P9-EEN-621

DATE DUE

DEMCO 38-296

Pittsburgh-Konstanz Series in
and History of Science

R

The Cosmos of Science

Essays of Exploration

EDITED BY John Earman and
John D. Norton

University of Pittsburgh Press / Universitätsverlag Konstanz

Riverside Community College
Library
4800 Magnolia Avenue
Riverside, CA 92506

Q175.3 .C69 1997
The cosmos of science :
essays of exploration

Published by the University of Pittsburgh Press,
Pittsburgh, Pa. 15261
Copyright © 1997, University of Pittsburgh Press
All rights reserved
Manufactured in the United States of America
Printed on acid-free paper
10 9 8 7 6 5 4 3 2 1

Library of Congress Cataloging-in-Publication Data
The cosmos of science : essays of exploration / edited by
    John Earman and John D. Norton.
        p.   cm. — (Pittsburgh-Konstanz series in the
    philosophy and history of science)
        Includes bibliographical references and index.
        ISBN 0-8229-3930-4 (cloth : alk. paper)
        1. Science — Philosophy.   2. Science — History.
    I. Earman, John.   II. Norton, John D., 1960–   .
    III. Series.
    Q175.3.C69   1996
    501 — dc20   95-52718
                                                                            CIP

Die Deutsche Bibliothek — CIP-Einheitsaufnahme
The cosmos of science : essays of exploration / ed. by
John Earman and John D. Norton. — Konstanz : UVK,
Univ.-Verl. Konstanz ; Pittsburgh : Univ. of Pittsburgh
Press, 1997
    (Pittsburgh-Konstanz series in the philosophy and
history of science ; 6)
    ISBN 3-87940-603-0
DBN: 95.025915.2 ⊗
            SG: 26

A CIP catalogue record for this book is available from
the British Library.

Contents

Induction, Scientific Methodology, and the Philosophy of Science

Action and Rationality

Preface

The accumulation of modern work in philosophy and history of science has brought us an understanding of the world of science of such richness and detail that it is no longer possible for a lone scholar to expect to add materially across all its parts. Each part has become a subdiscipline in its own right, with its own community of scholars and scholarly standards. The essays collected here document just how advanced are the explorations of the best work of the subdisciplines. Indeed, in many we find a sense that the subdisciplines have become so well developed that they contain standardized traditions that are now in need of reflection and correction. These essays bring us closer to the world we have made for ourselves, the cosmos of science.

History of Science

Bernard R. Goldstein's "What's New in Kepler's New Astronomy?" takes to task the easy caricature of the great astronomer Kepler to be found routinely in recent history of science: Kepler discovered his three laws of planetary motion, abandoning all earlier theories because of a slight discrepancy of 8′ of arc, in part through lucky guesses and in spite of pervasively confused metaphysics that enveloped all his work. Goldstein demonstrates that we come to a far deeper understanding of Kepler's work and achievements and avoid anachronistic misinterpretation if we place Kepler's work in its own context. Goldstein shows how and why we should take seriously Kepler's religious and broader metaphysical commitments and seek to understand the problems as he found them and posed them, as well as the methods he developed to solve them and the inner logic of his analyses.

In his "Experiment, Community, and the Constitution of Nature in the Seventeenth Century," Daniel Garber shows that much still remains to be understood about the historical development of that most common of notions in scientific discourse, objectivity. He finds in the early Royal Society a powerfully social dimension to facthood. Before an experimental outcome can enter into the register of attested facts, the experiment must be performed before competent witnesses. Garber argues that this social dimension of facthood is an innovation of the Royal Society and is not to be found in the methodological writings of the immediate predecessors of the Royal Society. These figures include René Descartes and, remarkably, Francis Bacon, whose vision of the practice of science was supposedly realized by the Royal Society.

Where Garber examines the emergence of social aspects of objectivity in seventeenth-century English science, William Harper, in "Isaac Newton on Empirical Success and Scientific Method," explains how, at the same time and place, a new and more rigorous standard of empirical success was being developed by Newton for the theory of a lone scholar within his research on gravitational physics. Newton required that the phenomena not merely be saved by his theory—this being the standard required in modern hypothetico-deductive confirmation. Rather he required that the theory's core parameters be deducible from the phenomena within the framework of the theory. This higher standard was painstakingly implemented in Newton's work on gravitation, as Harper shows in a series of detailed examples. Empirical success of this type results in a theory so solidly grounded in phenomena that Harper is able to use its solidity to deflect the skepticism of the Duhem-Quine thesis and of Kuhnian incommensurability. Harper's final irony is that this same standard of empirical success is now routinely exploited in work in relativitic physics that has decisively overturned Newton's theory.

Don Howard's "A Peek Behind the Veil of Maya: Einstein, Schopenhauer, and the Historical Background of the Conception of Space as a Ground for the Individuation of Physical Systems" begins with an innocent question. There is ample evidence that Einstein read Schopenhauer; what did Einstein find in Schopenhauer's writings that so attracted him? This question begins a fascinating journey which ends with the revelation that much of the modern discussion has missed the point of the classical debates over space and time. The point was identified lucidly by Schopenhauer who characterized space as a principle of individuation—that which would allow two things to be distinct individuals, even though they agree in all properties excepting their spatial positions. Whether such a function for space is legitimate lay at the heart of the famous Newton-Leibniz debate and became the focus of a tradition of analysis that was carried on by Kant, especially in his analysis of "incongruent counterparts," and by Schopenhauer. Howard portrays Einstein's deepest reservations about quantum theory as lying within this tradition and as deriving from an insistence that we preserve space as a principle of individuation. This was the real point of the celebrated but readily misunderstood Einstein-Podolsky-Rosen paper. The quantum theory, understood as a complete theory, requires that two systems, spacelike separated in spacetime, may nonetheless fail to be individuated by this separation into two distinct systems — a failure so fundamental that it leaves unclear just what are the systems that physics is supposed to describe.

Foundations of Mathematics and Physics

Geoffrey Hellman's "From Constructive to Predicative Mathematics" asks whether constructivism in mathematics is capable of underwriting the mathematics used in science. He answers first that a constructivism which restricts itself to intuitionist logic is too narrow since it implies, for example, that the assertion of the undecidability of a proposition is inconsistent. Next Hellman considers a radical form of constructivism which holds that proof-independent mathematical facts are not to be countenanced. There is an obvious analogy between this position and the logical positivists' verifiability theory of meaningfulness. And just as verifiability proved to be too strong a straitjacket for science so, Hellman argues, radical constructivism is too strong a straitjacket for the mathematics used in quantum mechanics. For example, Gleason's theorem, which is used to characterize the probability measures on closed subspaces of Hilbert space and also to rule out various hidden variable interpretations, is not constructively provable. Hellman finds more hope in a liberalized version of constructivism, such as predicativist analysis, that employs classical logic.

Prior to Newton, it was generally accepted that the laws governing the heavens are different from those that govern the terrestrial realm. In "Halfway Through the Woods: Contemporary Research on Space and Time," Carlo Rovelli notes that current physics finds itself in a similar state where the general theory of relativity governs large-scale objects, while quantum mechanics governs microscopic objects. Performing a marriage of these two theories which are enormously successful in their respective domains is, arguably, the most important task facing theoretical physics today. Rovelli, one of the leading research workers in quantum gravity, presents his perspective on the problems and prospects of such a marriage. This perspective takes in some of the most fascinating issues currently being discussed in the philosophy of space and time and the foundations of quantum mechanics.

In "What Superpositions Feel Like," David Z Albert addresses the measurement problem, the most vexing of problems in current philosophy of quantum mechanics. Quantum mechanics portrays the time development of systems as governed by a linear equation of motion, the Schrödinger equation. In order to explain why systems represented by superpositions of states display sharp values on measurement, it has become traditional to suppose that there is a second, somewhat perplexing type of time development, sometimes called the "collapse of the wave packet," that arises during a measurement operation. Albert reviews three attempts to do away with this addition and to construe quantum mechanics in such a way that the time development of the linear equations is all that is invoked. While initially promising, all three fail in the end to do justice to some mundane facts of everyday experience. The first attempt is the many-worlds interpretation of Everett; the second is a proposal of Gell-Mann, Hartle, and Zurek; the third is Albert's own. In it, Albert considers what it would be like for the quantum mechanical state of an observer to be entangled with a superposition of states of the system observed. While there would be no definite fact as to the result of the observation, the observer would still report that the observation resulted in a definite outcome. In this regard, superpositions would seem to feel quite ordinary.

Philosophers of science have assumed routinely that they can call up endless streams of ravens or emeralds, all of whose colors can be examined. Similarly, philosophers of physics routinely assume that they can be supplied with quantum systems prepared in precisely the states they specify. In "The Preparation Problem in Quantum Mechanics," Linda Wessels alerts us to the problems hiding in this innocent assumption. Among those who concern themselves with the problem of preparation, the presumption has generally been that preparation is essentially akin to measurement so that the preparation problem inherits all the vexation of the measurement problem. Wessels shows that this is not the case. While idealized measurement cannot be governed by the linear time development of the Schrödinger equation, idealized preparation is sufficiently different from measurement for it to be compatible with Schrödinger time development. Wessels shows, however, that this does not release preparation from all woe. Many, perhaps most, preparations envisaged in quantum mechanics actually fail to prepare. But at least the problem is not one of principle, as is the measurement problem; it is a de facto difficulty of the processes we happen to entertain as preparations.

One long-standing response to the foundational problems of quantum mechanics has been the assumption that the state of a system is not fully described by the state vector of standard quantum mechanics; more properties lie hidden beneath it. The best known of the resulting hidden variable theories is that of Bohm. We may expect that very many hidden variable theories may be constructible without directly contradicting the phenomena expected according to quantum theory. But that does not mean that results of great generality cannot be generated for this class. This is the primary burden of Jeffrey Bub's "Schrödinger's Cat and Other Entanglements of Quantum Mechanics." Using quite simple geometric methods, he shows how one can construct all possible theories in this class that solve the measurement problem formally in a sense he specifies. The chapter concludes by developing a proposal by Vink that allows the theories to be endowed with a Bohmian dynamics.

In "Chaos, Chance, and Determinism," John A. Winnie invites us to consider some of the deeper philosophical ramifications of the growing field of chaos theory. It is often said that chaos theory has deprived deterministic systems of their determinism in that sense that even simple chaotic systems exhibit random behavior at a broad level of description and thereby become empirically indistinguishable from stochastic systems. There are technical results that attempt to give precise formulation to this general claim. Winnie casts a critical eye on the claim and the technical results that underpin it. He finds that current analysis has failed to establish what is claimed. While a chaotic system may exhibit random behavior on some broad level of description, Winnie shows that we may, in some cases, only have to reduce the coarseness of our description slightly to recover the deterministic behavior.

As physicists grope their way to the Theory of Everything, R. I. G. Hughes reminds us in "Models, the Brownian Motion, and the Disunities of Physics" that as actually practiced, physics exhibits a number of disunities. It employs principles that are seemingly at odds with one another; it employs other principles that are consistent but only because they apply to different domains; and in bridging the gap between general principles and particular effects it makes use of a motley array of models. Hughes argues that these disunities favor the semantic view of theories (a view endorsed in different forms by Patrick Suppes, Fred Suppe, and Bas van Fraassen) over the statement view favored by the heirs of logical positivism. But Hughes also argues that this disunity calls for a more sophisticated version of the semantic view.

Induction, Scientific Methodology, and the Philosophy of Science

Carnap's famous "continuum of inductive methods," discovered independently by W. E. Johnson, applies to the following setup. There is a fixed number of kinds or species. Suppose that in a sample size n it is found that n_1 belong to the first species, n_2 belongs to the second, etc. Then the Carnap-Johnson continuum specifies that the probability that the next trial will yield a species of type i is a function of n, n_i, and an adjustable parameter, whose value measures (roughly) how fast we want to learn from experience. There is a glaring defect here and a limitation that chafes. The defect is that the Carnap-Johnson assignments of probabilities does not permit the confirmation of universal generalizations. The limitation is that the setup assumes that all of the species are known in advance. In "The Continuum of Inductive Methods Revisited," Sandy L. Zabell discusses a new three-parameter continuum in which the defect of the Carnap-Johnson continuum is removed and the limitation is overcome.

Judging from the lack of cross references in the literatures, epistemologists and philosophers of science have very little to say to one another, a very curious situation, since both groups are presumably concerned with understanding the sophisticated knowledge claims that are made in the sciences. Frederick's Suppe's "Science Without Induction" attempts to bring together the two literatures. He proposes a theory of knowledge for science that, in the current jargon, is externalist and nonreliabilist. The most novel feature of Suppe's view is that it allows knowledge of causal generalizations to be obtained noninductively. Just how this trick is pulled off is illustrated by the case study of J. J. Thomson's discovery of the electron.

David L. Hull's "That Just Don't Sound Right: A Plea for Real Examples" can be seen as furthering his agenda of making the philosophy of science as empirical as possible. As a first step, Hull recommends that philosophers of science restrict themselves to real examples, or at the very least to first exhaust actual examples before resorting to made-up, science fiction examples. Real examples come with a context. Fictitious examples, by construction, ignore context and the background knowledge that goes with it, thus removing the safe means of drawing out the implications of the example. The proponents of fictitious examples are thus forced to rely on intuition. But since our intuitions are formed in the actual world, we cannot rely on them when asked what we would say if confronted by the bizarre scenarios of made-up examples that transport us far from actuality. Hull discusses cases from the philosophy of biology where the "bug-eyed monsters" of fictional examples have led to confusion rather than enlightenment.

In the heyday of logical positivism, the philosophy of science was conceived as the logic of the sciences, that is, the investigation of the structure of the languages of science. Postpositivist philosophy of science has tended to deemphasize the philosophy of language. While this emphasis has its merits, it also has its downside, as demonstrated by J. Michael Dunn's "A Logical Framework for the Notion of *Natural Property*." Using an account of relevant predication developed in the framework of relevance logics, Dunn illuminates a wide range of issues in the philosophy of science, including the distinction between real and Cambridge change, Goodman's grue-bleen paradox, Hempel's ravens paradox, and the notion of natural kinds.

The received view of causation holds that for an event c to be a cause of the event e, there must be a (perhaps unknown) general regularity to the effect that events of type c are always conjoined with events of type e. Against the received view, David M. Armstrong's "Singular Causation and Laws of Nature" argues that causation can be present in singular instances without the support of regularities. Nevertheless, Armstrong maintains that there is a strong link between causes and laws; but for Armstrong laws must be conceived not as regularities but as relations among universals. This universals conception of laws has been challenged by van Fraassen in his *Laws and Symmetry*. Armstrong's essay offers his answers to van Fraassen's identification problem (identify the sort of facts about the world that give "law" its sense) and inference problem (explain why it is that L is a law should imply that L).

Action and Rationality

Philosophers of mind and action theorists have struggled to resolve a tension between seemingly plausible principles. (1) Explanatory Relevance: When beliefs explain behavior, it is their meaning (*what* is believed) that explains why the behavior occurred. (2) Extrinsicness of Meaning: A belief's meaning is an extrinsic property (i.e., it is a property that the belief possesses in virtue of the way other things are or were). (3) Causal Inertness: Extrinsic properties are causally inert. In his contribution, "Action and Autonomy," Fred Dretske proposes a stunning reversal of everything that is in the literature. Rather than trying to resolve the tension by rejecting these principles, he accepts it and proposes to use it to locate a sense in which intentional actions are autonomous — namely, the reasons for which we act are not causes of acting that way.

In "Explanations Involving Rationality," Peter Railton explores the prospects of achieving the nomothetic ideal in psychology — that is, finding more or less precise nomological generalizations that function as major premises in explanations of rational actions. That nothing approaches strict laws are to be currently found in desire-belief psychology is not in dispute. The question is what accounts for their absence. Donald Davidson is famous for arguing that this absence is not simply a reflection of the immaturity of the field but rather stems from a set of principled considerations that point to the impossibility of a nomothetic psychology. Railton provides deft and spirited responses to each of Davidson's considerations and in doing so points to a much more optimistic future for a genuine science of rational action.

John Earman
John D. Norton

History of Science

1

What's New in Kepler's New Astronomy?

Bernard R. Goldstein

Department of Religious Studies, University of Pittsburgh

Kepler's achievements are well known and can be stated succinctly, or so it seems. Yet he was a complex thinker who responded in unusual ways to many intellectual currents in astronomy and other disciplines. To set Kepler in his intellectual context I offer a few preliminary remarks on the historical development and the inner logic of the derivations of his laws of planetary motion and sketch some of his theological and methodological commitments that affected his astronomical arguments.

The Standard View of Kepler's Achievements in Astronomy

1. Kepler's contribution to astronomy consists in his three laws: the first two laws (the ellipse and the Area Law) that were derived in the *Astronomia Nova* (1609), and the third or Harmonic Law (relating planetary periods and heliocentric distances) in the *Harmonices Mundi* (1619). By introducing elliptical planetary orbits, Kepler "had overthrown for all time the 2000-year-old axiom . . . [of] uniform circular motion" (see Caspar 1962, 140).

2. Kepler abandoned all previous astronomical theories because of the discrepancy of 8 minutes of arc between Brahe's observations of Mars and values computed from the "best" equant model.

3. The ellipse and the third law were just lucky guesses or "approximations."

4. Kepler's views on matters other than his three laws can be ignored as "irrelevant to the progress of astronomy" (e.g., his introduction of regular solids between the "planetary orbs"; his use of analogy between the heavens and the trinity; and his appeal to principles of harmony, astrology, and magnetism). For example, Dijksterhuis (1961, 322) wrote that "the story of the discovery of Kepler's first two laws is significant not only historically but psychologically, because it clearly reveals the curious jumble of rational and irrational elements from which great discoveries tend to spring."

In addition to the anachronism of referring to Kepler's three laws (to which I have no objection insofar as they are understood as a summary of his contributions), the standard view fails to take into account the historical context and the conceptual framework in which Kepler operated. In other words, all four claims need to be nuanced.

Toward a Reevaluation of Kepler's Astronomical Contributions

1. Between 1596 and 1621 Kepler produced a series of books on astronomy, adhering to the program he had outlined in his first major publication, the *Mysterium Cosmographicum* (Duncan 1981), and he achieved his stated goal of finding what he took to be the physical causes of planetary motion. The content of these books as well as the rhetoric employed had many new features. For example, the full title of the *Astronomia Nova* includes the expression "Physics of the Heavens": Kepler had invented a new genre of astronomical writing — a technical treatise unifying astronomy and physics.

2. When Kepler wrote the *Mysterium*, he began by assuming that Copernicus's results were satisfactory and merely sought another way to reach the same conclusions. Kepler described Copernicus's method as a posteriori (i.e., based on observational data), and hoped to find the same results a priori (i.e., based on a set of first principles). Kepler's initial motivation for research in astronomy did not depend on observational data, and he only sought such data when he found an inadequate fit between his a priori reasoning and the Copernican distances that his teacher, Maestlin, derived for him.

The intellectual context in which Kepler operated was set in Aristotelian physics as understood in the late sixteenth century. In the *Mysterium*, chap. 1 (Duncan 1981, 77), Kepler wrote, "Nor do I hesitate to affirm that everything which Copernicus inferred a posteriori and derived from observations, on the basis of geometrical axioms, could be derived to the satisfaction of Aristotle, if he were alive (which Rheticus repeatedly wishes for), a priori without any evasions." In chap. 2 he added (ibid., 97):

For what could be said or imagined which would be more remarkable or more convincing than that what Copernicus established by observation, from the effects, a posteriori, by a lucky rather than a confident guess, like a blind man, leaning on a stick as he walks (as Rheticus himself used to say) and believed to be the case, all that, I say, is discovered to have been quite correctly established by reasoning derived a priori, from the causes, from the archetype of creation?

We can compare this to what Rheticus had proposed in the *Narratio Prima*, the first account of Copernican astronomy published in 1540 (Rosen [1939] 1959, 142): "Now in physics as in astronomy one proceeds as much as possible from effects and observations to principles" (this seems to be what Kepler meant by a posteriori in Copernicus). According to Buchdahl (1972, 275f), Kepler's a priori reasoning allowed him to convert the Copernican theory which only had the status of a "guess" into an "actual truth."

3. Kepler's debt to Tycho Brahe was not confined to observational data, for he used Brahe's models for the Sun (or the Earth, as Kepler preferred) and Mars as preliminary hypotheses in the *Astronomia Nova*. But Kepler also accepted Brahe's arguments, based on considerations of parallax, that the New Star of 1572 and the comet of 1577 were in the heavens rather than in the sublunary realm. Furthermore, from the path of the comet Brahe argued in his treatise published in 1588 that there were no solid planetary orbs, and concluded that Aristotle's distinction between the celestial and sublunary realms had to be abandoned. Kepler drew the conclusion that *models* for planetary motion were inappropriate since it had been assumed that the planet lay on a moving solid orb; for Kepler the planet was simply moving in a fluid. For some time Kepler held to the view that the planets were self-moved "like birds and fishes," but then withdrew intelligence from among the attributes of the planets and considered their motions as due simply to forces. It is surprising that Kepler was the only professional astronomer of the time to have noticed the consequence of Brahe's discovery, namely, that constructing planetary models no longer made sense. Indeed, even later in the seventeenth century this point was not understood by many members of the astronomical community.

4. We now come to Kepler's debt to his teacher, Michael Maestlin. Of particular interest is Maestlin's (1578) treatise on the comet of 1577–1578, ten years earlier than Brahe's major treatise on the same theme. Near the end of this treatise Maestlin presented a day by day ephemeris of the comet from 5 November 1577 to 10 January 1578 where the last three columns list the longitude and latitude of the comet followed by its distance in terrestrial radius from the center of the earth (pp. 52–53; see Brahe's treatment of the coordinates of the same comet in his publication of 1588 in Dreyer 1922, iv:177–79). Maestlin put the comet in a heliocentric orb just beyond Venus (i.e., between Venus and the Earth), but he did not suggest that such a three-dimensional approach should be applied to planets — nor had anyone else up to that time. (Note that Brahe's orbit for the comet in his publication of 1588 had many of the same features as that of Maestlin, but in a geostatic, rather than a heliostatic, framework; see Dreyer 1922, iv:160.) Nevertheless, Kepler may have concluded that if it was worth the effort to calculate the path in three dimensions of an ephemeral object (as a comet was thought to be), it would be reasonable to give as much attention to planetary paths — including the distances. On the title page of Maestlin's treatise, a figure displays the path of the comet in longitude and latitude against the background of the fixed stars (reproduced in Jarrell 1989, 25); similar figures had already been used by Apian for the comets of 1531 and 1532 (see Barker 1993). But surprisingly little attention was given to planetary distances in the period immediately preceding Kepler; this issue was not prominent in the debates over the Copernican system (except for the vexed question of the parallax of Mars). Yet here Maestlin treated a comet as if it were a planet with a heliocentric orb. It has been claimed (Westman 1972a, 23–24; Jarrell 1989, 26) that Maestlin failed to understand the equivalence of the Ptolemaic and Copernican models when he decided on a heliocentric orb for this comet. But Maestlin would have been aware that according to Ptolemy's nesting hypothesis there would be no room for a comet around the orb of Venus, whereas the Copernican orbs had some space between them (as Rheticus had already noted; see Rosen [1939] 1959, 147; Brahe's 1578 German treatise on the comet of 1577 in Dreyer 1922, iv:388 [translated in Christianson 1979, 136, but this passage seems to have been misunderstood]). In this sense the two systems were not equivalent. Indeed, in the preface to the first edition of the *Mysterium* (Duncan 1981, 63), Kepler considered adding a new

and invisible planet between Jupiter and Mars, and another between Venus and Mercury, taking advantage of the gaps in the Copernican system. Maestlin's discussion of the comet of 1577 was a key factor leading to Kepler's acceptance of the Copernican system, as he tells us in chapter 1 of the *Mysterium* (ibid., 79), even though he later rejected Maestlin's view on the path of comets. (See notes to the *Mysterium*, ed. 1621, in Duncan 1981, 87.)

Methodological Considerations

1. Kepler used the term *physics* in two senses which Westman (1972b, 247) has called "descriptive" and "causal" — I consider them both causal and distinguish between causes which for Kepler involve forces acting on bodies and causes which depend on the cosmic plan of creation. The failure to recognize the importance of this distinction has led to many misunderstandings in Kepler's arguments, since he does not always draw the distinction clearly. Kepler's appeal to a priori reasoning is based on his conviction that he can derive the details of planetary motion from an analysis of the plan of creation which is governed by theological considerations concerning the nature of God and His intervention in the world — for which no observational data are needed. The data, which are the starting points for a posteriori reasoning, only confirm Kepler's a priori reasoning or require modifications of it since, according to Kepler's methodology, a priori and a posteriori reasoning must agree.

2. Let us now turn to Kepler's view of God's plan of creation, which forms part of the groundwork for his astronomy. Though I have not found an explicit statement by Kepler of his indebtedness to Melanchthon, a leading theologian who worked with Luther, I suspect that Melanchthon's view of the Bible, directly or indirectly, had a serious impact on Kepler as a student of theology in a Lutheran seminary (see Kusukawa 1992, 44 and Westman 1975; 1980, 121, 142). According to Melanchthon, the Bible is a coherent document, and all its parts are equally authoritative as the word of God. Melanchthon was prepared to exclude the apocryphal books from the Bible because, in his view, they were incompatible with the coherence of the Bible (although he allowed such books to be studied in the same way that other non-Christian classics were studied). Moreover, Melanchthon accepted the view that God reveals himself in nature. Indeed, the pagan Greeks had

come to the knowledge of moral order from contemplating the order in nature, and from this order they arrived at knowledge of God, even if that knowledge was partial and inadequate (see, for example, Schneider 1990, 103f, 119f, 244). Kepler went further by asserting that the world, created in the image of God is also a coherent whole, just as Scripture is coherent as the word of God, and that the two are complementary. Kepler's conviction, that applying theological reasoning to the study of the natural world is appropriate, is not at all surprising in his intellectual environment. Although others at the time advocated such a fusion of disciplines, Kepler was the only one to seek *quantitative* results from it.

Kepler's image of the trinity in the plan of creation can be seen in the following passage from chapter 2 of the *Mysterium*, "The image of God the Three in One [is found] in a spherical surface, that is of the Father in the center, the Son in the surface and the Spirit in the regularity of the relationship between the point and the circumference. For what Nicholas of Cusa attributed to the circle . . . I reserve solely for a spherical surface" (Duncan 1981, 93). Although Kepler referred only to Cusa here, the application of this image to the heavens may be a response to Rheticus's *Narratio Prima*:

> First, . . . [Copernicus] established by hypothesis that the sphere of the fixed stars, which we commonly call the eighth sphere, was created by God to be the region which would enclose within its confines the entire realm of nature, and hence that it was created fixed and immovable as the place of the universe. Now motion is perceived only by comparison with something fixed. . . . Then in harmony with these arrangements, God stationed in the center of the stage His governor of nature, king of the entire universe . . . the sun. (Rosen [1939] 1959, 143)

As we shall see, Kepler also grounded astronomy in theology.

In the *Paralipomena* or the *Optical Part of Astronomy* of 1604, Kepler explicitly related the created world to God, "The creator in his great wisdom found nothing more perfect or more beautiful or more excellent than himself. This is why, thinking of the corporeal world, He gave it the form most like Himself" (Chevalley 1980, 107). This form turns out to be the sphere: defined by a surface, a center, and the interval between them which is everywhere symmetric and filled by straight lines. Later, in the *Harmonices Mundi* of 1619, Kepler replaced the sphere by the circle, but the correspondence remained. Kepler conceived of constructible polygons inscribed in an archetypal circle whose vertices define a set of arcs; harmonies are then defined by the ratios of those arcs. In this way Kepler succeeded in basing arithmetic ratios or harmonies on geometric figures.

This three-in-one-some is the "archetype" of the world, where there is also a "trinity" that serves as the background against which the planetary motions take place. In the preface to the 1596 edition of the *Mysterium* Kepler proclaimed "the splendid harmony of those things that are at rest, [namely], the Sun, the fixed stars, and the intermediate [space], with God, the Father, and the Son, and the Holy Spirit" (Duncan 1981, 63). The trinity is represented by the motionless parts of the heavens in contrast to the moving planets. In this context it is important for Kepler to reconcile Scripture with science:

Certainly God has a tongue, but he also has a finger. . . . Therefore, in matters which are quite plain, everyone with religious scruples will take the greatest care not to twist the tongue of God so that it refutes the finger of God in nature. (Notes to the *Mysterium*, chap. 1, ed. 1621, in Duncan 1981, 85)

3. For Kepler the job of the astronomer was to study the path of a planet in space, not to construct models (called "hypotheses" at the time) from which the path might be determined. This path was to be understood in three dimensions: heliocentric longitude, latitude, and distance (which could then be transformed to geocentric coordinates, if desired). In contrast, the observations discussed by previous astronomers were episodic, and they were used for determining the parameters of models, not paths.

Kepler claimed on the verso of the title page of the *Astronomia Nova* that he had met the challenge proclaimed by Ramus (d. 1572): to construct an astronomy without hypotheses (i.e., models). On the other hand, Brahe thought this challenge was meaningless because astronomers were supposed to construct hypotheses to account for planetary motions (see Blair 1990, 368). Already, in the *Mysterium*, Kepler paid special attention to the planetary paths, arguing that the planet's linear velocity along its path varies, as is the case in Ptolemy's equant model (considered by Copernicus to be a fault of that model). In the *Astronomia Nova*, chap. 1, Kepler illustrated the complexity of the geocentric path of Mars from 1580–1596 (see figure 1.1), displaying this trajectory in "depth" in the plane of the ecliptic—no such figure can be found in the previous astronomical literature (though it became a commonplace subsequently). By contrast, the heliocentric

DE MOTIB. STELLÆ MARTIS

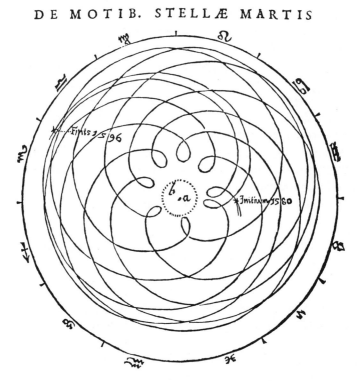

Figure 1.1 Kepler's figure to illustrate the geocentric trajectory of Mars for the period 1580–1596; *Astronomia Nova* ([1609] 1968, chap. 1).

path of a planet is an eccentric circle (or an oval) which is much simpler. In this way Kepler suggested a new reason to prefer a heliocentric over a geocentric system: on the basis of the planetary paths that are entailed by them.

4. In the *Apologia* (1600; see Jardine 1984), Kepler distinguished three tasks of an astronomer: (1) to record the apparent *paths* of the planets and their motions: the practical and mechanical part of astronomy; (2) to determine the true and genuine *paths*: the contemplative part of astronomy; and (3) to decide by what circles and lines certain images of these true motions may be depicted: the inferior tribunal of geometers. Again we see the crucial role of *paths* for Kepler (e.g., the apparent path is the "history" of a planet). Geometers provide tools for astronomers, and this is the role that Apollonius filled according to Kepler:

Apollonius was not an astronomer by profession, but a geometer. And he did not himself, as the office of astronomer requires, apply in practice what he demonstrated from a problem derived from astronomy in order, having adopted this hypothesis, to infer and demonstrate from observation the motion of some planet. Rather, he handed over to astronomers a merely geometrical demonstration, as a retailer hands over keys or an axe to an architect in case someone should need these things for his work. (Jardine 1984, 191f)

This was precisely the role that Viète (d. 1603) saw for himself as the "Apollonius Gallus" (see Swerdlow 1975).

5. Kepler also made a sharp distinction between astronomical and geometrical hypotheses in astronomy. He claimed, for example, that an oval shape for the moon's path would be an astronomical hypothesis — but when an astronomer shows by what circles this oval can be constructed he uses geometrical hypotheses (Jardine 1984, 153). Kepler presumably was referring to the oval shape of Ptolemy's deferent for the Moon; this fact was not mentioned by Ptolemy but it had already been noted in Reinhold's commentary on Peurbach's *Theoricae novae* (first ed. 1542; I consulted the 1557 Paris ed., fol. 38b–39a), a work that Kepler cited in a letter to Herwart von Hohenberg in 1599 (see Jardine 1984, 62) and in the *Astronomia Nova*, chap. 46 (see Donahue 1992, 467); Reinhold in turn probably depended on Brudzewo's commentary on Peurbach's *Theoricae novae* (dated 1482, printed 1495; ed. Birkenmajer 1900, 124, "Similiter etiam centrum epicycli Lunae infra unum mensem non circularem, sed etiam fere ovalem, propter descensum et ascensum suum, describit figuram"; this reference was kindly given to me by J. L. Mancha, Seville). Kepler was also aware that Copernicus's double epicycle for the Moon produced an oval (*Astronomia Nova*, chaps. 4 and 43), but he did not recognize that this oval is an ellipse (see Swerdlow and Neugebauer 1984, 197).

6. For Kepler the fundamental principle that governs planetary motion is the distance-velocity relation, that is, that the linear velocity of a planet varies inversely with its distance from the center of motion. This principle was not new with Kepler, indeed, Kepler cited Aristotle's *De Caelo* (see Duncan 1981, 197), and Copernicus (1543, 7v) referred to Euclid's *Optics* for it. But Kepler used it in a new way. It first served to support Ptolemy's equant models against Copernicus's models based on uniform circular motions, and then to support an equant model for the Earth. In Ptolemy's solar model (and correspondingly in Copernicus's terrestrial model), the variation in angular velocity is an optical

effect, due to the observer's location at a point eccentric to the center of motion. But Kepler insisted that the linear velocity of the Earth should vary with its distance from the Sun and that it is not an optical effect. Kepler drew an important distinction between angular velocity, a familiar concept in astronomy, and linear velocity along the path of a planet. For measuring angular velocity Kepler used angles per unit time, whereas for linear velocity he preferred times (*morae*) per unit distance (instead of distance per unit time, as we might expect). Indeed, for Kepler distance is prior to motion, and the distance-velocity relationship came as a natural consequence of this commitment. In the *Astronomia Nova* ([1609] 1968, chap. 33, 168; see Donahue 1992, 377), Kepler remarked:

Distance from the center is prior both in thought and in nature to motion over an interval. Indeed, motion over an interval is never independent of distance from the center, since it requires a space in which to be performed, while distance from the center can be conceived without motion. Therefore, distance will be the cause of intensity of motion (*causa vigoris in motu*), and a greater or lesser distance will result in a greater or lesser amount of time (*morae*).

It seems to me that Kepler's claim that the period of a planet varies with its distance from the Sun (which ultimately led to his Third Law) is also related to this commitment.

7. In sum, Kepler's basic tools for finding the laws of planetary motion in agreement with the observational data understood as the paths of the planets through space were (1) the plan of creation and the notion of archetypal reasoning associated with it, and (2) the distance-velocity relationship along with the priority of distance over motion.

Kepler's Laws of Planetary Motion

1. This section deals with the inner logic of Kepler's derivations of his three laws rather than the precise way he reached them. Several scholars have argued that Kepler misled many of his readers by introducing in some of his works (including the *Astronomia Nova*), for rhetorical purposes, a fictional order of discovery. Therefore, reconstructing the actual steps in his discoveries depends on a careful reading of his correspondence and manuscripts (see Donahue 1993, 1994). However, that reconstruction will not be discussed here because, in my view, Kepler's conceptual framework is sufficiently accessible through his published works.

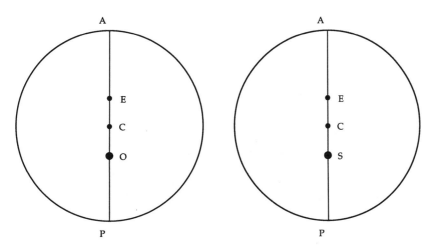

Figure 1.2a   Figure 1.2b

2. As is well known, in the early chapters of the *Astronomia Nova* Kepler showed the inadequacy of all three alternative planetary models that were worthy of consideration: the models of Ptolemy, Copernicus, and Brahe. Instead, Kepler's "vicarious hypothesis" which was based on Ptolemy's equant model, the very model that Copernicus had found so objectionable, best reproduced the observational data. Let us first consider Ptolemy's equant model for the deferent of the outer planets (Mars, Jupiter, and Saturn): Let C be the center of the deferent circle AP, E the equant point, and O the observer, such that $EC = CO$ (see figure 1.2a). Uniform motion takes place about E; hence the linear velocity on the circle is not uniform, as Ptolemy realized. In the *Astronomia Nova*, Kepler modified this model by replacing O with S (for the Sun), and allowing $EC \neq CS$. The models he derived from observations of Mars were two equant models: (1) where $EC = CS$ that accounted for the observed latitudes (see figure 1.2b), and (2) where $EC \neq CS$ that accounted well for the observed longitudes (to within 2′ of arc). No single equant model could account for all the observed data; the equant model with bisected eccentricity led to a discrepancy of 8′ of arc between theory and observations at the octant points, but it represented the distances from the Sun to the planet reasonably well (see Stephenson 1987, 42ff). The equant model with nonbisected eccentricity became Kepler's "vicarious hypothesis" (i.e., a substitute for the true theory yet to be discovered) because it was useful in computing the longitudes of Mars even though it was wrong in other respects.

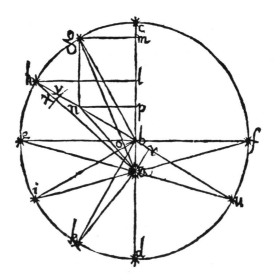

Figure 1.3 Kepler's figure to illustrate his derivation of the area law; *Astronomia Nova* ([1609] 1968, chap. 40, 193). The sun is located at point *a*, and the planet travels along circle *cedf* about center *b*.

3. Kepler's first derivation of the Area Law comes in the *Astronomia Nova*, chap. 40 (see Donahue 1992, 417ff), and his central argument referred to an eccentric circular orbit (see figure 1.3). Curiously, this derivation follows a chapter in which Kepler argued against the circular path for planetary motion.

To clarify Kepler's discussion we introduce figure 1.4: Consider the area A_1 defined by lines drawn from the Sun to the ends of a small arc *s*, representing the motion of a planet. Note that the Sun is not at the center of the circle on which the planet is supposed to move. Then, at aphelion and perihelion, where the motion of the planet is perpendicular to the radius vector drawn from the Sun, the area of this small triangle A_1 will be proportional to the product of the arc *s* and the length of the radius vector d_1. Strictly this relation will be valid only at the aphelion and perihelion (as Kepler was aware).

Next construct the motion of the planet around a portion of its orbit by adding small segments like that already defined (see figure 1.5). We then have *i* arcs of length *s*, and we seek the variable time intervals that correspond to each of these equal arc-lengths. Kepler calls these time intervals *morae*; in effect, they represent the time it takes a planet to move a unit distance along its trajectory. Following the proportionality already established, the sum of the areas A_1, A_2, A_3, ... A_i will be proportional to the sum of sd_1, sd_2, sd_3, ... sd_i, where d_1, d_2, d_3, ... d_i, are the lengths of the corresponding radius vectors; that is:

$$A_1 + A_2 + \ldots + A_i \propto sd_1 + sd_2 + \ldots + sd_i. \qquad (1.1)$$

It follows from expression (1.1) that the proportionality holds without the quantity s on the right side; we shall discuss the consequences of eliminating s later in this section. Now Kepler believed that the linear velocity of a planet is inversely proportional to its distance from the Sun. Hence, in each term in the above series we may replace the distance d_j (where $1 \leq j \leq i$) by the reciprocal of the corresponding velocity v_j, producing a quotient s/v_j. Each of these quotients represents the distance travelled by the planet along a small portion of its orbit divided by the velocity with which it traverses that portion of the orbit, and thus defines the time taken to traverse that portion of the orbit. That is:

$$A_1 + A_2 + \ldots + A_i \propto \frac{s}{v_1} + \frac{s}{v_2} + \ldots + \frac{s}{v_i}; \qquad (1.2)$$

$$\propto t_1 + t_2 + \ldots + t_i. \qquad (1.3)$$

Therefore, the ratio of the sum of the areas A_j making up a given segment of the orbit to the area of the whole orbit (A) will be equal to the ratio of the sum of the corresponding times t_j to the time required

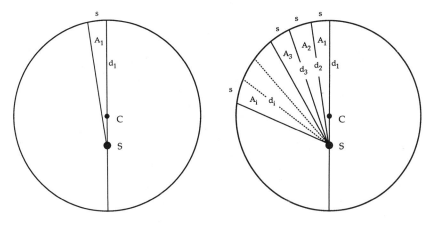

Figure 1.4   Figure 1.5

for the planet to complete one orbit that is the period of the planet (T). That is:

$$\frac{A_1 + A_2 + \ldots + A_i}{A} = \frac{t_1 + t_2 + \ldots + t_i}{T} \tag{1.4}$$

Now let us define

$$\alpha_i = A_1 + A_2 + \ldots + A_i$$

and

$$\tau_i = t_1 + t_2 + \ldots + t_i.$$

Then, we can rewrite equation (1.4) as

$$\frac{\alpha_i}{A} = \frac{\tau_i}{T} \tag{1.5}$$

The correlation established here between areas and time intervals is the same one we recognize, for the case of an elliptical orbit with the Sun at one focus, as the second law of planetary motion.

The derivation in chapter 40 suffers from a number of defects, some real, others alleged. The first defect is that the basis on which the small areas are calculated gives a good approximation only when the arc along which the planet is moving, and hence its velocity, is perpendicular to the radius vector drawn from the Sun. This condition is only satisfied at the apses, as Kepler recognized. Indeed, he gave an extensive analysis of the error introduced by this consideration, including a graphical representation of an exact solution. As Aiton (1969, 90), among others, has pointed out, Kepler presented a correction to the distance-velocity relation some years later in the *Epitome* of 1621: "The component of the velocity perpendicular to the radius vector is inversely proportional to the distance from the sun" (Aiton's paraphrase of the text in Kepler, *GW* 7:377).

A second defect is that Kepler seems to be appealing to infinitesimal values of s and summing an infinite number of radius vectors. Indeed, Kepler's language suggests this to be the case, and virtually all previous commentators on this chapter have pointed to this difficulty. But Kepler here explicitly appeals to Archimedes for his use of infinitesimals, and Archimedes took infinitesimals like s to be very small but still finite. In his only example, Kepler in fact took s to be 1°, far from an infinitesimal quantity. Moreover, when recasting this argument in the *Epitome* of 1621 (GW 7:377), Kepler was more careful and said that the orbit is to be divided into "the most minute equal parts" ("*in particulas minutissimas aequales*"). The difficulty in chapter 40 arises from Kepler's omission of some steps in his procedure whereby s could be eliminated altogether. Let us divide the entire circumference of the circle into n equal arcs of length s. Then, according to expressions (1.1) and (1.4),

$$\frac{A_1 + A_2 + \ldots + A_i}{A} = \frac{sd_1 + sd_2 + \ldots + sd_i}{sd_1 + sd_2 + \ldots + sd_i + \ldots + sd_n}$$

$$= \frac{s(d_1 + d_2 + \ldots + d_i)}{s(d_1 + d_2 + \ldots + d_i + \ldots + d_n)}$$

$$= \frac{d_1 + d_2 + \ldots + d_i}{d_1 + d_2 + \ldots + d_i + \ldots + d_n}. \tag{1.6}$$

Thus s has been eliminated by a proper procedure. We then combine equations (1.5) and (1.6) with the following result:

$$\frac{\tau_i}{T} = \frac{d_1 + d_2 + \ldots + d_i}{d_1 + d_2 + \ldots + d_i + \ldots + dn}. \tag{1.7}$$

In chapter 40 Kepler divided the circle into 360 equal parts and considered evaluating equation (1.7) for each value of i. Then, instead of pursuing this "tedious" method, he returned to equation (1.5), and took this as the principal result in Part III of the *Astronomia Nova*.

Let us return to the procedure in chapter 40. Kepler's goal at the outset was to approximate the motion on a circular orbit about an eccentric Sun according to the distance-velocity principle which he took to follow from physical causes. By the end of the chapter, this principle had been replaced by the Area Law as an approximation to it. Along the way, he proposed a method for taking the ratio of sums of radius vectors (see equation 1.7, above) to approximate the Area Law and then abandoned it as too cumbersome even though it appears to be much closer to the original distance-velocity relationship.

4. An argument similar to the one for the Area Law yields Kepler's third law (first presented in *Harmonices mundi* of 1619 and elaborated in the *Epitome* and in the notes to the *Mysterium* of 1621 [see Duncan 1981]). Let us begin by recalling that in a Copernican model, the length of the path of a planet about the Sun is simple to calculate, whereas in the Ptolemaic or Tychonic models the length of the planet's path about the Earth is no simple matter. The path of a planet in a circular orbit about the Sun is $2\pi R$, where R is its mean distance from the Sun. Let T be its period. Then its linear velocity

$$v = 2\pi R/T. \tag{1.8}$$

For two planets, P_1 and P_2, where P_2 is farther from the Sun than P_1,

$$\frac{v_2}{v_1} = \frac{T_1{}^* R_2}{T_2{}^* R_1}. \tag{1.9}$$

If $v_1 = v_2$, then $\dfrac{T_2}{T_1} = \dfrac{R_2}{R_1}$, $\tag{1.10}$

that is, $T \propto R$. Now let $v_2 < v_1$ (i.e., a planet farther from the Sun moves more slowly in linear velocity), and $T \propto f(R)$. For example, let $T \propto R^2$. Since both R and T are known, one can check whether this relationship holds. It is then easy to determine that T does not increase so rapidly. Hence the compromise is $T \propto R^{3/2}$ (Kepler's third law), which can be verified directly from the data.

It has been suggested that Kepler proceeded by trial and error (see Koyré 1961, 341), but that does not do justice to his method. Already in the *Mysterium* (ed. 1596, chap. 20), Kepler expressed his conviction that T varied with R and, realizing that the proportionality was not linear, he proposed a relationship based on the arithmetic mean. Then in the *Astronomia Nova*, chap. 39 ([1609] 1968, 186), he proposed that T varied with the square of R: "supposing the same planet to be in turn at two distances from the sun, remaining there for one whole circuit, the periodic times will be in the duplicate ratio [i.e., square] of the distances or magnitudes of the circle" (Donahue 1992, 407). In 1618 (the precise date is recorded in the *Harmonices Mundi* V,3), Kepler discovered the third law, but the notice about it in the *Harmonices Mundi* is brief. A fuller account appears in the notes to the *Myste-*

rium of 1621 (*ad* chap. 20; Duncan 1981, 205): "from the principles adopted [in chap. 20 of the *Mysterium*, ed. 1596] the geometric mean was the legitimate conclusion . . . but the arithmetic mean came closer to the mean according to the 3/2 power than the geometric mean, or that according to the square" (see Duncan 1981, 249–50, where it is shown that the substitution of a geometric mean for an arithmetic mean leads immediately to the law: $T \propto R^2$). In this way Kepler linked his arguments in the treatises of 1596, 1609, and 1619. It is clear that from the very beginning he sought a rule such that T would vary with R; he did not have to consider all possible exponents of R since information was gained after each trial. His reasoning was based on the distance-velocity relationship, and was confirmed by the data. This reasoning is "archetypal," which is "physical" for Kepler, for it depends on the plan of creation as the cause of planetary motion.

In the *Epitome* (IV, Part 2,4) Kepler attempted to produce another kind of physical argument to justify this rule, based on forces, and to some it has seemed ad hoc. But this view needs to be nuanced. In the *Astronomia Nova*, Kepler stated that the solar force diminishes linearly with the distance from the Sun; as the force diminishes, the linear velocity of a planet diminishes and hence its period increases. But the period must also increase as the orbit increases in length (which also depends on R). Therefore, combining both effects, $T \propto R^2$ (as in the *Astronomia Nova*, chap. 39). So far no account has been taken of the effect of the planetary body. Kepler believed that the density of the planets varied inversely with the square root of their distances from the Sun (which he took to be the densest of the heavenly bodies: *Epitome*, IV, Part 1,4; *GW*, 7:486ff); the denser a planet, the more "sluggish" it is; and the more "sluggish," the greater its period. This, together with the previous relationships, yields the third law:

$$T \propto R^*R/\sqrt{R}; \qquad (1.11)$$

$$\propto R^{3/2}. \qquad (1.12)$$

Kepler gave a priori reasons for the density to vary inversely with the distance, but the third law of planetary motion led to the introduction of the square root. Only in this restricted sense may his argument be called ad hoc.

Kepler also considered the volume of a planet (*moles*) and the amount of its matter (*copia*) separately: the volume (rather than the diameter or the surface) of a planet varies directly with its distance from the Sun; and the amount of its matter varies directly with the square root of that distance. The greater the volume (V), the greater is the effect of the solar force and the shorter the period — an inverse proportionality; while the greater the amount of matter (M), the stronger is the resistance to motion and the greater the period — a direct proportionality (see Gingerich 1975). Again, we reach the third law:

$$T \propto \frac{R*R*M}{V} \tag{1.13}$$

or

$$T \propto \frac{R*R*\sqrt{R}}{R}. \tag{1.14}$$

Thus

$$T \propto R^{3/2} \tag{1.15}$$

as before.

5. The derivation of the ellipse is the most complicated case, and the one for which Kepler provided the most information concerning the path that led him to discover it. But this wealth of information and calculation has tended to obscure the principles that underlie his procedure. At a very early stage Kepler was prepared to accept ovals, for he already mentioned an oval for the deferent of the Moon in the *Apologia* written in 1600, and there was a long tradition among Ptolemaic astronomers concerning the oval deferents of the Moon and Mercury. Brahe had even considered an oval path for the comet of 1577 (see Dreyer [1906] 1953, 366). Moreover, in the *Astronomia Nova*, chap. 4, Kepler remarked that in *De Rev.* V, 4, Copernicus indicated that the path of a planet is not circular but "goes outside the circle at the sides," whereas for Kepler the planetary ovals should go "inside" the circle at the sides (Donahue 1992, 136f). On the other hand, no mention of an ellipse is made in Kepler's early astronomical writings, or in the works on which he depended for planetary theory. Yet, we

would be wrong to conclude that only Kepler was capable of making such a "conceptual leap," for an unpublished manuscript by Viète includes a mathematical discussion of an ellipse in a planetary model. Kepler did not cite this work (though he does allude to a published work by Viète; see Donahue 1992, 256), and there is no reason to suppose that Kepler was aware of it. Moreover, Viète made no attempt to produce physical arguments in favor of his ellipses, and he did not seek observational evidence to confirm his insights. In sum, Viète approached planetary theory from the point of view of a geometer who could show astronomers (particularly Copernicus) how to construct elegant mathematical models (see Swerdlow 1975).

In his derivation of the elliptical orbit of Mars, Kepler seems to suggest that this result came as a surprise to him in the course of investigating various preliminary hypotheses that were later discarded. Despite the difficulties involved in reconstructing Kepler's procedures in arriving at the ellipse (see, e.g., Wilson 1968), I believe that he was guided once again by cosmic plan of creation and the distance-velocity relation, but detailed analysis must be left for another occasion.

Conclusion

Finally, we may note that Kepler's use of archetypal reasoning seems to have given more trouble to his modern readers than the complexity of his mathematical arguments. There is a related difficulty in interpreting his use of physical analogy which also deserves extensive treatment. Moreover, Kepler often interspersed archetypal reasoning with reasoning based on forces, adding to the confusion.

I have attempted to indicate that Kepler's religious and methodological commitments — as well as those that directly concern mathematics and natural philosophy — need to be exposed in order for us to appreciate fully his mode of thinking and to avoid anachronistic interpretations of his work. In the absence of such a global treatment of Kepler, even his technical achievements will remain unintelligible.

NOTES

This research was supported, in part, by grants from the National Science Foundation and the National Endowment for the Humanities, and it was undertaken in collaboration with Peter Barker. This paper was first presented at a workshop, "Kepler's Unification of Physics and Astronomy," held at the University of Groningen (The Netherlands) in June 1992 and, with some modifications, at the Center for Philosophy of Science (University of Pittsburgh), Annual Lecture Series, September 1992; the Joint City University of New York and Courant Institute–NYU History of Mathematical Sciences Seminar, May 1993; and the annual meeting of the History of Science Society (Santa Fe), November 1993. An earlier version was incorporated in "Distance and Velocity in Kepler's Astronomy," *Annals of Science* 51 (1994): 59–73 (with Peter Barker).

REFERENCES

Aiton, E. J . 1969. "Kepler's Second Law of Planetary Motion." *Isis* 60:75–90.

Barker, P. 1993. "The Optical Theory of Comets from Apian to Kepler." *Physis* 30:1–25.

Birkenmajer, L. A., ed. 1900. *Commentariolum super theoricas novas planetarum Georgii Purbachii in studio generali cracoviensi per Mag. Albertum de Brudzewo diligenter corrogatum A.D. MCCCCLXXXII.* Cracow: Jagellonian University.

Blair, A. 1990. "Tycho Brahe's Critique of Copernicus and the Copernican System." *Journal for the History of Ideas* 51:355–77.

Buchdahl, G. 1972. "Methodological Aspects of Kepler's Theory of Refraction." *Studies in the History and Philosophy of Science* 3:265–98.

Caspar, M. 1962. *Kepler: 1571–1630.* Trans. C. D. Hellman. New York: Collier Books.

Chevalley, C., trans. 1980. *Kepler: Les fondements de l'optique modern: Paralipomènes à Vitellion.* Paris: Vrin.

Christianson, J. R. 1979. "Tycho Brahe's German Treatise on the Comet of 1577." *Isis* 70:110–40.

Copernicus, N. 1543. *De revolutionibus orbium coelestium.* Nuremberg.

Dijksterhuis, E. J. 1961. *The Mechanization of the World Picture.* Oxford: Clarendon Press.

Donahue, W. H., trans. 1992. *Kepler: New Astronomy.* Cambridge: Cambridge University Press.

——. 1993. "Kepler's First Thoughts on Oval Orbits: Text, Translation, and Commentary." *Journal for the History of Astronomy* 24:71–100.

——. 1994. "Kepler's Invention of the Second Planetary Law." *British Journal for the History of Science* 27:89–102.

Dreyer, J. L. E. [1906]. 1953. *A History of Astronomy from Thales to Kepler.* Reprint ed. New York: Dover.

Dreyer, J. L. E., ed. 1922. *Tycho Brahe: Opera Omnia,* vol. 4. Copenhagen: Libraria Gyldendaliana.

Duncan, A. M., trans. 1981. *Kepler: Mysterium Cosmographicum: The Secret of the Universe.* New York: Abaris.

Gingerich, O. 1975. "The Origins of Kepler's Third Law." *Vistas in Astronomy* 18:595–601.

Jardine, N. 1984. *The Birth of History and Philosophy of Science.* Cambridge: Cambridge University Press.

Jarrell, R. A. 1989. "The Contemporaries of Tycho Brahe." In R. Taton and C. Wilson, eds., *Planetary Astronomy from the Renaissance to the Rise of Astrophysics; Part A: Tycho Brahe to Newton.* Cambridge: Cambridge University Press. Pp. 22–32.

Kepler, J. 1953. *GW 7.* Kepler: *Gesammelte Werke,* vol. 7. Edited by M. Caspar. Munich: C. H. Beck.

———. [1609] 1968. *Astronomia Nova.* Prague (facsimile ed. Brussels) Culture et Civilisation.

———. 1619. *Harmonices mundi libri v.* Linz.

Koyré, A. 1961. *La révolution astronomique.* Paris: Hermann.

Kusukawa, S. 1992. "Law and Gospel: The Importance of Philosophy at Reformation Wittenberg." *History of Universities* 11:33–57.

Maestlin, M. 1578. *Observatio et demonstratio cometae aetherae qui anno 1577 et 1578. . . .* Tübingen.

Reinhold, E. [1542] 1557. *Theoricae novae planetarum Georgii Purbachii. . . .* Paris.

Rosen, E. [1939] 1959. *Three Copernican Treatises.* Reprint ed. New York: Dover.

Schneider, J. R. 1900. *Philip Melanchthon's Rhetorical Construal of Biblical Authority: Oratio Sacra.* Lewiston N.Y.: Mellen Press.

Stephenson, B. 1987. *Kepler's Physical Astronomy.* New York and Berlin: Springer-Verlag.

Swerdlow, N. M. 1975. "The Planetary Theory of François Viète." *Journal for the History of Astronomy* 6:185–208.

Swerdlow, N. M. and O. Neugebauer. 1984. *Mathematical Astronomy in Copernicus's De Revolutionibus.* New York and Berlin: Springer-Verlag.

Westman, R. S. 1972a. "The Comet and the Cosmos: Kepler, Mästlin and the Copernican Hypothesis." *Studia Copernicana* 5:9–30.

———. 1972b. "Kepler's Theory of Hypothesis and the 'Realist Dilemma.'" *Studies in the History and Philosophy of Science* 3:233–64.

———. 1975. "The Melanchthon Circle, Rheticus, and the Wittenberg Interpretation of the Copernican Theory." *Isis* 66:165–93.

———. 1980. "The Astronomer's Role in the Sixteenth Century." *History of Science* 18:105–47.

Wilson, C. 1968. "Kepler's Derivation of the Elliptical Path." *Isis* 59:5–25.

2

Experiment, Community, and the Constitution of Nature in the Seventeenth Century

Daniel Garber
Department of Philosophy, University of Chicago

In his important and influential book, *How Experiments End*, Peter Galison (1987) discusses how scientists decide when a given experiment is finished, and when the supposed fact that it purports to establish can be accepted as fact, and not a mistaken reading of the apparatus, not a result of a malfunctioning piece of equipment, not a misinterpretation of a given observation, and so on. This epistemological question, the transition between individual observations, individual runs of a complex experiment, and the experimental fact that they are supposed to establish is a matter of some discussion in the recent literature in the history, philosophy, and sociology of science (see, e.g., Pickering 1984; Latour 1987; Labour and Woolgar 1986; and Shapin and Schaffer 1985). This is the question I explore in this essay.

While, in a sense, the question has been with us as long as people turned to experience to try and figure out how the world is, people were not always interested or aware of the question, and when they were, the answers that they suggested were not always the ones that we find most comfortable now. This essay addresses the history of the notion of an experimental fact, or, as Lorraine Daston has dubbed it, the prehistory of objectivity. In Boyle's generation in the Royal Society, as Shapin and Schaffer have emphasized, we have much of what we take for granted in experimental life, such as experiments performed on complex and temperamental equipment that often goes wrong, the centrality of the idea of reproducibility, and the idea of a community of scientists. But for a generation before Boyle, much of this familiar landscape is missing. Recent writers have emphasized the role played by the community of investigators in deciding what does and does not count as an experimental fact. This is a very prominent feature of the account of experimental facthood in the Royal Society. However, I argue that this is a very recent development.

Since I cannot tell the whole story in these few pages, I will begin with a brief discussion of experimental facthood in late Renaissance thought, before turning to Bacon, Descartes, and showing the extent to which their conception of experimental facthood is radically individualistic. I will then discuss the self-consciously social conception of experimental facthood found in the writings of the early Royal Society. After a digression about some recent issues concerning the rhetoric of scientific experiments in the period, I will end with some speculations about why the transition occurred when it did. The transformation in the philosophical view about the role of community in the establishment of experimental facts, I suggest, is closely connected with the emergence of a community entitled to make judgments necessary to establish such facts.

I should first address the distinction made between experiment and observation, that is, between information we get about the world from observing it as it follows its own natural course and information we get from torturing nature, as Bacon put it, setting up situations not normally found in nature and observing what happens. Important as this distinction is in the seventeenth century, it will not be relevant for this essay, so I will speak indifferently of observation, experiment, and experience.

Some Common Sense

People have turned to their senses for information about the world on which to ground their natural philosophy, their medicine and such, so long as there have been such disciplines. And so long as there have been such disciplines, halfway reflective people must have worried to at least some degree about how we can establish empirical facts about the world at the very lowest level, how we can be sure that the individual and particular observations we make on a given occasion are not misleading in some way, the product of chance or happenstance, malfunctioning equipment, a distracted observer, a nonrepresentative specimen, and so on. To this apparently simple question, we find in much early literature a relatively simple answer: When in doubt about a given observation or experiment, do it again.[1]

Peter Dear has recently found a very nice instance of this way of thinking about experiment in an obscure and generally unremarkable Jesuit textbook on optics published in 1613 by one Franciscus Aguilonius. Aguilonius writes:

A single sensory act does not greatly aid in the establishment of sciences and the settlement of common notions, since error can exist which lies hidden for a single act. But if the act is repeated time and again, it strengthens the judgement of truth until [that judgment] finally passes into common assent; whence afterwards they [i.e., the "common notions"] are put together, through reasoning, as with the first principles of a science. (Franciscus Aguilonius 1613, 215–16, quoted and translated in Dear 1991, 139)

Dear emphasizes that nothing is particularly original in Aguilonius's statement but that this statement in a way paraphrases Aristotle. In *Posterior Analytics* II.19, 100a, 5–7, Aristotle notes, "So from perception there comes memory, as we call it, and from memory (when it occurs often in connection with the same thing), experience; for memories that are many in number form a single experience" (translated in Barnes 1975, 81; see also *Metaphysics* I.1 980b, 28–30).

Aristotle's meaning here is by no means clear (for some indications of the complexities, see Barnes 1975, 253). But it is not too implausible to see Aristotle as standing behind Aguilonius's statement here. Aristotelian science is grounded not in individual events of sensory experience, particular observations made on particular occasions, but on the general course of experience, on common assent. It is not sufficient for an Aristotelian science that we have a particular observation that it snowed on the morning of 23 January 1979 in Chicago, Illinois, or that on 26 September 1664, a particular apple was observed to fall from a tree and hit Isaac Newton on the head. What is necessary for Aristotelian science is that it be generally accepted that it snows in northern climes in the winter months, or that heavy bodies fall; *this* is what constitutes experience as opposed to mere perception. To go from perception, the individual deliverance of the senses on a particular occasion, to what Aristotle and Aguilonius call experience, what we might call an experiential fact, requires the repetition of these individual perceptions. Should these individual perceptions speak with sufficient unanimity, then memory will transform them into experiential facts that can be acknowledged by common consensus and used as the foundation of a genuine body of knowledge. In this way an experimental fact can be regarded as a kind of low-level general statement established by repetition.

These perceptions can be repeated by many observers, but it is *sufficient* for them to be repeated by one observer alone; one observer, repeating the observation a sufficiently large number of times, is capable of constituting an experiential fact, on this conception. This conception of facthood is reflected in quite a number of figures in early modern science, and represents what might well be considered the commonsense view on the question as to how experimental facts are to be constituted. Consider, for example, the work of William Gilbert, who wrote in the preface of his *De Magnete* of 1600:

> Let whosoever would make the same experiments, handle the bodies carefully, skillfully and deftly, not heedlessly and bunglingly; when an experiment fails, let him not in his ignorance condemn our discoveries, for there is naught in these Books that has not been investigated and again and again done and repeated under our eyes. (Gilbert 1958, xlix)

Gilbert is aware that the complexity of the experiments he has performed, and the temperamental nature of the equipment he used, may cause difficulty for others to get the same outcomes that he did on his trials. Indeed, he begins the book proper by reporting on the mistaken results, corrected by his own more careful experiments, that others have gotten from antiquity to the present. Gilbert is convinced that his results are correct and that he has captured genuine experimental facts because he repeated his trials over and over again; "There is naught in these Books that has not been investigated and again and again done and repeated under our eyes," Gilbert writes, and these repetitions give him the authority to present his observations as fact.

Gilbert is hardly unusual here. Dear reports the same thread throughout a number of other writers of the period, including Galileo and Marin Mersenne: "I did the trial a hundred times, and it came out the same on every occasion" is a phrase that for these natural philosophers constitutes the ultimate justification for their confidence in a given experimental fact.

This is the same as the modern criterion for accepting experimental fact, but it is what is at issue here is not repeatability in general, but repeatability by the *individual experimenter*; to constitute a genuine fact, it must be possible for an experiment or observation to be reproducible, but to establish reproducibility, it is sufficient for the *individual investigator* to be able to reproduce the result a sufficiently large number of times. And so the individual investigator speaks with *complete authority*. If you the reader are not convinced, you can, of course, try the experiment yourself. But the benefits of this repetition accrue to you and you alone; as far as the investigator is concerned, the numerous repetitions that he did suffice to establish the result of his experiment as fact.

So much for common sense. Though natural philosophy and medicine had depended on observation of nature and experiment for many years before the new philosophers of the seventeenth century, with the new science, and with the increasing dependence on experience and the increasingly sophisticated forms that the appeal to experience took, came a new attention to the notion of experiment and experience.

Bacon and Experimental Facts

No seventeenth-century figure is more closely identified with the new experimental spirit in science than Francis Bacon. His program for science, *Instauratio Magna*, a plan for the revival and restoration of the sciences, has at its center the *Novum Organum* of 1620, a new logical instrument describing how to build a new science, that is, one more adequate than the Aristotelian science that still very much dominated the intellectual world in which Bacon grew up. And at the center of the new method outlined in the *Novum Organum* is observation and experiment, the collection of facts and their arrangement into natural histories.

The first step is simply the collection of experiments and observations, "For first of all we must prepare a natural and experimental history, sufficient and good; and this is the foundation of all, for we are not to imagine or suppose, but to discover, what nature does or may be made to do" (*Nov. Org.* II 10; in the *Novum Organum*, Bacon gives little guidance as to how to plan a series of experiments; on this see the discussion in the *De Augmentis* 1623, vol. 5, 2, Bacon 1863, vol. 9, 71f). But a natural history, a random collection of facts, is too unwieldy to work with directly, so, Bacon suggests, "we must therefore form tables and arrangements of instances, in such a method and order that the understanding may be able to deal with them" (*Nov. Org.* II 10). Consider Bacon's example on the investigation of the nature of heat. Bacon begins with what he calls the table of "Instances Agreeing in the Nature of Heat," or "Table of Essence and Presence," in which are listed a variety of circumstances in which heat may be found, including fiery meteors, quicklime sprinkled with water, iron dissolved in acid, and fresh horse dung (ibid., II 11). Bacon calls the second table "Instances in Proximity where the Nature of Heat is Absent," in which Bacon examines each entry in the table of essence and presence and tries to find similar circumstances in which heat is absent. So, for example, connected with the observation that iron in acid produces heat, Bacon notes that softer metals, like gold and lead, do not give off heat when dissolved in acid. Bacon calls the third table the "Table of Degrees," in which he observes things that contain the nature of heat, for example, in greater or lesser degree. And so he observes that while old dung is colder than fresh dung, it has a potential for heat insofar as it will produce heat when enclosed or buried. Similarly, Bacon observes that different substances burn with different degrees of heat (ibid., II 13).

Once we have compiled the natural history and arranged it into the proper tables, we are ready for the first inductive step, which Bacon calls the first vintage. Bacon writes that "the problem is, upon a review of the instances, all and each, to find such a nature as is always present or absent with the given nature, and always increases and decreases with it" (ibid., II 15). That is, in the case of heat, we want to find that which is always present when heat is present, and always absent when heat is absent. This proceeds in two stages. First Bacon uses his tables to exclude possible natures. For example, though Bacon thinks that heavenly bodies are hot, being a heavenly body cannot be part of the nature of heat since some terrestrial bodies are hot as well (ibid., II 18). Once we have thus excluded candidates for the nature of heat, we can then examine what is left and determine what all hot things have in common. Bacon suggests that heat is a particular kind of motion, which is an aspect common to all hot things found in the tables of our natural history: "Heat is a motion, expansive, restrained, and acting in its strife upon the smaller particles of bodies . . . not sluggish, but hurried and with violence" (ibid., II 20).

Presumably the first vintage is followed by successive vintages, in which we press more knowledge of nature from our initial observations. Furthermore, Bacon suggests, the knowledge we have derived from experiment will in some way suggest to us new experiments to perform, though he doesn't indicate how this might work. (See *Nov. Org.* II 10, where Bacon suggests that the interpretation of nature involves both deriving axioms from experience and deducing and deriving "new experiments from the axioms.)

The method of the *Novum Organum* is exemplified in the organization of the House of Salomon, the perfect scientific society that Bacon envisions in his science fiction story, the *New Atlantis*, published posthumously in 1627. At the bottom of the organization are those who form the tables of natural history, a total of 24 investigators. Twelve "Merchants of Light" sail into foreign countries under the names of other nations . . . [and] bring us the books and abstracts, and patterns of experiments of all other parts" (Bacon 1906, 273). Three "Depredators" collect experiments from books, three "Mystery-men" collect experiments from mechanical arts and liberal sciences, and three "Pioneers or Miners" try new experiments of their own devising. They are joined by three "Compilers" who arrange these observations and experiments into proper tables (ibid.). Twelve workers are employed at the next stage of the enterprise. Three "dowry-men or Benefactors" examine the initial tables compiled by the Compilers, and draw out both technological applications as well as the first theoretical conclusions that can be drawn from the tables, presumably what Bacon calls the First Vintage in the *Novum Organum*. Three "Lamps" then draw new experiments out of the work of the Compilers and Benefactors, which experiments are then performed by three "Inoculators." Finally, "we have three that raise the former discoveries by experiments into greater observations, axioms, and aphorisms. These we call Interpreters of Nature" (ibid., 274). In this way, Bacon's method for investigating nature is readily adopted to science as a social and cooperative enterprise. It is no wonder that organizations like the Royal Society looked back to Bacon for inspiration.

A great deal of attention has been given to the inductive stage in Bacon's method, of what it is and why it doesn't work. However, I will focus on the first and apparently less problematic stage, the collection and construction of natural histories, in particular, on the way Bacon thinks that the empirical facts contained in a natural history are to be established and checked.

In the *Advancement of Learning* of 1605, and later in the expanded and Latinized version of that work, the *De Augmentis* of 1623, Bacon offers a categorization of all human learning based on his conception of the mind: "The Best division of human learning is that derived from the three faculties of the rational soul, which is the seat of learning. History has reference to the Memory, poesy to the Imagination, and philosophy to the Reason" (1863, vol. 8; see also 1906, 75–76; see also the account in the *Descriptio Globio Intellectualis* in Bacon 1863, vol. 10, 404). In Bacon's conception of the category of history, he recognizes a number of different kinds of history: natural, civil, ecclesiastical, and literary. Unlike philosophy proper, which deals with abstractions and generalities, history deals with particulars, and in Bacon's conception, particular events in nature that happened at particular times. But Bacon suggests in the *De Augmentis* that these matters are complicated:

History is properly concerned with individuals, which are circumscribed by place and time. For though Natural History may seem to deal with species, yet this is only because of the general resemblance which in most cases natural objects of the same species bear to one another; so that when you know one, you know all. And if individuals are found, which are either unique in their species, like the sun and moon; or notable deviations from their species, like monsters; the description of these has as fit a place in Natural History as that of remarkable men has in Civil History. (1863, vol. 8, 407; see also vol. 10, 406)

History is the domain of atomic facts, as it were. But Bacon recognizes that some of these facts are more general than others. When we are dealing with knowledge about specific individuals, the sun, the moon, Julius Caesar, and so forth, then history deals with statements keyed to particular places and times: the sun or moon was observed to be at such-and-such a position in the sky from such-and-such a place at such-and-such a time; Julius Caesar was observed to have uttered such and such words at a particular place at a particular time. But when dealing with natural historical matters, a certain kind of generality can creep in. One can drop a certain piece of gold in a particular vat of aqua regia at a given time, and note that it dissolves. This, of course, might happen because of the particularities of the situation, the particular characteristics of the sample of gold or aqua regia, or, indeed, the observer may be mistaken in thinking that the gold dissolved on that occasion. But given the general similarity of samples of gold and aqua regia, and the general reliability of observers, at least with respect to events like this, "it would be a superfluous and endless labor to speak of [each individual case] severally" (Bacon 1863, vol. 10, 406). And so in compiling natural histories it is permitted to speak generally and include as a fact in our natural history that gold dissolves when put in aqua regia. In general, this is the sort of entry found in Bacon's natural histories (see, e.g., 1863, vol. 9, 399). Though the facts are based on observation and experiment, Bacon includes not the reports of the particular observations and experiments he (or others) might have made at some particular place and time, but the report of the general fact that came out of the particular events of observation or experiment. In this, I suspect that Bacon exemplifies what I called the commonsense conception of how experimental facts are to be established. For Bacon, as for the commonsense view, experimental facts seem to be just the unproblematic generalization of repeated experience, that is, similar instances repeated that constitute a general experience.

But Bacon's account has a further complexity. Bacon's natural histories are compiled from a number of sources, from his own observations and experiments, from those others have made and either published or related to him, from accounts travelers have brought back, from books, encyclopedias, ancient accounts, even from common says and proverbs: "In Britain, the east wind is considered injurious, insomuch that there is a proverb, When the wind is in the east, 'Tis neither good for man nor beast" (1863, vol. 9, 402; see other sources cited *passim* in this work, such as Acosta, Columbus, Aristotle, Knolle's *History of the Turks*, Gilbert, and Virgil). Bacon suggests that we be liberal in what we include in natural histories; the only thing he categorically excludes is "superstitious stories . . . and experiments of ceremonial magic," which he dismisses as "old wives tales" (1863, vol. 8, 360, 49). But among the sorts of things that Bacon does allow in his natural history, he recognizes that there will be differences in the degree of certainty. In the *Parasceve*, the portion of the *Instauratio Magna* in which Bacon discusses the preparation of the natural history, he notes that possible entries in the natural history will be of three sorts: certainly true, certainly false, and doubtful (ibid., 366–68). As for the first two categories, there is no particular problem; facts that are certainly true belong in, and commonly accepted "facts" that are generally accepted but false should be exposed and rejected as such. But Bacon's treatment of the third category is the most interesting. He suggests that we add them to the natural history, but with appropriate indication of their status,

"Nor is it of much consequence to the business in hand because . . . mistakes in experimenting, unless they abound everywhere, will be presently detected and corrected by the truth of axioms" (1863, vol. 8, 360; see also *Nov. Org.* I 118).

According to Bacon, we will inevitably find in the reports of others, or even in our own experimental work, that false statements are accepted as experimental facts. When too many of the entries in our natural history have that character, then we are obviously in trouble. But if our natural history is in general reliable, then we have a way of weeding out these nonfacts. For, Bacon suggests, *we can use the general statements derived by induction from our natural history to correct that natural history.* That is, once we have derived general statements from our experience using the careful method that Bacon outlines in the *Novum Organum*, then we are entitled to reject anything that does not conform to those general statements, or "axioms" as he calls them. In this way, Bacon writes in the *Instauratio Magna*, "the senses deceive, but then at the same time they supply the means of discovering their own errors" (1863, vol. 8, 43).

In this way Bacon seems to go beyond the common sense conception of facthood outlined earlier in this essay. For common sense, a fact is established as a fact through repetition alone; one goes from individual occurrence, the particular observation, the single run of an experiment by repeating the event until it is certain that there is no mistake of any sort. But to this Bacon adds another criterion, at least when we are dealing with doubtful results. Facts, as embodied in a natural history, determine theory. But, Bacon holds, theory determines fact as well; for a purported experimental or observational fact to enter the body of knowledge, it must conform to theory.

As interesting to me as the account Bacon hits upon is the one he misses. One presumes that in at least many of the doubtful cases that Bacon has in mind, at least *one* investigator has done the experiment in question numerous times, and has established to his own satisfaction that he has identified a genuine experimental fact. It would be a natural suggestion that the doubtful results could be checked by having *other* investigators try the experiment as well. But Bacon does not suggest this. In Bacon's House of Salomon, among the 36 investigators employed full time exploring nature, *not one is ever asked to redo* an experiment originally done by another investigator. As we shall see, matters are different when the House of Salomon is actually organized a generation later as the Royal Society. But first, let us turn to René Descartes, another important theorist of method in early seventeenth-century natural philosophy.

Descartes and Experimental Facts

While Bacon and Descartes are seeming opposites — the experimental-ist versus the rationalist — the two are not as distant from one another as we might think. Both are moderns from the point of view of the early seventeenth century, opponents of the sterile Aristotelian science of the schools, and both saw a new method of investigation as central in the attack against the old and in the establishment of a new science more adequate than the old. Descartes refers favorably to Bacon and his pro-gram in his correspondence, and his *Discours de la méthode* of 1637 echoes the very rhetoric of the *Instauratio Magna*, published while the young Descartes was working out his own ideas about scientific pro-cedure in the unfinished and unpublished *Regulae* of the 1620s. (See AT I 109, 195–96, 251. On the relation between Bacon's and Des-cartes's writings, see Lalande 1911. References in Descartes 1964–74 are abbreviated "AT"; volume and page are given in Roman and Ara-bic numerals. Translations for most of the material cited is found in Descartes 1984–92. Translations are not cited separately.) Descartes is certainly more circumspect about experiment than Bacon is, and trusts to reason more than Bacon does. But it is important to recognize that while experiment may play a somewhat more restricted role in Des-cartes's enterprise than in Bacon's, Descartes considered experiment crucial to the advance of his own program as well. Experiment appears prominently in his celebrated account of the rainbow in discourse 8 of the *Météores*, a discussion that he points to as a paradigm example of the method of the *Discours*. There, Descartes appeals to experiments done with prisms and flasks of water to support his conclusions about the cause of the rainbow. (See AT VI 325–44; it is identified as a product of the method on p. 325, line 7, which is the only reference to the method of the *Discours* in any of the three accompanying *Essais*. It is identified as "a brief sample of the method," the only example so identified, in a letter to Vatier, 22 February 1638, AT I 559). In re-sponse to a criticism of his *Principia Philosophiae* of 1644, transmitted by Huygens, that his views are insufficiently confirmed by experience or experiment, Descartes claims that there are "almost as many experiments as there are lines in my writings" (AT IV 224, possibly Descartes to Huygens, June 1645). Finally in part Vi of the *Discours*, Descartes's most prominent complaint is that the completion of his work is hindered by the lack of sufficient observations and experiments (see, e.g., AT VI 63, 65, 73).

Before turning to Descartes's conception of experiment and the constitution of experimental facts, let us take a brief and sketchy excursion into Descartes's conception of method, particularly as it is set out in the early *Regulae*. (For a fuller account of the *Regulae*, see Garber 1992, chap. 2, or Garber 1993, from which this discussion is borrowed.) Descartes ultimately wants to construct a deductive science. At bottom is what he calls intuition, the ability we have to immediately grasp certain truths with complete certainty. Descartes thinks that we can also see intuitive connections between some propositions known and others; this is what he calls deduction. All knowledge properly speaking, *scientia*, must come from intuition and deduction; completed science will have the structure of conclusions deduced from initially intuited premises. His method is a procedure for constructing such a science (see Rules 1–3).

Descartes's method is illustrated by his example of methodical investigation in Rule 8 of the *Regulae*. As illustrated in that example, Descartes's method has two parts: a reductive step, leading us from a question posed to an intuition, and a constructive step, in which a deduction of the answer to the question is presented. In Rule 8 Descartes poses the problem of finding the anaclastic line, that is, the shape of a surface "in which parallel rays are refracted in such a way that they all intersect in a single point after refraction" (AT X 394). Descartes notices — and this seems to be the first step in the reduction — that "the determination of this [anaclastic] line depends on the relation between the angle of incidence and the angle of refraction" (ibid.). But, Descartes notes, this question is still "composite and relative," that is, not sufficiently simple, and we must procede further in the reduction. Rejecting an empirical investigation of the relation in question, Descartes suggests that we must next ask how the relation between the angles of incidence and refraction is caused by the difference between the two media, for example, air and glass, which in turn raises the question as to "how the ray penetrates the whole transparent thing, and the knowledge of this penetration presupposes that the nature of the illumination is also known" (ibid., 394–95). However, Descartes claims, in order to understand what illumination is we must know what a natural power [*potentia naturalis*] is. Here the so-called reductive step of Descartes's method ends. At this point, Descartes seems to think that we can "clearly see through an intuition of the mind" what a natural power is (ibid., 395). Other passages suggest that this intuition is intimately connected with motion (see ibid., 402). Once we have such an intuition, we can begin the constructive step, and follow, in order, through the questions raised until we have answered the original question of the shape of the anaclastic line. This would involve understanding the nature of illumination from the nature of a natural power, understanding the ways rays penetrate transparent bodies from the nature of illumination, and the relation between angle of incidence and angle of refraction from all that precedes. Finally, once we know how angle of incidence and angle of refraction are related, we can solve the problem of the anaclastic line.

If we take the anaclastic line example as our guide, then methodological investigation begins with a question that, in turn, is reduced to questions whose answers are presupposed for the resolution of the original question posed (i.e., q_1 is reduced to q_2 iff we must answer q_2 before we can answer q_1). In a sense, then, the reduction leads us to more basic and fundamental questions from the anaclastic line to the law of refraction, and back eventually to the nature of a natural power and to the motion of bodies. Ultimately, Descartes thinks, when we follow out this series of questions, from the one that first interests us to the "simpler" and more basic questions on which it depends, we will eventually reach an intuition. When the reductive stage is taken to this point, then we can begin the constructive stage, turn the procedure on its head, and begin answering the questions that we have successively raised, in an order reverse of the order in which we have raised them. This *should* involve starting with the intuition that we have attained through the reductive step, and *deducing* down from there, until we have answered the question originally raised. Should everything work out as Descartes hopes it will (which it won't, but that's another story), then when we are finished we will have the certain knowledge he wants; an answer arrived at in this way will constitute a conclusion deduced ultimately from an initial intuition.

All of this is impressive, in a way. But where does experiment come in? How could Descartes have thought that experiment fits into his conception of scientific practice? There is not the space in this essay to enter into this question in the full detail that it deserves (for a fuller account, see Garber 1993). Briefly, experiment does not enter into the method proper. Rather, Descartes conceives of experiment as a kind of auxiliary in the reductive stage of the method that allows us to pass from one question to the next.

Consider the anaclastic line example, for instance. At one point in the argument Descartes says that the investigator must notice that the relation between the angles of incidence and refraction depends upon the changes in these angles due to the differences in the media through which the ray is passing (e.g., from air into glass, or water into air) and that these changes, in turn, depend upon the way in which the ray penetrates the transparent body (AT X 394). While this step may not require sophisticated optical experiments, it seems to require at the very least some minimal experience with light rays and lenses or other actual instances of refraction in order to see that light is typically bent by passing from one medium into another, and to come to the realization that in order to discover the law refraction obeys we must first understand how light passes through media of different sorts. Experiment that helps to perform the reduction and determine what question we should take up next in our investigation.

Descartes uses such appeals to experience more explicitly in his discussion of the rainbow. In that case he is interested in discovering how it is that colors arise in the rainbow. On the basis of experiments with a spherical flask of water, Descartes claims that the rainbow has two distinct bands of color, a primary and a secondary bow, and that the two bows of the rainbow derive from two combinations of reflection and refraction within a droplet of water. From this one might conjecture that the color might arise from the reflection, the refraction, or the fact that the droplets are spherical. But experiments with a prism show that color can arise from refraction alone. Reflection and the spherical surface of the droplet are thus judged irrelevant to the phenomenon, and in the next step of the reduction Descartes focuses on the question as to how refraction might produce colors in white light. Once again, experiment helps us to determine what our next question should be.

Descartes's use of experiment is quite different from Bacon's. Science, though experimental in a sense, remains deductive for Descartes; Baconian induction has no apparent role to play. But experiment seems to play a role in preparing the deduction. Insofar as it helps perform the reductive part of the method, the sequence of steps that leads from a question to an intuition, it helps determine the deduction, the same steps followed in reverse order that leads from intuition to the answer to the question posed. The deductive chain that the Cartesian scientist seeks in reason, the chain that goes from more basic to less, is *exemplified* in the connections we find in nature. Insofar as these latter connections are open to experimental determination, we can use experiment to sketch out the chain of connections in nature and find out what depends on what, *and thus we can use the connections we find in nature as a guide to the connections we seek in reason.* It may not be obvious to us at first just how we can go deductively from the nature of light to the rainbow, but poking about with water droplets, flasks, and prisms may suggest a path our deduction might follow.

Unlike Bacon's, Descartes's science is not directly grounded in natural history; yet the sorts of tables that Bacon recommends are not altogether irrelevant to Descartes's procedure. Writing to Mersenne on 10 May 1632, Descartes notes that "it would be very useful if some . . . person were to write the history of celestial phenomena in accordance with the Baconian method . . . without any arguments or hypotheses" (ibid., I 251). Such tables of phenomena and their correlations with one another, independent of any theory, is precisely what Descartes needs to determine the relations of dependence of one phenomenon upon another necessary to perform the reductive step of the method. But what status do the experimental facts that go into a natural history have for Descartes?

Descartes, of course, is well known for his distrust of the senses. He warns us that things are not at all as our senses tell us they are, that they are not red and green, sweet or salty, that our naive belief that all of our knowledge derives ultimately from our senses is a prejudice of sense- and body-bound youth, a prejudice that must be rejected before we will be able to penetrate to the true nature of things. However, he did not reject experience altogether.

Descartes's fullest, but complicated, account of the senses is in Meditation VI. Briefly, as with clear and distinct perceptions, Descartes deals with something that God gave us. As such, Descartes argues, they must be in some sense true: "It is doubtless true that everything that nature teaches me [and this includes the senses] has some truth in it" (ibid., VII 80). When it is truth about the nature of things that we are interested in, it is the light of reason, clear and distinct perceptions, that we must turn to first. And so, while some of the teachings of nature will turn out to be true, only the intellectual examination of them will establish this. In this way Descartes rejects the hyperbolic rejection of the senses that begins the *Meditations*, and, indeed, goes on to reject even the dream argument that is so prominent in Meditation I (ibid., 89–90). But though the teachings of nature, what we learn from our senses, are restored, they are subordinate to reason; they may be trusted to some extent and in some circumstances, but only after they have been given a clean bill of health by reason.

This can be illustrated by an example from the rainbow case. In the rainbow case, Descartes begins by observing that on his flask, the stand-in for the raindrop, there are two regions of color, at roughly 42° and 52° from the ray of sunlight; these two regions correspond to the primary and secondary bow of the rainbow. This observation is the starting place of his account, and we can presume that he repeated it often enough to convince himself that it is trustworthy.[2] Then, in the end, Descartes actually *deduces* from his law of refraction that parallel rays of light from the sun will converge at almost exactly those two angles after the appropriate number of reflections and refractions. After giving his account, Descartes notes that an earlier observer, Maurolicus, set the angles incorrectly at 45° and 56°, on the basis of faulty observations. Descartes remarks that "this shows how little faith one ought to have in observations which are not accompanied by the true reason" (AT VI 340). Only because we can *calculate* the angles of the primary and secondary bows from the account we have of the rainbow can we be sure of what they are, *despite* that the investigation *began* with an experimental determination of those angles. Though it is an observation that starts the ball rolling, *it is only through deduction that an experimental fact observed can actually enter the body of scientific knowledge, strictly speaking.* Descartes uses experiment here in the way we might in geometry. In geometry we might use carefully drawn diagrams and measurements made from them to suggest possible theorems. But still, we would want to hold, any geometrical facts found in this way are grounded in the geometrical demonstration, and not in the diagram that may have originally suggested the fact to the investigator. In this way, Descartes notes, "When [Pierre Petit] promises to refute my [laws of] refraction through experience, there is no more reason to listen than if he wanted to show that the three angles of a triangle aren't equal to two right angles by way of some faulty square rule" (Descartes to Mersenne, 9 February 1639, AT II 497). Observation and experiment may play an important role in establishing an experimental fact, but reason must confer the ultimate status of facthood on an observation. While there are important differences in detail, of course, Descartes's account here is not unlike Bacon's; for both there is an important sense in which theory must constitute experimental facts.

One further feature of Descartes's attitude toward experiment and experimental facts is that one of his basic commitments, indeed one of his obsessions, is the rejection of authority and the consequent centrality of the individual over community. In the *Regulae* Descartes emphasizes that only what an individual intuits and deduces is real knowledge for him; knowledge by authority is no knowledge at all (Rule 3).[3] The whole message of the *Discours de la méthode* is the rejection of authority and the importance of the individual building a world for himself. This is the project that is actually taken up in the *Meditations*, where the meditator begins by obliterating the world around him, and starting from scratch, building a world from the *cogito*, the thought of a solitary self. This radical individualism is also reflected in Descartes's attitude toward experimental science. Part VI of the *Discours de la méthode* is concerned with the need for additional experiments in order to complete Descartes's scientific program. Descartes begins by reporting the attitude he took in his youth. Originally, he reports, he believed that he should publish the details of his foundations for physics and the full system based on those in order to stimulate the work of others, to get others to build on his foundations, and make the new observations necessary to finish the job. So, Descartes thought, publishing his thoughts would convince others to "assist me in seeking those [observations] which remain to be made" (i.e., send money). At that time Descartes also hoped that others would "communicate to me the observations they have already made" (AT VI 65; see also 63). But motivated at least in part by the condemnation of Galileo in 1633, Descartes reports that he changed his mind about the wisdom of publishing his full system of physics (ibid., 60, 65). With that change, came others. He came to think, first of all, that others were not really in much of a position to advance his program; as the building metaphors of part II of the *Discours* suggest, that work is best done that is done by a single individual (ibid., 69, 11ff). Descartes also changed his mind about the value of experiments done by others. Descartes admits that "as regards observation . . . one man could not possibly make them all" (ibid., 72). But, he asserts, apart from paid assistants, "he could not usefully employ other hands than his own" (ibid.). Descartes continues:

And as for observations that others have already made, even if they were willing to communicate them to him . . . they are for the most part bound up with so many details or superfluous ingredients that it would be very hard for him to make out the truth in them. Besides, he would find almost all of these observations to be so badly explained or indeed so mistaken . . . that it would simply not be worthwhile for him to spend the time required to pick out those which he might find useful. (Ibid., 73)

Descartes concludes:

So, if there were someone in the world whom we knew for sure to be capable of making discoveries of the greatest possible importance and public utility, and whom other men accordingly were eager to help in every way to achieve his ends, I do not see how they could do anything for him except to contribute towards the expenses of the observations that he would need and, further, prevent unwelcome visitors from wasting his free time. (Ibid.)

The message is clear: send your money, *not* your observations, to R. Descartes, care of the publisher. (And don't visit either.) Experimental science is thus, for Descartes, a solitary activity that does not require a community and would be hindered by having to take place within a community.

While Descartes and Bacon may agree to some extent about the constitution of experimental facts, the contrast with Bacon here is dramatic. In response to the failure of the philosophy of the schools, their arguments from authority, their book learning, their disputations, Bacon turns to a new society and new forms of cooperative enterprise. Bacon's new society, the House of Salomon, is a society that institutionalizes his new experimental philosophy and exists outside of the schools altogether. Descartes, on the other hand, instead of trying to create a new society, sees inherent problems in any cooperative conception of the creation of new knowledge. Descartes thus chooses to place the scientist outside society altogether.

The Royal Society and the New Experimental Philosophy

When Thomas Spratt stepped forward in 1667 to defend the new Royal Society of London, a self-professed society for the promotion and perfection of the experimental approach to science, he turned to Bacon as a distinguished ancestor:

> I shall onely mention one great Man, who had the true Imagination of the whole extent of this Enterprize, as it is now set on foot; and that is, the Lord Bacon. In whose Books there are every where scattered the best arguments, that can be produc'd for the defence of Experimental Philosophy; and the best directions, that are needful to promote it. All which he has already adorn'd with so much Art; that if my desires could have prevail'd with some excellent Friends of mine, who engag'd me to this Work: there should have been no other Preface to the History of the Royal Society, but some of his Writings. (Spratt [1667] 1958, 35–36)[4]

But how was Bacon an inspiration? Certainly Spratt and his colleagues were attracted by his emphasis on experiment and natural history as the basis of all natural philosophy, and by his emphasis on the cooperative and communal nature of scientific investigation. Thus Joseph Glanvill writes in his *Plus Ultra*, a sympathetic though not-quite-authorized account of the Society:

> The deep and judicious Verulam [Bacon] . . . proposed . . . to reform and inlarge Knowledge by Observation and Experiment, to examine and record Particulars, and so to rise by degrees of Induction to general Propositions, and from them to take direction for new Inquiries, and more Discoveries, and other Axioms . . . So that Nature being known, it may be master'd, managed, and used in the Services of humane Life. This was a mighty Design, groundedly laid, wisely exprest, and happily recommended by the Glorious Author, who began nobly, and directed with an incomparable conduct of Wit and Judgment: But to the carrying it on, It was necessary there should be many Heads and many Hands, and Those formed into an Assembly, that might intercommunicate their Tryals and Observations, that might joyntly work, and joyntly consider. . . . This the Great Man desired, and form'd a SOCIETY of Experimenters in a Romantick Model; but could do no more: His time was not ripe for such Performances. (Glanvill [1668] 1958, 87–88; on the status of Glanvill's *Plus Ultra*, and its relation to the Royal Society, see Purver 1967, 13–14.)

The "Romantick Model" is, of course, the House of Salomon; though the time may not have been ripe for the realization of such a design in the 1620s, when Bacon envisioned it, Glanvill and his friends thought that the 1660s was just the time to realize Bacon's ambitious vision.

Though Bacon was the inspiration for the Society, and lauded for his great vision, he was not followed in every particular. Spratt, for example, appears to reject systematic rules of experimental method: "the true Experimenting has this one thing inseparable from it, never to be a fix'd and settled Art, and never to be limited by constant Rules" ([1667] 1958, 89). Though it is not entirely clear what exactly Spratt means to reject, and though Bacon is not mentioned by name here, it is not implausible to see in this a criticism of Bacon's fixed (though not rigidly so) methodology for experimental procedure. But most interesting for my purposes is another criticism Spratt directs at Bacon:

His Rules were admirable: yet his History not so faithful, as might have been wish'd in many places, he seems rather to take all that comes, then to choose; and to heap, rather, then to register. But I hope this accusation of mine can be no great injury to his Memory; seeing, at the same time, that I say he had not the strength of a thousand men; I do also allow him to have had as much as twenty. (Ibid., 36)

Though Bacon saw the importance of observation and experiment for the advancement of science, his natural histories are defective, Spratt argues. Spratt seems to recognize that Bacon does eventually sort through and reject some of the purported experimental facts that find their way into his natural histories; though he initially takes "all that comes," he eventually chooses what to base his induction on and eventually rejects observations that conflict with the general principles arrived at by induction. But, Spratt suggests, one must be more selective in the first place, and weed out bad observations before they even find their way into one's natural history. Such careful attention to the establishment of experimental facts is basic to the mission of the new Royal Society, Spratt argues; indeed, it is built into the very structure of that community.

The Royal Society was interested in gathering experimental facts from all who had them to contribute. But, Spratt writes, "I shall lay it down, as their Fundamental Law, that whenever they could possibly get to handle the subject, the Experiment was still perform'd by some of the Members themselves" (ibid., 83; when reporting this as "their Fundamental Law," Spratt is reporting what they *agreed* to do; what they actually did is another question. In what follows I limit myself to discussing what the Society thought of themselves as doing, their avowed practice, not what they did). That the experiment be performed not by one of the members, but by *some* of the members is crucial. When the Royal Society sponsored an experiment or series of experiments, as a matter of policy, Spratt reports, a number of different members were to be involved. Experiments were organized, Spratt writes:

. . . either by allotting the same Work to several men, separated one from another; or else by joyning them into Committees. . . . By this union of eyes, and hands there do these advantages arise. Thereby there will be a full comprehension of the object in all its appearances; and so there will be a mutual communication of the light of one Science to another: whereas single labours can be but as a prospect taken upon one side. And also by this fixing of several mens thoughts upon one thing, there will be an excellent cure for that defect, which is almost unavoidable in great Inventors. It is the custom of such earnest, and powerful minds, to do wonderful things in the beginning; but shortly after, to be overborn by the multitude, and weight of their own thoughts; then to yield, and cool by little and little; and at last grow weary, and even to loath that, upon which they were first the most eager. . . . For this the best provision must be, to join many men together. (Ibid., 84–85; see 100; see also Glanvill [1668] 1958, 108–09, 114)

The claim that experiments must be done by a number of *different* hands, and not by *one* experimenter's repetition, is explicit and carefully thought out.

In addition to the claim that experiments must be repeated by a variety of hands, Spratt further reports that facts must be established through the consensus of the community as a whole:

[After the performance of an experiment] comes in the second great Work of the Assembly; which is to judge, and resolve upon the matter of Fact. In this part of their imployment, they us'd to take an exact view of the repetition of the whole course of the Experiment . . . never giving it over till the whole Company has been fully satisfi'd of the certainty and constancy; or, on the otherside, of the absolute impossibility of the effect. This critical, and reiterated scrutiny of those things, which are the plain objects of their eyes; must needs put out of all reasonable dispute, the reality of those operations, which the Society shall positively determine to have succeeded. . . . [T]here is not any one thing, which is now approv'd and practis'd in the World, that is confirm'd by stronger evidence, than this, which the Society requires; except onely the Holy Mysteries of our Religion. ([1667] 1958, 99–100)

Experimental facts are now established by the community as a whole.

In this way, we have a new sense of reproducibility entering into the conception of an experimental fact. On the commonsense view, the repetition of an observation or experiment to the investigator's satisfaction was important, not *who* or *how many times* the repetition was done. But on the view of the Royal Society, to establish an experimental result as a genuine fact, it *must* be repeatable (and must be repeated) by a number of *different* persons, or at least it must have been repeated in their presence. (See Shapin and Shaffer 1985, 55f, on the practice of performing experiments in public for this purpose.) Repeatability in this sense is a considerably more stringent requirement for facthood.

Spratt's appeal to community is a repudiation of the Cartesian ideal of the solitary investigator, to be sure, but it is also a repudiation of Bacon's conception of experimental science. Bacon saw experimental philosophy as a cooperative venture. But for Bacon, the main advantage of the numerous investigators working together is that more facts can be collected for one's natural history, and consequences can be derived more expeditiously. As I pointed out earlier, not one of the thirty six investigators in the House of Salomon, Bacon's "Romantick Model" for the Royal Society, is involved in reproducing experiments done originally by others; experiments, done by individuals, working alone (with their servants and assistants acting only as extensions of themselves), enter into the natural histories directly. They do not discuss the experiments that some members of the House are deputed to perform. However, built into the very structure of the Royal Society are experiments to be performed by many hands, witnessed by many eyes, and certified as facts by the Society as a whole. For what I have called the commonsense conception of experiment, an experimental fact is established by an individual, through the senses. For Bacon, and in a different way for Descartes, an experimental fact is also established by an individual, though not directly through the senses. Although Bacon recognizes the importance of community to the advance of knowledge, in the establishment of particular experimental facts, he seems to be as much an individualist as Descartes is. However, on the new conception of the Royal Society, an experimental fact can be established through the senses, *but not by an individual.* Experiments end and experimental facts are constituted not when the individual investigator decides that it is time, but after an experiment is repeated by more than one investigator and when the community as a whole is satisfied that a fact has been established.

My account of the view of experimental facthood in the Royal Society is not particularly novel; in essence it is the backbone of Shapin and Shaffer's important study, *Leviathan and the Airpump* (1985, 55f).[5] However, I want to emphasize that this communitarian view of experimental facthood was new, a self-conscious innovation introduced by the Royal Society in the 1660s. One might possibly be able to find precedents for this, though I doubt it. It is an idea that is not found in the important theorists of scientific practice in the generations immediately preceeding the foundation of the Royal Society, and was regarded as an innovation by the Royal Society itself, a new and improved way of thinking about experiment, which is to say that *the social conception of experimental facthood is an idea with a history*. It arises at a particular time, in particular contingent circumstances.

The Rhetoric of Experimental Reports and the Constitution of Experimental Fact

Recent work on mid-seventeenth-century experimental science, particularly that of the Royal Society, has called attention to the way in which experimental results are reported. In the mid-seventeenth century, experimental reports become radically particular, in contrast to what they had been in earlier writers. In earlier writers, it is claimed, experimental reports are given in general terms, "such-and-such may be observed in such-and-such circumstances." In the Royal Society, however, reports characteristically describe exactly what was observed to happen in a particular place, at a particular time, with particular equipment, and particular people in attendance, both the successes and the failures. This seems to be a matter of conscious policy:

Whatever they have resolv'd upon; they have not reported, as unalterable Demonstrations, but as present appearances: delivering down to future Ages, with the good success of the Experiment, the manner of their progress, the Instruments, and the several differences of the matter, which they have apply'd: so that, with their mistake, they give them also the means of finding it out. (Spratt [1667] 1958, 108)

Dear (1985, 1987, 1991) is particularly insistent on this point in a series of penetrating articles (see also Shapin and Shaffer 1985, 60ff.). Dear relates this change to the rejection of an Aristotelian conception of natural philosophy:

"Experience" as an element of scholastic natural philosophical discourse took the form of generalized statements about how things usually occur; as an element of characteristically seventeenth-century, non-scholastic natural philosophical discourse it increasingly took the form of statements describing specific events. . . . For the scholastic natural philosopher, writing his commentaries on Aristotle, the grounding in experience of the physical facts debated in his discussions was guaranteed by their generality as experiential statements — "heavy bodies fall" is a statement to which all could assent, through common experience embodied in authoritative texts. . . . The new "experience" of the seventeenth century . . . established its legitimacy in historical reports of events, often citing witnesses. (1987, 134; Dear offers this passage as a summary of the main argument of Dear 1985.)

This apparently stylistic difference between the old and the new is actually substantive, Dear argues. When experience functions as the illustration of the universal statements that constituted the starting place of a scientific syllogism, as it does in Aristotelian science, there is little reason to expect controversy; all will agree that stones fall and fire rises. But in the new experimental science, particularly as practiced in the Royal Society, experiment functions to create novel facts. And here the situation is different. When one deals with novel facts, there is a possibility for controversy that simply didn't exist in earlier Aristotelian science; "Controversy, however, or the threat of controversy, demanded more radical measures, and at the same time placed greater emphasis on discrete events as justification for assertions" (Dear 1987, 169). When experiment makes novel claims, Dear argues, then the reporting of an observation or an experiment has a new function, not *reminding* the reader of something already known, but actually *convincing* the reader that the conclusion reported actually happened. This was done, Dear claims, by making the report that of a particular witness on a particular occasion, a procedure that bears an obvious relation to legal reasoning, as noted by Shapin and Schaffer (1985, 56–57). Shapin and Schaffer argue that an important point of this new style of presentation is to give readers faith in the truth of the outcomes reported by giving them faith in the scientist producing those outcomes. In the case of Boyle, they argue, "It was the burden of Boyle's literary technology to assure his readers that he was such a man as should be believed. He therefore had to find the means to make visible in the text the accepted tokens of a man of good faith" (ibid., 65). (On the notion of a literary technology, see 25; the style also has the function of bringing readers into the community of experimenters, thus making the readers "virtual" witnesses to the experiment, contributing to its success in constituting an experimental fact—see 60, 63.) Boyle was a credible witness because detail upon detail made the story credible as a report of something that actually happened in the world in a particular place and at a particular time. (One should not overestimate the degree of detail in Boyle's experimental reports. In the Proëmical Essay to his *Certain Physiological Essays and Other Tracts* (1661),

Boyle notes that he often leaves out important particulars in reporting his experiments, for various reasons; see Boyle [1772] 1965, I,315–16.)

Although earlier writers appear to present direct reports of actual events, and later writers present their experimental results in general terms, a general trend certainly runs in the experimental literature toward more and more particularity in reporting the results of experiments. Dear makes the case that in some circumstances, at least, this increased particularity is connected with the problem of convincing an audience to accept novel and unexpected results (see, e.g., 1987, 169; 1991, 163n). However, novelty and the rejection of an Aristotelian conception of the function of experience in natural philosophy is not the only factor at work.

The use of general statements in reporting the outcomes of experiments is not necessarily connected with either an Aristotelian conception of the use of experience, or with the reporting of non-novel facts. For example, Bacon takes great pains to emphasize that his use of experience is different from Aristotle's. His is a new Organon, a completely different way of going from experience to a knowledge of the world; for Bacon, as for his later followers, his science will clearly be built from a collection of facts, many of which will be novel. Yet, as we have seen, the statements of fact that make up the bulk of Bacon's natural histories will be general rather than particular. Although "history is properly concerned with individuals, which are circumscribed by place and time," he writes in the *De Augmentis* II 1, "because of the general resemblance which in most cases natural objects of the same species bear to one another," natural history most often deals with general statements about species of things: "when you know one, you know all" (1863, vol. 8, 407; see also vol. 10, 406).

In Gilbert's case, Gilbert takes great pains to emphasize the originality of his exploration of the magnet: "This natural philosophy [*physiologia*] is almost a new thing, unheard-of before; a very few writers have simply published some meagre accounts of certain magnetic forces. . . . Our doctrine of the loadstone is contradictory of most of the principles and axioms of the Greeks" (1958, 1). Despite the self-conscious novelty of the experiments that Gilbert reports, the form of the reports is decidedly general. Often Gilbert simply reports the properties he has observed (numerous times, presumably) in his loadstones: "Iron rubbed and excited by a loadstone is seized at the fitting ends by a loadstone more powerfully than iron not magnetized" (ibid, 159). Even when more complicated and directly experimental facts are related, they are given in a relatively nonparticular way:

A concave hemisphere of thin iron, a finger's width in diameter, is applied to the convex polar superficies of a loadstone and properly fastened; or an iron acorn-shaped ball rising from the base into an obtuse cone, hollowed out a little and fitted to the surface of the stone, is made fast to the pole. . . . Fitted with this contrivance, a loadstone that before lifted only 4 ounces of iron will now lift 12 ounces. (Ibid., 137)

While the important details are there, Gilbert gives only the results of the experiment, and those in general terms; in his report there are no indications of time or place, who was performing the experiment or observing it, how many times the experiment was performed, what difficulties there may have been in constructing the apparatus, and so on — all of the features found in later Royal Society experiments, like those of Boyle. Yet what Gilbert was relating was decidedly novel, as he fully recognized.

The new importance of novel facts in science cannot completely explain the new forms that experimental reports took; the importance of novel facts was recognized without necessarily resulting in any changes in the way in which experimental results were reported. What other factors are relevant here? Why did the Royal Society find it necessary to couple novelty of results with a new form of presentation for those results? My suggestion is that we look to the change in the conception of experimental facthood that I have been developing in this essay.

I have tried to show that with the Royal Society, we have a new conception of experimental facthood. For earlier investigators, it was possible for an individual working entirely alone to establish an experimental fact, through either simple repetition of a trial, or through reasoning. And so, when a Gilbert or a Bacon or a Descartes report the outcome of an experiment, they can report it *as fact*; others may challenge what they claim to have established, but the epistemology of experimental facthood does not in any way demand the concurrence of others to constitute a fact. But, I have argued, matters are entirely different with respect to the conception of experimental facthood in the Royal Society. There it is essential that others perform the experiment and witness the results before a purported experimental fact can enter the register of attested facts. And so, when an experimenter reports the outcome of an experiment, or even a series of experiments, he is *not* reporting anything that could *possibly be* an experimental fact; facts cannot be established in that way. And so, the best that can be reported is, as Spratt puts it, "present appearances," the way things looked to an individual at a given time in a given place. Only by putting this together with the observations of others can we constitute a fact. And so, I suggest, it is no surprise that new conventions for reporting the outcomes of experiments come at the same time as the Royal Society is explicitly rethinking how it is that experimental facts are to be established.

Community and Fact

I have traced the development of a social conception of experimental facthood, in particular, the explicit recognition of the social character of experimental facts. This is an interesting claim in the history of the philosophy of science, perhaps. But something even more interesting is happening. The social criterion of experimental facthood in connection with the Royal Society presupposes certain *social structures*. Consider the strong notion of reproducibility. As embodied in the communal conception of facthood, to be a candidate for an experimental fact, a given experimental result must be capable of being reproduced by *different* hands and eyes and requires the consensus of the scientific community as a whole. Not just anyone can participate in this enterprise.[6] If an experiment is performed by a member of the community, and I, for example, cannot reproduce it, that would not necessarily count against it; standing outside the community, I am not competent to cast my vote for or against a purported experimental fact. However, if others within the community could not reproduce an experiment, that might count against it. The very standard of strong reproducibility would seem to presuppose some criterion for membership in the community of peers. Similarly, it presupposes various kinds of social structures that are relevant to doing experiments and evaluating their outcomes in an appropriately public way.

The social structures necessary for one to be able to adopt the Royal Society's conception of experimental facthood were not always present in society. The community necessary to support such a conception of science was created only in the mid-seventeenth century, and then explicitly to enable its members to realize such a communal conception of scientific activity. While communities, schools and universities, academies and scientific societies, existed before the mid-seventeenth century, they were not organized in a way appropriate for the performing and reperforming of experiments, or for the communal judging of the outcome of experiments. This is not to say that such communities *could not* have arisen before then. Descartes might perhaps have transformed the Jesuit fathers of La Flèche or the Collège de Clermont or the members of the Mersenne Circle into such a group, but neither he nor anyone else did either. Such a community might also have come with Bacon, but dreamed of a community of gatherers of facts, and gave it many tasks and an elaborate organization, he never dreamed that they would cooperate with the production of facts, and the structure he proposed assumed that the many workers in the House of Salomon would work alone. This suggests that we must view the rise of the new communal conception of experimental facthood, a feature of the way practitioners thought about their natural philosophy, as intimately connected with the social transformation of the institutional structure in which science (natural philosophy) is done. I don't know which came first, if either, the social transformation or the philosophical transformation, but the two must go hand in hand. While this does not explain why the social conception of experimental facthood arose when it did, it does suggest a direction in which we might look for an answer: the rise of the social conception of facthood must go hand in hand with the emergence of the institutions appropriate to its support.

This leads me to a final moral. It has recently become fashionable to press the social factors in experimental facthood, and the role that the community plays in the establishment of experimental facts. Indeed, the importance of social factors in recent experimental science has lead some to view that the establishment of experimental facts can be explained entirely in sociological terms. On their view, establishing an experimental fact is simply a matter of social negotiation among members of the relevant community. Concerning the concept of experimental facthood in the Royal Society, Shapin and Schaffer write that "the objectivity of the experimental matter of fact was an artifact of certain forms of discourse and certain modes of social solidarity" (1985, 77–78; see 25). Indeed, they go so far as to claim that matters of fact are "social conventions," the result of "negotiations between experimenters" (ibid., 226; see also 281–82). "A fact," Bruno Latour writes in a similar spirit, "is what is collectively stabilized from the midst of controversies when the activity of later papers does not consist only of criticism or deformation but also of confirmation (1986, 243; see also

Latour 1987, 42; for a general account, see Latour and Woolgar 1986, 174–83, 236–52; Latour 1987, 41–44).[7]

While I have considerable sympathy for the view, I think that it has some historical limitations. One might take it for an almost a priori truth: belief in experimental facts, as in everything, simply *must* be a function of some communal agreement or other, explicit or tacit; belief, one might claim, as with the language in which it is framed, is by its nature social, and whatever Descartes or Bacon or anyone else might have thought about it, they, too, were caught up in the invisible web of social structure. Understood in this way, the thesis would seem to be grounded in very, very general facts about language, belief, and society, largely independent of any particularities about history and circumstance. Regarded in this way, though, the thesis is a general philosophical claim, one largely without any special interest to the historian or philosopher of science. But if the sociological claim is taken to be a thesis with real content and relevance for the historian of philosophy and science, then I think that *at best* it can only be an account that holds for experimental science as practiced in the last 350 years or so, since the appropriate social (and intellectual) structures were simply missing before then.

But even when the social constructivist is suitably historicized, I have my doubts. The thesis that the world of facts established by science is simply a matter of social agreement has an obvious deflationary consequence for the whole enterprise of science, turning what was thought to be objective fact into the collective illusion of a particular community. It would be a great irony if the social criterion of experimental facthood that, in a sense, marks the beginning of modern experimental science also marks the beginning of its demise.

NOTES

Chapter 2 appeared in a slightly different form in *Perspectives on Science* 3, no. 2 (1995): 173–205. Earlier versions were read at the University of Chicago; University of Pittsburgh; Johns Hopkins University; University of Wisconsin, Milwaukee; Canadian Society for History and Philosophy of Science; King's College, University of London; Princeton University; Western Canadian Philosophical Association; and the History of Science Society. I thank my audiences for challenging discussions that greatly improved this essay, and thanks to Peter Dear and Lorraine Daston for special help.

1. This, of course, will not work for astronomy, where the events observed are radically unique: the observation of a particular heavenly body in a particular position in the sky at a given time. Different strategies evolved for dealing with the fallibility of astronomical observations, generally involving numerous observations made over long periods of time. See, for example, discussions of the determination of mean motions of heavenly bodies in Swerdlow and Neugegauer 1984.

2. See the letter to Mersenne, 29 January 1640, AT III 7, where he suggests that to have complete assurance, a given observation with respect to the declinations of a magnet should be performed "a thousand times" rather than only three, as another investigator, John Pell, had done. We can presume that he would have adopted this standard for his own work, in principle if not in practice.

3. In part I of the *Discours*, for example, Descartes elaborately goes through what he learned at school, only to argue that there is little value in it; instead, he concludes, he must leave school and the traditions it embodies and find out how things are for himself. In part II, he employs city planning and architectural metaphors to advance the view that the best cities and houses arise not from the accidents of history but from the careful planning of a single individual. It is no accident that the *Discours* is written in first person, a single individual giving an account of the world he builds for himself.

4. The "excellent Friends" Spratt mentions in this passage are the other members of the Royal Society. Spratt's *History* was closely supervised by the Society and can be fairly read as a representation of the members' collective views; see Purver 1967, 9–19.

5. Shapin and Shaffer question the history of experimental philosophy as such, why experimental philosophy *as such* arose in England when it did, and how and why it came to triumph over a different and nonexperimental conception of science, such as that represented by Hobbes. They appeal to the political context of the debates and how Hobbes's and Boyle's positions fit into that context. They seem to take it for granted that the very idea of experimental science carries with it a social criterion of experimental facthood. See for example Shapin and Shaffer 1985, 25f, 77–78, 225–26, 281–82. My interest is in the circumstances under which this criterion first arose.

6. Spratt ([1667] 1958, 344) claims that virtually anyone, no matter how idle or industrious, how learned or ignorant, can participate in the program of experimental science. Of course, in practice this was not so.

7. This is at one extreme of those who call themselves social constructivists. A wide variety of such views can be found in the literature.

REFERENCES

Anguilonus, F. 1613. *Opticorum libri sex*. Antwerp.

Bacon, F. 1863. *The Works of Francis Bacon*, edited by J. Spedding, R. L. Ellis, and D. D. Heath. Cambridge, England: Riverside Press.

——. 1906. *The Advancement of Learning and New Atlantis.* Oxford: Oxford University Press.

Barnes, J. 1975. *Aristotle's Posterior Analytics.* Oxford: Oxford University Press.

Boyle, R. [1772] 1965. *The Works of the Honorable Robert Boyle in Six Volumes.* 2d edition. Edited by T. Birch. Hildesheim: Georg Olms Verlag.

Dear, P. 1985. "*Totius in verba*: Rhetoric and Authority in the Early Royal Society." *Isis* 76: 145–61.

——. 1987. "Jesuit Mathematical Science and the Reconstitution of Experience in the Early Seventeenth Century." *Studies in History and Philosophy of Science* 18: 133–75.

——. 1991. "Narratives, Anecdotes, and Experiments: Turning Experience into Science in the Seventeenth Century." In P. Dear, ed., *The Literary Structure of Scientific Argument: Historical Studies.* Philadelphia: University of Pennsylvania Press, pp. 135–63.

Descartes, R. 1964–74. *Oeuvres de Descartes.* vols. 1–11. Edited and translated by J. Cottingham, R. Stoothoff, D. Murdoch, and A. Kenny. Cambridge: Cambridge University Press.

Galison, P. 1987. *How Experiments End.* Chicago: University of Chicago Press.

Garber, D. 1992. "Descartes and Experiments in the *Discourse* and *Essays.*" In S. Voss, ed., *Essays on the Philosophy and Science of René Descartes.* New York: Oxford University Press, pp. 288–310.

Gilbert, W. 1958. *De Magnete.* Translated by P. Fleury Mottelay. New York: Dover Books.

Glanvill, J. [1668] 1958. *Plus Ultra: or, the Progress and Advancement of Knowledge Since the Days of Aristotle.* Reprint. St. Louis: Washington University Studies.

Lalande, A. 1911. "Sur quelques textes de Bacon et Descartes." *Revue de métaphysique et de morale* 19: 296–311.

Latour, B. 1987. *Science in Action.* Cambridge: Harvard University Press.

Latour, B., and S. Woolgar. 1986. *Laboratory Life: The Construction of Scientific Facts.* 2d edition. Princeton: Princeton University Press.

Pickering, A. 1984. *Constructing Quarks.* Chicago: University of Chicago Press.

Purver, M. 1967. *The Royal Society: Concept and Creation.* London: Routledge & Kegan Paul.

Shapin, S., and S. Schaffer. 1985. *Leviathan and the Air-Pump: Hobbes, Boyle, and the Experimental Life.* Princeton: Princeton University Press.

Spratt, T. [1667] 1958. *The History of the Royal-Society of London for the Improving of Natural Knowledge.* Reprint. Gainesville, Fla.: Scholars' Facsimiles and Reprints.

Swerdlow, N., and O. Neugebauer. 1984. *Mathematical Astronomy in Copernicus's* De revolutionibus, vols 1–2. New York: Springer-Verlag.

Wood, P. B. 1980. "Methodology and Apologetics: Thomas Sprat's *History of the Royal Society.*" *British Journal for the History of Science* 13: 1–26.

3

Isaac Newton on Empirical Success and Scientific Method

William Harper
Department of Philosophy, University of Western Ontario

Several closely related methodological themes from Newton were central to the transformation of natural philosophy into natural science as we now know it. These themes continue to be exemplified in research in gravitational physics. One such theme is an attempt to turn theoretical questions into ones that can be answered empirically by measurement. Another is an ideal of empirical success according to which a successful theory has its parameters accurately measured by the phenomena it purports to explain. A third is that phenomena are to be projectable generalizations that fit open-ended bodies of data. All of these themes play an important role in Newton's attempt to distinguish propositions that count as established scientific facts from merely conjectured hypotheses.

In this essay, Newton's argument for universal gravitation and work by his successors on perturbation theory are used to inform my interpretation of these themes and their implications for philosophy of science. Newton's inferences from phenomena are measurements backed up by systematic dependencies that make them more able to generate theoretical information from approximate phenomena than some of Clark Glymour's bootstrap confirmations. This provides resources to answer objections based on the fact that universal gravitation requires corrections to Keplerian phenomena. Newton's ideal of empirical success also answers a skeptical objection associated with the Quine-Duhem thesis.

More recent work in gravitational physics is used to back up implications I draw from Kuhn's classic investigation of scientific revolutions. Kuhn's discussion of how meaning changes make Newtonian limit theorems refer to parameters of the new theory shows how general relativity can take over the empirical successes of universal gravitation. This supports my suggestion that the most revolutionary outcome of the original triumph of Newton's theory was having his ideal of empirical success become a higher standard which was later used to overturn that theory. Recent tests of general relativity show that the methodological themes associated with this role for Newton's ideal of empirical success continue to inform the practice of gravitational physics.

Classic Deductions from Phenomena

Consider Newton's inference to a centripetal force from the phenomenon that an orbit satisfies Kepler's area law.[1] Theorem 1 of *Principia* Book 1 asserts that if the force deflecting a body into an orbit is directed at a center, the orbit will lie in a plane and the radius will sweep out equal areas in equal times. According to Theorem 2, if a body moves in a plane orbit, and satisfies the area law with respect to a center, then it is deflected into that orbit by a force directed at the center. Corollary 1 of Theorem 2 asserts that an increasing areal rate corresponds to a deflecting force that is off-center in the direction of the motion, while a decreasing rate corresponds to a force that is off-center in the opposite direction. These systematic dependencies make a constant areal rate measure the centripetal direction of a force deflecting a body into a plane orbit.

Newton's appeal to these systematic dependencies made it possible for him to turn the theoretical question of the direction of the force deflecting a planet into its orbit into one that is answered empirically by measurement. Given his assumptions, the accuracy with which Kepler's law of areas is known to hold fixes a correspondingly accurate measurement of the centripetal direction of the force.[2] Here we see all three themes at work in Newton's argument from the area law to establish the proposition that a planet is held in its orbit by a force directed toward the sun.

Given that bodies are being deflected into their orbits by centripetal forces, Newton can use theorems about motion under centripetal forces to make phenomena, such as stable orbits, measure the inverse square variation of those forces. According to Corollary 1 of Proposition 45 of Book 1, the centripetal force f is as the $(360/360 + p)^2 - 3$ power of the distance, just in case p is the number of degrees of precession per revolution.[3] For a stable orbit $p = 0$, which makes stability measure the force to vary as the -2 power of the distance.[4] Newton's inference to inverse square variation from Kepler's harmonic law for a system of orbits is also backed up by systematic dependencies which would make alternatives to the phenomenon measure alternative powers of distance.[5]

Comparison with Bootstrap Confirmation

Newton's inferences from phenomena are examples of Clark Glymour's (1980, 203–14) bootstrap confirmation. Glymour's work challenged the adequacy of hypothetico-deductive accounts of confirmation. Like bootstrap confirmations, Newton's inferences are based on theorems from background assumptions, according to which the exact truth of the phenomenon would deductively entail that the theoretical magnitude has the value inferred from that phenomenon. The systematic dependencies backing up Newton's inferences, which go beyond the requirements of bootstrap confirmation, make Newton's inferences more robust with respect to approximations. This can be illustrated with Newton's basic theorems about elliptical orbits.

According to Proposition 11 of Book 1, the law of the centripetal force directed to the focus of an elliptical orbit produced by it is inverse square. This makes an elliptical orbit satisfying the area law with respect to a focus of the ellipse sufficient to require the inverse square variation of the centripetal force deflecting a body into that orbit. If an elliptical orbit satisfied the area law with respect to the center of the ellipse, rather than a focus, then, according to Proposition 10 of Book 1, the centripetal force would be directly as the distance, instead of inversely as the square of the distance. Adding this to Proposition 11 is sufficient to make an elliptical orbit satisfying the area law with respect to a focus bootstrap confirm the inverse square variation of the centripetal force deflecting a body into that orbit.[6] It is not, however, sufficient to provide the systematic dependencies that back up Newton's actual inferences to inverse square variation.[7] In each of these actual cases there is a phenomenal magnitude with a whole interval of alternative values corresponding to alternative powers of distance.

One advantage of the systematic dependencies backing up Newton's inferences is realized when the phenomenon is established only up to some approximation. These dependencies make whatever tolerances to which the relevant value of the phenomenal magnitude is established carry over to corresponding tolerances on point estimates of the theoretical magnitude in question. In the case of bootstrap confirmations, which are not backed up by such dependencies, allowing that the phenomenon has not been established to hold exactly may undercut the inference to a corresponding value of the theoretical magnitude (Harper and DiSalle 1996).

The Argument for Universal Gravitation

In the argument for universal gravitation in *Principia*, these classic deductions from phenomena were used to establish that the moons of Jupiter (Proposition 1) and the primary planets (Proposition 2) are deflected into their orbits by inverse square centripetal forces. In Proposition 3 Newton argued that the moon is held in its orbit of the earth by an inverse square centripetal force. In this case the approximation involved in the assumption of a stable orbit actually conflicted with the available data.[8] Newton pointed out that the known slow precession of the lunar orbit, on average about 3°3′ per revolution, would measure a centripetal force that was inversely as the 2 and 4/243 power of distance, rather than as the inverse square; but, he suggested that this precession may be neglected as it is a perturbation due to the action of the sun on the moon's orbit about the earth. Newton's suggestion points out how the precession theorem can measure inverse square variation in the presence of precession due to perturbations. In order to realize it, however, one must be able to account for all observed precession as due to perturbations. This was not clearly achieved for the lunar precession until Clairaut's pioneering work in 1749, long after Newton wrote *Principia* (Whiteside 1976, 320; Wilson 1980, 133–45).

In Proposition 4 Newton argued to identify the force holding the moon in its orbit with terrestrial gravity. This inference appeals to the first two of what he formulated as rules of reasoning for natural philosophy.[9] The identification makes the centripetal acceleration of the lunar orbit and the length of a seconds pendulum at the surface of the earth into two phenomena which give agreeing measurements of the same inverse square force of gravitation toward the center of the earth. It is, therefore, motivated by Newton's ideal of empirical success.[10] In

Proposition 5 he generalizes the identification of inverse square centripetal orbital forces with gravitation toward the primary and extends gravity to those planets without moons.[11]

In Proposition 6 Newton argued for the direct proportion of gravitation on a body to the inertial mass of that body. Among the arguments he gives are two that can be regarded as measurements backing up the identification of passive gravitational and inertial mass. One is a frequently cited set of pendulum experiments in which equal weights of different materials are measured to have equal inertial masses.[12] According to Clifford Will (1993), these experiments establish that the identification of passive gravitational mass with inertial mass holds for laboratory-sized test bodies to one part in a thousand, because they establish corresponding limits on the measure of relative acceleration that we would now represent by the Eötvös ratio.[13] In the other argument, Newton cites the absence of polarization with respect to the sun of the orbits of Jupiter's moons as evidence that gravitation toward the sun of Jupiter and of each of its moons are respectively proportional to their inertial masses.[14] One aspect of Newton's conception of gravitation as a universal force of interaction is that the pendulum experiments and the absence of polarization of the orbits of Jupiter's moons can both be regarded as measurements which put limits on Eötvös ratios that would correspond to differences between inertial and passive gravitational mass.

In arguing for Proposition 7, Newton applied the third law of motion to identify the reaction corresponding to gravitation of one body toward another directly with a corresponding gravitation of the second to the first.[15] If gravitation is such an interaction between bodies, this argument extends the identification of gravitational mass with inertial mass to include active, as well as passive, gravitational mass. It also leads directly to universal gravitation. According to Stein:

With this interpretation of what the third law of motion requires — combined with what I have called the far-reaching hypothesis that gravity is such an interaction *between* the heavy body and the central body towards which it has weight — Newton's short and simple argument for proposition 7 leads directly to universal gravitation. (1991, 219)

This far-reaching hypothesis, however, had not been established by the argumentation of Propositions 1–7. Stein's comments (1991, 217, 219) suggest that it was sufficiently open to doubt that the argument to

Proposition 7 cannot be regarded as having established universal gravitation as something that should have been counted as a scientific fact.

Beyond Hypothetico-Deductive Success?

According to Stein (1991, 220), the argument from phenomena to universal gravitation does not end at Proposition 7, but includes the whole of *Principia* Book 3. The fact that Newton could not directly establish his hypothesis by measurement illustrates how some background assumptions receive empirical support only indirectly by the empirical successes they lead to.[16] I want to emphasize, however, that the additional work in Book 3 includes substantial efforts on Newton's part to realize what I have called his *new ideal of empirical success*. According to this ideal, what counts as success is accurate measurement of a theory's parameters by the phenomena it purports to explain. This ideal goes beyond hypothetico-deductive success, for which accurate prediction of the phenomena would be sufficient.

Corollary 1 of Proposition 8, the very next proposition,[17] exhibits Newton's efforts to realize this stronger ideal of success. In this corollary Newton exploits Proposition 7 to measure the masses of the sun and the planets with moons, from the harmonic law ratios of their orbiting satellites. We can say, therefore, that immediately after the argument for universal gravitation Newton began using the assumption that it holds, not just to make predictions, but to make measurements which contributed toward the theory's realization of his new, stronger ideal of empirical success.

Earlier remarks by Stein (1967, 180), made when hypothetico-deductivism dominated philosophy of science, may suggest that Newton's appeal to the rest of Book 3 is merely hypothetico-deductive.

when Newton responds to criticism of *Principia*, he appears to rest his case for universal gravitation, not upon the *argument leading to* the theory, but upon the *extremely detailed agreement of its consequences with observed phenomena*.

Stein (ibid.) backed up this remark with the following passage quoted from a letter from Newton to Leibniz dated 16 October 1693:

What that very great man Huygens has remarked on my work is acute. . . . But . . . since all the phenomena of the heavens and of the sea follow accurately, so far as I am aware, from gravity alone acting in accordance with the laws described by me, and nature is most simple; I myself have judged that all other causes are to be rejected.

This passage does suggest a methodology based on hypothetico-deductive success, as it seems to endorse inference to the best explanation based on such success.[18] The following passage, from the general scholium added to the second edition of 1713, illustrates the more prominent role given to inferences which realize Newton's stronger ideal of empirical success in later editions of *Principia*.

Hitherto we have explained the phenomena of the heavens and of our sea by the power of gravity, but have not yet assigned the cause of this power. This is certain, that it must proceed from a cause that penetrates to the very centers of the sun and planets, without suffering the least diminution of its force; that operates not according to the quantity of the surfaces of the particles upon which it acts (as mechanical causes used to do), but according to the quantity of the solid matter which they contain, and propagates its virtue on all sides to immense distances, decreasing always as the inverse square of the distances. Gravitation towards the sun is made up of the gravitations towards the several particles of which the body of the sun is composed; and in receding from the sun decreases accurately as the inverse square of the distances as far as the orbit of Saturn, as evidently appears from the quiescence of the aphelion of the planets: nay, and even to the remotest aphelion of the comets, if those aphelion are also quiescent. (Cajori, 546)

These properties of gravitation are what Newton takes to be the scientifically established facts that any hypothesis about the cause of gravity must account for. Prominent among them is the inverse square variation of gravitation toward the sun measured by the stability of planetary orbits.

Challenge from the Mechanical Philosophy

The mechanical philosophy made an a priori commitment to explanation by mechanical contact action.[19] This commitment reflected the outcome of what Kuhn (1963, 104–05) described as a hard-won paradigm shift that had overthrown Aristotelian explanation and led to the investigations of collisions upon which Newton was able to build his laws of motion. Universal gravitation was criticized for introducing occult qualities, because it seemed impossible to provide an explanation by contact action for what appeared to be the long-range action at a distance of gravitation as a universal force of interaction. In the

1730s the Bernoullis and others were still trying to find ways "to combine the technical advantages of the English physics with the philosophical superiority of the French physics" (Aiton 1972, 239).

I have suggested that it was not clear that universal gravitation really could realize Newton's ideal of empirical success until work by his successors on perturbation theory was developed. Salient among these developments was Clairaut's solution to the lunar precession problem, whereby he was able to completely account for the known precession upon recalculating the action of the sun on the lunar orbit.[20] In the period from Clairaut and Euler to Laplace, the Newtonian corrections of Keplerian phenomena became increasingly precise, projectable generalizations that accurately fit open-ended bodies of increasingly precise data.[21] These developments were accompanied by, and depended upon, increasingly accurate measurements of relative masses of bodies in the solar system. Newton's estimate of the mass of the sun in earth masses in the second edition of *Principia* was 32 percent less than the value accepted in 1976.[22] By 1796 Laplace was giving estimates within 1 or 2 percent of the 1976 value (Hufbauer 1993, 29).[23]

These developments exhibited Newton's new ideal of success as an achievable aim. I suggest that this eventually led to a Newtonian revolution that was more radical than the overthrow of the commitment to explanation by contact action. This may be regarded as the transformation of natural philosophy into natural science. Central to this transformation was having Newton's ideal of empirical success become a higher standard that could override even the most cherished of the lower standards from paradigms on which Kuhn focuses much of his discussion. On my suggestion, later changes of standard through paradigm shift, such as the transition from particle to wave conceptions of light, can be understood as applications of this higher standard to changing circumstances of research.[24]

In the light and color debates, Newton had distinguished what he called his "experimental philosophy" from the mechanical philosophy of his continental critics by arguing that the proper method in natural philosophy is first to establish properties of things by experiments and only afterwards to attempt to find hypotheses to explain them.[25] The experimentally established properties are what the explanations must account for.[26] In the *Principia*, what he still called "experimental philosophy" was extended by his extraordinary new resources for empirically establishing propositions on the basis of measurement from phenomena. Among the revisions designed to address continental critics of universal gravitation is a rule of reasoning added to the third (1726) edition.

Rule 4   In experimental philosophy, propositions gathered from phenomena by induction should be considered either exactly or very nearly true notwithstanding any contrary hypotheses, until yet other phenomena make such propositions more exact or liable to exceptions. (Cohen and Whitman, 488)

The arguments we have been examining suggest that what count as "propositions gathered from phenomena by induction" are those sufficiently backed up by measurements from phenomena, while what makes something a mere "contrary hypothesis" (not allowed to undercut inference to such a proposition) is that it is not sufficiently backed up to count as a serious rival. This makes Rule 4 endorse inference to (at least) approximate truth of any proposition backed up by an explanation that clearly counts as a better realization of Newton's ideal of empirical success than explanations available to back up alternative hypotheses.

Objections from Philosophers of Science

In his classic *The Aim and Structure of Physical Theory*, Pierre Duhem pointed out:

The principle of universal gravity, very far from being derivable by generalization and induction from the observational laws of Kepler, formally contradicts these laws. If Newton's theory is correct, Kepler's laws are necessarily false. (1962, 193)

Karl Popper (1972, 200–01) pointed out that universal gravitation requires correction to Kepler's harmonic law even in a two-body system consisting of the sun and a single planet. For Paul Feyerabend (1970, 164), such incompatibilities undercut Newton's claims to use Kepler's laws as phenomena on which to base knowledge of universal gravitation. He claims, "Used in this way phenomena are no longer a basis for knowledge." Imre Lakatos (1978, 212) comments:

The schizophrenic combination of the mad Newtonian methodology, resting on the *credo quid absurdum* of "experimental proof" and the wonderful Newtonian *method* strikes one now as a joke. But from the rout of the Cartesians until 1905 nobody laughed. Most textbooks solemnly claimed that first Kepler "*deduced*" his laws "from the accurate observations of Tycho Brahe," then Newton "*deduced*" his laws from "Kepler's laws and the laws of motion" but also "added" perturbation theory as a crowning achievement. The philosophical bric-a-brac, hurled by Newtonians at their contemporary critics to defend their "proofs" by hook or crook, were taken as pieces of eternal wisdom instead of being recognized for the worthless rubbish that they really were.

These comments are based on the point, noted by Duhem and Popper, that universal gravitation is inconsistent with Kepler's laws because it requires interactions that produce perturbations of Keplerian motion.

Just as Newton took the corrections he was able to achieve as further evidence for universal gravitation, rather than as developments that undercut his inferences from Keplerian phenomena, so did later figures such as Laplace take their very impressive improvements to further underwrite Newton's theory. Lakatos suggests that while this attitude toward corrections due to perturbations may be appropriate to what he calls "the wonderful Newtonian method," it is incompatible with the Newtonian methodology expressed in claims to have deduced inverse square centripetal forces from Keplerian phenomena. We have been using Newton's method, as revealed in the details of his argument for universal gravitation, to inform our interpretation of his methodological remarks. Based on this interpretation, Newton's claims to have deduced inverse square variation of gravitation from orbital phenomena are claims that the phenomena measure this inverse square variation.

The fact that perturbation theory provides the resources to carry out these measurements in the presence of precession due to perturbations draws the sting from Lakatos's criticism.[27] We have already seen this illustrated by measurement of the inverse square variation of gravitation to the earth made possible by Clairaut's solution to the lunar precession problem. It may be worth adding some remarks about measurement of the inverse square variation of gravitation toward the sun. For each planet other than Mercury, the zero leftover precession, after accounting for precession due to perturbations, measures its gravitation toward the sun to be the -2 power of distance. The 43 seconds per century unaccounted-for precession of Mercury would measure gravitation toward the sun to be as the -2.000000157 power of distance.[28]

When the Keplerian phenomena are taken as approximations, Newton's argument can be construed as one using this first level of approximation to establish the existence of the basic inverse square centripetal forces. The reasoning behind universal gravitation introduces interactions that lead to corrections that count as more accurate, higher-level approximations, but the Keplerian phenomena continue to hold at the first level of approximation. Indeed, it was only long after *Principia* that perturbation theory led to appreciable improvement over Keplerian phenomena (Wilson 1985, 16). For example, the two-body correction to Kepler's harmonic law, pointed out by Popper, leads to a correction of about 3 parts in 10,000 for Jupiter's mean distance, which was well below what could be detected by data available to Newton (Harper 1993, 178).

The Quine-Duhem Thesis

Duhem's criticism (1962) of Newtonian methodology did not rest just on the inconsistency he pointed out between Keplerian phenomena and universal gravitation.

Therefore, if the certainty of Newton's theory does not emanate from the certainty of Kepler's laws, how will this theory prove its validity? It will calculate, with all the high degree of approximation that the constantly perfected methods of algebra involve, the perturbations which at each instant remove every heavenly body from the orbit assigned to it by Kepler's laws; then it will compare the calculated perturbations with the perturbations observed by means of the most precise instruments and the most scrupulous methods. (Ibid., 193–94)

Here we see that Duhem's positive account of how perturbation methods contribute toward establishing validity for Newton's theory is hypothetico-deductive. Calculating perturbations from the theory precisely and having the resulting predictions accurately fit the most precise observations is what counts as the empirical success that supports the theory.

Duhem goes on to articulate the first part of what has become a classic argument purporting to establish that empirical support can only be holistic.

Such a comparison will not only bear on this or that part of the Newtonian principle, but will also involve all the principles of dynamics; besides it will call in the aid of all the propositions of optics, the statics of gases, and the theory of heat, which are necessary to justify the properties of telescopes in their construction regulation, and correction, and in the elimination of the errors caused by diurnal or annual aberration and by atmospheric refraction. It is no longer a matter of taking, one by one, laws justified by observation, and raising each of them by induction and generalization to the rank of a principle; it is a matter of comparing the corollaries of a whole group of hypotheses to a whole group of facts. (Ibid.)

Here Duhem points out that the inferences to establish that the predicted perturbations accurately fit the most precise observations require appeal to wide-ranging background assumptions. These include the laws of motion and theories correcting instrument observations, in addition to the propositions Newton inferred from the uncorrected Keplerian phenomena.

The next part of the argument may be suggested by the following remark about implications of the fact that Kepler's laws are only approximations.

Thus the translation of Kepler's laws into symbolic laws, the only kind useful for a theory, presupposed the prior adherence of the physicist to a whole group of hypotheses. But, in addition, Kepler's laws being only approximate laws, dynamics permitted giving them an infinity of different symbolic translations. Among these various forms, infinite in number, there is one and only one which agrees with Newton's principle. The observations of Tycho Brahe, so felicitously reduced to laws by Kepler, permit the theorist to choose this form, but they do not constrain him to do so, for there is an infinity of others they permit him to choose. (Ibid., 195)

Examples of the sort of symbolic translations of Kepler's laws, which Duhem refers to here, are trajectories in accord with the one body theorems that, according to Newton, make those laws measure inverse square centripetal forces. The corrections generated by taking into account Newtonian perturbations are other examples. For Duhem the set of acceptable alternative hypotheses would include any formulae which could be made to generate trajectories compatible with the approximations allowed by the data. Duhem's apparent commitment to merely hypothetico-deductive criteria suggests that quite wide-ranging hypothetical adjustments of background assumptions could be appealed to in counting an alternative acceptable.[29]

The theme that alternative hypotheses are not ruled out because they could be protected by adjusting the background assumptions is added to the theme that appeal to background assumptions cannot be avoided, generating what has become known as the Quine-Duhem thesis. According to the Quine-Duhem thesis, theoretical hypotheses do not face trial by data in isolation, so that only the holistic body of all one's assumptions together ever gets tested. This suggests the skeptical implications that empirical support cannot differentiate among assumptions.[30]

As we have seen, Newton's measurements of parameters by phenomena do require the sort of appeal to background assumptions that Duhem called attention to. What does not follow from this, however, is the idea that such appeal to background assumptions prevents measurement by phenomena to differentially support propositions of a theory. This should be especially obvious in cases, such as the inverse square variation of gravitation toward the sun, that are supported by agreeing measurements from many separate phenomena.

The Newtonian ideal of empirical success requires that a viable alternative hypothesis achieve comparably precise and well-supported measurements of its parameters from corrections to the phenomena that comparably accurately fit the best data. This makes finding viable alternative hypotheses far more difficult than it would be if the only criterion for empirical success were the hypothetico-deductive one that predictions fit the data. Consider attempting to undercut Newton's inference to the identification of the force holding the moon in its orbit with the earth's gravity. It would not be enough to for an alternative hypothesis to predict the phenomena, or to provide believable alternatives to the phenomena that would fit believably appropriate corrections to the data as well as Newton's phenomena fit the actual data. In addition, an alternative hypothesis would have to receive enough extra success to offset the success provided by having the length of a seconds pendulum and the centripetal acceleration of the lunar orbit count as agreeing measurements of the same inverse square acceleration field. I suggest that, to achieve this, the hypothesis would have to be supported by something that would count as positive evidence that the identification was mistaken. Similar problems would arise every time a proposed alternative involved changing the value of any parameter measured by other phenomena.

Kuhn on Scientific Revolutions

Our remarks about the developments in celestial mechanics and its application to planetary astronomy suggest that Newton's ideal of empirical success led to, and was exemplified in, the development of what Kuhn might call the "normal science" stage of gravitational physics. I want to argue that Newton's ideal of empirical success was also ex-emplified in Einstein's revolution which overthrew Newton's theory of gravity.

Kuhn (1970, 101–03) criticizes attempts to argue from Newtonian limit results of relativistic dynamics to the claim that Newtonian dy-namics remains approximately true.

Can Newtonian dynamics really be *derived* from relativistic dynamics? What would such a derivation look like? Imagine a set of statements, $E_1, E_2, \ldots, E_n$, which together embody the laws of relativity theory. These statements contain variables and parameters representing spatial position, time, rest mass, etc. From them, together with the apparatus of logic and mathematics, is deducible a whole set of further statements including some which can be checked by ob-servation. To prove the adequacy of Newtonian dynamics as a special case, we must add to the E_i's additional statements, like $(v/c)^2 << 1$, restricting the range of the parameters and variables. This enlarged set of statements is then manip-ulated to yield a new set, $N_1, N_2, \ldots N_n$, which is identical in form with New-ton's laws of motion, the law of gravity, and so on. Apparently Newtonian dy-namics has been derived from Einsteinian, subject to a few limiting conditions.

Yet the derivation is spurious, at least to this point. Though the N_i's are a special case of the laws of relativistic mechanics, they are not Newton's Laws. Or at least they are not unless those laws are reinterpreted in a way that would have been impossible until after Einstein's work. The variables and parameters that in the Einsteinian E_i's represented spatial position, time, mass, etc., still occur in the N_i's; and they there still represent Einsteinian space, time, and mass. But the physical referents of these Einsteinian concepts are by no means identical with those of the Newtonian concepts that bear the same name. (Newtonian mass is conserved; Einsteinian is convertible with energy. Only at low velocities may the two be measured the same way, and even then they must not be conceived to be the same.)

According to Kuhn some changes, such as the new laws making mass velocity-dependent in relativity theory, count as meaning changes, which make the referents of such terms in the Newtonian limits of relativity theory incompatible with the referents of the corresponding terms in Newtonian theory.

The realization of Newton's methodological themes in the revolu-tionary transition to relativity theory does not depend on challenging Kuhn's claims about meaning and reference of scientific terms. Accord-ing to Newton's ideal of empirical success, what matters for relativity to take over the empirical successes of Newtonian gravitation is that the phenomena which accurately measured Newtonian parameters now accurately measure the corresponding parameters of relativity theory.

This is exactly what, according to Kuhn, the Newtonian limit results can provide. Consider measurements of the mass of the sun from the harmonic law ratios of planets which orbit it. The harmonic law ratios are determined for a rest frame with respect to the sun. It turns out, therefore, that, according to relativity theory, it was rest mass that had been measured all along.

This suggests that the transition to relativity theory may have been more like normal science than the following famous passage suggests paradigm choices could be.

Like the choice between competing political institutions, that between competing paradigms proves to be a choice between incompatible modes of community life. Because it has this character, the choice is not and cannot be determined merely by the evaluative procedures characteristic of normal science, for these depend in part upon a particular paradigm, and that paradigm is at issue. When paradigms enter, as they must, into a debate about paradigm choice, their role is necessarily circular. Each group uses its own paradigm to argue in that paradigm's defence. (Ibid., 94)

The contribution of the changes introduced in Einstein's theories to the choice of those theories over Newton's may have been due more to new resources they provide for realizing Newton's ideal of empirical success than to any question begging appeal to new standards of theory evaluation.

Newtonian phenomena are projectable generalizations that correspond to relative trajectories that can be generated by applications of universal gravitation. Laubscher's (1981) account of the motion of Mars is an example of such a phenomenon. It illustrates that corrections of Keplerian orbits to take account of Newtonian perturbations can generate an ephemeris which fits 303 data on right ascensions taken from meridian observations of Mars from 1954 to 1969 with a mean error of .29 seconds of arc (Laubscher 1981, 375, 399). Laubscher estimates the probable error for each of these observations to be .4 seconds (ibid., 375). Newtonian limit results allow general relativity to recover such phenomena as approximations good enough to fit such data.[31] These limit results also, recover for parameters of general relativity the accurate measurements that such phenomena provide for parameters of universal gravitation (DiSalle and Harper 1996). This suggests that general relativity may be able to take over all of the Newtonian empirical success of universal gravitation.

In addition to recovering Newtonian phenomena as approximations, general relativity provides for post-Newtonian corrections for many such phenomena. The classic tests provide examples where such corrections improve the fit with data. In what we may regard as its original post-Newtonian correction, general relativity generated the 43 seconds per century of Mercury's precession that Newtonian perturbations had been unable to account for (Einstein 1915; Roseveare 1982; Earman and Janssen 1993). A post-Newtonian correction of the Newtonian trajectory of light passing near the sun was regarded as a better fit than the Newtonian trajectory to the somewhat rough data provided by the classic light-bending test.[32]

Recent tests have provided more precise data that greatly increase the improvement of fit generated by the light-bending correction (Will 1993, 333). They have also exhibited many more post-Newtonian corrections that count as relativistic phenomena (Will 1993, 320–52). These post-Newtonian corrections improve on Newtonian phenomena in the same way that the Newtonian perturbations, which Duhem called attention to, improved on Keplerian phenomena. They are projectable generalizations that more accurately fit open-ended bodies of what, in many cases, have now become very much more precise data.[33] These post-Newtonian corrections provide measurements of parameters of general relativity that clearly distinguish it from universal gravitation. By now, it has become overwhelmingly clear that general relativity meets the higher standard, corresponding to Newton's ideal of empirical success better than universal gravitation.

Before the classic light-bending test, Einstein had provided the limit results that make general relativity recover the empirical successes of universal gravitation[34] and the post-Newtonian correction accounting for Mercury's perihelion precession. He had already successfully challenged Newtonian dynamics with special relativity dynamics. The appeal to the equivalence principle in his theoretical argument can exploit an empirical advantage to be gained by dropping Newton's distinction between inertial and gravitationally accelerated trajectories.[35] Even in 1919, when the new empirical evidence was limited to that provided by Eddington's light-bending test, there was no need to appeal to any standard other than Newton's to choose Einstein's theory over Newton's on the basis of empirical success.

Testing General Relativity[36]

In 1963, the year after the first edition of Kuhn's book was published, Robert H. Dicke gave a course of lectures in which he criticized the experimental support produced for general relativity (Dicke 1964). He pointed out that the red shift experiment, suggested by Einstein, was not so much a test of general relativity as of a more general idea corresponding to Einstein's equivalence principle — that gravitation be represented by space-time curvature (ibid., 25). The Einstein equivalence principle (EEP) is the conjunction of the weak equivalence principle (WEP, which Newton's pendulum experiment had measured), local Lorentz invariance (LLI), and local position invariance (LPI, which the gravitational red shift measured). Dicke provided a testing framework that would allow tests of these principles to count as measurements that would constrain a very wide range of possible alternative theories of gravity (ibid., 50).

Dicke's work led to extensive efforts to test EEP. Results include limits on Eötvös ratios corresponding to violations of WEP of one part in 10^{12} (Will 1993, 25). For LLI there are limits respectively of 10^{-23} for strong nuclear interactions, 10^{-22} for electrostatic, 10^{-22} for hyperfine, and 10^{-18} for weak nuclear interactions (ibid., 31). Red shift experiments have measured limits of 2×10^{-4} for violations of LPI (ibid., 35). Dicke's work and the extensive testing it has led to count as very impressive realizations of Newton's methodological theme, recommending attempts to turn theoretical questions into ones that can be answered empirically by measurement (DiSalle, Harper, and Valluri 1994a, 1994b). Unlike the defensive appeal to question-begging assumptions, which according to Kuhn characterizes defenders of paradigms, Newton's methodology recommends striving to open assumptions to scientific test.

In his lectures Dicke also presented a competing theory that had been published jointly with Carl Brans (Brans-Dicke 1961, reprinted as Appendix 7 of Dicke 1964). This theory was motivated by criticisms that Berkeley and Mach had raised against Newton.

As pointed out already in 1710 by Bishop Berkeley, and later elaborated by E. Mach, the Newtonian scheme has logical difficulties. The acceleration is to be reckoned with respect to "absolute space," a concept emphasized by Berkeley to be without an observational basis. What one actually measures is always an accleration relative to other, generally distant matter in the universe. (Dicke 1964, x)

According to Berkeley, thought experiments and unbiased reflection are sufficient to reveal that only relative motions can be empirically established (*Principles*, 110–13, *De Motu*, 57–60). For Newton, whether or not true motions can be empirically determined is, itself, an empirical question to be decided by the empirical successes of supporting background assumptions such as the laws of motion. Insofar as his laws of motion are taken to be adequate, the bucket experiment shows that one can distinguish true from merely relative accelerations by dynamical effects.[37]

Brans and Dicke offer the following comment on a thought experiment designed to show that general relativity does not realize the operationalist ideas provided by Mach's and Berkeley's criticisms of Newton.

It is clear that what is being described here is more nearly an absolute space in the sense of Newton rather than a physical space in the sense of Berkeley and Mach. (ibid., 78)

Unlike those who would make such ideas a priori commitments, Dicke sees that their realization requires an alternative theory able to overcome the empirical success of general relativity.

To give a modern interpretation to Mach's ideas it is necessary to interpret inertial forces as acceleration dependent interactions of a particle with a field generated by the distant matter in the universe. (ibid., x)

For Dicke, just as for Newton, whether or not only relative motions can be empirically distinguished is a question to be settled by empirical success of theories.

Like general relativity (GR), the Brans-Dicke theory satisfies EEP, so that it represents gravitation by a metric corresponding to space-time curvature. It differs by postulating a scalar field that represents the contribution of distant masses to local curvature. According to this theory, the precession of Mercury is

$$[(4 + 3\omega)/(6 + 3\omega)] \times (\text{value of GR})$$

where ω is an adjustable constant representing the scalar field (ibid., 88). In 1966, experiments by Dicke and Goldenberg on solar oblateness suggested values of the solar quadrupole moment that would have contributed about 4 seconds per century to the precession (Will 1993, 181–82). This would have favored a version of Brans-Dicke with ω set at about 5, over general relativity (ibid., 182). Experiments by Hill and collaborators in 1973 on solar oblateness produced conflicting results that agreed with GR's account of Mercury's precession (ibid., 182).

Other sorts of estimates of the quadrupole moment also have failed to support any anomaly for GR (Shapiro 1990, 316–17).

Work by Kenneth Nordtvedt and Clifford Will (Nordtvedt 1968, Will 1971, Will and Nordtvedt 1972), extending earlier work by Eddington (1922), led to development of parametrized post-Newtonian (PPN) formalism. The assumptions of PPN formalism are approximations detailed enough to make solar system phenomena measure parameters on which competing metrical gravitation theories can differ (Will 1993, 86–99). Consider the phenomenon of the time delay of light and the PPN parameter γ which specifies how much space curvature is produced by unit rest mass. According to GR, $\gamma = 1$. According to the Brans-Dicke theory

$$\gamma = (1 + \omega)/(2 + \omega),$$

(ibid., 117). The results of Shapiro's time-delay experiments measure $\frac{1}{2}(1 + \gamma)$ to be within .2% of 1, which limits allowable assignments of the Brans-Dicke parameter to $\omega \geq 250$ (Shapiro 1990, 316). Accordingly, viable versions of Brans-Dicke theory cannot diverge far from GR. The time delay experiments are not alone. For example, Will (1993, 335) cites Lunar laser ranging experiments which force ω over 600. More recent Lunar laser ranging results (Dickey et al. 1994, 485–86) force it over 900.[38] Such experiments have greatly reduced the prospect of finding evidence that would clearly favor Brans-Dicke over GR.

Metrical theories, those which, like Brans-Dicke and general relativity, satisfy EEP and represent gravity by the space-time metric, are committed to the PPN assumptions for solar system tests. This makes the contribution of the PPN formalism to such tests an outstanding realization of Newton's methodology. The outcomes of the measurements count as the scientifically established facts that acceptable versions of the theories under consideration must account for.

The Brans-Dicke theory has not been the only alternative proposed. There has been quite lively activity generating alternative theories (Will 1993, 19, 117). Moreover, there has been widespread agreement that general relativity will turn out to be only an approximation of a more fundamental theory of quantum gravity (Wald 1984, 378). This fits Kuhn's description of the transition from normal to extraordinary research.

The proliferation of competing articulations, the willingness to try anything, the expression of explicit discontent, the recourse to philosophy and to debate over fundamentals, all these are symptoms of a transition from normal to extraordinary research. It is upon their existence more than upon revolutions that the notion of normal science depends. (Kuhn 1970, 91)

So far, however, no alternative theory of gravity has successfully challenged general relativity. If Newton's methodology continues to guide gravitational physics, we can expect that transition to a successor theory will not happen unless it can count as progress according to Newton's ideal of empirical success.

NOTES

I am grateful to George Smith for critical remarks about an early draft, and to Howard Stein, Curtis Wilson, Michael Nauenberg, and Kenneth Nordtvedt for critical remarks on a later draft.

1. See Harper 1991 for a more detailed account of these classic deductions from phenomena and Harper 1993 for a discussion of the illumination these inferences can provide to Newton's controversial rules of reasoning. This essay uses Newton's ideal of empirical success to further illuminate the applications of his rules of reasoning in the argument for universal gravitation.

2. Propositions 1 and 2 are one-body idealizations with respect to centers which can be treated as inertial (In the case of moons about a planet, as well of planets about the sun, Newton determined rotation of radius vectors relative to the fixed stars). Proposition 3, which applies to centers that are themselves accelerating, appeals to Corollary 6 of the laws of motion, according to which equal and parallel accelerations on all bodies in a system can be ignored.

3. Newton has the force as the $(n/m)^2 - 3$ power of distance, where $n = 360$ and m is the total angular motion with which the body returns to the same apsis (Cohen and Whitman 1987, 192). So, $m = 360 + p$ for p degrees of precession, which may be positive, zero, or negative.

4. His proof of Proposition 45 uses a limiting case in which the orbital eccentricity approaches zero. This raises the question of how (and whether) it applies to elliptical orbits with appreciable eccentricities. Ram Valluri (Valluri, Wilson, Harper 1997) has shown, however, that the equivalence holds more generally. As the eccentricity varies (even approaching one), the formula relating precession to the power of the distance also varies, but in each case, zero precession corresponds to inverse-square variation.

5. A system of orbits satisfies Kepler's harmonic law just in case the ratio R^3/t^2, where R is the distance and t is the period, is constant. That is, to have $t \propto R^{3/2}$ (the periods be as the ³⁄₂ power of distance). According to Corollary 7 of Proposition 4 Book 1, for a system of concentric circular orbits under a centripetal force f, $t \propto R^n$

iff $f \propto R^{(1-2n)}$, so that the harmonic law also measures the centripetal force to be as the -2 power of the distance. Theorem 4 and Corollaries 1–7 of it are proved for concentric circular orbits (see Harper 1993, 151–52 for discussion).

6. Clark Glymour (1980) proposed bootstrap confirmation as an alternative to what was then the dominant hypothetico-deductive account of confirmation. The central idea is that background assumptions may be appealed to in order to make empirical data entail a hypothesis. The basic requirement is that the background assumptions be compatible with alternatives to the data that would, relative to those background assumptions, be incompatible with the hypothesis (Glymour 1980, 127). Let T be Newton's laws of motion together with the assumption that a body is orbiting under a centripetal force. Let E be the phenomenon that the body orbits in an ellipse that satisfies the area law with respect to a focus. Let H be the hypothesis that the centripetal force deflecting the body into the orbit is inverse square. Let E' be the alternative phenomenon in which the body moves in an elliptical orbit but satisfies the area law with respect to the center of the ellipse. This example clearly satisfies Glymour's basic requirement. For each formulation of bootstrap confirmation (ibid., 130–31; 1983, 5–8) there is a corresponding formulation of this example according to which E bootstrap confirms H relative to T.

7. Newton did not include the ellipse among the explicitly cited phenomena in *Principia*, Book 3, and Proposition 11 of Book 1 is not cited to back up any inference from an elliptical orbit to inverse-square variation of a centripetal force. Perhaps these omissions are motivated by a lack of systematic dependencies to back up such an inference from an ellipse. As shown below, this motivation is further supported insofar as elliptical orbit phenomena are established only up to approximations.

8. This is one example compatible with Ronald Laymon's conjecture that "Newton's descriptions of the phenomena were typically incompatible with the then accepted observational data" (1983, 187). Contrary to Laymon, however, most of what Newton cited as phenomena were within what would count as reasonable limits of error on the data available to him (Harper 1989, note 12).

9. Here are these rules as translated by Cohen and Whitman (1987, 485–86).

Rule 1   No more causes of natural things should be admitted than are both true and sufficient to explain their phenomena.

Rule 2   Therefore, the causes assigned to natural things of the same kind must be, so far as possible, the same.

As pointed out below, the application of these rules to identify the force on the moon with gravity realizes Newton's ideal of empirical success.

See Harper 1993, 159–62; 1989, 135–40 for other discussion of the role of Newton's rules in his reasoning from phenomena. Stein 1991, 210–15 provides an excellent account of the role of Newton's definitions of measures of centripetal force in backing up this application of the rules of reasoning to identify the inverse square centripetal force on the moon with terrestrial gravity. Newton's definitions are appropriate when, as in the case of the earth's gravity, a centripetal force can count as an acceleration field.

10. According to Newton's ideal, a successful theory has its parameters accurately measured by the phenomena it purports to explain. On the identification of

the force holding the moon in its orbit with the earth's gravity, the data on the moon supports Huygen's measurement of the rate of fall at the surface of the earth, even though the measurement it provides is far less precise. The lunar data backs up the pendulum measurement by making alternative values, differing by great enough amounts, more improbable than they would be on the pendulum data alone.

11. See Harper 1989, 140–42; 1993, 162–63 and esp. Stein 1991, 214–15 for more detail.

12. In editions 2 and 3, Newton added a corollary in which the inference to Proposition 6 from the equality of gravitation of bodies toward the earth, as revealed by this experiment and by many terrestrial phenomena, is backed up by what he formulated as his third rule for natural philosophy. Here is Rule 3 from Cohen and Whitman's translation:

Rule 3   Those qualities of bodies that cannot be intended and remitted [that is, qualities that cannot be increased and diminished] and that belong to all bodies on which experiments can be made should be taken as qualities of all bodies universally.

See Harper 1989, 140–47 for connections between Newton's third rule and natural kind inferences, such as the inference to the mass for all electrons from measurements on one or a few.

Insofar as the phenomena all can be construed as agreeing measurements of a single parameter that represents the direct proportionality of gravitation on a body to its inertial mass, this application of this rule is supported by Newton's ideal of empirical success (Harper and Smith, Sec. 4c). The Eötvös ratio (see note 13) can represent such a magnitude (DiSalle and Harper 1996).

13. Where a_1 is the gravitational accleration produced on one of the bodies and a_2 is that produced on the other, the Eötvös ratio is

$$2|a_1 - a_2|/|a_1 + a_2|$$

(Will 1993, 25). To have the gravitational accelerations produced on the two bodies at equal distances from the center of the earth be equal is to have the gravitational force on each be proportional to its inertial mass. To the extent that the pendulum experiments measure general constraints on Eötvös ratios to one part in a thousand, they establish to three decimal places what we now call the weak equivalence principle (WEP) (ibid., 22). To whatever accuracy WEP is established, so is the identification of passive gravitational mass with inertial mass for laboratory-sized test bodies.

14. This argument is much less often cited, but, as Damour (1987, 143–44) points out, it is especially interesting as Jupiter and its moons are massive enough to count as having non-negligible self-gravitational energy. That passive gravitational mass for such bodies should equal inertial mass is what is now called the gravitational weak equivalence principle (GWEP) (Will 1993, 184). The polarizations toward the sun of orbits of moons of planets (where moon and planet gravitate toward the sun in different proportions to their inertial masses) are among what are now called Nordtvedt effects. Nordvedt (1968) established that such

effects could distinguish alternative metrical gravitation theories such as the Brans-Dicke theory from general relativity (see Will 1986, 139–46 for a vivid historical account). Laser ranging experiments have established limits on polarization of our moon's orbit that measure limits on violations of GWEP of 7 parts in 10^{12} (Will 1993, 190).

Newton claimed that the absence of such polarization of the orbits of Jupiter's moons measured agreement in proportion of gravitation toward the sun of Jupiter and its moons to their respective inertial masses to about one part in a thousand. Damour (1987, 144) criticizes details of Newton's calculation, but he admires Newton's insight in backing up his pendulum experiments with measurements based on such orbital phenomena. Damour and Vokrouhlický (1996, 4199) credit Laplace with having calculated limits on polarization of Jupiter's moons which measure GWEP to one part in 10^7.

15. Newton refers to Proposition 69 Book 1, which applies the third law to argue that if bodies in a system attract each other by inverse-square accelerative forces, then the attraction toward each will be as its inertial mass. The interpretation of the forces as attractions between the bodies is what allows the third law to be applied.

16. I have not attempted to explicate how accurate measurement of parameters by phenomena generates support for a theory. Rosenkrantz's (1983) discussion of Glymour's bootstrap confirmation suggests that appropriate Bayesian priors will make agreeing measurements of the same parameter by several phenomena better confirmation than mere prediction of those phenomena would be. Jon Dorling (1991) has argued that Newtonian deductions from phenomena support a theory, according to the simplicity-based priors developed by Solomonoff, when they reveal its simplicity relative to rivals. A recent paper by Forster and Sober (1994) suggests that an account of predicted fit developed by Akaike may help to show circumstances under which phenomena which accurately measure theoretical parameters may offer better supported predictions of future fit than alternative generalizations which better fit the data so far.

My interpretation of Newton's fourth rule of reasoning suggests that direct empirical support for a theory from measurements of its parameters can be undercut when a rival theory clearly does better. This would make it much harder to construct counter examples where clearly inadequate theories are being supported by approximate phenomena which would give fairly accurate measurements of its parameters. For example, it should help avoid Kuhn's (1970, 99) worry about being committed to say that the Phlogiston theory should still count as approximately true. Whether it would also offer support for Kuhn's contention that the approximate truth of Newtonian phenomena need not underwrite the approximate truth of Newtonian gravitation theory is not clear without a more detailed investigation of scientific approximation.

17. Proposition 8 shows that the idealization of treating bodies as point masses, which was implicit in the theorems backing up Newton's inferences, can be replaced by more realistic descriptions of spherically shaped bodies. This illustrates the role of what I. B. Cohen (1980, 52–64) called Newton's "mathematical style." It also illustrates what Laymon (1983, 195) sees as the especially important em-

pirical virtue of maintaining fit with data as idealized assumptions are replaced by more accurate ones.

18. The following comment from the General Scholium makes a similar appeal to results which appear later in Book 3, but is not quite so explicitly based on hypothetico-deductive success.

And to us it is enough that gravity does really exist, and act according to the laws which we have explained, and abundantly serves to account for all the motions of celestial bodies, and of our sea. (Cajori, 547)

A passage Stein quotes from Newton's discussion of Rule 3 makes a somewhat more detailed appeal to results added in the rest of Book 3.

Lastly, if it universally holds by experiments and astronomical observations that all bodies about the earth gravitate toward the earth, and that in proportion to the quantity of matter in each, and the moon gravitates towards the earth in proportion to her quantity of matter, and that our sea in turn gravitates towards the moon, and all the planets gravitate mutually towards one another, and the comets in like manner gravitate towards the sun: it is to be asserted by this rule [i.e., Rule III itself], that all bodies gravitate mutually towards one another. For the argument from the phenomena will be even stronger for universal gravitation, than for the impenetrability of bodies: for which among heavenly bodies in particular we have no experiment, no observation whatever. (Quoted in Stein 1991, 220)

This passage is much less obviously suggestive of a methodology based on hypothetico-deductive success, since many of the cited properties of gravitation are backed up by measurements of parameter values rather than mere prediction of the phenomena.

19. See Harper and Smith (1995) for discussion of the relation between Newton's "experimental philosophy" and the rival "mechanical philosophy," both as it developed in the earlier debate on light and colors, and as it continued to develop in Newton's revisions of *Principia* in response to criticism from the mechanical philosophers.

20. Newton accounted for only about half of the known precession of the moon by his treatment, in *Principia*, of the action of the sun on the lunar orbit. Clairaut was one of a new generation of ardent Newtonians for whom work by others, including Malebranche, had paved the way (Guerlac 1981, 61). When, in his initial calculation, Clairaut found that he could account for only half of this precession, a result also obtained in calculations by Euler and d'Alembert, he proposed that gravitation toward the earth was not inverse square, but also depended upon a higher power term (Waff, 1976).

Clairaut's successful recalculation involved a more complex treatment of inverse square perturbational terms (Wilson 1980, 138). Euler's reaction, after duplicating the result with a somewhat different method (ibid., 139–43), expresses the enormous support it was believed to provide for the inverse square law.

The more I consider this happy discovery, the more important it seems to me, and in my opinion it is the greatest discovery in the Theory of Astronomy, without which it would be absolutely impossible ever to succeed in knowing the perturbations that the planets cause in each others motions. For it is very certain that it is only since this discovery that one can regard the law of attraction reciprocally proportional to the squares of the distances as solidly established; and on this depends the entire theory of astronomy (Ibid., 143)

That Clairaut's solution to the lunar precession problem contributed to the final demise of the vortex theory is suggested by Bouguer's appeal to Clairaut's earlier proposal (to alter the inverse square law) to help argue that attraction need not be an independent principle (Aiton, 248).

This suggestion implies that Kuhn's (1962, 81; 1970, 81) remark that "no one took these proposals [to alter the inverse square law] very seriously" may give a somewhat misleading impression of how well accepted Newton's theory had become before Clairaut's solution to the precession problem. Both the first (1962) and second (1970) editions of Kuhn's classic book were written before the work by Aiton, Waff, and Wilson on reactions to the precession problem had become available. It may be worth noting that one of the two passages Kuhn explicitly changed between the first and second editions (Kuhn 1970, 30–31) was modified to take into account new work, such as Truesdell's (1967) account of resistance to *Principia* based on differences between its dynamics and the dynamical techniques that had been developed by the Bernoullis and others.

21. Kepler's orbit for Mars gave an account of its motion with respect to the sun that, when combined with his account of the motion of the earth, produced an ephemeris that predicted the geocentric angular position of Mars at any time. Using this ephemeris, the fit of these predictions to Brahe's data had mean errors of about 2 minutes of arc, which was more or less the precision to which Brahe's data could be trusted (Harper, Bennett, and Valluri 1993, 138–39). Laubscher's *The Motion of Mars 1751–1969* illustrates the dramatic improvement of fit obtained (with much more precise data) when Keplerian phenomena are corrected by introducing perturbations due to gravitational interactions. Laubscher's ephemeris is compared with over 5,000 right ascensions and declinations of Mars taken between 1751 and 1969. For the earliest group of this data (1751–1822), the mean errors in right ascension are under 2 seconds of arc, while in later groups they are under ½ second (ibid., 140).

22. Newton's estimate in the first edition was 91 percent less, and in the third edition was 41 percent less. These errors were due to the values Newton accepted for the solar parallax, 20″ in 1687, 10″ in 1713, and 10.5″ in 1726. The modern value for solar parallax is 8.7942″, which makes the solar distance 23,455 earth radii, rather than the 10,000 Newton assumed in 1687, the 20,000 he assumed in 1713, or the 19,600 he assumed in 1726. See Hufbauer 1993, 28–29.

23. As corrections to Keplerian motion for gravitational interactions led to better and better fit, with data that became more and more precise, the idea that measurement from phenomena could progress toward exact values for parameters may even have made Laplace's conjecture about determinism plausible.

If an intelligence, for one given instant, recognizes all the forces which animate Nature, and the respective positions of the things which compose it, and if that intelligence is also sufficiently vast to subject these data to analysis, it will comprehend in one formula the movements of the largest bodies of the universe as well as those of the minutest atom: nothing will be uncertain to it, and the future as well as the past will be present to its vision. (This is a somewhat more vivid translation of a passage from Laplace quoted by Earman 1986, 7)

This conjecture of Laplace's, however, is an extrapolation that goes far beyond what, according to the Newtonian ideal of empirical success, should count as scientifically established fact.

Recent calculations exhibiting chaotic Newtonian perturbations for all the planets (Sussman and Wisdom, 1992) undercut Laplace's conjecture that the empirical successes of planetary celestial mechanics could be expected, in principle, to be extended to measure magnitude values precise enough to support determinism. The existence of chaos in the solar system, however, does not undercut Newton's methodology or his ideal of empirical success. Arguments, such as that of Ekeland 1990, suggesting that chaos requires giving up Newton's methodology confuse that methodology with Laplace's conjecture about the prospects for future research. The efforts to estimate sizes of chaotic zones in work on chaos (e.g., Laskar 1990, as well as Sussman and Wisdom 1992) is very much in line with what I have been suggesting is Newton's methodology.

24. This suggestion would augment, at least for parts of physics, the five characteristics of a good scientific theory discussed by Kuhn (1977, 321–99) by adding to what he calls accuracy, which appears to be (321–23) mainly hypothetico-deductive prediction, Newton's ideal of empirical success as accurate measurement of a theory's parameters by phenomena.

25. Newton's reply (1958, 106) to Pardies' second letter.

26. Newton makes it clear that any proposition he counts as conclusively established from experiment is subject to challenge either by showing defects in the reasoning purporting to establish it on the basis of the cited experiments or by providing further experiments that go against it (1958, 94). See Harper and Smith, Sec. I(f).

27. Harper 1991 argued that corrections of Keplerian phenomena introduced to accommodate perturbations corresponding to gravitational interactions with other bodies cannot undercut the inferences to inverse square centripetal forces from the uncorrected phenomena, since the corrected motion is understood to result from composing additional forces of interaction with the inverse square centripetal force inferred from the Keplerian phenomena.

28. This value was proposed by Hall 1894 as an alternative force law to account for mercury's precession. Valluri, Wilson, and Harper 1997 argue that it is appropriate, even when taking into account the actual eccentricity of Mercury's orbit.

29. An account of scientific approximations based on measurement of theoretical parameters would share many of the features that Duhem discusses. We have seen, however, that Newton's ideal of empirical success goes beyond hypothetico-deductive criteria. This suggests that it would motivate some modifications to Duhem's account of scientific approximation.

30. Kuhn's (1977, 321–22) criterion of consistency — "not only internally or with itself, but also with other currently accepted theories applicable to related aspects of nature" — provides an argument against this skeptical implication suggested by the Quine-Duhem thesis. As Jon Dorling argues (1979), plausible Bayesian priors for real examples of scientific research also provide an answer. As will be seen below, Newton's ideal of empirical success provides an additional argument for theories that realize it.

31. Einstein's original Newtonian limit result (1916) is for one-body accelerations as are the results of Troutman discussed by Glymour (note 35). Newtonian perturbation theory was used to account for interactions in the original applications of GR to solar system phenomena such as those provided by the Schwart-

schild solution. Recently, however, applications of GR to solar system phenomena use n-body Lagrangians, which can be given numerical solutions to very precise approximations even though exact solutions to even two-body problems in GR have not been developed (e.g., Damour 1994, Nordtvedt 1994, Brumberg 1991).

32. Earman and Glymour (1980, 79–80) criticize Dyson et al. (1919, 291) for limiting the alternatives to three — zero deflection for light uninfluenced by gravitation, 0.87″ deflection for Newtonian gravity with gravitation on light in accordance with mass-energy equivalence, and 1.75″ for general relativity. It is not at all obvious that one could not make a case justifying Dyson and Eddington's limitation of alternatives to what they had reason to regard as the serious rivals. Moreover, other alternatives are not relevant to my claim that general relativity does better than Newtonian gravity theory. Earman and Glymour (1980, 75) also criticize Dyson et al. for dismissing the Sobral astrograph plates, while giving some weight to the astrograph plates at Principe. Collins and Pinch (1994, 48–52) describe the criticisms of Glymour and Earman in such a way as to suggest that the evidence from the data of the 1919 eclipse expeditions was even more equivocal than Glymour and Earman suggest.

It is not at all obvious that these suggestions or the Glymour-Earman criticisms are accurate reflections of the evidential implications of the data generated by these expeditions. The Sobral 4-inch telescope had seven good plates (Dyson et al. 1919, 302), not eight, as cited by Collins and Pinch (1994, 49). If we take the standard deviations Collins and Pinch cite from Glymour and Earman the Einstein value (1.75, Dyson et al. 1919, 291 — Collins and Pinch 1994, 48, give this as 1.7″) is 1.35 sd below the mean (1.98″, ibid., 49), while the Newton value (.87″) is 6.23 sd below. If we use sd$^+$ = $(n/n - 1)^{1/2}$ sd where n is 7 (the number of good plates) the Einstein value is 1.21 sd$^+$ below, while the Newton value is 5.84 sd$^+$ below. This suggests that even if both sets of astrograph data were dropped the Einstein prediction fits much better than the Newton prediction, 5 or 6 standard deviations is a clear rejection. The worst case for the Einstein prediction, the Sobral astrograph data (mean is .86, almost right on the Newton prediction), which Dyson et al. argue are defective, would make the Einstein prediction 1.75 sd.$^+$ too high, which (unlike the case with the Newton prediction on the much more solid Sobral 4-inch telescope data) is not a rejection. The Principe astrograph data (mean is 1.62) gives .2 sd$^+$ for the Einstein prediction and 1.19 sd$^+$ for the Newton prediction. This suggests that the Einstein prediction would remain a clear winner over the Newton prediction even if all the astrograph data were included along with the data from the Principe 4-inch.

The criticisms of Earman and Glymour would need to be backed up by a far more detailed investigation to support the skeptical assessment of the evidential implications of the 1919 eclipse expeditions which Collins and Pinch purport to draw from them. The points raised above strongly suggest that the claim that general relativity did better than Newtonian gravity theory can be expected to hold up in such an investigation.

33. The following quotation from the preface to a 1991 text on relativistic celestial mechanics illustrates the sort of precision expected for ephemerides based on general relativity.

In recent years the general theory of relativity (GTR) in its simplest applications in celestial mechanics and astrometry is no longer seen as a theory to be proved, but rather serves as a necessary framework for the construction of accurate dynamical ephemerides (10^{-8} to 10^{-9} with respect to the main Newtonian terms) and in the discussion of high precision observations ($0.001''$ in angular distance, 1 microsecond in time, 10^{-14} in frequency). (Brumberg 1991, vii)

34. As pointed out in Note 31, Einstein's Newtonian limit results are one-body theorems. Given that Newtonian perturbation theory is appropriate to calculate the harmonic law ratio for a given planet, which would measure the mass of the sun in Newtonian gravitation theory, it allows that same harmonic law ratio to accurately measure the corresponding component of the stress-energy tensor in the Schwarzschild metric of general relativity representing the space-time curvature generated by the sun (DiSalle and Harper 1996).

35. Glymour (1977; 1980, 358–64) compares the Newtonian space-time formulation of universal gravitation with an alternative space-time theory of gravity that allows exactly the same trajectories (the accelerations agree in all relevant local coordinates), but represents these trajectories as geodesics corresponding to a space-time connection that directly represents the gravitation generated by a distribution of matter. The Newtonian space-time formulation of universal gravitation represents trajectories as effects of a gravitational potential, generated by the same distribution of matter, operating against a flat background space-time. Glymour argued that the second theory is better tested by bootstrap confirmation from trajectories than the first, even though the trajectories they allow are the same.

The equations of motion make it clear that, by dropping the unmeasured flat background space-time, the second theory does better than the first at realizing Newton's ideal of empirical success (DiSalle, Harper, and Valluri 1996a). This second theory is a Newtonian limit of general relativity that keeps its key idea of representing gravitational trajectories as geodesics in curved space-time, though the space at any given time is flat (Malament 1986). This suggests that, even on Newtonian phenomena alone, general relativity can do better than universal gravitation by Newton's standard for empirical success.

36. Some of the material in this section is covered in more detail in DiSalle, Harper, and Valluri 1994c. DiSalle, Harper, and Valluri (1994b) is a shorter presentation for physicists. The Shapiro time delay test is also discussed in ibid. 1994b, which also includes discussion of Newtonian limits of general relativity. (This paper was presented at the 7th Marcel Grossman Conference on General Relativity held at Stanford in July 1994 and will appear in the proceedings of that conference.)

37. DiSalle 1992 argues that realism about space-time structure is not a belief in an additional unobservable entity that explains dynamical laws. Rather, it is no more than a belief in dynamical laws that allow distinctions such as Newton's to be made objectively.

38. The lunar laser ranging results that limit Nordtvedt effects to $(2 \pm 5) \times 10^{-13}$ limit the Nordtvedt parameter η ($\eta = 4\beta - 3 - \gamma$ in the PPN framework and is zero in GR) to -0.0005 ± 0.0011 (Dickey, et al. 1994, 485–86). According to Brans-Dicke theory, the PPN parameter $\beta = 1$ (Will 1993, 117). This makes $\eta =$

$1 - \gamma$ and makes constraints on η constrain the Brans-Dicke ω, since $\gamma = (1 + \omega)/(2 + \omega)$. (Taking $\eta = -0.0005 \pm 0.0011$ would put γ between .9994 and 1.0016 which would require ω to be over 1000.) Following the usual practice (where null results are compatible with the data) by taking η as 0 ± 0.0011 makes ω at least 908.

REFERENCES

Aiton, E. J. 1972. *The Vortex of Planetary Motions*. New York: American Elsevier.

Ashby, N., D. Bartlett, and W. Wyss, eds. 1990. *General Relativity and Gravitation, 1989: Proceedings of the 12th International Conference on General Relativity and Gravitation*. Cambridge: Cambridge University Press.

Berkeley, G. 1965. *Berkeley's Philosophical Writings*. London: Collier-Macmillan.

Brans, C., and R. H. Dicke. 1961. "Mach's Principle and a Relativistic Theory of Gravitation." *Physical Review* 124: 925–35.

Brumberg, V. A. 1991. *Essential Relativistic Celestial Mechanics*. New York: Adam Hilger.

Cohen, I. B. 1980. *The Newtonian Revolution*. Cambridge: Cambridge University Press.

Damour, T. 1987. "The Problem of Motion in Newtonian and Einsteinian Gravity." In S. W. Hawking and W. Israel, eds., 128–98.

Damour, T., and D. Vokrouhlický. 1996. "Equivalence Principle and the Moon." *Physical Review* 53: 4177–4201.

Dicke, R. H. 1964. *The Theoretical Significance of Experimental Relativity*. New York: Gordon and Breach.

Dickey, J., P. L. Bender, J. E. Faller, Newhall, R. L. Ricklefs, J. G. Ries, P. J. Shelus, C. Veillet, A. L. Whipple, J. R. Wiant, J. G. Williams, C. F. Yoder. 1994. "Lunar Laser Ranging: A Continuous Legacy of the Apollo Program." *Science* 265: 482–90.

DiSalle, R. 1992. "Einstein, Newton, and the Empirical Foundation of Spacetime Geometry." *International Studies in the Philosophy of Science* 6: 181–89.

DiSalle, R., and W. L. Harper. 1996. "Empirical Success and Scientific Revolution: General Relativity and the Newtonian Limit." Manuscript.

DiSalle, R., W. L. Harper, and S. R. Valluri. 1994a. "General Relativity and Empirical Success." *Proceedings of the 7th Marcel Grossman Meeting on General Relativity*, July 1994.

———. 1994b. "Reasoning from Phenomena in General Relativity." In R. B. Mann and R. G. McLenaghan, eds., *Proceedings of the 5th Canadian Conference on General Relativity and Relativistic Astrophysics*. Singapore: World Scientific, 209–14.

———. 1994c. "Theory and Evidence in Relativistic Gravitation." Manuscript.

Dorling, J. 1979. "Bayesian Personalism, the Methodology of Scientific Research Programs, and Duhem's Problem." *Studies in the History and Philosophy of Science* 10: 177–87.

——. 1991. "Reasoning from Phenomena: Lessons from Newton." *PSA 1990* 2: 197–208.

——. 1995. "Einstein's Methodology of Discovery was Newtonian Deduction-form-the-Phenomena." In J. Leplin, ed., *The Creation of Idea in Physics*. Dordrecht: Kluwer.

Duhem, P. 1962. *The Aim and Structure of Physical Theory*. Trans. P. Wiener. New York: Atheneum.

Dyson, F. W., A. S. Eddington, and C. Davidson. 1920. "A Determination of the Deflection of Light by the Sun's Gravitational Field." From observations made at the total eclipse of 29 May 1919. Royal Society. *Philosophical Transactions* 220: 291–333.

Earman, J., ed. 1983. *Testing Scientific Theories*. Minneapolis: University of Minnesota Press.

Earman, J., ed. 1986. *A Primer on Determinism*. Dordrecht: D. Reidel.

Earman, J., and C. Glymour. 1980. "Relativity and Eclipses: The British Eclipse Expeditions of 1919 and Their Predecessors." *Historical Studies in the Physical Science* 11: 49–85.

Earman, J., and M. Janssen. 1993. "Einstein's Explanation of the Motion of Mercury's Perihelion." In J. Earman, M. Janssen, and J. D. Norton, eds., *The Attraction of Gravitation: New Studies in the History of General Relativity*. Boston: Birkhäuser, 129–72.

Eddington, A. S. 1922. *The Mathematical Theory of Relativity*. Cambridge: Cambridge University Press.

Ekeland, I. 1990. *Mathematics and the Unexpected*. Chicago: University of Chicago Press.

Feyerabend, P. K. 1970. "Classical Empiricism." In R. E. Butts and J. W. Davis, eds., *The Methodological Heritage of Newton*. Toronto: University of Toronto Press, 150–70.

Forster, M., and E. Sober. 1994. "How to Tell When Simpler, More Unified, or Less *Ad Hoc* Theories Will Provide More Accurate Predictions." *British Journal of Philosophy of Science* 45: 1–35.

Glymour, C. 1977. "The Epistemology of Geometry." *Nous* 11: 227–51.

——. 1980. *Theory and Evidence*. Princeton: Princeton University Press.

——. 1983. "On Testing and Evidence." In Earman 1983, 3–26.

Guerlac, H. 1981. *Newton on the Continent*. Ithaca, N.Y.: Cornell University Press.

Hall, A. 1894. "A Suggestion in the Theory of Mercury." *Astronomical Journal* 319: 49–51.

Harper, W. L. 1989. "Consilience and Natural Kind Reasoning in Newton's Argument for Universal Gravitation." In J. R. Brown and J. Mittelstrass, eds., *An Intimate Relation, Studies in the History and Philosophy of Science*. Dordrecht: Kluwer Academic Publisher, 115–52.

——. 1991. "Newton's Classic Deduction from Phenomena." *PSA 1990* 2: 183–96.

——. 1993. "Reasoning from Phenomena: Newton's Argument for Universal Gravitation and the Practice of Science." In P. Theerman and A. F. Seef, eds.,

Action and Reaction. Proceedings of a Symposium to Commemorate the Tercentenary of Newton's Principia. Newark: University of Delaware Press, 144–82.

Harper, W. L., B. H. Bennett, and S. Valluri. 1994. "Unification and Support: Harmonic Law Ratios Measure the Mass of the Sun." In D. Prawitz and D. Westerstahl, eds., *Logic and Philosophy of Science in Uppsala, Synthese Library* 236. Dordrecht: Kluwer, 131–46.

Harper, W. L., and R. Disalle. 1996. "Inferences from Phenomena in Gravitational Physics." *PSA 1996.*

Harper, W. L., and G. Smith. 1995. "Newton's New Way of Inquiry." In J. Leplin, ed., *The Creation of Ideas in Science.* Dordrecht: Kluwer, 113–66.

Hawking, S. W., and W. Israel. 1987. *300 Years of Gravitation.* Cambridge: Cambridge University Press.

Hufbauer, K. 1993. *Exploring the Sun, Solar Science Since Galileo.* 2nd ed. Baltimore: Johns Hopkins University Press.

Kuhn, T. S. [1962] 1970. *The Structure of Scientific Revolutions.* 2nd ed. Chicago: University of Chicago Press.

———. 1977. *The Essential Tension.* Chicago: University of Chicago Press.

Lakatos, I. 1978. *The Methodology of Scientific Research Programes.* Cambridge: Cambridge University Press.

Laskar, J. 1990. "The Chaotic Behavior of the Solar System: A Numerical Estimate of the Size of the Chaotic Zones." *Icarus* 88: 266–91.

Laubscher, R. E. 1981. "The Motion of Mars 1751–1969." *Astronomical Papers prepared for the use of the American Ephemeris and Nautical Almanac,* vol. 22, 363–494. Washington: U.S. Government Printing Office.

Laymon, R. 1983. "Newton's Demonstration of Universal Gravitation." In Earman 1983, 179–99.

Malament, D. 1986. *Newtonian Gravity, Limits, and the Geometry of Space.*

Newcomb, S. 1897. *The Elements of the Four inner Planets and the Fundamental Constants of Astronomy, Supplement to the American Ephemeris and Nautical Almanac for 1897.*

Newton, I. 1934. *Principia,* vol. 2. Trans. A. Motte and F. Cajori. Berkeley: University of California Press.

———. 1958. *Papers and Letters on Natural Philosophy and Related Documents.* Ed. I. B. Cohen and R. E. Schofield. Cambridge: Cambridge University Press.

———. 1987. *Mathematical Principles of Natural Philosophy.* Trans. I. B. Cohen and A. Whitman. Los Angeles: University of California Press, forthcoming.

Nordtvedt, K. 1968. Equivalence Principle for Massive Bodies. II. Theory. *Physical Review* 169: 1017–25.

———. 1972. "Gravitation Theory: Empirical Status from Solar System Experiments." *Science* 178: 1157–64.

Nordtvedt, K., and C. M. Will. 1972. "Conservation Law and Preferred Frames in Relativistic Gravity. II. Experimental Evidence to Rule Out Preferred-Frame Theories of Gravity." *Astrophysical Journal* 177: 775–92.

Popper, K. R. 1972. *Objective Knowledge.* Oxford: Oxford University Press.

Rosenkrantz, R. 1983. "Why Glymour Is a Bayesian." In Earman 1983, 69–97.

Roseveare, N. T. 1982. *Mercury's Perihelion from Le Verrier to Einstein.* Oxford: Carendon Press.

Shapiro, I. I. 1990. "Solar System Tests of General Relativity: Recent Results and Present Plans." In Ashby et al. 1990, 313–30.

Stein, H. 1967. "Newtonian Space-Time." *Texas Quarterly*, Autumn, 174–200.

———. 1991. "From the Phenomena of Motions of the Forces of Nature: Hypothesis or Deduction?" *PSA 1990* 2: 209–22.

Sussman, G. J., and J. Wisdom. 1992. "Chaotic Evolution of the Solar System." *Science* 257: 56–62.

Truesdell, C. 1967. "Relations of Late Baroque Mechanics to Success, Conjecture, Error, and Failure in Newton's *Principia.*" *Texas Quarterly*, Autumn, 281–97.

Valluri, S., C. Wilson, and W. L. Harper. 1997. "Newton's Apsidal Precession Theorem and Eccentric Orbits." *Journal for the History of Astronomy* 28: 13–27.

Waff, C. B. 1976. *Universal Gravitation and the Motion of the Moon's Apogee: The Establishment and Reception of Newton's Inverse Square Law, 1687–1749.* PhD. diss., Johns Hopkins University.

Wald, R. M. 1984. *General Relativity.* Chicago: University of Chicago Press.

Whiteside, D. T. 1976. "Newton's Lunar Theory: From High Hope to Disenchantment." *Vistas in Astronomy* 19: 317–28.

Will, C. M. 1971. "Theoretical Frameworks for Testing Relativistic Gravity. II. Parametrized Post-Newtonian Hydrodynamics and the Nordtvedt Effect." *Astrophysical Journal* 163: 611–28.

———. 1986. *Was Einstein Right? Putting General Relativity to the Test.* New York: Basic Books.

———. 1993. *Theory and Experiment in Gravitational Physics.* 2d. ed. Rev. ed. Cambridge: Cambridge University Press.

Will, C. M., and K. Nordtvedt. 1972. "Conservation Laws and Preferred Frames in Relativistic Gravity. I. Preferred-frame Theories and an Extended PPN Formalism." *Astrophysical Journal* 177: 757–74.

Wilson, C. 1980. "Perturbations and Solar Tables from Lacaille to Delambre: The Rapprochement of Observation and Theory. Part I." *Archive for History of Exact Sciences* 22: 53–188.

———. 1985. "The Great Inequality of Jupiter and Saturn: From Kepler to Laplace." *Archive for History of Exact Sciences* 33: 15–290.

4

A Peek Behind the Veil of Maya

Einstein, Schopenhauer, and the Historical
Background of the Conception of Space as a
Ground for the Individuation of Physical Systems

Don Howard

Department of Philosophy, University of Notre Dame

> We find among their many efforts to bring to light the analogy between all the
> phenomena of nature, many attempts, although unfortunate ones, to derive
> laws of nature from the mere laws of space and time. However, we cannot
> know how far the mind of a genius will one day realize both endeavors.
> —Arthur Schopenhauer, *Die Welt als Wille und Vorstellung*

Einstein's Berlin Portrait Gallery

According to Einstein's son-in-law and biographer, Rudolf Kayser, portraits of three figures hung on the wall of Einstein's Berlin study in the
late 1920s: Michael Faraday, James Clerk Maxwell, and Arthur Schopenhauer (Reiser-Kayser 1930, 194).[1] One can guess why Faraday and
Maxwell, the inventors of field theory, were there. But I have long been
puzzled about what Schopenhauer was doing in this august company.

Something else puzzled me as well. From what source did Einstein
draw the idea of "spatiotemporal separability," — that a non-null spatiotemporal separation is a sufficient condition for the individuation of
physical systems and their states, an idea fundamental both to his
conception of field theory and to his famous reservations about the
quantum theory? This idea makes its first appearance in Einstein's
1905 paper on the photon hypothesis and gradually finds ever-clearer
expression. By the late 1940s, Einstein has disentangled it from other,
related conceptions of independence, such as the concept of "locality,"

or "local action," and has come to regard it as the most essential aspect of his understanding of the very concept of physical reality. But, search as one will, this idea is not to be found anywhere in the scientific and philosophical literature normally regarded as having influenced Einstein's world view in his early years. It will not be found in Mach, Maxwell, Lorentz, Boltzmann, Hume, Poincaré, or Hertz.[2] It will not be found in period textbooks, either elementary texts, such as Violle (1892–1893), or more advanced ones, such as Föppl (1894). It will not be found in treatises on phoronomy, as in Aurel Voss's article in the *Encyklopädie der mathematischen Wissenschaften* (1901). And it will not be found in the popular science literature of the day, such as the works by Aaron Bernstein (1853–1857) or Ludwig Büchner (1855) that Einstein read as a youth. The idea may, of course, have been Einstein's own. Still, one must ask whether there was a source for it elsewhere in Einstein's readings.

I suggest that the solutions to the two puzzles may be related. Surprising as it may seem, Schopenhauer may well have been the source for the idea of spatiotemporal separability. Given how fundamental that idea was to Einstein's conception of a field theory, this may explain Schopenhauer's rather exalted place next to Faraday and Maxwell in Einstein's little Berlin gallery.

The argument for this conjecture will be circumstantial at many crucial places; I have found no smoking gun. All the more reason to be skeptical, as I was at first (and still am, to some small extent) about the possibility that a thinker like Schopenhauer, famous as "the philosopher of pessimism," could have had an important influence on Einstein's thinking about fundamental questions of the ontology of spacetime and field theories. Schopenhauer was certainly widely read in the late nineteenth and early twentieth centuries. His aesthetics is said to have been a crucial influence on Wagner's notion of the *Gesammtkunstwerk*. He was an important source for Nietzsche's concept of the will. And he influenced several generations of literary figures, Thomas Mann being only the most famous of many.[3] But such influences are different in kind from those I suggest he had on the young physicist, Einstein. Could Schopenhauer *really* have had an influence on Einstein's thinking about fundamental questions of the ontology of spacetime?

Patience and forbearance are required in order to permit this argument to get off the ground. One must put on hold a strong disinclination to believe in the influence suggested here. For part of that initial skepticism results from the fact that we have all grown to philosophical maturity in an antimetaphysical age (at least the philosophers of science among us), an experience that leaves us ill-prepared, if not positively ill-disposed, to believe that Schopenhauer could have had this kind of influence on Einstein. Indeed, I suspect that one reason why better documentation for this influence is lacking is that, whereas many ardent young positivists were eager to ask Einstein about his reading of Mach and Hume, almost no one bothered to ask him about Schopenhauer.

Fortunately, a few people did ask, and Einstein himself on at least a few occasions wrote about what he learned from Schopenhauer. So we can say some things with reasonable certainty about his reading of Schopenhauer, his estimation of Schopenhauer as a writer, and the way in which Schopenhauer influenced his world view.

What I will argue, more specifically, is that several crucial features of Einstein's world view, in addition to the idea of spatiotemporal separability, could easily have been derived from his reading of Schopenhauer. At the very least, they would have found important confirmation in Schopenhauer. In particular, we will find in Schopenhauer a unique view, a critical reaction to Kant, about the equally fundamental importance of, on the one hand, space and time as the "*principium individuationis*," the "ground of being," and, on the other hand, causality, the "ground of becoming," for the constitution of representations of empirical objects in the understanding, both causality and space and time being seen by Schopenhauer as forms of the principle of sufficient reason.

There is a context for Schopenhauer's development of these themes. It goes back at least to Newton, if not still earlier, to ancient and medieval discussions of the problem of individuation. The specific issue of space and time as principles of individuation comes to the fore in the disputes between Leibniz and the Newtonians over absolute versus relational conceptions of space and time, as in the Leibniz-Clarke correspondence. An important chapter in the story concerns Kant's turn away from Leibniz and back to Newton, with his invention of the "incongruent counterparts" argument at the time of his *Inaugural Dissertation*. But it is Schopenhauer, more than anyone else — more even than the Marburg neo-Kantians — who brings the theme of space and time as principles of individuation into nineteenth-century discussions of space and time.

That Schopenhauer's philosophy could have been read in this way by philosophers of science and philosophically sophisticated physicists of Einstein's generation will be shown by looking at what thinkers as diverse as Mach, Schlick, Schrödinger, Weyl, Pauli, and Cassirer did in fact say about Schopenhauer. As one begins to appreciate *who* it was who read Schopenhauer in this way, and exactly *how* they read him, a picture begins to emerge in which Schopenhauer's distinctive views on the importance of space and time as the *principium individuationis* arguably form the background and provide the vocabulary for the early twentieth-century discussion of the way in which the spatiotemporal manner of individuating physical objects is thrown into doubt by the development of the quantum theory. Thus, Einstein's ardent defense of spatiotemporal separability as something fundamental to general relativity or any field theory, and the equally forceful critiques of separability in the work of Schrödinger and Pauli (and perhaps also Bohr) as something explicitly denied in the quantum-mechanical theory of interactions, must all be seen against this background informed by Schopenhauer.

Finally, when Einstein's defense of spatiotemporal separability is seen against this Schopenhauerian background, its place in his understanding of the ontology of general relativity assumes a new significance, inasmuch as it helps us to locate Einstein squarely within a tradition regarding the nature of space or spacetime going all the way back to Newton. What defines this tradition is not one's position in the dispute over absolute versus relative conceptions of space and time, or one's position in that more recent debate between substantival and relational conceptions of spacetime — both of which debates miss something essential in the Newtonian conception of space. What defines this tradition is, rather, a commitment to the idea that spatiotemporal separation is an objective feature of spacetime sufficient to serve as a ground for the individuation of systems and their states.

It was this characteristically Newtonian idea to which Leibniz really objected, arguing that systems are individuated not by extrinsic spatiotemporal determinations, but by intrinsic, qualitative determinations. It was this Newtonian idea that Kant reaffirmed with his incongruent counterparts argument. It was this idea that Schopenhauer bequeathed to the nineteenth century. And it was this idea that was preserved, in its most elementary form, stripping away all of the baggage of absolutist ideas, in Einstein's making the infinitesimal metric interval the fundamental invariant of the general theory of relativity.

Einstein's Reading of Schopenhauer

Schopenhauer was born in 1788 and died in 1860. His major published works include *Ueber die vierfache Wurzel des Satzes vom zureichenden Grunde* (On the fourfold root of the principle of sufficient reason, 1813); *Die Welt als Wille und Vorstellung* (The world as will and representation, 1st ed. 1819, 2nd ed., including first publication of vol. 2, 1844); *Ueber den Willen in der Natur* (On the will in nature, 1836); *Die beiden Grundprobleme der Ethik* (The two fundamental problems of ethics, 1841b); and *Parerga und Paralipomena* (1851). Schopenhauer began to win an audience for his writings only near the end of his life, but in the latter half of the nineteenth century and the first decades of the twentieth he had become perhaps the most widely read philosopher in German-speaking Europe, both among academic philosophers and among the broader public; there was also a large audience for his works in translation (see Laban 1880 and Hübscher 1981 for details on the editions and translations of Schopenhauer's works). Here is how a recent biographer explains Schopenhauer's posthumous popularity:

In Schopenhauer, readers found an encomium of a sober sense of reality, of materialistic explanation, and a justification, based on Kant, of why our empirical curiosity must follow this road. They found a confirmation of what they were, materialistically, doing. Simultaneously, however, they found in Schopenhauer the empirical proof that this approach to reality was not the only one. Even the materially visualized world still remains an idea. Schopenhauer inaugurated a new renaissance of Kant and opened up the possibility of a "materialism as if." One could endorse strictly empirical science, one could surrender to the materialistic spirit, but one need not be totally captured by it. With Schopenhauer's "Beyond" of the self-experienced will one could now withstand the pull of a materially interpreted immanence.

Even more effective than this "as if" materialism was the "as if" ethics which Schopenhauer sketched out in his "philosophy for the world." After 1850 his "Aphorisms on Practical Wisdom" [*Aphorismen zur Lebensweisheit*, part of *Parerga und Paralipomena* (1851)] rapidly became the Bible of the educated bourgeoisie. (Safranski 1990, 334)

Einstein, born in 1879 to a father just starting out in the new electro-technology business, was a child of this educated bourgeoisie. (See Einstein 1987, xlviii–lxvi, on Einstein's early life and family circumstances.)

Rudolf Kayser, the son-in-law and biographer who gave us the story of the portraits in Einstein's Berlin study, describes Einstein's reading of Schopenhauer during his student years at the ETH in Zurich:

Despite all the progress in understanding and knowledge achieved by the youthful physicist and mathematician, almost a dislike for science and its intellectual technique remained with him after he had finished his course of study. He overcame it only a long time afterwards. He approached the broader aspects of thought through philosophical studies, chiefly through his readings in Kant and Schopenhauer, and later through his study of Hume, with whom he felt a special kinship. (Reiser-Kayser 1930, 55)

This being the only biography of himself that Einstein read before publication, it deserves to be trusted in such matters.[4] But we have Einstein's own words to the same effect. Responding to a question about his early readings from another, later biographer, Carl Seelig, Einstein wrote on 20 April 1952:

As a young man (and even later) I concerned myself little with poetical litera-ture and novels. . . . I preferred books whose content concerned a whole world view [*Bücher weltanschaulichen Inhalts*] and, in particular, philosophical ones. Schopenhauer, David Hume, Mach, to some extent Kant, Plato, and Aristotle. (EA 39-019)[5]

Moreover, there is every reason to believe that this interest in Schopen-hauer continued in Einstein's later years. Kayser says about Einstein in the late 1920s, "He has read the most important works of classical philosophy, and in his hours of leisure he returns with especial pleasure to Plato, Hume and Schopenhauer" (Reiser-Kayser 1930, 197). And Konrad Wachsmann, the architect of Einstein's summer house in Ca-puth, recalls of that same time, the late 1920s and early 1930s, "The philosophers occupied him more than *belles lettres*. Above all, he read a lot of Schopenhauer. On Haberlandstraße and also in Caputh, he often sat with one of the already well-worn Schopenhauer volumes, and as he sat there reading, he seemed so pleased, as if he were engaged with a serene and cheerful work" (Grüning 1989, 243). Wachsmann adds, "He always insisted that the engagement with Schopenhauer, Kant, the Greeks, or even Locke and Hume, gave him far and away greater plea-sure than, for example, Goethe's — as it is so nicely known — great epic of human kind [*großes Menschheitsgedicht*]" (Grüning 1989, 247).

Remarks such as these suggest that Einstein read widely in Scho-penhauer's works. A complete 1894–1896 edition of Schopenhauer's works, answering Wachsmann's description of the "already well-worn

Schopenhauer volumes," survives in Einstein's private library.[6] But what, exactly, did Einstein read? Here, unfortunately, the documentation is thin. The only explicit reference I have found is in a letter from Einstein to his friend, ETH classmate, and future collaborator, Marcel Grossmann, on 6 September 1901: "What kinds of things do you do with your free time these days? Have you too already looked at Schopenhauer's *Aphorismen zur Lebensweisheit*? It's part of *Parerga & Paralipomena* and I liked it very much" (Einstein 1987, Doc. 122, 316). The reading of the *Aphorismen* evidently affected Einstein deeply enough that an analogy with Schopenhauer's professional isolation readily came to mind when he wrote to his future first wife, Mileva Marić, on 17 December 1901, describing the peculiar situation in which he found himself as a private tutor living and working in the home of Carl Baumer, a teacher in Schaffhausen:

It is really a very funny life I live here, precisely in Schopenhauer's sense. That is to say that, aside from my pupil, I speak with no one the whole day long. Even Herr Baumer's company seems to me boring and insipid. I always find that I am in the best company when I am alone, except when I am together with you. (Ibid., Doc. 128, 325)

When he wrote this, one day before applying for the position at the Swiss Federal Patent Office that he occupied from 1902 to 1909 (ibid., Doc. 129), Einstein was nearing the end of a long period of despair over his failure to find a regular professional position. He could, no doubt, empathize with Schopenhauer's failure ever to win a regular academic appointment, after his failed attempt as a lecturer in Berlin in 1820 (see Safranski 1990, 250–63).

Indeed, Schopenhauer's acid cynicism about professional philosophy and the academy more generally would have found a sympathetic audience in the young Einstein. Einstein was a headstrong young man, sure enough of his own abilities that, according to legend, he did not bother with many of the required lectures during his years at the ETH (1896–1900), borrowing Grossmann's notes if necessary and otherwise reading on his own the works of Mach, Helmholtz, Maxwell, Hertz, Boltzmann, and Lorentz (see Einstein 1946, 15; 1955, 10–11; Kollros 1955, 22). Einstein did not like the rigid structure of university instruction. His recollections of this time echo Kayser's remark about the young Einstein's "distaste for science and its intellectual technique" at the time when he started reading Schopenhauer:

For people like me, with a brooding interest, university study is not an unqualified blessing. If forced to eat so many good things, one can do lasting damage to one's appetite and stomach. The light of holy curiosity can be forever extinguished. Happily, in my case, this intellectual depression endured for only one year after the successful completion of my studies. (Einstein 1955, 12)

At the end of that year, Einstein was recommending Schopenhauer to Grossmann.

Einstein's cynicism about the university and professional academic life was reinforced by the fact that his hopes for a position as *Assistent* were thwarted at every turn (see Einstein 1987, esp. letters to Marić in the spring of 1901) forcing him to take unrewarding positions as a temporary teacher at the *Technikum* in Winterthur and as a private tutor in Schaffhausen, before finally going to work at the Patent Office. What Schopenhauer says in the *Aphorismen* about finding happiness in "what one is," as opposed to "what one has" or "what one represents" (one's fame, reputation, or standing) would surely have appealed to an Einstein who lacked a steady income and whose talents were not being recognized:

A tranquil and serene temperament, the result of perfect health and successful organization, a clear, lively, penetrating, and sure intellect, a measured, gentle will, and, thus, a clear conscience, these are advantages that no fame or riches can replace. For what a person is for himself, what abides with him in his loneliness and isolation, and what no one can give or take away from him, this is obviously more essential for him than everything that he possesses or what he may be in the eyes of others. A gifted man, in complete isolation, enjoys splendid converse with his own thoughts and fancies, whereas for an obtuse stump of a man even a continuing variety of companions, plays, excursions, and amusements cannot protect against torturous boredom. . . . Thus, for one's happiness in this life, that which one *is*, one's personality, is absolutely the first and most essential thing. (Schopenhauer 1851, vol. 1, 348–49)

But was this all that Einstein took from Schopenhauer? Was providing a voice and a validation for Einstein's late-adolescent alienation enough to earn Schopenhauer a place of honor in the portrait gallery? Was this why the mature and world-wise Einstein of 1930 was still reading Schopenhauer avidly? I think not. Einstein, himself, tells us in several places what else he learned from Schopenhauer.

Consider, first, Einstein's solution to the problem of free will. Einstein's commitment to determinism is well known. Equally important to his larger world view, however, was the extension of this metaphysics to the consideration of human freedom. One of his first public discussions of this issue was in an essay entitled, "The World as I See It":

I do not at all believe in human freedom in the philosophical sense. Everybody acts not only under external compulsion but also in accordance with inner necessity. Schopenhauer's saying, "A man can do what he wants, but not want what he wants," has been a very real inspiration to me since my youth; it has been a continual consolation in the face of life's hardships, my own and others', and an unfailing well-spring of tolerance. This realization mercifully mitigates the easily paralyzing sense of responsibility and prevents us from taking ourselves and other people all too seriously; it is conducive to a view of life which, in particular, gives humor its due. (Einstein 1931, 8–9)

As Friedrich Herneck has pointed out—in the only essay I know of that takes seriously the question of Schopenhauer's influence on Einstein—that exact quotation, "Ein Mensch kann zwar tun, was er will, aber nicht wollen, was er will," is not to be found in Schopenhauer's writings (Herneck 1969, 204), although many similar formulations are scattered throughout his works. Most relevant is his essay, "Ueber die Freiheit des menschlichen Willens" (1841). Schopenhauer begins by suggesting that we recast the question of free will not as a question about *physical freedom*—whether our doing what we will may be blocked by physical impediments ("Frei bin ich, wenn ich *thun* kann, *was ich will*")—but as a question about *moral freedom*, which he understands as a question about our capacity to want that which we will—"Kanst du auch *wollen*, was du willst?" (ibid., 364). Schopenhauer's answer, which rests on his doctrine of the will as the thing-in-itself, is that we are obviously free in the physical sense—that one does what one wills being almost an analytic truth—but that we are not free in the moral sense. Hence Einstein's paraphrase, and his remark that we act "in accordance with inner necessity." Our actions are controlled outwardly by physical causality—Einstein and Schopenhauer are both strict determinists as regards the course of physical events, such physical determinism being the first of Schopenhauer's four forms of the principle of sufficient reason. But we are also constrained inwardly—our psychological motives are no more under voluntary control than our outward actions, psychological determinism being the fourth form of the principle of sufficient reason. The will itself, as thing-in-itself, is not under the control of the principle of sufficient reason in any of its forms. It is only the objectifications of the will in phenomenal objects and events, both outer and inner, to which the principle of sufficient reason pertains.

How much of the metaphysical foundation to Schopenhauer's position on the freedom of the will Einstein accepted is hard to say. But another hint that he accepted enough to discomfit even the most tolerant logical empiricist is contained in a second essay from about this same time, a brief piece entitled "Religion and Science," first published in the *New York Times Sunday Magazine* for 9 November 1930. Einstein's main thesis is that while science and religion, viewed historically, may appear to be "irreconcilable antagonists," at a deeper level the "cosmic religious feeling," which distinguishes the "religious geniuses of all ages," is in fact "the strongest and noblest motive for scientific research," this same feeling being what gave Kepler and Newton "the strength to remain true to their purpose in spite of countless failures." And what is this "cosmic religious feeling"? It is a "third stage of religious experience" common to all historical religions, after the religion of fear and moral religion. Einstein explains:

It is very difficult to elucidate this feeling to anyone who is entirely without it, especially as there is no anthropomorphic conception of God corresponding to it.

The individual feels the futility of human desires and aims and the sublimity and marvelous order which reveal themselves both in nature and in the world of thought. Individual existence impresses him as a sort of prison and he wants to experience the universe as a single significant whole. The beginnings of cosmic religious feeling already appear at an early stage of development, e.g., in many of the Psalms of David and in some of the Prophets. Buddhism, as we have learned especially from the wonderful writings of Schopenhauer, contains a much stronger element of this. (Einstein 1930, 38)

One will not find this precise conception of "cosmic religion" anywhere in Schopenhauer, but the quoted lines echo a number of characteristically "Schopenhauerian" themes. Thus, for example, Schopenhauer writes that the historical religions and science are "natural enemies," adding that "to want to speak of peace and reconciliation between the two is most laughable; it is a *bellum ad internecionem*" (1851, vol. 2, 431).

More importantly, however, Schopenhauer distinguishes philosophy and religion as two kinds of metaphysics, the first having "its verification and credentials *in itself*, the other *outside itself*," this external verification being revelation (1859, vol. 2, 164). They are further distinguished by philosophy's being obliged to be true *sensu stricto et proprio*, whereas religion is obliged only to be true *sensu allegorico*, neglect of religion's allegorical nature being the source of error and confusion, especially among those who seek to rationalize religion and to reconcile it with science (ibid., 166–68). Schopenhauer adds about Buddhism:

The value of a religion will depend on the greater or lesser content of truth which it has in itself under the veil of allegory; next on the greater or lesser distinctness with which this content of truth is visible through the veil, and hence on the veil's transparency. It almost seems that, as the oldest languages are the most perfect, so too are the oldest religions. If I wished to take the results of my philosophy as the standard of truth, I should have to concede to Buddhism pre-eminence over all the others. In any case, it must be a pleasure to me to see my doctrine in such close agreement with a religion that the majority of men on earth hold as their own, for this numbers far more followers than any other. (Ibid., 169)

In this light, Einstein's conception of "cosmic religion," a "much stronger element" of which is contained in Buddhism, has a distinctly Schopenhauerian cast. At the heart of Schopenhauer's philosophy is the idea that human individuality is a kind of illusion, the necessary form in which the will objectifies itself in space and time, but not an aspect of the will in itself. The momentary overcoming of one's individuality, the recognition of "the futility of human desires and aims," a glimpse into "the sublimity and marvelous order which reveal themselves both in nature and in the world of thought," the experience of "the universe as a single significant whole" — all of these are, for Schopenhauer, marks of the way genius apprehends the world:

Genius is the ability to leave entirely out of sight our own interest, our willing, and our aims, and consequently to discard entirely our own personality for a time, in order to remain *pure knowing subject*, the clear eye of the world; and this not merely for moments, but with the necessary continuity and conscious thought to enable us to repeat by deliberate art what has been apprehended, and "what in wavering apparition gleams fix in its place with thoughts that stand forever." (1859, vol. 1, 185–86; the quotation is from Goethe's *Faust*)

It is essential to the expression of genus, for Schopenhauer, that one apprehends the "Idea" standing behind the plurality of its phenomenal manifestations. This "Idea" is not subject to the forms of space and time, the *principium individuationis*, and is thus a unity, a whole, different in kind from individual objects in space and time.

From an early age, Einstein was captivated by the prospect of a mode of comprehension that would carry one out of oneself, and he linked this yearning with his scientific ambitions. In his "Autobiographical Notes," he writes thus about his turn away from organized religion:

It is quite clear to me that the religious paradise of youth, which was thus lost, was a first attempt to free myself from the chains of the "merely personal," and from an existence dominated by wishes, hopes, and primitive feelings. Out yonder there was this huge world, which exists independently of us human beings and which stands before us like a great, eternal riddle, at least partially accessible to our inspection and thinking. The contemplation of this world beckoned as a liberation, and I soon noticed that many a man whom I had learned to esteem and to admire had found inner freedom and security in its pursuit. The mental grasp of this extra-personal world within the frame of our capabilities presented itself to my mind, half consciously, half unconsciously, as a supreme goal. Similarly motivated men of the present and of the past, as well as the insights they had achieved, were the friends who could not be lost. The road to this paradise was not as comfortable and alluring as the road to the religious paradise; but it has shown itself reliable, and I have never regretted having chosen it. (1946, 5)

Such remarks breathe the spirit of Schopenhauer, a connection made explicitly by Einstein himself in his 1918 address in honor of Max Planck's sixtieth birthday:

In the temple of science are many mansions, and various indeed are they that dwell therein and the motives that have led them thither. Many take to science out of a joyful sense of superior intellectual power; science is their own special sport to which they look for vivid experience and the satisfaction of ambition; many others are to be found in the temple who have offered the products of their brains on this altar for purely utilitarian purposes. Were an angel of the Lord to come and drive all of the people belonging to these two categories out of the temple, the assemblage would be seriously depleted, but there would still be some men, of both present and past times, left inside. Our Planck is one of them, and that is why we love him. . . . Now let us have another look at those who have found favor with the angel. Most of them are somewhat odd, uncommunicative, solitary fellows, really less like each other, in spite of these common characteristics, than the hosts of the rejected. What has brought them to the temple? . . . To begin with, I believe with Schopenhauer that one of the strongest motives that leads men to art and science is escape from everyday life with its painful crudity and hopeless dreariness, from the fetters of one's own ever shifting desires. A finely tempered nature longs to escape from personal life into the world of objective perception and thought; this desire may be compared with the townsman's irresistible longing to escape from his noisy, cramped surroundings into the silence of high mountains, where the eye ranges freely through the still, pure air and fondly traces out the restful contours apparently built for eternity. (1918, 224–25)

Talk of escape from "personal life" into the world of "objective perception" is vintage Schopenhauer; the expression, "objective perception," is Schopenhauer's term for the kind of knowledge achieved when the genius frees himself from the will and from his own individuality enough to apprehend the Ideas (in a quasi-Platonic sense) standing behind the will's objectification in phenomenal, empirical objects and events. Ordinary scientific knowledge is termed, by contrast, "subjective" (see Schopenhauer 1859, vol. 1, book 3). Even the image of the mountaintop is a Schopenhauer hallmark (see Schopenhauer 1985, 14, for one example of many).

It has been said that Einstein appreciated Schopenhauer only as a writer, not as a thinker. Thus, an early interviewer, Alexander Moszkowski, writes: "To Schopenhauer and Nietzsche he assigns a high position as writers, as masters of language and moulders of impressive thoughts. He values them for their literary excellence, but denies them philosophic depth" (Moszkowski 1921, 237). And Philipp Frank observes:

Einstein read philosophical works from two points of view, which were sometimes mutually exclusive. He read some authors because he was actually able to learn from them something about the nature of general scientific statements, particularly about their logical connection with the laws through which we express direct observations. These philosophers were chiefly David Hume, Ernst Mach, Henri Poincaré, and, to a certain degree, Immanuel Kant. Kant, however, brings us to the second point of view. Einstein liked to read some philosophers because they made more or less superficial and obscure statements in beautiful language about all sorts of things, statements that often aroused an emotion like beautiful music and gave rise to reveries and meditations on the world. Schopenhauer was pre-eminently a writer of this kind, and Einstein liked to read him without in any way taking his views seriously. In the same category he also included philosophers like Nietzsche. Einstein read these men, as he sometimes put it, for "edification," just as other people listen to sermons. (1947, 51)

It is difficult to reconcile these characterizations with the evidence of Einstein's own words. I think that Moszkowski and Frank are simply wrong. It is worth recalling that the Moszkowski volume is notoriously unreliable, Einstein having been persuaded by close friends to try to block its publication (see Born 1969, 62–70), and that Philipp Frank was an ardent logical empiricist whose characterization of Einstein's way of reading philosophers deserves to be dismissed as nothing more than a piece of positivist propaganda, part of the positivist-revisionist reading of Einstein that, unfortunately, gained wide acceptance in the middle decades of this century in spite of Einstein's careful, repeated attempts to distance himself from positivism.[7]

If we put aside positivist prejudices, and let Einstein speak for himself, it appears that he did see in Schopenhauer a certain "philosophic depth." After all, Einstein, if not Frank, put Schopenhauer first in the list of his early philosophical readings, *before* Hume, Mach, and Kant. It is, again, only an antimetaphysical, positivist prejudice that would lead one to say that Einstein could have taken his solution to the problem of free will and key ingredients of his concept of cosmic religion from Schopenhauer "without in any way taking his views seriously."

Do such metaphysical doctrines have a place in Einstein's world view? Recall what Einstein wrote to Schlick in a letter that sealed Einstein's turn away from Schlick's brand of positivism:

Generally speaking, your presentation does not correspond to my way of viewing things, inasmuch as I find your whole conception, so to speak, too positivistic. Indeed, physics *supplies* relations between sense experiences, but only indirectly. For me *its essence* is by no means exhaustively characterized by this assertion. I put it to you bluntly: Physics is an attempt to construct conceptually a model of the *real world* as well as of its law-governed structure. To be sure, it must represent exactly the empirical relations between those sense experiences accessible to us; but *only* thus is it chained to the latter. . . . You will be surprised by Einstein the "metaphysician." But in this sense every four-and two-legged animal is, de facto, a metaphysician. (EA 21-603)[8]

This was written on 28 November 1930, roughly the time when various sources, such as Kayser and Wachsmann, report Einstein's intense preoccupation with Schopenhauer and immediately before the period when we find Einstein citing Schopenhauer in discussions of free will and cosmic religion. Had he perhaps just read or reread, in one of those "well-worn Schopenhauer volumes," chapter 17 of volume 2 of *Die Welt als Wille und Vorstellung*, the chapter entitled "Über das metaphysische Bedürfniß des Menschen" (On man's need for metaphysics), where Schopenhauer wrote:

[Man] is an *animal metaphysicum*. At the beginning of his consciousness, he naturally takes himself also as something that is a matter of course. This, however, does not last long, but very early, and simultaneously with the first reflection, appears that wonder which is some day to become the mother of metaphysics. (1859, vol. 2, 160)[9]

We begin to appreciate why Schopenhauer's portrait was in Einstein's study. But still one must ask, why in the company of Faraday and Maxwell?

Einstein on Spatiotemporal Separability

Important though they may be to Einstein's larger world view, cosmic religion and the problem of free will are not among our main concerns here. Our concern is rather with the question of whether or not Schopenhauer's doctrine of space and time as the *principium individuationis*, the ground of individuation for empirical objects, was the source for Einstein's doctrine of separability, according to which non-null spatiotemporal separation is a sufficient condition for the individuation of physical systems. To begin to try to answer this question, let us review what Einstein had to say about separability (see Howard 1990, 1985).

Einstein's concern with the problem of separability first emerges in his 1905 paper on the light quantum hypothesis. Einstein explains there that what he means by the assumption that light quanta behave like spatially localized, "independent" particles is that, in a two-particle system, the joint probability for the two particles occupying given cells in phase space is simply the product of the separate occupation probabilities. This factorizability is a necessary and sufficient condition for Boltzmann's principle, $S = k \cdot \log W$, to hold (1905, 140). But he adds that this way of conceiving light quanta leads not to Planck's law for black-body radiation, but to Wien's law, which is valid only in the limit of large ν/T, implying that this independence and factorizability must be *denied* in order to get the universally valid Planck formula (ibid., 143).

The fact that photons behave like independent particles in the Wien regime, but *not* like independent particles more generally was a source of great puzzlement to Einstein in subsequent years. This puzzlement led first to his speculations about wave-particle duality in 1909, when he found that expressions for the mean-square fluctuations of the energy in a radiation-filled cavity and the mean-square fluctuations in radiation pressure can each be written as a sum of two terms. Einstein interpreted one as corresponding to behavior typical of "independent," localized, particlelike light quanta appearing in the Wien regime, and the other as corresponding to nonindependent, wavelike behavior appearing in the limit of small ν/T (Einstein 1909a, 1909b).

The next step came in late 1924 and early 1925, as Einstein pondered Satyendra Nath Bose's (1924) new idea for deriving the Planck formula from a non-Boltzmannian statistics, based on a different set of assumptions about what count as distinct configurations in the particle's phase space. The key idea here is that, in a two-particle phase space, for example, mere exchange of spatial location of the two particles does not yield a different configuration, implying that the different spatial locations of the two particles do not suffice to endow them with discernible identities. In a series of three papers, Einstein (1924, 1925a, 1925b) applied Bose's idea to a quantum gas of material particles. In the second paper, prompted by a question from Paul Ehrenfest, Einstein for the first time drew attention to the way in which the application of the new statistics to material particles throws into question our classical assumptions about particle independence: "Thus, the formula [for the entropy] indirectly expresses a certain hypothesis about a mutual influence of the molecules — for the time being of a quite mysterious kind — which determines precisely the equal statistical probability of the cases here defined as 'complexions' " (1925a, 6).

Erwin Schrödinger's reading of these three papers was an important stimulus to his development of wave mechanics. As we now understand, one of his main aims was to provide an explanation for the failure of quantum systems to behave like independent classical particles. The issue of independence is at the heart of an extensive correspondence between Einstein and Schrödinger during 1925 and 1926, beginning with Einstein's explaining to Schrödinger in a letter of 28 February 1925, "In the Bose statistics employed by me, the quanta or molecules are not treated as being *independent of one another.*" Einstein added in a postscript that, in relatively dense gases, where the difference between Boltzmann and Bose statistics should be especially noticeable: "There the interaction between the molecules makes itself felt, — the interaction which, for the present, is accounted for statistically, but whose physical nature remains veiled" (EA 22-002).

Schrödinger eventually solved this problem by showing that, if we employ configuration space, rather than physical space, for the representation of many-particle systems, we can construct there joint wave functions that are not factorizable into separate wave functions for the constituent systems, and that these nonfactorizable wave functions are necessary to account correctly for interference effects in quantum mechanical interactions. But precisely this feature of the Schrödinger formalism troubled Einstein deeply, as he made clear first in a letter to Schrödinger of 16 April 1926 (EA 22-012), and then, even more clearly, in a note "Added in Proof" to the never published manuscript of a talk to the Prussian Academy of Sciences on 5 May 1927. In his talk to the Academy, entitled "Does Schrödinger's Wave Mechanics Determine the Motion of a System Completely or Only in the Statistical Sense?" Einstein had attempted a kind of hidden-variables interpretation of the Schrödinger formalism.[10] The attempt failed, in Einstein's eyes (presumably why the talk was never published), because of nonseparability. Einstein wrote in the note "Added in Proof":

I have found that the schema does not satisfy a general requirement that must be imposed on a general law of motion for systems.

Consider, in particular, a system Σ that consists of two energetically independent subsystems, Σ_1 and Σ_2; this means that the potential energy as well as the kinetic energy is additively composed of two parts, the first of which contains quantities referring only to Σ_1, the second quantities referring only to Σ_2. It is then well known that

$$\Psi = \Psi_1 \cdot \Psi_2,$$

where Ψ_1 depends only on the coordinates of Σ_1, Ψ_2 only on the coordinates of Σ_2. In this case we must demand that the motions of the composite system be combinations of possible motions of the subsystems.

The indicated scheme does not satisfy this requirement. In particular, let μ be an index belonging to a coordinate of Σ_1, ν an index belonging to a coordinate of Σ_2. Then $\Psi_{\mu\nu}$ does not vanish. (EA 2-100)

From this time on, Einstein ceased being an active contributor to the development of the quantum theory he had been helping to shape since 1905. Instead, he turned his energies both to the development of a unified field theory, which he believed would satisfy the mentioned "general requirement that must be imposed on a general law of motion for systems" — in essence the separability principle — and to the progressive refinement of his argument for why the quantum theory's failure to satisfy this requirement is so objectionable.

The next phase of this story, namely, Einstein's encounters with Niels Bohr at the 1927 and 1930 Solvay conferences, has been told many times by many authors. However, most of them, including, most notably, Max Jammer (1974, 1985) and Bohr himself (1949), get the story wrong. The old view — encouraged by both Jammer and Bohr — was that Einstein first sought to show the quantum theory incorrect, as with the famous "photon-box" thought experiment, purportedly designed to exhibit violations of the Heisenberg uncertainty principle, and that he shifted to arguing for the theory's incompleteness only after Bohr had cleverly refuted all of his attempts to prove it incorrect. We now know that from the start Einstein was trying instead to formulate thought experiments that would bring to the fore the peculiar consequences of quantum nonseparability. Even the photon-box thought experiment was intended to show that, *if* one assumed the separability of the systems involved in two spacelike separated measurement events (the box and the emitted photon), *then* the quantum mechanical account of these measurements would be incomplete. (See Howard 1990, 98–100, for a discussion of Ehrenfest's letter to Bohr, 9 July 1931.)

Arthur Fine was the first to draw our attention to the way Einstein's concern with the separability problem informed his critique of the quantum theory, especially in his pioneering reanalysis of the Einstein-Podolsky-Rosen (EPR) argument for the incompleteness of quantum mechanics (Fine 1981; see also Howard 1985). Fine showed that, in correspondence with Schrödinger starting in June 1935, immediately after publication of the EPR paper (Einstein, Podolsky, and Rosen 1935), Einstein repudiated the published EPR paper, noting, "For reasons of language, this was written by Podolsky after many discussions. But still it has not come out as well as I really wanted; on the contrary, the main point was, so to speak, buried by the erudition" (Einstein to Schrödinger, 19 June 1935, EA 22-047). Einstein went on to explain what that "main point" was, elaborating a very different incompleteness argument similar in concept to the earlier photon-box thought experiment and based explicitly upon what Einstein there dubs the "separation principle."

Einstein begins with a simple example: two boxes and a single ball always located in one or the other of the boxes. We make an "observation" simply by lifting the lid on a box and looking inside. Consider now a state description: "The probability that the ball is in the first box

is 1/2." Is it a complete description? Einstein says that one who subscribes to the "Born" interpretation must answer no, since, from that
point of view, a complete description would be a categorical assertion
that the ball *is* (or *is not*) in the first box. One who subscribes to the
"Schrödinger" interpretation, by contrast, would say yes, claiming that
before we make an observation, the ball is not really in either box, and
that this "being in a definite box" is, in fact, the result of the observation, so that the state of the first box before we make an observation is
described completely by the probability 1/2.

At least "Born" and "Schrödinger" will talk about the *real* state of
the system. Not so Bohr, who is too much a positivist in Einstein's eyes:
"The talmudic philosopher doesn't give a hoot for 'reality,' which he
regards as a hobgoblin of the naive, and he declares that the two points
of view differ only as to their mode of expression." What is Einstein's
view? He writes:

My way of thinking is now this: properly considered, one cannot get at the
talmudist if one does not make use of a supplementary principle: the "separation principle." That is to say: "the second box, along with everything having
to do with its contents, is independent of what happens with regard to the first
box (separated partial systems)." If one adheres to the separation principle,
then one thereby excludes the second ("Schrödinger") point of view, and only
the Born point of view remains, according to which the above state description
is an *incomplete* description of reality, or of the real states. (EA 22-047)

One cannot "get at" the "talmudist" — the antirealist Bohr — without
the "separation principle."[11]

The actual argument for the incompleteness of quantum mechanics,
developed in the 19 June 1935 letter to Schrödinger, proceeds as follows. We start with a definition of the "completeness" of a state description, Ψ: "Ψ is correlated one-to-one with the real state of the real
system. . . . If this works, then I speak of a complete description of
reality by the theory. But if such an interpretation is not feasible, I call
the theoretical description 'incomplete'" (EA 22-047). Einstein then
sketches the typical EPR-type thought experiment involving spatially
separated, but previously-interacting systems A and B. According to
the orthodox quantum-mechanical formalism, depending upon the
kind of measurement we choose to perform on system A (choice of observable, not actual outcome), we ascribe different state functions, Ψ_β
or Ψ_β, to system B. It follows that quantum mechanics is incomplete:

Now what is essential is exclusively that Ψ_β and $\Psi_{\underline\beta}$ are in general different from one another. I assert that this difference is incompatible with the hypothesis that the Ψ description is correlated one-to-one with the physical reality (the real state). After the collision, the real state of (AB) consists precisely of the real state of A and the real state of B, which two states have nothing to do with one another. *The real state of B thus cannot depend upon the kind of measurement I carry out on A.* ("Separation hypothesis" from above.) But then for the same state of B there are two (in general arbitrarily many) equally justified Ψ_β, which contradicts the hypothesis of a one-to-one or complete description of the real states. (EA 22-047)

Among the many important outgrowths of the Einstein-Schrödinger correspondence from the summer of 1935 was the famous Schrödinger "cat paradox" (Schrödinger 1935a; see Fine 1986a). Perhaps the most important result, however, was Schrödinger's now classic formulation of the quantum mechanical interaction formalism, in which the non-separability of the formalism — in the form of the nonfactorizability of the post-interaction joint state — is so prominently highlighted (1935b, 1936).

Over the next fourteen years, Einstein repeated and refined this argument in a number of publications (for details, see Howard 1985). Most important for our purposes is the way in which he clarified his understanding of the kind of independence intended in the "separation principle," gradually disentangling what is now more commonly designated the "separability principle" from the "locality principle." The separability principle is a fundamental ontological principle according to which any two systems separated by a non-null spatiotemporal interval, regardless of their history of interaction, possess their own separate real states that completely determine the joint state. In the language of the orthodox quantum mechanical formalism, this is simply the requirement (not generally satisfied in the quantum mechanical account of interactions) that $\Psi_{AB} = \Psi_A \otimes \Psi_B$. The "locality principle" — essentially a statement of the special relativistic prohibition on superluminal influences — says that, given any two space-like separated systems, A and B, the separate real state of one cannot be influenced by events (such as the choice of an observable to measure, for example, by rotating a Stern-Gerlach apparatus plus detectors) in the vicinity of the other.

Einstein explained the distinction in a restatement of the incompleteness argument for a 1948 special issue of the Swiss journal *Dialec-*

tica, edited by Pauli, that was devoted to the interpretation of quantum mechanics:

If one asks what is characteristic of the realm of physical ideas independently of the quantum-theory, then above all the following attracts our attention: the concepts of physics refer to a real external world, *i.e.*, ideas are posited of things that claim a "real existence" independent of the perceiving subject (bodies, fields, *etc.*), and these ideas are, on the other hand, brought into as secure a relationship as possible with sense impressions. Moreover, it is characteristic of these physical things that they are conceived of as being arranged in a space-time continuum. Further, it appears to be essential for this arrangement of the things introduced in physics that, at a specific time, these things claim an existence independent of one another, insofar as these things "lie in different parts of space." Without such an assumption of the mutually independent existence (the "being-thus") of spatially distant things, an assumption that originates in everyday thought, physical thought in the sense familiar to us would not be possible. Nor does one see how physical laws could be formulated and tested without such a clean separation. Field theory has carried out this principle to the extreme, in that it localizes within infinitely small (four-dimensional) space-elements the elementary things existing independently of one another that it takes as basic, as well as the elementary laws it postulates for them.

For the relative independence of spatially distant things (*A* and *B*), this idea is characteristic: an external influence on *A* has no *immediate* effect on *B*; this is known as the "principle of local action," which is applied consistently only in field theory. The complete suspension of this basic principle would make impossible the idea of the existence of (quasi-) closed systems and, thereby, the establishment of empirically testable laws in the sense familiar to us. (Einstein 1948, 321–22)

These are very rich paragraphs. Separability, "the mutually independent existence (the 'being thus') of spatially distant things," is distinguished from locality, "the principle of local action." Both are tied in special ways to field theory, which is said to carry out the separability principle "to the extreme" and to be the only place where the locality principle is applied "consistently."

Elsewhere I have speculated that what Einstein meant by saying that field theory, like general relativity, provides an extreme embodiment of the separability principle is that, by assigning a well-defined value of the fundamental field quantity — such as the metric tensor in general relativity — to every point of the spacetime manifold, and by taking the joint reality associated with any two such points to be completely determined by the values of the field quantities associated with those two points, a field theory implicitly regards every point of the manifold as representing a separable system, a separable bit of reality (Howard 1985, 1989). Here we need only attend to the specific manner in which Einstein discusses the first principle, separability, including even the vocabulary he employs.

It is characteristic of physics that its concepts "refer to a real external world," which is to say that "ideas are posited of things that claim a 'real existence' independent of the perceiving subjects (bodies, fields, *etc.*)." Beyond that, the elements of our ontology, "these physical things," are represented "as being arranged in a space-time continuum." And it is "essential" that "these things claim an existence independent of one another, insofar as these things 'lie in different parts of space.'" This "assumption of the mutually independent existence (the 'being-thus') of spatially distant things . . . originates in everyday thought." Without it, "physical thought in the sense familiar to us would not be possible." Keep these formulations in mind when, in the next section, we learn how Schopenhauer discussed space and time as the *principium individuationis*, the ground for the individuation of empirical objects.

To conclude this survey of what Einstein said about separability and individuation, consider one final remark from handwritten comments he added to the manuscript of Max Born's 1949 Waynflete Lectures, *Natural Philosophy of Cause and Chance* (Born 1949), answering Born's request to respond to his representation of their discussions of quantum mechanics. The subject is once again reality and the fundamental ontology of any possible, future, fundamental physical theory. Einstein writes:

I just want to explain what I mean when I say that we should try to hold on to physical reality. We are, to be sure, all of us aware of the situation regarding what will turn out to be the basic foundational concepts in physics: the point-mass or the particle is surely not among them; the field, in the Faraday-Maxwell sense, might be, but not with certainty. But that which we conceive as existing ("actual") should somehow be localized in time and space. That is, the real in one part of space, *A*, should (in theory) somehow "exist" independently of that which is thought of as real in another part of space, *B*. If a physical system stretches over the parts of space *A and B*, then what is present in *B* should somehow have an existence independent of what is present in *A*. What is actually present in *B* should thus not depend upon the type of measurement carried out in the part of space, *A*; it should also be independent of whether or not, after all, a measurement is made in *A*.

If one adheres to this program, then one can hardly view the quantum-theoretical description as a *complete* representation of the physically real. If one attempts, nevertheless, so to view it, then one must assume that the physically real in *B* undergoes a sudden change because of a measurement in *A*. My physical instincts bristle at that suggestion.

However, if one renounces the assumption that what is present in different parts of space has an independent, real existence, then I do not at all see what physics is supposed to describe. For what is thought to be a "system" is, after all, just conventional, and I do not see how one is supposed to divide up the world objectively so that one can make statements about the parts. (Einstein to Born, March 1948, in Born 1969, 223–24)

The point-particle and the field are dispensable. The one thing that cannot be given up is the idea of separability: "The real in one part of space, *A*, should (in theory) somehow 'exist' independently of that which is thought of as real in another part of space, *B*." Deny this assumption, and "I do not at all see what physics is supposed to describe." Why? Because "what is thought to be a 'system' is, after all, just conventional, and I do not see how one is supposed to divide up the world objectively so that one can make statements about the parts." But it is precisely nonseparability that is the distinguishing, nonclassical feature of the quantum mechanical interaction formalism. And there's the rub. Einstein says that separability is indispensable, and quantum mechanics denies it.[12]

Schopenhauer on Space, Time, Causality, and Individuation

Schopenhauer represents the first generation of the nineteenth-century neo-Kantian reaction to the idealism of Hegel, Fichte, and, to some extent, Schelling (see Köhnke 1986; Wiley 1978). What he says about space, time, and causality can only be understood against this background, although he is no strict Kantian.

Consider first the doctrine as developed in Schopenhauer's major work, *Die Welt als Wille und Vorstellung* (1859). According to Schopenhauer, the will is the thing-in-itself, and, as such, is not knowable through either experience or reason. A kind of knowledge of the will is possible through aesthetic experience, however, in moments when the subject, slipping the bonds of its individuality and freeing itself momentarily from the urgings of the will, manages a fleeting insight, as much a feeling as anything else, into the non-empirical Ideas that are, as it were, the first stage of the objectification of the will in nature. What we know through intuition and experience — empirical objects — are those things in nature that are the ultimate objectifications of the will. The most basic necessary forms of this objectification are space and time, the *principium individuationis*, through which empirical objects are originally distinguished from one another and thus represented to us in experience: "It is only by means of time and space that something which is one and the same according to its nature and the concept appears as different, as a plurality of coexistent and successive things. Consequently, time and space are the *principium individuationis*" (1859, vol. 1, 113). Or, as we read in a later passage: "We know that *plurality* in general is necessarily conditioned by time and space, and only in these is conceivable, and in this respect we call them the *principium individuationis*" (ibid., 127). Or, again, "We have called time and space the *principium individuationis*, because only through them and in them is plurality of the homogenous possible. They are the essential forms of natural knowledge" (ibid., 331). What we know through reason are concepts, all of them formed by abstraction from intuited empirical objects thus individuated in space and time.

If space and time are the *principium individuationis*, how do they differ from one another, aside from the obvious fact that, as for Kant, space is the form of outer intuition and time the form of inner intuition — space being thus the ground for the possibility of geometry, time the ground for the possibility of arithmetic? Schopenhauer says that space is "absolutely nothing else but the possibility of the reciprocal determinations of its parts by one another, which is called *position*" and "succession is the whole essence and nature of time" (ibid., 8). What does this imply? "Space renders possible the persistence of substance" and "time renders possible the change of accidents" (1859, vol. 2, 50). What Schopenhauer seems to mean by this, to employ a modern idiom, is that space individuates *objects* or *systems* in nature, while time individuates the *states* of those objects. In the more scholastic vocabulary that Schopenhauer often employs, space is that against which we distinguish individual substances, while time is that against which we distinguish the changing accidents or properties born by those substances. Hence there can be no ontology of empirical objects and their changing characteristics without space and time as the *principium individuationis*.

Many things are therefore individuated in space and time: tables and chairs, planets and stars, chemical atoms and birds in the sky. But most importantly, for Schopenhauer, the human body, the individuated, objective correlate of the knowing subject, exists in space and time:

That which knows all things and is known by none is the *subject*. . . . Everyone finds himself as this subject. . . . But his body is already object, and therefore from this point of view we call it representation. For the body is object among objects and is subordinated to the laws of objects, although it is immediate object. Like all objects of perception, it lies within the forms of all knowledge, in time and space, through which there is plurality. (1859, vol. 1, 5)

Thus, the very *object-ivity* of human knowledge is bound up with the fact that the human body, the phenomenal side of the human subject, is an object individuated in space and time like all of the other objects in space and time that are the objects of human knowledge.

Added to space and time is the one category of the understanding that Schopenhauer takes over from Kant, namely, *causality*. More so than Kant, Schopenhauer sees causality standing in an intimate relation with space and time, uniting them together in the perception of *matter*, in a manner presupposing the individuating character of space and time. The result is the *reality* that is investigated by empirical science:

For matter is absolutely nothing but causality. . . . Thus, its being is its acting; it is not possible to conceive for it any other being. Only as something acting does it fill space and time. . . . Thus cause and effect are the whole essence and nature of matter; its being is its acting. . . . The substance of everything material is therefore very appropriately called in German *Wirklichkeit* [i.e., activeness], a word much more expressive than *Realität*. . . . Time and space . . . each by itself, can be represented in intuition even without matter; but matter cannot be so represented without time and space. The form inseparable from it presupposes *space*, and its action, in which its entire existence consists, always concerns a change, and hence a determination of *time*. But time and space are not only, each by itself, presupposed by matter, but a combination of the two constitutes its essential nature, just because this, as we have shown, consists in action, causality. (Ibid., 8–9)

According to Schopenhauer, causality derives its meaning and necessity from the fact that change is not mere variation. Instead, the essence of change

consists in the fact that, at the *same place in space*, there is now *one* condition or state and then *another*, and at *one* and the same point of time there is *here* this state and *there* that state. Only this mutual limitation of time and space by each other gives meaning, and at the same time necessity, to a rule according to which change must take place. What is determined by the law of causality is therefore not the succession of states in mere time, but that succession in respect of a particular space, and not only the existence of states at a particular place, but at this place at a particular time. Thus change, i.e., variation occurring according to the causal law, always concerns a particular part of space and a particular part of time, *simultaneously* and in union. Consequently, causality unites space and time. (Ibid., 9–10)

Finally, causality's unification of space and time grounds our a priori knowledge of matter:

On this derivation of the basic determinations of matter from the forms of our knowledge, of which we are *a priori* conscious, rests our knowledge *a priori* of the sure and certain properties of matter. These are space-occupation, i.e., impenetrability, i.e., effectiveness, then extension, infinite divisibility, persistence, i.e., indestructibility, and finally, mobility. On the other hand, gravity, notwithstanding its universality, is to be attributed to knowledge *a posteriori*, although Kant in his *Metaphysical Rudiments of Natural Science* . . . asserts that it is knowable *a priori*. (Ibid., 11)

It was in his first book, *Ueber die vierfache Wurzel des Satzes vom zureichenden Grunde* (1813), that Schopenhauer originally developed the idea of space and time as principles of individuation. They are distinguished as the third of the four forms of the principle of sufficient reason — causality is the first — and are together designated the "ground of being," or the "principle of sufficient reason of being":

Space and time are so constituted that all their parts stand in mutual relation and, on the strength of this, every part is determined and conditioned by another. In space this relation is called *position*, in time *succession*. These relations are peculiar and differ entirely from all other possible relations of our representations. Therefore neither the understanding nor the faculty of reason by means of mere concepts is capable of grasping them, but they are made intelligible to us simply and solely by means of pure intuition a priori. . . . Now the law whereby the parts of space and time determine one another as regards these relations is what I call the *principle of sufficient reason of being, principium rationis sufficientis essendi*. (1813, 194)

As later, he characterizes matter as "the perceptibility of time and space" and "causality that has become objective" (ibid., 193), but in this earlier text we find some helpful additional detail about why space and time are needed as a ground for the individuation of objects in experience, and about the manner in which such individuation, especially in space, is effected.

Why are space and time, as pure intuitions, necessary for individuation? The answer harkens back to Kant's famous "incongruent counterparts" argument, which we will consider more fully in the next section:

It is impossible to explain clearly from mere concepts what are above and below, right and left, front and back, before and after. Kant quite rightly confirms this by saying that the difference between the right and left gloves cannot possibly be made intelligible except by means of intuition. (Ibid., 194)

As Kant argued, the only way to distinguish "incongruent counterparts," like a right- and left-handed glove, is by their different situations with respect to absolute space. Assuming an otherwise empty space, and assuming two such objects that are qualitatively identical, their internal and external relational properties would not suffice to tell them apart. That there are two objects, rather than one, we might establish via their mutual relations with one another, but we could not determine, in this way, which is which. It is only against the background of absolute space, so argues Kant, that we can tell them apart; hence, space is necessary as a ground for the individuation of physical objects.

How, according to Schopenhauer, does space do the work of distinguishing physical objects? The answer comes in a discussion of how we know the truth of geometrical axioms not on the basis of rational demonstration, but of pure intuition. Space and time themselves are pure intuitions. Figures and numbers in space and time, respectively, are dubbed "normal intuitions" (*Normalanschauungen*), and most of Euclid's axioms are said to depend for their truth on one fact about these "normal intuitions":

What Plato says of his Ideas would hold good of these normal intuitions, even in geometry, as well as of concepts, namely that two cannot exist exactly alike because such would be only one. I say that this would hold good also of normal intuitions in geometry if it were not that, as exclusively *spatial* objects, they differ through mere juxtaposition and hence through *place*. (Ibid., 198–99)

Such spatial individuation is surely necessary to ground the possibility of congruent, but not identical figures, but otherwise it is not clear how this fact about "normal intuitions" can, by itself, assure the truth of Euclid's axioms. Schopenhauer merely says:

> Now the mere view that such a difference of place does not abolish the rest of the identity seems to me to be capable of replacing those nine axioms, and of being more suitable to the true nature of science whose purpose is to know the particular from the general, than is the statement of nine different axioms that are all based on one view. (Ibid., 199)

However Schopenhauer imagined the grounding of the axioms to work, the significance of these remarks will loom larger when, again in the next section, we try to locate Schopenhauer's views on space, time, and individuation in a larger historical context, including, most importantly, the Leibniz-Clarke debate. For Schopenhauer asserts here that otherwise qualitatively identical objects (remember that material objects are just figures made perceptible by activity or causality) may be distinguished from one another solely by virtue of their different spatial locations, their separation in space. This is precisely what Leibniz denies.

It should be stressed that Schopenhauer is no philosopher of geometry, no philosopher of physics. He does not write about space and time as the *principium individuationis* out of an interest in phoronomy or the rational foundations of mechanics. On the contrary, Schopenhauer's larger aims concern aesthetics, ethics, and social philosophy. By the time of *Die Welt als Wille und Vorstellung* (1819), he is interested in space and time as the *principium individuationis* mainly because of what it can tell us about the way in which the embodied human subject lives in the phenomenal world as a human individual, subject to the law of causality, and ontologically separated, by necessity, from all other human individuals. By contrast, the will, the thing-in-itself, not being subject to the pluralizing forms of space and time, is a kind of unity. Much of what is most interesting in Schopenhauer's philosophy — in his ethics, his aesthetics, and his social philosophy — turns around the tension between the unity of the will and the plurality of objectifications of the will. It is here, at this point of creative tension, that Schopenhauer makes such masterful use of the image of the Veil of Maya, an idea borrowed from Vedic thought as a representation of the way in which space and time as the *principium individuationis* draw a veil across the deeper unity of the world as will, forcing us to apprehend that world only in the will's dispersed objectifications in space and time, except for those brief moments when, freeing ourselves from the will's urgings and transcending our individuality, we catch a glimpse behind the Veil of Maya. Let me illustrate by quoting at length from section 63 of *Die Welt als Wille und Vorstellung*, where the issue is "eternal justice":

The eyes of the uncultured individual are clouded, as the Indians say, by the veil of Maya. To him is revealed not the thing-in-itself, but only the phenomenon in time and space, in the *principium individuationis*, and in the remaining forms of the principle of sufficient reason. In this form of his limited knowledge he sees not the inner nature of things, which is one, but its phenomena as separated, detached, innumerable, very different, and indeed opposed. For pleasure appears to him as one thing, and pain as quite another; one man as tormentor and murderer, another as martyr and victim. . . . He sees one person living in pleasure, abundance, and delights, and at the same time another dying in agony of want and cold at the former's very door. He then asks where retribution is to be found. . . . He often tries to escape by wickedness, in other words, by causing another's suffering, from the evil, from the suffering of his own individuality, involved as he is in the *principium individuationis*, deluded by the veil of Maya. . . . This separation, however, lies only in the phenomenon and not in the thing-in-itself; and precisely on this rests eternal justice. . . . The person is mere phenomenon, and its difference from other individuals, and exemption from the sufferings they bear, rest merely on the form of the *principium individuationis*. According to the true nature of things, everyone has all the sufferings of the world as his own; indeed, he has to look upon all merely possible sufferings as actual for him, so long as he is the firm and constant will-to-live, in other words, affirms life with all his strength. For the knowledge that sees through the *principium individuationis*, a happy life in time, given by chance or won from it by shrewdness, amid the sufferings of innumerable others, is only a beggar's dream, in which he is a king, but from which he must awake, in order to realize that only a fleeting illusion has separated him from the suffering of his life. (1859, vol. 1, 352–53)

What passages like this should help to make clear is that the conception of space and time as the *principium individuationis* runs like a Leitmotiv throughout Schopenhauer's writings. Until one has read *Parerga und Paralipomena* or *Die Welt als Wille und Vorstellung* in their entirety, one cannot appreciate just how ubiquitous the idea is — how it tends to pop up every few pages, as a sort of reminder, even in places where one might not expect to find it. One cannot read even a small part of Schopenhauer's oeuvre without encountering it many times over.[13] The regular emphasis on space and time as the *principium individuationis* is thus one of the chief distinguishing features of Schopenhauer's philosophy. It sets him apart from virtually all other nineteenth-century philosophers, including even the later generation of Marburg neo-Kantians. Although they derived this idea from Kant, none of them highlights it in the same way as Schopenhauer.

Of course, all of this is still no proof that Einstein borrowed the idea of spatiotemporal separability from Schopenhauer, or even that he was influenced in some small way by what Schopenhauer wrote about space and time as the *principium individuationis*. After all, on the face of it, there are significant differences between the two conceptions, not the least of which is the fact that Schopenhauer, writing in the early nineteenth century, assumes the classical distinction between space and time, whereas Einstein sees them as aspects of a single, four-dimensional spacetime. But there is a context in which Schopenhauer was writing, a long-standing controversy about space, time, and individuation. An appreciation of the significance of that context will make it appear more likely that Einstein could have been influenced by Schopenhauer's position in this controversy.

From Suárez to Schopenhauer: The Historical Background

Schopenhauer tells us himself where to look for the earliest roots of this controversy:

As we know, time and space belong to this principle [of sufficient reason], and consequently plurality as well, which exists and has become possible only through them. In this last respect I shall call time and space the *principium individuationis*, an expression borrowed from the old scholasticism, and I beg the reader to bear this in mind once and for all. For it is only by means of time and space that something which is one and the same according to its nature and the concept appears as different, as a plurality of coexistent and successive things. Consequently, time and space are the *principium individuationis*, the subject of so many subtleties and disputes among the scholastics which are found collected in Suárez (*Disp.* 5, sect. 3). (1859, vol. 1, 112–13)

The reference is to the fifth of Francisco Suárez's *Disputationes Metaphysicae* (1597), "De unitate individuali ejusque principio," a compendious summary of the various medieval debates on individuation (see Gracia 1982).

The medieval debate over individuation, summarized by Suárez, had its origins in the mid-thirteenth century, in Aquinas's *Commentary on Boethius' "De Trinitate"* (see Maurer 1987, xxiii–xxxv). The problem was to understand how the Father, the Son, and the Holy Spirit could be one God, not three. For Boethius, things can be different in three ways: in genus (like human and stone), in species (like human and horse), or in number (like Cato and Cicero). To show that none of these differences is present in God, Boethius must explain in what they consist. For our purposes, difference in number is the important case, and Boethius says that this is the result of the diversity of the accidental characteristics. But what about differences in accidents that do not affect the identity of individual substances, such as the difference between a person sitting and standing? Does this mean that accidents are not relevant to individuation? No. "For if we mentally remove all [other] accidents, still each occupies a different place, which we cannot possibly conceive to be the same" (Thomas Aquinas 1987, 57).

Aquinas takes up the Question in Question 4, Article 4 of his *Commentary*, "Does a Difference in Place Have Some Bearing on a Difference in Number?" Aquinas was writing in a very different intellectual climate, one in which Aristotle loomed far larger than he did for the neo-Platonist, Boethius. He therefore could not simply endorse Boethius's position, holding instead that the ultimate cause of individuation is "matter, itself, as existing under dimensions." He seems to mean that matter as that which is susceptible to division is the cause of individuation, of diversity in number. But it is equally the cause of difference in place, making difference in place a *sign* of difference in number — indeed, the surest sign: "Taken in this sense, difference of place plays the greatest role because it is the sign most closely related to diversity in number" (ibid., 109–10).

Suárez's fifth *Disputation*, which Schopenhauer cites, reviews the "subtleties and disputes" that arose in response to Aquinas's *Commentary*. We need not follow this history in detail here, except to note that by the time Suárez was writing, in the late sixteenth century, the issue had become quite complicated, owing to the intrinsic difficulty of even posing the question clearly in the vocabulary available to the Scholastics. For, happily, the nature of the debate was about to change, thanks largely to the revival of atomism in Renaissance natural philosophy and to related late-Renaissance developments in the conception of space and the void.

Among the first generation of classical Greek atomists, Leucippus and Democritus, the void was not simply the infinite, all-pervading arena or container within which the atoms moved; which is to say that it was not yet identified with the space of the geometers, as was to become the case by the time of Epicurus and, later, Lucretius. Instead, the void, the empty (κενὸν) is that which separates the atoms; it consists of the intervals (διαστήματα) between the atoms. The function of the void, then, is to be the principle making possible plurality (see Bailey 1928, 69–76). It retains this function when it later becomes the space of the geometers. For without a void or a space capable of thus separating, there would be no basis for distinguishing, one from another, the otherwise qualitatively identical atoms.

The way was made easier for the revival of classical Greek atomism in Renaissance natural philosophy by the growing tendency to regard *quantitative*, as opposed to *qualitative*, properties as ontologically primary. This tendency was especially in evidence in the neo-Platonic, Averroistic atmosphere of Padua, starting from the fifteenth-century debate between Biagio Pelacani and Gaetano of Thiene over whether the primary accident of a substance is quantitative or qualitative, to the late-sixteenth century continuation of this debate between Galileo and Cremonini (Dijksterhuis 1961, 235). Galileo gave the classic statement of the primacy of quantitative, mathematical properties when he described natural philosophy as being written "in the language of mathematics," explaining that "its characters are triangles, circles, and other geometric figures, without which it is humanly impossible to understand a single word of it; without these, one wanders about in a dark labyrinth" (*Il Saggiatore*; as quoted in ibid., 362). Of course, Descartes carried this tendency to its logical extreme by identifying matter with extension itself. In such a metaphysical context, where the only properties, or at least the primary qualities, are geometrical properties, the only possible basis for distinguishing otherwise identical (congruent) parts of the universe is difference in place, or spatial separation. There could be no diversity, no plurality, no change, no motion, no physics, were it not for the most basic metaphysical fact of individuation by spatial separation.

From this point of view, the otherwise large differences between the Cartesians and the Newtonians appear relatively insignificant. For whether one identifies matter with extension, or distinguishes space from matter, individuation via spatial separation remains a necessary part of one's world view, if one takes quantitative geometrical properties to be the metaphysically fundamental properties of material substance.

In Newton's case, a helpful text in this regard is his *De Gravitatione et aequipondio fluidorum*. (Hall and Hall 1962, 89–156), an unpublished manuscript that the Halls date to between 1664 and 1668, evidently at the time of Newton's first serious critical reaction to Descartes. Newton's principal aim in the heart of this manuscript is to refute the Cartesian identification of material substance with extension. Toward that end, he sets forth his own conception of the nature of extension and body.

From the start, the radical break with Scholasticism is clearly marked. "Perhaps now it may be expected that I should define extension as substance or accident or else nothing at all. But by no means, for it has its own manner of existence which fits neither substances nor accidents" (ibid., 131–32). What manner of existence is this? To begin with, there is the universal divisibility of space:

1. In all directions, space can be distinguished into parts whose common limits we usually call surfaces. . . . Furthermore spaces are everywhere contiguous to spaces, and extension is everywhere placed next to extension, and so there are everywhere common boundaries to contiguous parts. . . . And hence there are everywhere all kinds of figures, everywhere spheres, cubes, triangles, straight lines, everywhere circular, elliptical, parabolical and all other kinds of figures, and those of all shapes and sizes, even though they are not disclosed to sight. For the material delineation of any figure is not a new production of that figure with respect to space, but only a corporeal representation of it, so that what was formerly insensible in space now appears to the senses to exist. (Ibid., 132–33)

Thus the first, most fundamental property of space is that it can be divided anywhere, in any way, into parts. The second property of space is infinity. The third is that the parts of space are motionless. And to this point Newton comments as follows:

Moreover the immobility of space will be best exemplified by duration. For just as the parts of duration derive their individuality from their order, so that (for example) if yesterday could change places with today and become the later of the two, it would lose its individuality and would no longer be yesterday, but today; so the parts of space derive their character from their positions, so that if any two could change their positions, they would change their character at the same time and each would be converted numerically into the other. The parts of duration and space are only understood to be the same as they really are because of their mutual order and position; nor do they have any hint of individuality apart from that order and position which consequently cannot be altered. (Ibid., 136)

The individuality of the parts of space consists solely in their order and position. The same will be true of bodies, because, as we shall see, they are nothing more than the aforementioned "material delineations" of parts of space, "*determined quantities of extension which omnipresent God endows with certain conditions.*" These include mobility — God successively determines immediately adjacent, congruent parts of space; impenetrability — two different determined quantities of extension cannot coincide; and the capacity to excite "perceptions of the senses" (ibid., 140). This is the meaning of the fourth fundamental property of space:

4. Space is a disposition of being *qua* being. No being exists or can exist which is not related to space in some way. God is everywhere, created minds are somewhere, and body is in the space that it occupies; and whatever is neither everywhere nor anywhere does not exist. And hence it follows that space is an effect arising from the first existence of being, because when any being is postulated, space is postulated. And the same may be asserted of duration: for certainly both are dispositions of being or attributes according to which we denominate quantitatively the presence and duration of any existing individual thing. (Ibid., 136)

Space is a disposition of being *qua* being in the sense that it is the ground of individuation, that "according to which we denominate quantitatively the presence . . . of any existing individual thing." The fifth property of space is that "the positions, distances and local motions of bodies are to be referred to the parts of space," a fact that "will be more manifest if you conceive that there are vacuities scattered between the particles," and the sixth and last property is that space is "eternal in duration and immutable in nature" (ibid., 137).

Whatever absolute space becomes for Newton in the *Principia* and later, whatever dynamical properties it must possess in order to make sense of the first and second laws — in order, that is, to ground the concept of acceleration and to make possible the identification of inertial motions — the more basic role of space as the ground of individuation remains.

This was not lost on Newton's contemporaries. We find John Locke, in 1694, writing in the second edition of the *Essay Concerning Human Understanding* (1694):

When therefore we demand whether anything be the *same* or no, it refers always to something that existed such a time in such a place, which it was certain, at that instant was the same with itself, and no other. From whence it follows, that one thing cannot have two beginnings of existence, nor two things one beginning; it being impossible for two things of the same kind to be or exist in the same instant, in the very same place; or one and the same thing in different places. That, therefore, that had one beginning, is the same thing; and that which had a different beginning in time and place from that, is not the same, but diverse. . . . From what has been said, it is easy to discover what is so much inquired after, the *principium individuationis*; and that, it is plain, is existence itself; which determines a being of any sort to a particular time and place, incommunicable to two beings of the same kind. (1894, 439–41)

This passage is from book II, chapter 27, "Of Identity and Diversity," a chapter added in the second edition at the urging of William Molyneux (Molyneux to Locke, 2 March 1693, in De Beer 1979, vol. 4, 647–52; see also Locke to Molyneux, 23 August 1693, ibid., 719–23; and 8 March 1695, ibid., vol. 5, 284–88).

Whether or not Molyneux's motivation was a concern about Leibniz's reaction, the latter did react, leaving no uncertainty as to where he differed from Locke. Leibniz writes in the *Nouveaux essais sur l'entendement humain* (1704–1705), with specific reference to the quoted passage from chapter 27 of the *Essay*:

In addition to the difference of time or of place there must always be an internal *principle of individuation*: although there can be many things of the same kind, it is still the case that none of them are ever exactly alike. Thus, although time and place (i.e. the relations to what lies outside) do distinguish for us things which we could not easily tell apart by reference to themselves alone, things are nevertheless distinguishable in themselves. Thus, although diversity in things is accompanied by diversity of time or place, time and place dc not constitute the core of identity and diversity, because they impress different states upon being. To which it can be added that it is by means of things that we must distinguish one time or place from another, rather than *vice versa*; for times and places are in themselves perfectly alike, and in any case they are not substances or complete realities. . . . The "principle of individuation" reduces, in the case of individuals, to the principle of distinction of which I have just been speaking. If two individuals were perfectly similar and equal and, in short, *indistinguishable* in themselves, there would be no principle of individuation. I would even venture to say that in such a case there would be no individual distinctness, no separate individuals. That is why the notion of atoms is chimerical and arises only from men's incomplete conceptions. For if there were atoms, i.e. perfectly hard and perfectly unalterable bodies which were incapable of internal change and could differ from one another only in size and in shape, it is obvious that since they could have the same size and shape they would then be indistinguishable in themselves and discernible only by means of external denominations with no internal foundation; which is contrary to the greatest principles of reason. . . . One can see from these considerations, which have until now been overlooked, how far people have strayed in philosophy from the most natural notions, and at what a distance from the great principles of true metaphysics they have come to be. (1981, 230–31)

But even though the new chapter 27 of the 1694 second edition of Locke's *Essay* was the immediate occasion of these remarks, the basic idea is of older vintage in Leibniz's thinking, as evidenced by the following remarks from a 1696 note "On the Principle of Indiscernibles":

A consideration which is of the greatest importance in all philosophy, and in theology itself, is this: that there are no purely extrinsic denominations, because of the interconnexion of things, and that it is not possible for two things to differ from one another in respect of place and time alone, but that it is always necessary that there shall be some other internal difference. So there cannot be two atoms which are at the same time similar in shape and equal in magnitude to each other; for example, two equal cubes. Such notions are mathematical, that is, they are abstract and not real. For all things which are different must be distinguished in some way, and in the case of real things position alone is not a sufficient means of distinction. This overthrows the whole of purely corpuscularian philosophy. (1696, 133)

Leibniz's critique of Locke on individuation looms in the background of the Leibniz-Clarke correspondence of 1715–1716. Whatever else might be the points of difference separating Leibniz and the Newtonians regarding the nature of space, this much is without doubt at the very heart of the controversy: The Newtonians want to defend the idea that space is the principle of individuation, that spatial separation is a sufficient condition for marking two bodies — even two qualitatively and geometrically identical atoms — as different individuals, with their own, separate identities, whereas Leibniz and his followers, like Christian Wolff, want to deny this.

Consider one of the most famous and oft-quoted passages from the Leibniz-Clarke correspondence, this being Leibniz's main argument against the Newtonian concept of space in his "Third Paper" of 25 February 1716:

I have many demonstrations, to confute the fancy of those who take space to be a substance, or at least an absolute being. But I shall only use, at the present, one demonstration, which the author here gives me occasion to insist upon. I say then, that if space was an absolute being, there would something happen for which it would be impossible there should be a sufficient reason. Which is against my axiom. And I prove it thus. Space is something absolutely uniform; and, without the things placed in it, one point of space does not absolutely differ in any respect whatsoever from another point of space. Now from hence it follows, (supposing space to be something in itself, besides the order of bodies among themselves,) that 'tis impossible there should be a reason, why God, preserving the same situations of bodies among themselves, should have placed them in space after one certain particular manner, and not otherwise; why every thing was not placed the quite contrary way, for instance, by changing East into West. But if space is nothing else, but that order of relation; and is nothing at all without bodies, but the possibility of placing them; then those two states, the one such as it now is, the other supposed to be the quite contrary way, would not at all differ from one another. Their difference therefore is only to be found in our chimerical supposition of the reality of space in itself. But in truth the one would exactly be the same thing as the other, they being absolutely indiscernible; and consequently there is no room to enquire after a reason of the preference of the one to the other. (Alexander 1956, 26)

Yes, the issue is an absolute versus a relative conception of space. But what is it about Newton's absolute space that so troubles Leibniz? It is not — or certainly it is not *only* — that absolute space is endowed with dynamical properties sufficient to distinguish inertial motions. It is rather that, all the other characteristics of absolute space aside, Newton uses his absolute space as a ground of individuation. That is what really troubles Leibniz. Notice how he begins this passage: "I have many demonstrations, *to confute the fancy of those who take space to be a substance, or at least an absolute being*" (emphasis added). It is neither the substantival character of space — which, recall, Newton *denied* ("for it has its own manner of existence which fits neither substances nor accidents") — nor the dynamical properties of absolute space. What is objectionable is simply the claim that position or spatial separation is a sufficient condition for individuation.

Make Leibniz's argument simpler by considering a universe containing just two atoms, *A* and *B*. Now let God change "East into West." Why would it be impossible to distinguish the two states of the universe? Because, the spatial separation of *A* and *B* not being a sufficient condition for establishing their separate identities, we could not, even before the change of "East into West" tell *which* atom was *A* and *which* atom was *B*. That there were *two* atoms we may be able to determine by their mutual order; but *which* is *which* we cannot say. And therefore, there is no discernible difference between the two states of the universe before and after changing "East into West." It is worth pondering the similarity between Leibniz's position, vis-à-vis Newton's, and the position of Bose-Einstein statistics, vis-à-vis Boltzmann statistics. In both the Leibnizian ontology and in Bose-Einstein statistics, the exchange of the positions of the two particles, the changing of "East into West," makes no difference at all.

Let me venture at this point a strong historiographical claim. Too much of the historical literature on the Leibniz-Clarke correspondence misses this absolutely central point in Leibniz's critique of the Newtonian conception of space because of the error of anachronism. Recent interest in the Leibniz-Clarke correspondence arose mainly because the development of relativity theory awakened our curiosity about the controversy over absolute versus relative conceptions of space. We went back looking for anticipations of this twentieth-century controversy in the seventeenth and eighteenth centuries, not pausing to consider that what really mattered to Newton in the conception of space, as well as what really bothered Leibniz about Newton's position, was perhaps quite different from what our twentieth-century categories led us to expect. More recently, we have picked over the Leibniz-Clarke correspondence again, interested now in the more subtle question of substantival versus relational conceptions of space. But the anachronistic error has been repeated: Even Newton's explicit *denial* that space is something substantial is not enough to deter scholars from trying to turn him into a substantivalist.

Anachronism has similarly infected the literature on Leibniz's most penetrating critic, Immanuel Kant. Much has been written on the "incongruent counterparts" argument, but again, too much of this literature sees the argument as part of the absolute-relative or substantival-relational controversies. In fact, Kant was trying to defend Newton on *precisely* the point we have been tracking: that space is the ground of individuation.

The argument appeared first in Kant's 1768 essay, "Von dem ersten Grunde des Unterschiedes der Gegenden im Raume." It was used again in the 1770 *Inaugural Dissertation,* and is mentioned both in the *Prolegomena* (1783) and in the *Metaphysische Anfangsgründe der Naturwissenschaft* (1786), although, curiously, it is not mentioned in either the first (1781) or second (1787) editions of the *Kritik der reinen Vernunft.* There are genuine puzzles about the place of this argument in the development of Kant's thought. It puzzled some commentators that, when the argument first appears in 1768, it is employed as an argument for the reality of absolute space, whereas already at the time of the *Inaugural Dissertation* and thereafter it appears as an argument for the transcendental ideality of space (Buroker 1981, 3–4). But this much is beyond dispute. The argument first appears at the crucial moment of Kant's final turn away from Leibniz and Wolff, a time when, according to Cassirer, Kant was intensely engaged with a careful reading of the Leibniz-Clarke correspondence (Cassirer 1918, 111).

What is the argument? Kant asks us to consider two objects, such as our right and left hands or a right- and left-handed screw, objects that are "perfectly like and similar and yet . . . in themselves so different that the limits of the one cannot at the same time be the limits of the other." Such objects are called "incongruent counterparts" (1768, 41). What does the existence of such objects demonstrate?

It is already clear from the everyday example of the two hands that the figure of a body can be completely similar to that of another, and that the size of the extension can be, in both, exactly the same; and that yet, however, an internal difference remains: namely, that the surface that includes the one could not possibly include the other. As the surface limiting the bodily space of the one cannot serve as a limit for the other, twist and turn it how one will, this difference must, therefore, be such as rests on an inner principle. This inner principle cannot, however, be connected with the different way in which the parts of the body are connected with each other. For, as one sees from the given example, everything can be perfectly identical in this respect. (Ibid., 42)

What, then, is this "inner principle"? Kant explains:

If one accepts the concept of modern, in particular, German philosophers, that space only consists of the external relations of parts of matter, which exist alongside one another, then all real space would be, in the example used, simply that *which this hand takes up.* However, since there is no difference in the relations of the parts to each other, whether right hand or left, the hand would be completely indeterminate with respect to such a quality, that is, it would fit on either side of the human body. But that is impossible.

From this it is clear that the determinations of space are not consequences of the situations of the parts of matter relative to each other; rather are the latter consequences of the former. It is also clear that in the constitution of bodies differences, and real differences at that, can be found; and these differences are connected purely with *absolute and original space*, for it is only through it that the relation of physical things is possible. (Ibid., 43)

Or, as Kant announced the conclusion at the beginning of the essay, *"absolute space has its own reality independently of the existence of all matter and that it is itself the ultimate foundation of the possibility of its composition"* (ibid., 37).

What is it about absolute space that Kant means here to defend against Leibniz? Again, it has nothing whatsoever to do with the dynamical properties of Newtonian absolute space. It is rather that more basic property of space that it is, first, independent of the existence of matter, and is, second, *"the ultimate foundation of the possibility of its composition,"* that is to say, that space is the ground of individuation. It was this that Newton and Locke held to be the most fundamental property of space; it was this that Leibniz attacked; and it is this that Kant defends.

Realizing that individuation is the issue actually helps us to resolve some of the puzzles about the role of the "incongruent counterparts" argument in the development of Kant's thought. For I think it wrong to see a major difference between the 1768 employment of the argument and its later employments. Already in 1768, where Kant says that he is arguing that "absolute space has its own reality independently of the existence of all matter," he nevertheless concludes:

It is also clear that since absolute space is not an object of external sensation, but rather a fundamental concept, which makes all these sensations possible in the first place, we can only perceive through the relation to other bodies that which, in the form of a body, purely concerns its relation to pure space. (1768, 43)

This is not so very different from the way the "incongruent counterparts" argument is used two years later in the *Inaugural Dissertation* to prove that "the concept of space is therefore a pure intuition" (Kant 1770, sec. 15, 69). Indeed, what is common to each of the occasions when Kant invokes the argument is the idea that we require a ground for the individuation of objects in experience. In the *Prolegomena*, Kant puts the point this way:

Space is the form of the external intuition of this sensibility, and the internal determination of every space is possible only by the determination of its external relation to the whole of space, of which it is a part (in other words, by its relation to the outer sense). That is to say, the part is possible only through the whole, which is never the case with things in themselves, as objects of the mere understanding, but which may well be the case with mere appearances. Hence the difference between similar and equal things which are not congruent (for instance, two symmetric helices) cannot be made intelligible by any concept, but only by the relation to the right and left hands which immediately refers to intuition. (1783, sec. 13, 34)

That space is a form of intuition follows from what is required for individuation, for the conditions for the possibility of the parts of space. He says later in the *Prolegomena*: "The mere universal form of intuition, called space, must therefore be the substratum of all intuitions determinable to particular objects; and in it, of course, the condition of the possibility and of the variety of these intuitions lies" (ibid., sec. 38, 68–69; see also 1786, 23–24).

We saw above that, when Schopenhauer first introduced the concept of space and time, the *principium individuationis*, as the "ground of being," the third form of the Principle of Sufficient Reason, in *Ueber die vierfache Wurzel*, it was precisely Kant's "incongruent counterparts" argument that he cited to begin the explanation of how spatial objects differ "through mere juxtaposition and hence through *place*" (1813, 194, 199; see also 40). When Schopenhauer writes, thus, about space and time as the *principium individuationis*, he writes in a well-defined historical context, in which the problem of space as the ground of individuation was the central issue in two centuries of controversy over the nature of space. Schopenhauer follows Kant in affirming the Newtonian, Lockean view in the face of the Leibnizian critique. It is in this context also that we must situate Einstein's reading of Schopenhauer.

Not that it is essential to Einstein's understanding of Schopenhauer, but could Einstein have understood that Schopenhauer was writing in this context? Newton's *De Gravitatione* would not have been available to Einstein. It is a good guess that he was familiar with the Leibniz-Clarke correspondence, although the issue of individuation is not mentioned in the discussion of the correspondence in Ferdinand Rosenberger's *Isaac Newton und seine physikalischen Principien* (1895), a copy of which was in Einstein's personal library. Einstein read Kant's first *Kritik* at least once, as a youth (see Howard 1994), but the "incongruent counterparts" argument, as noted above, is not explicitly mentioned there. It is, however, discussed at length, with generous quotations from the 1768 essay in which it first appeared, in the published version of August Stadler's lectures on Kant for which Einstein enrolled as a student at the ETH in the summer semester of 1897 (Stadler 1912, 104–05, 189–92).[14] Einstein also read the *Prolegomena* at least once, in 1918 (see Einstein to Max Born, summer 1918, in Born 1969, 25). And, finally, as noted above, there is evidence that Einstein read Locke (Grüning 1989, 247). Still, this does not mean that Einstein knew the history of the controversies over space as the ground of individuation; his reading of Schopenhauer could very well stand on its own.

Schopenhauer in the Eyes of Einstein's Contemporaries: Philosopher of Pessimism or Philosopher of Physics?

We have documented Einstein's extensive reading of Schopenhauer. We have examined what Schopenhauer had to say about space and time as the *principium individuationis*. We have explored the historical context in which Schopenhauer developed this doctrine. But we lack explicit documentation that Einstein's distinctive conception of spatiotemporal separability was influenced, in any way, by his reading of Schopenhauer. It is possible that Einstein did not read Schopenhauer as a thinker with interesting and important things to say about fundamental questions in physics. Lacking any specific documentation on this point, how might we proceed? What we can do is to ask how others read Schopenhauer at this time and, most important, how other physicists and philosophers of science contemporary with Einstein read Schopenhauer.

Mention was made above of many people who were, without question, deeply influenced by their reading of Schopenhauer, such as Wagner, Nietzsche, and Thomas Mann. But what about thinkers closer to Einstein in their interests and outlook? Robert Musil, author of *Der Mann ohne Eigenschaften*, is an interesting case, standing as he does between the philosophical and literary communities.[15] In his youth, he read Schopenhauer with keen interest. As David Luft puts it, "Musil's generation (from Hermann Broch to Thomas Mann) received the Kantian doctrine and romantic aesthetics via Schopenhauer" (1980, 39). Or consider the novelist and critic Max Brod, and his Prague friend, the philosopher Hugo Bergmann. They were at the core of the Fanta-Kreis, the Jewish, intellectual salon with which Einstein associated during his year in Prague, 1911–1912. Brod and Bergmann are among the many Jewish intellectuals who had an enthusiasm for Schopenhauer, according to Henry Brann (1975, 60–61).[16] Closer still to Einstein is Ludwig Wittgenstein, whose debt to and respect for Schopenhauer has been widely discussed (see, for example, Janik and Toulmin 1973; Magee 1983, esp. 286–315; Monk 1990).

What about Schopenhauer's reception among philosophers of science and physicists? Was he taken seriously as a thinker with interesting and important things to say about fundamental questions in physics? Lest we forget, Ludwig Boltzmann took Schopenhauer seriously enough to do him the honor of a harsh, witty, at times almost sarcastic, critique, "Über eine These Schopenhauers," before the Vienna Philosophische Gesellschaft in January 1905. Boltzmann's main target was Schopenhauer's ethical pessimism, but along the way he devoted special attention to Schopenhauer's views on the a priori status of space and time (Boltzmann 1905, 387–88). Ernst Mach had a different attitude toward Schopenhauer. Among many references to Schopenhauer, we find Mach writing the following in *Erkenntnis und Irrtum*, as a footnote to the remark that "the foundation of all knowledge is . . . *intuition*": "As it appears to me, after Kant, Schopenhauer accorded the highest value to the significance of intuition" (1905, 310).

Moritz Schlick went out of his way on many occasions to criticize Schopenhauer's conception of metaphysics, especially the possibility of a metaphysical knowledge different in kind from scientific knowledge (see, for example, Schlick 1932, 167–68). He nevertheless saw the relevance of Schopenhauer in other areas. Thus, in a discussion of causality and reality in his *Allgemeine Erkenntnislehre*, which Einstein read with care (see Howard 1984), Schlick writes:

If previously we said that we call real whatever is the cause of experiences, we can now give up the relation to experience and still maintain the position that everything real is a *cause*. Anything that does not make itself noticeable in some way, never manifests itself, is in fact not *there*, is not real; whether we experience the manifestation of a thing, however, is accidental. Thus we capture the essential as opposed to the accidental if we accept the formulation: the real is that which *has an effect* (*wirklich ist, was wirkt*).

Even our language seems to exert pressure in behalf of this interpretation and to demonstrate that it has caught the sense of the popular view. In German, the word "real" ("wirklich") is derived from the verb "to have an effect" ("wirken"). In Aristotle the concept *energeia* coincides with that of reality. And Leibniz, too, declared: "*quod non agit, non existit.*" The best known advocate of this conception is no doubt Schopenhauer. Of matter, he said: "its being is its acting on something; it is impossible even to think of its having any other being." In another passage, he wrote that matter is "causality itself, objectively conceived." The reality of things, he explained, is their materiality: reality is thus the "efficacy of things generally." Today we find the same definition in many thinkers. (Schlick 1918, 181–82)

What about the reception specifically of Schopenhauer's doctrine of space and time as the *principium individuationis*? Was he taken seriously by physicists and philosophers of science on this topic as well? Schlick was no doubt echoing Schopenhauer when he wrote, also in the *Allgemeine Erkenntnislehre*: "In material reality, space and time are the great uniters and dividers. In the end, all the determinations by which we mark off an object of the external world and distinguish it as an individual thing from other individual things consist of specifications of time and place" (1918, 53). But what of those who were most involved in the debates about space, time, and individuation, the debates in which Einstein deployed his principle of spatiotemporal separability? I would count among the members of this core group, aside from Einstein, Erwin Schrödinger, Wolfgang Pauli and Niels Bohr (see Howard 1990). Evidently Bohr did not read Schopenhauer (according to a personal communication from David Favrholdt), but Schrödinger and Pauli did.

Karl von Meyenn, the editor of Pauli's scientific correspondence, tells us that Pauli "admired [Schopenhauer] greatly" (Pauli 1985, 586). In a 1954 essay on scientific and epistemological aspects of the unconscious, written on the occasion of Carl Jung's eightieth birthday, Pauli wrote sympathetically about extrasensory perception, noting approvingly that "even such a thoroughly critical philosopher as Schopenhauer not only regarded parapsychological effects going far beyond what is secured by scientific evidence as possible, but even considered them as a support for his philosophy," and adding: "Schopenhauer speaks metaphysically of the 'Will' that breaks through space and time, the 'principium individuationis,' as he calls them, and opposes the 'nexus metaphysicus' to the customary 'nexus physicus'" (Pauli 1954, 124, referring to the chapter "Animalischer Magnetismus und Magie" in Schopenhauer 1836). This is the same Pauli who, in discussing the EPR argument with Werner Heisenberg, in a letter of 15 June 1935 (four days *before* the letter to Schrödinger in which Einstein first enunciated his "separation principle"), cut right to the essential point about separability:

Quite independently of *Einstein*, it appears to me that, in providing a systematic foundation for quantum mechanics, one should *start* more from the composition and separation of systems than has until now (with Dirac, e.g.) been the case. — This is indeed — as Einstein has *correctly* felt — a very fundamental point in quantum mechanics, which has, moreover, a direct connection with your reflections about the *cut* and the possibility of its being shifted to an arbitrary place. (Pauli 1985, 404)

We glimpse here a difference in the way different thinkers read Schopenhauer. If I might venture a shaky generalization, I would say that those thinkers broadly sympathetic to quantum mechanics, with its nonseparable way of accounting for interacting systems, tended to be drawn toward the holism that lies on the side of the Will, the thing-in-itself, in Schopenhauer; whereas a thinker critical of quantum nonseparability, like Einstein, tended to emphasize the more Kantian aspect of Schopenhauer's thought — in particular, the insistence that objective knowledge of empirical objects is possible only as those objects are individuated in space and time. Thus, Schrödinger, a friend of quantum nonseparability, is drawn, like Pauli, to the mystical side of Schopenhauer, the Schopenhauer who was the prophet of Vedanta.

Schrödinger's biographer, Walter Moore, details the lifelong influence of Schopenhauer on Schrödinger, starting with the fact that "Erwin read everything written by Schopenhauer," and adding that "his direct influence on Schrödinger was considerable, but equally important was the introduction he provided to Indian philosophy" (Moore 1989, 111, 112). Schrödinger himself later recalled that in 1918 he was "with great enthusiasm becoming familiar with Schopenhauer and, through him, with the doctrine of unity taught by the Upanishads" (ibid., 109). The traces of this influence are everywhere in Schrödinger, such as in the very Schopenhauerian epilogue, "On Determinism and Free Will," that concludes his famous *What Is Life?* (1944), or the metaphysical fourth chapter, "The Arithmetical Paradox: The Oneness of Mind," of *Mind and Matter* (1958), or the Schopenhauerian label that Schrödinger put on one folder of papers in his files: "Sammlung der Gedanken über das physikalische Principium individuationis" (AHQP, Erwin Schrödinger Papers, 44, 9).

Most interesting, however, is the essay, "Seek for the Road," which Schrödinger wrote during the summer of 1925 when he was first coming to grips with the failure of classical assumptions about the individuality of spatially separated particles brought to light by Einstein's work on Bose-Einstein statistics in the winter and spring of 1925, and immediately before the conceptual breakthrough that led to wave mechanics in December 1925 (Moore 1989, 191–209), the wave mechanics that Schrödinger saw as "a method of dealing with the problem of many particles" (1950, 206), the problem being the new statistics required by the failure of individuality.

"Seek for the Road" is thoroughly Schopenhauerian in spirit. Its main point is to argue for the unity of all consciousness, in spite of "the spatiotemporal plurality of individuals" (1925, 31). "For philosophy," Schrödinger writes, "the real difficulty lies in the spatial and temporal multiplicity of individuals." The difficulty is not to be solved by logical means, he claims, "but it is quite easy to express the solution in words, thus: the plurality that we perceive is only *an appearance; it is not real*" (ibid., 18). Applied to the problem of consciousness, this "solution" implies that "this life of yours which you are living is not merely a piece of the entire existence, but is in a certain sense the *whole*" (ibid., 21). Schrödinger summarizes this view by quoting the very same mystic formula of the Brahmins, called the *Mahavakya*, "*Tat tvam asi*" ["This living thing art thou"], that Schopenhauer regularly quoted for the purpose of expressing the idea of the unity of the will (see, for example, Schopenhauer 1859, vol. 1, 220, 355, 374). What follows from the unity of all consciousness? Schrödinger says: "It is the vision of this truth (of which the individual is seldom conscious in his actions) which underlies all morally valuable activity" (1925, 22).

To the end of his life, a belief in the unity of consciousness was central to Schrödinger's philosophy. In the chapters appended to "Seek for the Road" at the time of its preparation for publication in 1960, just months before his death in January 1961, Schrödinger wrote: "We still have the lovely thought of unity, of belonging unqualifiedly together, of which . . . Schopenhauer said that it was his comfort in life and would be his comfort in death"; and "Schopenhauer's books are still beautiful" (1964, 104, 110).[17]

There is surely no straight logical line leading from a belief in the unity of consciousness to the invention of entangled quantum states. But Schrödinger himself was no friend of a misplaced logic ("In a considerable number of cases logical thinking brings us up to a certain point and then leaves us in the lurch" [1925, 19]), and it seems not unlikely that the metaphysical vision of unity preached in "Seek for the Road" helped prepare the ground for Schrödinger's construction in wave mechanics of a theoretical framework within which entangled many-particle physical states could be realized.

So Pauli and Schrödinger, both of whom focused on the failure of classical notions of spatial separability in quantum mechanics, were avid readers of Schopenhauer, much influenced by his holism, his glimpse of unity on the side of the Will, the thing-in-itself. Still, we have not yet found any of Einstein's philosophical and scientific contemporaries making *explicit* the connection between Schopenhauer's views on space and time as the *principium individuationis* and the failure of separability in quantum mechanics. Did anyone make this point?

Hermann Weyl came very close. To begin with, he clearly understood the role of space (if not also time) as the *principium individuationis*. Thus, he opens his seminal 1922 Barcelona and Madrid lectures, *Mathematische Analyse des Raumproblems*, as follows:

With respect to reality, we follow Kant in distinguishing the *qualitative content* from its *Form*, the spatio-temporal extension that first makes possible a distinguishing of the qualitative. Without changing the nature of its content, in that it remains exactly such as it was, a body can find itself, instead of being *here*, at any other place in space. In this way it is possible, in the extensive medium of the external world (in which, in addition to space, we count time), for things that are, in their essences, in their natures, the same as one another, to be individually distinguished. (1923, 1)

And five years later, in his *Philosophie der Mathematik und Naturwissenschaft* (1927), he characterized "the essence of space" in this way:

The penetration of the This (here-now) and the Thus is the general form of consciousness. A thing exists only in the indissoluble unity of intuition and sensation, through the superimposition of continuous extension and continuous quality. . . . Since the mere Here is nothing by itself that might differ from any other Here, space is the *principium individuationis*. It makes the existence of numerically different things possible which are equal in every respect. That is why Kant contradistinguishes it as the *form of intuition* from "the matter of phenomena, i.e. that which corresponds to sensation." Here lies the root of the concepts of similarity and congruence. (1927, 130–31)

Schopenhauer may not here be mentioned by name (he is discussed elsewhere by name; see ibid., 34, 210), but Weyl's use of the telltale Schopenhauerian expression, *"principium individuationis,"* leaves little doubt as to the source. In the appendices added to the 1949 translation of this work, we find extensive discussions of quantum nonseparability (1949, 261–62), as well as of the failure of classical conceptions of particle individuation resulting from Bose-Einstein and Fermi-Dirac statistics (ibid., 245–47). Here again, the vocabulary he uses is interesting: "The upshot of it all is that the electrons satisfy Leibniz's *prin-

cipium identitatis indiscernibilium. . . . In a profound and precise sense physics corroborates the Mutakallimûn; neither to the photon nor to the (positive and negative) electron can one ascribe individuality" (ibid., 247). The reference is to certain practitioners of *kalam* (speculative theology) in early Islam whose way of denying becoming involved a denial of the enduring identities of individual things. But even Weyl does not make the explicit connection we seek.

The one thinker I have so far found who does make the connection in the most unmistakable terms is Ernst Cassirer. The place is Chapter II.ii, "Zum Problem des 'materiellen Punktes,' " of *Determinismus und Indeterminismus in der modernen Physik*. Cassirer writes:

Quantum mechanics has emphasized again and again that within it a strictly mathematical schema exists, but that this schema is not to be imagined as a simple interconnection of things in space and time. If this is so, however, some very definite conclusions follow concerning the "individuality" of the elements with which quantum mechanics constructs nature. Schopenhauer declares that space and time are to be regarded as the real *principium individuationis*. In other places also in the philosophical history of the problem of individuation, we encounter this determination frequently. "It is easy to discover," Locke observed, "what is so much inquired after, the *principium individuationis*; and that, it is plain, is existence itself, which determines a being of any sort to a particular time and place incommunicable to two beings of the same kind. Let us suppose an atom . . . existing in a determined time and place; it is evident that, considered in any instant of its existence, it is, in that instant, the same with itself." Conversely it follows, however, that, when an object is no longer determinable by means of a "here" and "now," when it is not denotable as a τόδε τι, its "individuality" can no longer be maintained in the conventional sense, valid for things in space and time. . . . If then we continue to talk about the individuality of single particles, this can only be done indirectly; not insofar as they themselves, as individuals, are given, but so far as they are describable as "points of intersection" of certain relations. If we scrutinize the development of modern quantum mechanics, we will find this assumption fully confirmed. In de Broglie's wave theory of matter and in Schrödinger's wave mechanics the concepts of proton and electron are maintained, but they are defined no longer as "material points" in the sense of classical mechanics, but instead, as centers of energy. We may thus continue to talk of the electron as a definite object but it no longer possesses that individuation that could be designated by a simple "here" and "now." Waves are not tied to a single spatiotemporal point; they enjoy a kind of omnipresence. Each extends through the entire space — which, however, is no longer to be considered as an empirical space but as a configurational space. (1936, 180–81)

Cassirer goes on to note that the situation regarding individuation is the same in Heisenberg's matrix mechanics and in Born's statistical interpretation, adding, in a long footnote:

The impossibility of delimiting different electrons from one another, and of ascribing to each of them an independent individuality, has been brought into clear light through the evolution of the modern quantum theory, and particularly through the considerations connected with the Pauli exclusion principle. Considered solely from the standpoint of its methodological significance in the construction of the quantum theory, Pauli's exclusion principle is strangely analogous to the general principle introduced into philosophy by Leibniz under the name of *principium identitatis indiscernibilium*. This principle states that there cannot be two objects which completely correspond to each other in every determining characteristic, and thus are indistinguishable except by mere number. There are no things that differ from one another *solo numero*; rather every true difference must be definable as a qualitative difference, a distinction of the attributes and conditions that constitute the object. Cf. Leibniz, *Briefwechsel mit Clarke*, Letter 4, sec. 4; Letter 5, 6, etc. The Pauli principle is, as it were, the *principium identitatis indiscernibilium* of quantum theory. (Ibid., 184–85)

And, finally, Cassirer relates all of this to the problems of nonseparability and non-Boltzmann statistics:

A system consisting of two electrons determines, from the point of view of quantum mechanics, the state of these electrons, but the reverse does not follow. A knowledge of the states of the two parts does not determine the state of the joint system, and a derivation of the latter from the former is out of the question. The question, how, within a given whole, the separation into parts may be accomplished and how a certain aggregate may be differentiated and "individualized," accordingly always constitutes a difficult problem for quantum theory. The ordinary method of counting, which presupposes that it is known from the beginning what is to constitute one thing and what two or more things, is here insufficient. Individual things are not separate from each other in as simple a manner as in the sensuous-spatial view; complicated theoretical considerations are thus always required in order to determine precisely what is to be treated as an individual, what is to be counted as "one." According to the premises chosen, entirely different forms of quantum statistics may arise. . . . Here also we see clearly that the determination of the individual, or that which truly figures as "one" being, is not the *terminus a quo*, but always only the *terminus ad quem* for quantum theory. (Ibid., 187–88)

The large conclusion that Cassirer draws from this analysis is that we need not regard the quantum theory as threatening to impugn the universal validity of the causal principle, as long as we get clear about what counts as an individual in the ontology of the quantum theory. In other words, we can save causality, if we are willing to abandon the classical, spatiotemporal mode of individuating physical systems.

Einstein read Cassirer's *Determinismus und Indeterminismus in der modernen Physik* in March 1937. He recorded his reaction in a letter to Cassirer of 16 March:

> I have read your book carefully and with sincere admiration. I do not know whether one should admire more the sagacity, the skill in presentation, or the depth of your knowledge of the subject. Moreover, it dawned on me for the first time in reading this book what a towering intellect Leibniz was. That he did not find the hypothesis of forces acting at a distance satisfactory is not so surprising; even Newton, himself, did not believe that this postulate was to be conceived as final, irreducible. His rejection of an absolute space is even more admirable. But that he realized that the atomic theory was to be rejected as a foundation for physics, because it is, in principle, irreconcilable with a representation by means of continuous functions (the law of impact as an elementary law), that required at that time true genius. (EA 8-394)

(The discussion of Leibniz is in the chapter immediately preceding that from which the above quotations were taken.) Einstein then goes on to rehearse for Cassirer the version of the argument for the incompleteness of quantum mechanics that he had first presented in correspondence with Schrödinger in June 1935, where he introduced his "separation principle." This time he concluded that one either recognizes that the quantum theoretical description of separated subsystems in the EPR-type experiment is incomplete, or "one would make up one's mind to believe that there is some kind of 'telepathic' interaction between the separated masspoints 1 and 2, something to which no theorist known to me could subscribe" (EA 8-394).

Did physicists and philosophers of science contemporary to Einstein take Schopenhauer seriously as someone with interesting and important things to say about fundamental questions in physics? Yes. Did they understand the relevance of what Schopenhauer said about space and time as the *principium individuationis* to the problems of nonseparability in quantum theory? Yes. Could Einstein have seen this connection? How could he not have seen it if he read Schopenhauer as thoroughly as the evidence suggests, and if, besides, he read Cassirer's *Determinismus und Indeterminismus in der modernen Physik* as carefully as he told Cassirer he had? The evidence is, of course, still only circumstantial. But it is not wildly implausible to assume that Einstein understood the connection. If he did, it would help us to understand why he accorded Schopenhauer's portrait a place of honor alongside those of Faraday and Maxwell.

Conclusion

What, then, have we learned? We have learned that, whether or not Einstein learned about space and time as the *principium individuationis* from Schopenhauer, his invocation of the separability principle in the course of explicating his concerns about the quantum theory occurs in a context. The context is a centuries-old tradition, which found clear expression in the writings of thinkers like Newton, Locke, and Kant, a tradition that regards space as the ground against which physical systems are individuated, a tradition according to which "in material reality, space and time are the great uniters and dividers" (Schlick 1918, 53). And we have seen that it was Schopenhauer, more than anyone else, who made this idea a commonplace in the minds of early twentieth-century physicists and philosophers of science.

Understanding this context makes it plausible that Einstein, like many of his contemporaries, learned the lesson about space and time as the *principium individuationis* from Schopenhauer. At the very least, understanding the context should help us better to understand why Einstein clung so tenaciously to his separability principle. And it should also help us to understand why Einstein regarded field theory — which provides the most extreme embodiment of the separability principle — as the proper framework for fundamental physics.

Einstein clung to separability and the field concept because, in this tradition, space and time as *principium individuationis* function as a priori conditions for the very possibility of objective scientific knowledge of empirical objects. Notice that I said "space and time as *principium individuationis.*" It is not the absolute space and time of Newton that have this a priori status. It is not space as described by Euclidean geometry. It is not the space and time of the turn-of-the-century neo-Kantianism that was so offended by general relativity. Instead, it is space and time as *principium individuationis*. Remember what Einstein said in his comments to Born in March 1948:

If one renounces the assumption that what is present in different parts of space has an independence, real existence, then I do not at all see what physics is supposed to describe. For what is thought to be a "system" is, after all, just conventional, and I do not see how one is supposed to divide up the world objectively so that one can make statements about the parts. (Born 1969, 224)

Einstein was certainly no friend of the Kantian a priori. He believed, with Kant, that we necessarily bring something of our own to our construction of scientific theories of the world — at least that our theories are not logical deductions from experience — but he denied that what we contribute is somehow fixed by the very nature of the human mind or knowledge itself (see Howard 1994). I would not want to overread a passage such as this. Still, the sentiment expressed is a strong one. Give up separability "and I do not see how one is supposed to divide the world objectively so that one can make statements about the parts."

From the point of view that regards separability as the fundamental metaphysical fact about space, there is a continuity in the evolution of our understanding of space since the seventeenth century (and earlier) that is largely overlooked in the standard histories. Thus, if it is separability, space as *principium individuationis*, that is the basic property of space, then it makes little difference whether one regards space as something absolute or relative. The space or spacetime of Einstein individuates physical systems just as well as the space of Newton. Of course the relative space of Leibniz does not. In asserting a relative conception of space in opposition to Newton's absolute space, Leibniz meant to deny something much more fundamental about Newtonian space than Einstein ever meant to deny. And that something is, again, separability. To put it in different words, Einstein's relativity is not Leibniz's relativity.

It also makes no difference whether one takes the material particle or the field as ontologically primary. The idea of separability, space as *principium individuationis*, first grew up in the context of atomism. But Einstein understood that it was field theory that gives the most extreme expression of the idea of separability, by its treating every point of space (or spacetime) as, in effect, a separable physical system endowed with its own, separable, real physical state. Even allowing the material contents of space to be swallowed up, as it were, by the geometry of space (spacetime) makes no difference in the capacity of space to act as the ground of individuation.

And, finally, it makes no essential difference whether one distinguishes space and time as Newton did, or unites them into one, four-dimensional spacetime manifold, as Einstein did. The only change this brings about in the way we individuate things in nature is that, whereas classically space is the ground for the individuation of *systems, bodies, things, or substances,* and time is the ground for the individuation of the *states* of those systems, bodies, things, or substances, in Einsteinian spacetime the distinction between *system* and *state* lapses.

All that is necessary for a unified spacetime to fulfil its inherited role as *principium individuationis,* is that, in it, the basic idea of *separation* retains its objective significance. In the case of a four-dimensional continuum, the minimum necessary condition for the possibility of objective individuation is the objectivity, which is to say the invariance under arbitrary transformations, of the infinitesimal metric interval:

$$ds^2 = g_{ik} dx_i dx_k.$$

But this is precisely the fundamental invariant of the general theory of relativity. So one could say that general relativity is, simply, the most general way of doing physics that retains the basic idea of space and time, or spacetime, as the *principium individuationis.*

With this insight, history closes upon itself. For the principle underlying the definition of the infinitesimal metric interval is the Pythagorean theorem. It is the most abstract representation of the idea of *distance* and was the great secret of the Pythagoreans. But it was those same Pythagoreans to whom, in the end, we owe our idea of space. As F. M. Cornford reminds us, in "The Invention of Space":

Our first glimpse of the Void in philosophic literature we owe to a passage in Aristotle's *Physics,* recording a feature of the primitive Pythagorean cosmology: "The Pythagoreans too asserted that Void exists and that it enters the Heaven itself, which, as it were, breathes in from the boundless a sort of breath which is at the same time the Void. This keeps things apart, as if it constituted a sort of separation or distinction between things that are next to each other. . . ." (*Phys.,* IV, 6, 213b, 23). (Cornford 1928, 8)

There is more to be said about Einstein's reading of Schopenhauer. For example, the characteristic way in which Schopenhauer combines space and time with causality, as equally essential a priori conditions for the possibility of objective scientific knowledge, suggests Einstein's equally firm commitments to both separability and causality. Of course, one learns a similar lesson from Kant himself, but it is not at all impossible that, like Robert Musil, and so many other young thinkers of that day, Einstein learned his Kant filtered through the lens of Schopenhauer. More and more I see the ghost of Schopenhauer in the whole debate over how the quantum theory might force us to choose between spacetime representation and causality, or at least to regard them as complementary modes of description. But this deserves a fuller study in its own right.

Finally, we should, perhaps, draw a lesson about how an Einstein must be regarded as a whole thinker. We should resist the positivist prejudice (so much in evidence in Frank's characterization of Einstein's reading of Schopenhauer) that regards some thinkers from the past, such as Hume, Mach, and (grudgingly) Kant, as legitimate candidates for being influences on an Einstein's scientific development, and consigns thinkers such as Schopenhauer to the category of the merely literary. Schrödinger, for example, would have been offended at our denying the importance of Schopenhauer to his intellectual development. Who are we to say, by the lights of our own puny prejudices, how a thinker of Einstein's stature is to be allowed to weave together, from a hundred different sources, an original picture of the natural world? We should be content to be visitors in the gallery, not bluenose censors of the life of the mind.

NOTES

The intellectual debts that I hope in some small measure to repay with this essay are old ones. Two of my graduate school teachers, Milič Čapek and Judson Webb, will see important parts of themselves in what I write here. They may not want to claim credit for it, and I do not want to implicate them in any errors, oversights, or misinterpretations the essay may contain. But I do want, sincerely and generously, to express my thanks to them for the seeds they planted long ago. More specific thanks are owing to a variety of other colleagues, including Mara Beller, David Cartwright, Catherine Chevalley, Bob Cohen, Dan Breazeale, Arthur Fine, Dan Frank, Don Giles, John Inglis, Michel Janssen, John Norton, Ze'ev Rosnkranz, Tom Ryckman, Henry Schankula, Robert Schulmann, Abner Shimony, John Stachel, and Jim Wilkinson. I received many helpful suggestions from audiences at the University of Pittsburgh, Northwestern University, the University of Notre Dame, Boston University, the Universität Göttingen, and the 1994 meeting of the North American Division of the Schopenhauer Society, where versions of this paper were read.

My sincere thanks go to the University of Kentucky for granting me a University Research Professorship for the 1992–1993 academic year, during which time this essay was written. Finally, I thank the Hebrew University of Jerusalem, which holds the copyright, for permission to quote from Einstein's unpublished correspondence.

1. This report is seconded by Konrad Wachsmann, who also explicitly disputes the assertion by Helen Dukas and Banesh Hoffmann (Hoffmann-Dukas 1972, 46) that the portraits were of Newton, Faraday, and Maxwell; see the interview with Wachsmann in Grüning 1989, 144. On the other hand, in an undated photograph of Einstein in his study reproduced in Grüning 1989, 136, one sees only a portrait of Newton, on the wall (the same photo is also reproduced in Sugimoto 1989, 102).

2. For Einstein's early reading of these individuals and others named in this paragraph, see the respective references in Einstein 1987, 1989.

3. For a representative sampling of testimony to this influence, see the many essays and "testimonials" conveniently collected in Haffmans 1977, including Nietzsche's "Schopenhauer als Erzieher," originally published in 1874 as part of the *Unzeitgemäße Betrachtungen*, and Mann's "Schopenhauer," originally written in 1938 as an introduction to *Living Thoughts of Schopenhauer* (New York: Longmans, Green, 1939).

4. Einstein wrote in the preface: "The author of this book is one who knows me rather intimately in my endeavor, thoughts, beliefs — in bedroom slippers. I have read it to satisfy, in the main, my own curiosity. . . . I found the facts of the book duly accurate, and its characterization, throughout, as good as might be expected of one who is perforce himself, and who can no more be another than I can."

5. The question actually came from Einstein's second son, Eduard, then housed in a Swiss mental institution, where he was befriended by Seelig; see Seelig 1960, 191–92.

6. Einstein owned the twelve-volume 1894–1896 edition by Cotta, which was based on the Julius Frauenstädt edition of 1873–1874 (see Hübscher 1981, 37) and carried an introduction by Rudolf Steiner. My thanks to Ze'ev Rosenkranz, director of the Albert Einstein Archives, for locating and identifying this edition.

7. For a healthy antidote to Frank's reading of Einstein, see Holton 1968. For a critical response to Holton, see Howard 1993. Frank is also wrong about Einstein's attitude toward Kant, a thinker whom Einstein took seriously, even while disagreeing with him (see Howard 1994). What is convincingly "Einsteinian" in Frank's account is the use of the musical image to disparage a second-rate thinker. But I know of only one case where Einstein himself deployed this image for such a purpose, and Hegel, not Schopenhauer, was the target. On 16 May 1951, Einstein wrote to G. Broggi: "In my eyes, a philosopher of Hegel's type is a man who juggles words that correspond to no clear concepts, a kind of word-music" (EA 25-428). I thank Arthur Fine for drawing my attention to this remark.

8. In a very different intellectual tradition, Schopenhauer's influence on Einstein's world view was long taken seriously, but for all the wrong reasons. In 1951, the Soviet philosopher M. M. Karpov wrote the following in a contribution to a long-running discussion of Einstein's philosophy in *Voprosy filosofii*: "Einstein's views and opinions were formed under the influence of such idealist philosophers as Hume, Mach, and Schopenhauer. That could not help influencing his philosophical views. Einstein answers the basic question of philosophy idealistically" (1951, 130, quoted in Gribanov 1987, 38).

9. Schopenhauer says that this need for metaphysics arises from wonder experi-

enced by the will made objective in man: "And its wonder is the more serious, as here for the first time it stands consciously face to face with *death*, and besides the finiteness of all existence, the vanity and fruitlessness of all effort force themselves on it more or less" (1859, vol. 2, 160). It is perhaps no coincidence that Einstein's first really serious physical collapse, in the form of an attack of pericarditis, had occurred only shortly before, in the spring of 1928, forcing him to endure a long period of bed rest (see Clark 1971, 348–50).

10. For a reconstruction of the hidden variables model attempted by Einstein, see Cushing 1994, chap. 8, sec. 3, and app. 3.

11. It is clear from a letter of 9 August 1939 that the "talmudist" is intended to be Bohr. After once again distinguishing the "Born" and "Schrödinger" interpretations of the Ψ-function, Einstein writes: "There are also, however, the mystics, who altogether prohibit, as unscientific, any question about something existing independently of experience (Bohr). Then the two conceptions flow together in a soft haze, in which I do not feel any better, however, than in one of the aforementioned conceptions, which take a position on the reality concept" (EA 22-060).

12. There is a way of reading the implications of Bell's theorem and the Bell experiments that implies that Einstein was wrong, that, in order to avoid violations of the locality principle, which, recall, encapsulates special relativistic constraints on superluminal signals, one must give up separability. And there is the further argument that, if separability is so intimately connected with the concept of a field theory like general relativity, then giving up separability threatens also the very foundations of general relativity (see Howard 1989, 1993).

13. See the lengthy entries under "Individuation," "Individuum," and "Individualität," in Wagner 1909, 184–86.

14. For Einstein's enrollment in Stadler's lectures, see Einstein 1987, 46, 364. As noted above, however, there is some question about Einstein's actual attendance at lectures.

15. In 1908, Musil wrote a doctoral dissertation on Ernst Mach, *Beitrag zur Beurteilung der Lehren Machs*, under Karl Stumpf in Berlin; Luft 1980, 81–88.

16. For Brod's enthusiasm for Schopenhauer, see the references in Brod 1947, 1966, 1979.

17. The unity of *consciousness* is, to be sure, a more specifically Vedantic theme; Schopenhauer insists on the unity of the *will*. Schrödinger's emphasis on the unity of consciousness thus may reflect his reading of Paul Deussen's *Das System der Vedanta* (1906) and other works on Indian philosophy and religion (see Moore 1989, 113), as much as it reflects his reading of Schopenhauer, Schrödinger's interpretation of Schopenhauer being influenced, perhaps, by his interest in the Vedanta. My thanks to David Cartwright on this point.

REFERENCES

Alexander, H. G., ed. 1956. *The Leibniz-Clarke Correspondence. With Extracts from Newton's "Principia" and "Opticks."* Manchester: Manchester University Press; New York: Barnes & Noble.

Bailey, C. 1928. *The Greek Atomists and Epicurus. A Study.* Oxford: Clarendon Press.

Bernstein, A. 1853–1857. *Aus dem Reiche der Naturwissenschaft. Für Jedermann aus dem Volke.* 12 vols. Berlin: Besser's Verlagsbuchhandlung. Reissued as *Naturwissenschaftliche Volksbücher.* 20 vols. Berlin: F. Duncker, 1867–1869.

Bohr, N., 1949. "Discussion with Einstein on Epistemological Problems in Atomic Physics." In Schilpp 1949, 201–41.

Boltzmann, L. 1905. "Über eine These Schopenhauers." In *Populäre Schriften.* Leipzig: Johann Ambrosius Barth, 385–402.

Born, M. 1949. *Natural Philosophy of Cause and Chance.* Oxford: Oxford University Press.

Born, M., ed. 1969. *Albert Einstein — Hedwig und Max Born. Briefwechsel 1916–1955.* Munich: Nymphenburger.

Bose, S. N. 1924. "Plancks Gesetz und Lichtquantenhypothese." *Zeitschrift für Physik* 26: 178–81.

Brann, H. W. 1975. *Schopenhauer und das Judentum.* Bonn: Bouvier Verlag Herbert Grundmann.

Brod, M. 1947. *Diesseits und Jenseits.* 2 vols. Zurich: Mondial-Verlag.

———. 1966. *Der Prager Kreis.* Stuttgart: Kohlhammer.

———. 1979. *Streitbares Leben. Autobiographie 1884–1968.* Frankfurt am Main: Insel Verlag.

Büchner, L. 1855. *Kraft und Stoff. Empirisch-naturphilosophische Studie.* Frankfurt am Main: Meidinger Sohn & Cie.

Buroker, J. V. 1981. *Space and Incongruence: The Origin of Kant's Idealism.* Dordrecht: D. Reidel.

Cassirer, E. 1918. *Kants Leben und Lehre.* Vol. 11 (Ergäzungsband) of *Immanuel Kants Werke.* Ernst Cassirer, ed. Berlin: Bruno Cassirer. Page numbers from *Kant's Life and Thought,* trans. James Haden. New Haven and London: Yale University Press, 1981.

———. 1936. *Determinismus und Indeterminismus in der modernen Physik. Historische und systematische Studien zum Kausalproblem.* Vol. 42, pt. 3, *Göteborgs Högskolas Ärsskrift.* Separatum: Göteborg: Elanders Boktryckeri Aktiebolag, 1937. Page numbers and translations from *Determinism and Indeterminism in Modern Physics: Historical and Systematic Studies of the Problems of Causality,* trans. O. Theodor Benfey. New Haven and London: Yale University Press, 1956.

Clark, R. W. 1971. *Einstein: The Life and Times.* New York and Cleveland: World Publishing.

Cornford, F. M. 1928. "The Invention of Space." In *Essays in Honor of Gilbert Murray.* London: Allen & Unwin. Page numbers from the partial reprinting in *The Concepts of Space and Time: Their Structure and Their Development.* Milič Čapek, ed. Dordrecht: D. Reidel, 1976.

Cushing, J. T. 1994. *Quantum Mechanics: Historical Contingency and the Copenhagen Hegemony.* Chicago: University of Chicago Press.

De Beer, E. S., ed. 1979. *The Correspondence of John Locke.* 8 vols. Oxford: Clarendon Press.

Deussen, P. 1906. *Das System des Vedanta. Nach den Brahma-Sutra's des Badara-*

yana und dem Kommentare des Cankara über dieselben als ein Kompendium der Dogmatik des Brahmanismus vom Standpunkte des Cankara aus dargestellt. 2nd ed. Leipzig: F. A. Brockhaus.

Dijksterhuis, E. J. 1961. *The Mechanization of the World Picture.* Trans. C. Dikshoorn. London: Oxford University Press.

Einstein, A. 1905. "Über einen die Erzeugung und Verwandlung des Lichtes betreffenden heuristischen Gesichtspunkt." *Annalen der Physik* 17: 132–48.

———. 1909a. "Zum gegenwärtigen Stand des Strahlungsproblems." *Physikalische Zeitschrift* 10: 185–93.

———. 1909b. "Über die Entwickelung unserer Anschauungen über das Wesen und die Konstitution der Strahlung." *Deutsche Physikalische Gesellschaft. Verhandlungen* 11: 482–500. Reprint. *Physikalische Zeitschrift* 10 (1909): 817–25.

———. 1918. "Motive des Forschens." In *Zu Max Plancks sechzigstem Geburtstag. Ansprachen, gehalten am 26. April 1918 in der Deutschen Physikalischen Gesellschaft.* Karlsruhe: C. F. Müller, 29–32. Page numbers from "Principles of Research," in Einstein 1954, 224–27.

———. 1924. "Quantentheorie des einatomigen idealen Gases." *Preussische Akademie der Wissenschaften. Physikalisch-mathematische Klasse. Sitzungsberichte*: 261–67.

———. 1925a. "Quantentheorie des einatomigen idealen Gases. Zweite Abhandlung." *Preussische Akademie der Wissenschaften. Physikalisch-mathematische. Klasse. Sitzungsberichte*: 3–14.

———. 1925b. "Zur Quantentheorie des idealen Gases." *Preussische Akademie der Wissenschaften. Physikalisch-mathematische Klasse. Sitzungsberichte*: 18–25.

———. 1930. "Religion and Science." *New York Times Sunday Magazine,* 9 November 1930, 1–4. A German version, "Religion und Wissenschaft," was published in the *Berliner Tageblatt,* 11 November 1930, suppl., p. 1, and in Einstein 1934, 36–45. Page numbers from Einstein 1954, pp. 36–40.

———. 1931. "The World as I See It." *Forum and Century* 84: 193–94. Reprinted in *Living Philosophies.* New York: Simon & Schuster, 1931, 3–7. Page numbers from Einstein 1954, 8–11.

———. 1934. *Mein Weltbild.* Amsterdam: Querido.

———. 1946. "Autobiographisches — Autobiographical Notes." In Schilpp 1949, 1–94. Page numbers and translations from the corrected version, *Autobiographical Notes: A Centennial Edition.* Trans. and ed. Paul Arthur Schilpp. La Sable, Ill.: Open Court, 1979.

———. 1948. "Quanten-Mechanik und Wirklichkeit." *Dialectica* 2: 320–24.

———. 1954. *Ideas and Opinions.* New York: Bonanza Books.

———. 1955. "Erinnerungen-Souvenirs." *Schweizerische Hochschulzeitung* 28 (*Sonderheft*): 145–53. Page numbers from the reprinting as "Autobiographische Skizze," in Seelig 1956, 9–17.

———. 1987. *The Collected Papers of Albert Einstein.* Vol. 1, *The Early Years, 1879–1902.* John Stachel et al., eds. Princeton: Princeton University Press.

———. 1989. *The Collected Papers of Albert Einstein.* Vol. 2, *The Swiss Years:*

Writings, 1900–1909. John Stachel et al., eds. Princeton: Princeton University Press.

Einstein, A., B. Podolsky, and N. Rosen. 1935. "Can Quantum-Mechanical Description of Physical Reality Be Considered Complete?" *Physical Review* 47: 777–80.

Fine, A. 1981. "Einstein's Critique of Quantum Theory: The Roots and Significance of EPR." In *After Einstein: Proceedings of the Einstein Centennial Celebration at Memphis State University, 14–16 March 1979.* Peter Barker and Cecil G. Shugart, eds. Memphis, Tenn.: Memphis State University Press, 147–58. Reprinted in Fine 1986b, 26–39.

———. 1986a. "Schrödinger's Cat and Einstein's: The Genesis of a Paradox." In Fine 1986b, 64–85.

———. 1986b. *The Shaky Game: Einstein, Realism, and the Quantum Theory.* Chicago: University of Chicago Press.

Föppl, A. 1894. *Einführung in die Maxwell'sche Theorie der Elektrizität.* Leipzig: B. G. Teubner.

Frank, P. 1947. *Einstein: His Life and Times.* Trans. George Rosen. Shuichi Kusaka, ed. New York: Alfred A. Knopf.

Gracia, J. E., trans. and ed. 1982. *Suárez on Individuation. Metaphysical Disputation V: Individual Unity and Its Principle.* Milwaukee: Marquette University Press. The text is based on Carolo Berton's edition of Disputation V in Suárez's *Opera omnia,* vol. 25. Paris: Vivès, 1861, 145b–201a.

Gribanov, D. P. 1987. *Albert Einstein's Philosophical Views and the Theory of Relativity.* H. Campbell Creighton, trans. Moscow: Progress Publishers.

Grüning, M. 1989. *Ein Haus für Albert Einstein. Erinnerungen • Briefe • Dokumente.* Berlin: Verlag der Nation.

Haffmans, G., ed. 1977. *Über Arthur Schopenhauer.* Zurich: Diogenes.

Hall, A. R., and M. B. Hall, eds. and trans. 1962. *Unpublished Scientific Papers of Isaac Newton: A Selection from the Portsmouth Collection in the University Library, Cambridge.* Cambridge: Cambridge University Press.

Herneck, F. 1969. "Einstein und die Willensfreiheit." *Physik in der Schule,* no. 4: 145–49. Page numbers from the revised and expanded version, "Einstein und Schopenhauer." In *Einstein und Sein Weltbild. Aufsätze und Vorträge.* Berlin: Buchverlag der Morgen, 1976, 199–210.

Hoffmann, B., and H. Dukas. 1972. *Albert Einstein: Creator and Rebel.* New York: Viking Press.

Holton, G. 1968. "Mach, Einstein, and the Search for Reality." *Daedalus* 97: 636–73. Reprinted in *Thematic Origins of Scientific Thought: Kepler to Einstein.* Cambridge: Harvard University Press, 1973, 219–59.

Howard, D. 1984. "Realism and Conventionalism in Einstein's Philosophy of Science: The Einstein-Schlick Correspondence." *Philosophia Naturalis* 21: 618–29.

———. 1985. "Einstein on Locality and Separability." *Studies in History and Philosophy of Science* 16: 171–201.

———. 1989. "Holism, Separability, and the Metaphysical Implications of the Bell Experiments." In *Philosophical Consequences of Quantum Theory: Reflec-*

tions on Bell's Theorem. James T. Cushing and Ernan McMullin, eds. Notre Dame, Ind.: University of Notre Dame Press, 224–53.

———. 1990. "'Nicht sein kann was nicht sein darf,' or the Prehistory of EPR, 1909–1935: Einstein's Early Worries about the Quantum Mechanics of Composite Systems." In *Sixty-two Years of Uncertainty: Historical, Philosophical, and Physical Inquiries into the Foundations of Quantum Mechanics.* Arthur I. Miller, ed. New York and London: Plenum, 61–111.

———. 1993. "Was Einstein Really a Realist?" *Perspectives on Science: Historical, Philosophical, Social* 1: 204–51. Revised version of "Einstein era realamente un realista?" In *Realismo/Antirealismo: Aspetti del dibattito epistemologico contemporaneo.* Alessandro Pagnini, ed. Florence: La Nuova Italia, 1995, 93–141.

———. 1994. "Einstein, Kant, and the Origins of Logical Empiricism." In *Language, Logic, and the Structure of Scientific Theories: The Carnap-Reichenbach Centennial.* Wesley Salmon and Gereon Wolters, eds. Pittsburgh: University of Pittsburgh Press; Konstanz: Universitätsverlag, 1994, 45–105.

Hübscher, A. 1981. *Schopenhauer-Bibliographie.* Stuttgart-Bad Cannstatt: Frommann-Holzboog.

Jammer, M. 1974. *The Philosophy of Quantum Mechanics: The Interpretation of Quantum Mechanics in Historical Perspective.* New York: John Wiley.

———. 1985. "The EPR Problem in Its Historical Context." In *Symposium on the Foundations of Modern Physics.* P. Lahti and P. Mittlestaedt, eds. Singapore: World Publishing, 129.

Janik, A., and S. Toulmin. 1973. *Wittgenstein's Vienna.* New York: Simon and Schuster.

Kant, I. 1768. "Von dem Ersten Grunde des Unterschiedes der Gegenden im Raume." *Königsberger Frag- und Anzeigungsnachrichten.* In Kant 1900–1942, vol. 2, pp. 375–84. Translations and page numbers from "Concerning the Ultimate Foundations of the Differentiation of Regions in Space." In Kant 1968, pp. 36–43.

———. 1770. *De mundi sensibilis atque intelligibilis forma et principiis.* Königsberg: Königl. Hof- und akadem. Druckerei. In Kant 1900–1942, vol. 2, 385–419. Translations and page numbers from "On the Form and Principles of the Sensible and Intelligible World." In Kant 1968, 45–92.

———. 1781. *Kritik der reinen Vernunft.* Riga: Johann Friedrich Hartknoch. In Kant 1900–1942, vol. 4, 1–252.

———. 1783. *Prolegomena zu einer jeden künftigen Metaphysik, die als Wissenschaft wird auftreten können.* In Kant 1900–1942, vol. 4, 253–384. Page numbers from *Prolegomena to any Future Metaphysics.* Lewis White Beck, trans. Indianapolis: Bobbs-Merrill, 1950.

———. 1786. *Metaphysische Anfangsgründe der Naturwissenschaft.* In Kant 1900–1942, vol. 4, pp. 465–566.

———. 1787. *Kritik der reinen Vernunft,* 2nd ed. Riga: Johann Friedrich Hartknoch. In Kant 1900–1942, vol. 3, pp. 1–552.

———. 1900–1942. *Kants gesammelte Schriften.* 22 vols. Berlin: Preussische Akademie der Wissenschaften.

———. 1968. *Kant: Selected Pre-Critical Writings and Correspondence with Beck.* G. B. Kerferd and D. E. Walford, trans. Manchester: Manchester University Press; New York: Barnes & Noble.

Karpov, M. M. 1951. "On the Philosophical Views of A. Einstein." *Voprosy filosofii 1.*

Köhnke, K. C. 1986. *Entstehung und Aufstieg des Neukantianismus. Die deutsche Universitätsphilosophie zwischen Idealismus und Positivismus.* Frankfurt am Main: Suhrkamp.

Kollros, L. 1955. "Erinnerungen-Souvenirs." *Schweizerische Hochschulzeitung* 28 (*Sonderheft*): 169–173. Page numbers from translation as "Erinnerungen eines Kommilitonen," in Seelig 1956, 17–31.

Laban, F. 1880. *Die Schopenhauer-Literatur. Versuch einer chronologischen Uebersicht derselben.* Reprint, New York: Burt Franklin, 1970.

Leibniz. G. W. 1696. "On the Principle of Indiscernibles." In *Leibniz: Philosophical Writings.* G. H. R. Parkinson, ed. London: J. M. Dent, 133–35.

———. 1981. *New Essays on Human Understanding.* Peter Remnant and Jonathan Bennett, trans. and eds. Cambridge: Cambridge University Press.

Locke, J. 1894. *An Essay Concerning Human Understanding.* 2 vols. Alexander Campbell Fraser, ed. Oxford: Oxford University Press. Reprint, New York: Dover, 1959.

Luft, D. S. 1980. *Robert Musil and the Crisis of European Culture, 1880–1942.* Berkeley: University of California Press.

Mach, E. 1905. *Erkenntnis und Irrtum. Skizzen zur Psychologie der Forschung.* Leipzig: Johann Ambrosius Barth.

Magee, B. 1983. *The Philosophy of Schopenhauer.* Oxford: Clarendon Press.

Maurer, A. 1987. "Translator's Introduction." In Thomas Aquinas 1987, vii–xxxviii.

Monk, R. 1990. *Ludwig Wittgenstein: The Duty of Genius.* New York: Penguin.

Moore, W. 1989. *Schrödinger: Life and Thought.* Cambridge: Cambridge University Press.

Moszkowski, A. 1921. *Einstein. Einblicke in seine Gedankenwelt. Gemeinverständliche Betrachtungen über die Relativitäts-Theorie und ein neues Welt-System entwickelt aus Gesprächen mit Einstein.* Hamburg: Hoffmann und Campe; Berlin: F. Fontane. Translated as *Einstein the Searcher: His Work Explained from Dialogues with Einstein.* Henry L. Brose, trans. New York: E. P. Dutton, 1921. Page numbers from *Conversations with Einstein.* New York: Horizon Press, 1970.

Pauli, W. 1954. "Naturwissenschaftliche und erkenntnistheoretische Aspekte der Ideen vom Unbewußten." *Dialectica 8.* Reprinted in *Aufsätze und Vorträge über Physik und Erkenntnistheorie.* Braunschweig: Friedrich Vieweg & Sohn, 113–28.

———. 1985. *Wissenschaftlicher Briefwechsel mit Bohr, Einstein, Heisenberg u.a.* Vol. 2, *1930–1939.* Karl von Meyenn, ed. Berlin: Springer-Verlag.

Reiser, A. [Rudolf Kayser, pseud.] 1930. *Albert Einstein: A Biographical Portrait.* New York: Albert and Charles Boni.

Rosenberger, F. 1895. *Isaac Newton und seine physikalischen Principien. Ein*

Hauptstück aus der Entwickelungsgeschichte der modernen Physik. Leipzig: Johann Ambrosius Barth.

Safranski, R. 1990. *Schopenhauer and the Wild Years of Philosophy.* Trans. Ewald Osers. Cambridge: Harvard University Press.

Schilpp, P. A., ed. 1949. *Albert Einstein: Philosopher-Scientist.* Evanston, Ill.: Library of Living Philosophers.

Schlick, M. 1918. *Allgemeine Erkenntnislehre.* Berlin: Julius Springer. Translations and page numbers from *General Theory of Knowledge.* Albert E. Blumberg, trans. La Sable, Ill.: Open Court, 1985.

———. 1932. "Gibt es ein materiales Apriori?" In *Wissenschaftlicher Jahresbericht der Philosophischen Gesellschaft an der Universität zu Wien — Ortsgruppe Wien der Kant-Gesellschaft für das Vereinsjahr 1931/32.* Vienna, 55–65. Page numbers from "Is There a Factual *A Priori?*" In *Philosophical Papers,* vol. 2 [1925–1936]. Henk L. Mulder and Barbara F. B. van de Velde-Schlick, eds. Trans. Peter Heath, Wilfrid Sellars, Herbert Feigl, and May Brodbeck. Dordrecht: D. Reidel, 1979, 161–70.

Schopenhauer, A. 1813. *Ueber die vierfache Wurzel des Satzes vom zureichenden Grunde. Eine philosophische Abhandlung.* Rudolstadt: in Commision der Hof-Buch-und-Kunsthandlung. Translations and page numbers from *On the Fourfold Root of the Principle of Sufficient Reason.* Trans. E. F. J. Payne. La Sable, Ill.: Open Court, 1974.

———. 1819. *Die Welt als Wille und Vorstellung.* Leipzig: F. A. Brockhaus.

———. 1836. *Ueber den Willen in der Nature. Eine Erörterung der Bestätigungen, welche die Philosophie des Verfassers, seit ihrem Auftreten, durch die empirischen Wissenschaften erhalten hat.* Frankfurt am Main: Siegmund Schmerber.

———. 1841a. "Ueber die Freiheit des menschlichen Willens." In Schopenhauer 1841b, 361–458.

———. 1841b. *Die beiden Grundprobleme der Ethik.* Frankfurt am Main: Joh. Christ. Hermann'sche Buchhandlung, F. E. Suchsland, 1–100. Page numbers from the reprinting in *Arthur Schopenhauers Werke in fünf Bänden.* Ludger Lütkehaus, ed. Vol. 3, *Kleinere Schriften.* Zurich: Haffmanns Verlag, 1988.

———. 1844. *Die Welt als Wille und Vorstellung.* 2 vols. 2nd ed. Leipzig: F. A. Brockhaus.

———. 1851. *Parerga und Paralipomena. Kleine philosophische Schriften.* 2 vols. Berlin: A. W. Hayn. Page numbers and quotations from the edition supported by the Schopenhauer-Gesellschaft and the Schopenhauer-Archiv, *Parerga und Paralipomena. Kleine philosophische Schriften.* 2 vols. Zurich: Diogenes, 1977, which is based on the 3d ed. of Arthur Hübscher's critical-historical edition of the original text, Wiesbaden: Brockhaus, 1972.

———. 1859. *Die Welt als Wille und Vorstellung,* 3d impr. and enl. ed. Leipzig: F. A. Brockhaus. Translations and page numbers from *The World as Will and Representation.* 2 vols. E. F. J. Payne, trans. Indian Hills, Colo.: Falcon's Wing, 1958; reprint, New York: Dover, 1969. The Payne translation is based on Arthur Hübscher's German edition, Leipzig: F. A. Brockhaus, 1937.

———. 1874. *Parerga und Paralipomena. Kleine philosophische Schriften.* 2 vols.

3rd ed. Julius Frauenstädt, ed. Leipzig: Brockhaus. [4th ed., 1878; 5th ed., 1888; 6th ed., 1888; 7th ed., 1891.]

———. 1886. *Aphorismen zur Lebensweisheit. Separatausgabe aus "Parerga und Paralipomena."* 2 vols. Wilhelm Gwinner, ed. Leipzig: Brockhaus.

———. 1985. *Der handschriftliche Nachlass in fünf Bänden.* Vol. 1, *Frühe Manuskripte (1804–1818).* Arthur Hübscher, ed. Munich: Deutscher Taschenbuch Verlag. Translations and page numbers from the English translation: *Manuscript Remains in Four Volumes.* Vol. 1, *Early Manuscripts (1804–1818).* E. F. J. Payne, trans. Oxford, New York, and Hamburg: Berg, 1988.

Schrödinger, E. 1925. "Seek for the Road." In Schrödinger 1964, 1–58.

———. 1935a. "Die gegenwärtige Situation in der Quantenmechanik." *Die Naturwissenschaften* 23: 807–12, 824–28, 844–49.

———. 1935b. "Discussion of Probability Relations between Separated Systems." *Cambridge Philosophical Society. Proceedings* 31: 555–63.

———. 1936. "Probability Relations between Separated Systems." *Cambridge Philosophical Society. Proceedings* 32: 446–52.

———. 1944. *What Is Life? The Physical Aspect of the Living Cell.* Cambridge: Cambridge University Press.

———. 1950. "What Is an Elementary Particle?" *Endeavor* 7. Page numbers from the reprinting in *Science, Theory, and Man.* New York: Dover, 1957, 193–223.

———. 1958. *Mind and Matter.* Cambridge: Cambridge University Press.

———. 1964. *My View of the World.* Cecily Hastings, trans. Cambridge: Cambridge University Press, 1964; reprint, Woodbridge, Conn.: Ox Bow Press, 1983. English trans. of *Meine Weltansicht.* Hamburg: P. Zsolnay, 1961.

Seelig, C. 1960. *Albert Einstein. Leben und Werk eines Genies unserer Zeit.* Zurich: Europa Verlag.

Seelig, C., ed. 1956. *Helle Zeit—Dunkle Zeit. In Memoriam Albert Einstein.* Zurich: Europa Verlag.

Stadler, A. 1912. *Kant. Akademische Vorlesungen.* Leipzig: Johann Ambrosius Barth.

Sugimoto, K. 1989. *Albert Einstein: A Photographic Biography.* Barbara Harshav, trans. New York: Schocken.

Thomas Aquinas. 1987. *Faith, Reason, and Theology: Questions I–IV of his Commentary on the "De Trinitate" of Boethius.* Trans. Armand Maurer. Toronto: Pontifical Institute of Medieval Studies, 1987.

Violle, J. 1892–1893. *Lehrbuch der Physik.* E. Gumlich et al., eds. Berlin: Julius Springer.

Voss, A. 1901. "Die Prinzipien der rationellen Mechanik." In *Encyklopädie der mathematischen Wissenschaften, mit Einschluss ihrer Anwendungen.* Vol. 4, *Mechanik,* pt. 1. Felix Klein and Conrad Müller, eds. Leipzig: B. G. Teubner, 1901–1908, 3–121.

Wagner, G. F. 1909. *Encyklopädisches Register zu Schopenhauer's Werken, nebst einem Anhange, der den Abdruck der Dissertation von 1813, Druckfehlerverzeichnisse u.a.m. enthält.* Karlsruhe im Breisgau: G. Braunsche Hofbuchdruckerei und Verlag. Reprinted as *Schopenhauer-Register.* Arthur Hübscher, ed. Stuttgart-Bad Cannstatt: Günther Holzboog, 1960.

Weyl, H. 1923. *Mathematische Analyse des Raumproblems. Vorlesungen gehalten in Barcelona und Madrid*. Berlin: Julius Springer.

———. 1927. *Philosophie der Mathematik und Naturwissenschaft*. Munich and Berlin: R. Oldenbourg. Page numbers and translations from Weyl 1949.

———. 1949. *Philosophy of Mathematics and Natural Science*. Olaf Helmer, trans. Princeton: Princeton University Press; reprint, New York: Athenaeum, 1963.

Wiley, T. E. 1978. *Back to Kant: The Revival of Kantianism in German Social and Historical Thought, 1860–1914*. Detroit: Wayne State University Press.

Foundations of Mathematics and Physics

5

From Constructive to Predicative Mathematics

Geoffrey Hellman
Department of Philosophy, University of Minnesota

Constructive mathematics takes many forms, representing a variety of philosophical perspectives, but a common thread has been a dissatisfaction with the platonist commitments of classical analysis and set theory and the desire to provide an epistemically more secure alternative. But by what standard can the adequacy of any such alternative be judged? Apart from a priori philosophical principle, surely we are well advised to take seriously Hermann Weyl's dictum that "it is the function of mathematics to be at the service of the natural sciences" (Weyl 1949, 61). For if a proposed constructivist framework should prove incapable of recovering portions of mathematics genuinely needed in carrying out important theoretical or experimental work of a science such as physics, that framework cannot reasonably be taken seriously as an alternative to classicism. There are obvious practical considerations behind this stance, but there is also a theoretical one that deserves explicit mention: so bound up with material applications has mathematics been throughout its history — and here we mean to include ordinary as well as scientific applications, such as simple counting and measuring, etc. — that a mathematics without such applications is difficult even to conceive. Such a "mathematics" would have the character of a formal game, and the question of truth would not even seem to arise. Whatever philosophical problems ensue upon taking mathematical truth seriously in virtue of its material applicability cannot be resolved by creating a new subject lacking in that range of applicability.

Modern platonism, especially in its Quinean version (1953; cf. Putnam 1971), and to some extent also as Gödel (1947) defended it, goes further, seeing in claims of indispensability for scientific (and finitary mathematical) applications the key to justifying abstract mathematical axioms. By analogy with physical laws, such axioms enter into theoretical reasoning leading to experimentally (or computationally) confirmed conclusions and thus can receive the same sort of indirect, broadly empirical confirmation as those laws. If we are able to show that indeed such powerful axioms are genuinely required for the deductions involved, the confirmation is correspondingly strengthened (in much the way that a theoretical physical postulate gains by elimination of rival hypotheses). To be sure, the force of this analogy and the scope of these methods require much further investigation. Meanwhile, however, as suggested by Weyl's dictum and my reflections above, one may fruitfully pursue questions of indispensability for their limitative or negative as well as positive implications. If we are able to show that some piece of classical mathematics or logic is genuinely indispensable for an (important?) application, we thereby rule out as *unviable* any constructivist framework which denies itself access to that mathematics. (Such an enterprise might be called "deconstructive mathematics.")

In what follows, I shall sketch some recent steps in this direction pertaining especially to intuitionism and Bishop constructivism. It will be argued that *there is a strong case for the need to transcend any variety of constructivism which restricts itself to intuitionistic logic* as that has been developed and explained by Brouwer, Heyting, Kreisel, Troelstra, et al. In fact, as will emerge, despite the radical philosophical standpoint associated with intuitionism (especially as Dummett 1977 expounds it), virtually all varieties of *constructivist mathematical practice* employ "liberal" methods incompatible with such philosophical pronouncements. This leads naturally to a next stage in constructive (or antiplatonist) mathematics, predicativist analysis of Weyl (1918) and Feferman (1988). Finally, we shall examine how this standpoint may be combined with what we have called a "modal-structuralist" alternative to traditional objects-platonism (Hellman 1989a). Surprisingly, perhaps, a partial adaptation is possible that enables a recovery of Dedekind's categorical treatment of the natural number system. This is welcome, as the modal-structural view provides a rather attractive justification of predicativist set theory. Whether such a system is really adequate for the sciences we do not attempt to address here, but this then becomes a natural focal point for further investigation.

"Never" Again!

Let us begin by recalling the standard explanations (due to Brouwer, Heyting, and Kreisel [BHK], cf. Troelstra 1977) of the intuitionistic logical connectives. These tell us what the "proof conditions" of a complex statement are in terms of those of its components, as contrasted with classical "truth conditions." To know such proof conditions — what it is for an "idealized human mathematician" to have a method or construction for proving the statement in question — is to know the content or meaning of the statement. This, of course, reflects the intuitionist's replacement of the classical conception of a mind-independent mathematical reality determining the truth or falsity of mathematical propositions (independently of our capacity to prove or refute them) with a constructivist conception according to which mathematics ultimately concerns mental capacities to produce suitable "constructions." Given that atomic sentences are already so understood (e.g., as asserting something demonstrable about constructive objects such as natural numbers), the inductive clauses of the BHK explanations give us the assertability conditions, hence the meaning, of arbitrary sentences built up from the atomic ones using the intuitionist logical connectives. Using a subscript i to indicate intuitionistic connectives, the clauses are as follows:

A proof of $p \mathbin{\&}_i q$ is a pair (c_1, c_2) of constructions such that c_1 proves p and c_2 proves q.

A proof of $p \vee_i q$ is a pair (c_1, c_2) such that c_1 proves p or c_2 proves q.

A proof of $p \rightarrow_i q$ is a constructive operation c on constructions such that it can be recognized of c that, when applied to any proof c' of p, it yields a proof $c(c')$ of q.

A proof of $\exists_i x A$ is a construction c such that c proves $A(o)$, where o denotes some constructive object o from the domain of quantification.

A proof of $\forall_i x A$ is a constructive operation c on constructions of which it can be recognized that, when applied to any construction c' of an object o of the domain, it yields a proof $c(c')$ of $A(o)$.

Negation, $\neg_i p$, is explained as $p \rightarrow_i 0 = 1$, or some other absurdity.

Now, there are a number of interesting interrelated problems concerning these explanations, such as the problem of circularity, the problem of impredicativity in the clauses for the conditional and the universal quantifier, the problem of whether to countenance a universe of constructions, and the problem of decidability of the proof relation. (Cf., e.g., Weinstein 1983, Beeson 1980, 1985, Hellman 1989b). Here we wish to emphasize a further problem generated by these explanations together with minimal assumptions about the proof relation. That is a problem of *drastic expressive incompleteness* of this intuitionistic logical apparatus, with ramifications in many areas of applied mathematics, and even within the domain of pure intuitionistic mathematics.

The problem may be brought out by observing that, if we attempt to express within intuitionist language that an arbitrary proposition A is (absolutely) undecidable, we actually contradict ourselves. More precisely, let "A is undecidable" abbreviate the following intuitionistic formula:

$$\forall_i c(\neg_i c \, Pr \, A \, \&_i \, \neg_i c \, Pr \, \neg_i A), \qquad (UA) \qquad (5.1)$$

where 'Pr' is the intuitionistic proof relation and 'c' is a variable ranging over intuitionistic constructions. The formula A is arbitrary. Then we have the following

Metatheorem: "A is undecidable," UA, is intuitionistically inconsistent.

This rests on the following iteration principle governing 'Pr', which we take to be axiomatic:

Iteration Principle: $\exists_i a$ constructive operation $'$ such that if $d \, Pr \, A$, then $d' \, Pr \, 'd \, Pr \, A'$.

According to this, intuitionistic provability satisfies the first Löb condition on a good proof predicate, $Prov(A) \to_i Prov('Prov(A)')$. Its justification rests on the self-certifying character of intuitionistic proofs: by inspecting a proof c of A, we become convinced that indeed c is a proof of A; this "procedure" yields a proof that c proves A. (Of course, we need not invoke it explicitly — there is no epistemic regress!)

Proof of Metatheorem: From UA we have,

$$\forall_i c(\neg_i c \, Pr \, A). \qquad (5.2)$$

From the BHK interpretations of the connectives $\forall_i$ and $\neg_i$, this yields,

$$\exists_i m \forall_i b \forall_i c(b \, Pr \, 'c \, Pr \, A' \to_i m(b) \, Pr \, '0 = 1'), \qquad (*) \qquad (5.3)$$

i.e., we have a method m which transforms any proof b of '$c\ Pr\ A$' into a proof of absurdity. (The new quantifiers in (*) are presumably to be read intuitionistically; they represent the metalinguistic quantifiers occurring on the right sides of the BHK explanations.) We claim that this m can be used to construct an m' such that $m'\ Pr\ \neg_i A$. Instantiating on this m' in UA will then yield a contradiction.

To prove the claim, let $d\ Pr\ A$. By the iteration principle, we have

$$d'\ Pr\ \text{'}d\ Pr\ A\text{'};\ \text{by}\ (*),\ m(d')\ Pr\ \text{'}0\ =\ 1\text{'}. \qquad (5.4)$$

Now set $m' = mo'$; then

$$m'(d)\ Pr\ \text{'}0\ =\ 1\text{'},\ \text{any}\ d\ \text{such that}\ d\ Pr\ A,\ \text{i.e.,}\ m'\ Pr\ \neg_i A. \quad (5.5)$$

This proves the claim and the metatheorem.

This means that in intuitionistic language we can never even contemplate the *possibility* that a proposition A is (absolutely) undecidable: the attempt leads to "possibly UA," which asserts the possibility of a contradiction! Matters are actually worse, for, reflecting on the meaning of the combination '$\forall_i c\ \neg_i$', we see that it cannot be used even to entertain the possibility that a proposition A *can never be proved*. Due to the strong existential import of $\forall_i$, a formula of the form,

$$\forall_i c\ \neg_i(c\ Pr\ A), \qquad (5.6)$$

says that a method is available for refuting (any claim to have a proof of) A. But that is far stronger than the possibility envisioned, that one will forever simply lack a proof of A. (Note that we may assume the language already enriched with a suitable modal operator for possibility. The problem persists due to the interpretations of '$\forall_i$' and '$\neg_i$'. Moreover, the problem of unintended constructive existential commitment of '$\forall_i$' persists even if the language is expanded to include classical negation. For a fuller discussion of this and related points, see Hellman 1989b.)

At this point it is natural to ask what difference such expressive limitations make for intuitionistic mathematics. Every language, after all, is expressively incomplete in some respects. So what if such uses of *never* transcend intuitionistic language?

As already suggested, there are serious ramifications, both in connection with (*i*) the internal workings of intuitionistic logic and mathematics and (*ii*) myriad ordinary and scientific applications of mathematics.

Under (*i*), consider, for example, intuitionism's stance regarding "the law of the excluded middle" ('LEM$_i$', where the subscript reminds us that it is the version, $p \vee_i \neg_i p$, formulated with intuitionistic connectives that is at issue, not the classical law). Indeed, intuitionism wishes to remain agnostic with respect to this law. It cannot claim to refute it, on pain of contradiction, and it certainly does not claim to demonstrate it, as that would involve possession of a method for deciding all mathematical questions it can formulate. There is surely no problem with intuitionism's refusal to include this law in its logic *today*. However, suppose we wish to inquire whether intuitionistic logic is *stable* in this respect, whether it will, at any stage in the future, remain weaker than classical logic in this way. Surely, we would like to say that, at any stage of mathematical inquiry, there will be exceptions to LEM$_i$, i.e., there will be p such that p is neither provable nor refutable *at that stage*. (Here we may assume that 'p is provable at stage s' is a decidable relation, that there is no problem in saying 'p is undecided at s'.) In other words, we should like to say that we may *never* be in a position to decide all mathematical questions. But if we attempt to express this (finally dropping our subscript i as understood throughout) as,

$$\text{(possibly) } \forall s \exists p (\neg \vdash_s p \vee \neg p), \tag{5.7}$$

where '$\vdash_s$' means "is assertible at stage s," we inadvertantly assert the possibility of a method for *producing* undecided questions p as a function of stages, which is not at all intended. And if we write the weaker claim,

$$\text{(possibly) } \forall_s \neg \forall p (\vdash_s p \vee \neg p), \tag{5.8}$$

we still inadvertantly assert the possibility of having a general method (as a function of s) of refuting any method of deciding any p at s, which is also more than we intend to assert. In order to assert simply that possibly at any stage we will lack a method for deciding all p, we seem to require nonconstructive universal quantification over an unbounded, noncanonically generable totality (of stages, states of information, or constructions).

Does this mean that intuitionism cannot even "express its rejection of LEM"? This is ambiguous. It can certainly remain agnostic, silent and uncommitted. At the opposite extreme, it can, in a logic with quantification over propositions, easily express the *refutability of quantified* LEM ('QLEM'), $\forall p(p \vee \neg p)$, by simply writing

$$\neg \forall p(p \vee \neg p). \qquad (\neg QLEM) \qquad (5.9)$$

However, neither the intuitionist nor anyone else is in a position to assert either QLEM or $\neg$QLEM: we certainly lack any method of deciding all p, but also we lack any method of refuting the having of any such method, although we (along with intuitionists) certainly regard it as extremely unlikely that any such decision method could ever be found. On further reflection, we also doubt that we will *ever* have a method of refuting such a decision method. So our actual epistemic situation is best expressed, not by $\neg$QLEM, which we are in no position to assert, but rather by saying that we believe QLEM to be *undecidable*. But, as shown above, in order to say *this*, we must employ nonintuitionistic logical connectives. In short, *the stability of intuitionistic logic seems eminently reasonable, but it cannot be formulated in intuitionistic language.*

Before taking leave of the LEM, let us pause to infer a curious corollary of the above metatheorem. Consider a more explicit version of QLEM, namely,

$$\forall p \exists c(c \; Pr \; p \vee c \; Pr \; \neg p), \qquad (5.10)$$

where c ranges over constructions. This, as just discussed, is reasonably regarded as undecidable. But what of its weakening,

$$\forall p \neg \neg \exists c(c \; Pr \; p \vee c \; Pr \; \neg p)? \qquad (WQLEM) \qquad (5.11)$$

This says that a general method (as a function of p) is available for refuting any refutation of the decidability of p. Well, suppose for arbitrary p a refutation of the decidability of p were available, i.e., that we could prove $\neg \exists c(c \; Pr \; p \vee c \; Pr \; \neg p)$. This implies $\forall c(\neg(c \; Pr \; p) \; \& \; \neg(c \; Pr \; \neg p))$. But, as the reasoning of the metatheorem shows, this leads to a contradiction. Since that reasoning is intuitionistically acceptable, it actually then provides a refutation of the assumption that we could prove $\neg\exists c(c \, Pr \, p \, \text{v} \, c \, Pr \, \neg p)$, and it is perfectly general in p. Thus, we have as a

Corollary: The weakened quantified law of the excluded middle, WQLEM, is an intuitionistic theorem.

On reflection, this should come as no surprise, as it is just a fancier version of $\neg\neg(p \, \text{v} \, \neg p)$, which is also a theorem. Such is intuitionistic logic.

Closely related to these considerations is a second use of *never* which both intuitionistic and Bishop-constructivist mathematics would seem to require (and which its practitioners often employ) but which it strictly cannot express. That is bound up with the method of "weak counterexamples," in which one demonstrates the constructive un-provability of a given statement by showing—e.g., by a clever con-struction of a real number—that its provability would yield a method of solving a problem of a general form—e.g., an existence problem in number theory such as whether Goldbach's conjecture has a coun-terexample, or whether its least such, if it exists, has some decidable property—*which is assumed to be unsolvable*. (For a clear exposition with examples, see, e.g., Troelstra 1977. For its importance in con-structive mathematics, see, e.g., Bridges 1979, 3–5.) Without this prior assumption of unsolvability of the reduced problem (e.g., the halting problem, or (its "lesser" variant) whether an arbitrary real $x \leq 0$ or ≥ 0), the method would lose its force, which is commonly understood as a method for demonstrating that certain statements are "essentially non-constructive" (Bridges 1979), that "they will never be intuitionistically provable" (Dummett 1977). But, as already argued, this prior assump-tion—that a reduced problem is unsolvable—cannot even be articu-lated within intuitionistic language. As Dummett puts it:

[The method] gains its force from the fact that we have a uniform way of constructing similar 'counterexamples' for each unsolved problem of the same form. Since we can be virtually certain that the supply of such unsolved prob-lems will never dry up, we can conclude with equal certainty that the . . . statement will never be intuitionistically provable. Such a recognition that a . . . statement is unprovable does not amount to a proof of its negation. (1977, 45)

Indeed it does not! But, intuitionistically formalized, that is precisely what it would mean. Moreover, the very same problem of inadvertant commitment to a method of generating unsolved problems (as a function of times or states of information, say) that we encountered in connection with the LEM confronts us here.

Once again, a minimalist, agnostic stance is of course not ruled out. One can employ the method of weak counterexamples to conclude, "I should not adopt this statement as an axiom today." But it is difficult even to infer, "I should not attempt to prove this statement," for one cannot even assert the possibility, let alone the probability, of coming up empty-handed no matter how long one were to search. Again, the *stability* of intuitionistic or Bishop-constructivist mathematics would seem to be constructively inexpressible.[1]

When we turn to the second category (*ii*), applications of mathematics to the material world, there are abundant examples of infinitistic quantification that resist constructive formalization, even if appropriate adjustments to the notion of "method of proof" are made. Even if we understand "constructive method" to mean "constructive method for evaluating as empirically confirmed (or probable, or what have you)," we cannot express our epistemic predicament in a great variety of cases, *that we may never know whether p or not*, where *p* is, say, an empirical existence statement, such as, "there are extraterrestrial mathematicians," or "there are violations of symmetry *S* (especially beyond an event horizon on some cosmological hypothesis)," or "there are macroscopic quantum interference effects of type T," etc. For such cases, constructive universal quantifiers are quite inappropriate, as we are attempting to express that we may remain forever ignorant in the relevant sense. The general point here is that, whatever method of evaluation of statements may be built into such a (quasi-constructive) quantifier, we can find many examples of reasonable claims or conjectures of perpetual ignorance in the sense of perpetual lack of just *that sort of method* (for affirming or denying the statement).

These considerations confront radical constructivism with a dilemma: *either* it remains within the confines of intuitionistic logic, renouncing all these applications of logical and mathematical concepts as "deficient in cognitive significance" (as is characteristic of *radical* constructivism), in which case, we submit, it binds itself to an untenable position; *or* it expands its horizons to incorporate the logical concepts requisite to expressing such things as the stability of its own logic and the nonconstructive uses of *never* we have sketched, in which case it allows for the intelligibility of nonconstructive quantification over infinite, not effectively generable domains; then it has embraced an essential component of the classicism from which it sought to distance itself. It must then renounce its radical critique and acknowledge itself as at best a part of the wider world of classical mathematics.

From Radical to Liberal Constructivism

As just indicated, a major philosophical divide within constructivism is marked by a verificationist doctrine concerning the meaningfulness of mathematical concepts and statements. That doctrine may be stated thus:

A mathematical concept is not meaningfully applicable apart from an idealized mathematician's having a constructive method that shows that it applies. Proof-independent mathematical facts are not countenanced. (PCCM)

PCCM stands for "proof-conditional criterion of meaningfulness."[2] Much as early versions of logical positivism sought to distinguish meaningful scientific statements from meaningless metaphysical ones by requiring (conclusive) verifiability or falsifiability, *radical* constructivism seeks to demarcate the domain of meaningful constructive mathematical practice from the rest, which is rejected on grounds of deficiency in cognitive significance due to proof-independent (hence "metaphysical"?) truth commitments. One may view, for instance, the BHK explanations of the intuitionistic logical connectives as an effort to demarcate these limits for the logical apparatus. Classical infinitistic quantification transcends these limits and so, according to the PCCM doctrine, cannot be used to form (fully?) meaningful statements.

Moreover, it should be emphasized that the radical constructivist, in rejecting the legitimacy of classical practice on the basis of a principle such as the PCCM, is adopting a "philosophy first" stance with regard to mathematical practice.[3] If practice fails to conform to philosophy principle, it is the practice rather than the principle that the radical says must yield. Perhaps few practicing constructivist mathematicians (as opposed to philosophers) would subscribe to such a stance, at least in relation to *constructivist practice*. Certainly this writer does not.

Thus, it must not be thought that, merely in confining oneself to intuitionistic logic in one's reasoning, one is thereby embracing the radical position. There are other reasons for the confinement: one may believe that constructive proofs are more informative; that they yield valuable insights otherwise unobtainable; that they confer a higher or better grade of mathematical certainty; and that a constructive logical apparatus can be used to mark many distinctions of independent interest (such as that between $\neg \forall x \varphi$ and $\forall x \neg \varphi$, brought out in differences over Markov's principle, etc.) normally obscured in a classical treatment. (Cf., e.g., Richman 1987.) These are characteristic of what may be called *liberal constructivism*, which pursues constructive mathematics in the manner of Bishop et al., but without embracing anything like the PCCM, or even a "philosophy first" stance.[4] Unlike the radical, the liberal constructivist may (cheerfully or despondently) admit the need, argued for above, to transcend intuitionistic logic for a variety of purposes, without thereby at all impugning the value of constructive mathematics. Both for the classical mathematician and for the philosopher of mathematics, it is of great interest to know the actual limits of the various constructivist approaches to mathematics.

Above I have concentrated on expressive limitations of a very elementary character, but, as I have suggested, with far-reaching ramifications. Still, they pertain to matters (such as the stability of logic, and various epistemic limitations) somewhat removed from the core of mathematics proper. What is the situation closer to home? Can the mathematics used and needed in the empirical sciences proper be constructivized along Bishop's lines? Since Bishop's landmark work (1967, Bishop and Cheng 1972, cf. Bridges 1979), of course, we know that a great deal of such mathematics — including large portions of functional analysis, the theory of metric spaces, and even measure theory — can indeed be constructivized. But that does not settle the matter, for "a great deal" is not all, and there may be important exceptions with far-reaching philosophical implications. It is well known, for example, that extremal value problems and classical uses of the calculus of variations present difficulties for constructivization (see, e.g., Beeson 1985, 19–22). Such problems arise in classical mechanics. Here we will review in brief some more recent findings of our own, bearing on the constructivization of the mathematics of quantum mechanics, an area known to present challenges.[5] For details and a fuller treatment, we refer the reader to Hellman 1993a, 1993b.

There are two important pieces of functional analysis for quantum mechanics that present special problems for constructive mathematics. The first concerns the problem of characterizing the *probability measures* that can, in principle, be introduced into the theory; the second concerns the theory of *unbounded closed linear operators* in Hilbert space, including the *spectral theorem, Stone's theorem*, etc. Such operators play a prominent role in representing a great many of the physical observables of quantum mechanics (e.g., position, linear momentum, total energy, etc.).

Concerning the first problem, the central classical result is Gleason's theorem (1957), which solved a problem, posed by Mackey, to characterize the measures on the (closed) subspaces of a (separable) Hilbert space $\mathcal{H}$ (forming what geometers call "projective space," also known as a quantum lattice $\mathcal{L}(\mathcal{H})$), which, since von Neumann's seminal work (1955, 1932), has been generally taken to be in one-one correspondence with the system of experimental propositions about a quantum system, hence, the appropriate domain of definition of quantum probability measures. (Cf. Mackey 1963.) Gleason's theorem states that, for $\mathcal{H}$ of dimension ≥ 3, every such measure μ (additive with respect to sequences of pairwise orthogonal subspaces) is given by a (unique) positive self-adjoint operator W of trace class via the formula, $\mu(A) = Tr(WP_A)$, where P_A is the projection operator of $\mathcal{H}$ onto the subspace A. (Here 'subspace' means topologically closed linear submanifold of $\mathcal{H}$.) In the case of probability measures, the trace $Tr(W) = 1$ and W is called a "density matrix." Then Gleason's theorem implies that every mathematically possible probability measure on $\mathcal{L}(\mathcal{H})$ (dim($\mathcal{H}$) ≥ 3) is given by the quantum mechanical pure and mixed states. (Cf. Jordan 1969, chap. 5.) (In the case of pure states, W is a projector P_φ onto a one-dimensional subspace ("ray") of $\mathcal{H}$ (spanned by unit vector φ), and the formula $Tr(WP_A)$ reduces to the familiar $(\varphi, P_A\varphi)$. In the case of mixed states, W can be taken as a convex linear combination of such ray projectors.) It follows as a corollary to Gleason's theorem that no measure on $\mathcal{L}(\mathcal{H})$ can be dispersion-free ($\{0,1\}$-valued), thereby ruling out a major category of hidden-variables theories for quantum mechanics.

Known proofs of Gleason's theorem are non-constructive, making heavy reliance on the nonconstructive principle of sequential compactness (Every bounded infinite sequence includes a convergent subsequence), which goes back to the nonconstructive Bolzano-Weierstrass theorem (Every bounded infinite set of reals contains an accumulation point). (Cf. Gleason 1957, also the "elementary" but highly nonconstructive proof of Cook, Keane, and Moran 1985.) As it turns out, this is unavoidable: it can be proved, by the method of (weak) counterexamples, that even the weakest relevant special case of Gleason's theorem is not constructively provable (Hellman 1993a). That case concerns the three-dimensional Hilbert space $\mathbb{R}^3$ and the class of so-called bounded frame functions on the unit sphere S of $\mathbb{R}^3$, functions f such that for any orthogonal triple (*frame*) of points, (p,q,r), in S, $f(p) + f(q) + f(r)$ has the same fixed value (called the *weight* of f, w_f). Gleason's theorem in this case implies that for each such f, there exists a frame, (p,q,r), in terms of which $f(s)$ for any $s \in S$ can be computed, via,

$$f(s) = Mx^2 + \gamma y^2 + mz^2,$$

where (x,y,z) are the coordinates of s in the frame (p,q,r), M is the maximum of f, m is its minimum, and $\gamma = w_f - M - m$. A constructive proof of this would thus provide a method of computing some point at which f assumes its maximum, and this can readily be shown to reduce the (lesser) halting problem. Even if the restriction to *normalized* (weight = 1) non-negative frame functions is imposed, the reduction goes through (Hellman 1993a).

It is worth adding that, if we take Church's thesis (CT) in any form which implies,

> "Constructive reals x are recursive" & "Effective decidability of $x \leq 0 \lor x \geq 0 \to \exists e$ (e an index of a recursive function in numerical parameters & $\{e\}(x) = 1$ iff $x \leq 0$ & $\{e\}(x) = 0$ iff $x \geq 0$)",   (CT$\to$)   (5.12)

then the above mentioned proof by reduction yields the following

Corollary: The constructive provability of Gleason's theorem (in any of the restricted forms of Hellman 1993a) is inconsistent with Church's thesis.[6]

This follows because the proof by reduction implies that constructive provability of Gleason's theorem provides a decision method for $x \leq 0 \lor x \geq 0$, for constructive reals x; by (CT$\to$), there would then be a recursive function (in numerical parameters) to decide this; but this contradicts outright the recursive counterpart of the (lesser halting problem) reduction which shows categorically, by invoking a pair of recursively inseparable r.e. sets, that there can be no such recursive function (cf. Troelstra 1977, 997–98).

Thus, any framework that accepts (CT→) cannot prove Gleason's theorem on pain of contradiction. Internally, it will conclude that Gleason's theorem is absurd. (Such is constructive mathematics.)

Turning to the second topic, the theory of unbounded linear operators in Hilbert space, we encounter once again the problem of *expressive limitations* of constructivist mathematics. For an analysis of a key theorem of Pour-El and Richards (1983) shows that a wide class of unbounded linear operators T (between Banach spaces X and Y) fail to preserve constructivity of inputs and hence are nonconstructive as functions. (See Hellman 1993b and Feferman 1984 for independent routes to essentially the same conclusion.) The reasoning is classical, but it pertains to any framework of constructivity accepting minimal closure axioms on a class of "constructive sequences of Banach space elements" and recognizing the constructive analogue of functions generating recursively enumerable but nonrecursive sets of natural numbers. The only conditions on unbounded operators T, besides linearity, are that the graph of T be topologically *closed* (the next best thing after continuity [= boundedness in fact]),[7] and that Te_n be a constructive sequence for some computable sequence e_n whose span is (in fact, not necessarily constructively) dense in X. Most operators of physical interest (e.g., in quantum mechanics) satisfy these conditions. (In particular, every self-adjoint operator is closed. See Jordan 1969, Theorem 11.3, 32–33.) They and their restrictions to constructive Hilbert space elements lie outside the constructivist universe of discourse. As I spell out in greater detail elsewhere (Hellman 1993b), central theoretical portions of quantum mechanics — including the canonical commutation relations (between position and momentum operators); the derivation of the dynamical law (abstract Schrödinger equation) from symmetry consideration; the correspondence principle and its applications expressing deep connections between quantum and classical physics; and so forth — are not constructively recoverable, at least not in a form that applies to the standard (classically closed) QM operators or their restrictions to constructive arguments. Although particular computations can, of course, be performed employing unbounded operators (e.g., a differential operator such as $-id\psi(x)/dx$), the *theory* of such operators (quantifying over them as objects) on which results such as these depend lies beyond the constructive frontier.[8]

Above I claimed that, in fact, constructivist mathematical practice is properly viewed as liberal rather than radical. The present topic provides an example. According to the radical PCCM, a "slippery slope" argument leads naturally to the requirement that the domains of definition of constructive operators (functions) should be decidable. Just as the (meaningful mathematical) question, "Is the value of $T(f) = g$?" must be constructively answerable, so must the (meaningful mathematical) question, which arises all the time in the theory of (unbounded) operators T, "Is T defined at f?" Otherwise a realm of proof-independent fact—whether T is or is not defined at f—is implicitly being recognized. If a rule is required for definitely answering the first kind of question, why not also the second kind? Presumably, such a rule must yield a recognizable answer in a finite number of steps. Thus, the domain of definition of T must be decidable. However, it is clear from consulting, for example, Bishop 1967 or Bridges 1979 that constructive analysis does not conform to this requirement, nor does it claim to. On the contrary, it recognizes sets so long as they are specifiable by a (presumably predicative) formula, or at least this is one standard to which one is reasonably led by Bishop's rather terse remarks on the matter. (Cf. Bridges 1979, 5, Beeson 1980, Hellman 1992.) And, given any set X of constructive objects, one may trivially introduce a "constructive function" f on X by specifying that $f(x) = x$ if $x \in X$, undefined otherwise.

Suppose the application of the PCCM to the question of definedness were accepted. Then the above conclusion from the Pour-El and Richards theorem concerning unbounded operators could be reached quite a bit more simply. By the (classical) *closed graph theorem*, a closed linear operator T defined on all of a Hilbert space $\mathcal{H}$ is necessarily bounded; so unbounded such operators cannot be defined everywhere. Indeed, most of the unbounded operators of physical interest are defined on a dense proper subdomain of $\mathcal{H}$. It is not difficult to show that such a domain cannot form a decidable totality (on pain of solving the halting problem, see Hellman 1993b, sec. 3). This already shows that the theory of these operators lies beyond the bounds of *radical constructivism* as we have defined it. But to arrive at the parallel conclusion vis-à-vis *liberal constructivism*, the more elaborate argument based on the Pour-El and Richards theorem seems required.

In any case, neither variety of constructivism appears to be capable of recovering this major piece of functional analysis for quantum mechanics in a manner which preserves the applicability of the QM operators in constructive domains. This reinforces the conclusion of the previous section that a viable constructivism will have to expand beyond the confines of intuitionistic logic. How far it must expand remains a major open question in this area.

Predicativist Analysis and Modal-Structuralism

If intuitionistic logic cannot be viewed as marking a boundary of intelligibility or rational belief, it becomes instead a valuable research tool: in addition to whatever independent insights it may yield, it is at least a useful bookkeeping device for demarcating those portions of classical mathematics susceptible to constructive treatment of this particular kind. However, other more liberal notions of constructivity — employing classical logic — have been developed and are worth pursuing. In particular, predicativist analysis forms a next natural "stopping point" after Bishop constructivism, in the route back to full classical analysis and set theory. As already noted, predicative comprehension principles already provide the sets invoked in Bishop constructivism; the main change, of course, comes in allowing the law of the excluded middle and proofs of existence by reductio, although it is indeed possible (as illustrated by Feferman 1984, cited in connection with Pour-El and Richards 1983) to distinguish constructively acceptable existence proofs from the rest within a classical framework. Indeed, once we move to classical predicativist analysis — especially, say, Feferman's system W (1988) designed to fulfill Weyl's program — the expressive limitations pointed to in the previous sections of this paper no longer apply, at least not to the same extent. ("Never" of course is no longer a problem, but whether Gleason's theorem and the general analysis of unbounded operators are recoverable in say Feferman's W is, I believe, an open question.)

Constructivism has, of course, not been the only reaction to the "theological" aspects of platonism. Apart from formalism and instrumentalism (e.g., Field 1980, cf. Burgess 1984), one may pursue modal-structural alternatives (as suggested in Putnam 1967 and developed in Hellman 1989a). These, in effect, seek to purge platonism of its more extravagant ontological commitments and to bypass the problems of reference to abstract particulars that they engender. Instead, just enough abstract machinery is employed to characterize the relevant types of mathematical structures of interest (ω-sequences for number theory, separable ordered continua for real analysis, etc.), and then all mathematical statements S proper to the (classical) theories of those structures are systematically translated as hypotheticals of the form,

"Were $\mathcal{M}$ any structure (of the relevant type), S would hold in $\mathcal{M}$."

For all of standard classical mathematics, a system of second-order modal logic suffices for these purposes. (For details, see Hellman 1989a.) That system was frankly impredicative, employing unrestricted *extensional second-order comprehension* for classes and relations (of first-order objects), although (*i*) containing only actualist universal quantifiers, and (*ii*) limited to modal-free formulas, that is,

$$\exists R \forall x_1 \ldots \forall x_n [R(x_1 \ldots x_n) \equiv A], \qquad \text{(CS)} \qquad (5.13)$$

where A is modal free and lacks free R. (The necessitation of this is then available.) Since A may contain quantification over classes, it is relatively straightforward to recover Dedekind's analysis of the natural number structure (what he called "simply infinite systems" and we call ω-sequences), and, moreover, to prove that that analysis provides a categorical characterization. (An ω-sequence is defined explicitly as a pair $[X,f]$ satisfying the usual Peano-Dedekind axioms, including full second-order induction, f behaving as the successor function, and X the domain to which all quantifiers in the axioms are relativized. One then proves in second-order logic that any two such structures are isomorphic. The comprehension scheme CS plays its usual role in guaranteeing the existence of an isomorphism as the intersection of all maps defined on one of the two structures with values in the other, associating their zeroes and preserving succession.) All this is central to the modal-structural interpretation, serving to justify it as accurate and adequate.

It would thus appear that impredicativity is essential to the modal-structural approach. Although, of course, one could weaken the system (restricting CS), one would expect to pay a heavy price, renouncing the possibility of justifying the translation scheme, of showing that Dedekind's definition indeed captures the intended type of structures. This would be unfortunate, for (as was spelled out in Hellman 1989, 61ff.) modality can be used in a natural way to introduce talk of abstract sets (as possible results of selection procedures). Predicative comprehension principles themselves can readily be justified in this manner by recognizing the semantic relation of *application* (of predicates to objects) as a selection procedure. If these were the only sets we needed, one could sustain nominalism, at least in a modal form: all talk of sets would be shown derivative in this way. It is difficult to see, however, how impredicative comprehension principles could be so justified, for the predicates involved presuppose a domain of sets or classes if they are to apply even to individuals. Ultimately, of course, this may be unavoidable, but it is probably too soon to conclude that.

In any case, as it happens, there *is*, surprisingly, a way in which the modal-structural interpretation for number theory (at least) can be carried out without employing the full impredicative comprehension scheme. The method adapts a kind of trick (attributed by Wang 1964 to Michael Dummett, discussed in Isaacson 1987), enabling Dedekind's analysis to go through in a manner that is predicatively justifiable. Apart from the motivation already given, it is interesting in its own right, as it shows that the predicativist need not simply take the natural numbers for granted (as is usually done), but can recover much of Dedekind's achievement (1888) without violating predicativist principles.

The trick is to modify the second-order induction axiom slightly:

$$\forall F[F(0) \;\&\; \forall y\{F(y) \;\&\; y \neq x \to F(s(y))\} \to F(x)]. \quad \text{(DT)} \quad (5.14)$$

"DT" stands for "Dummett's Trick." This says that x belongs to every class of 0 which contains the successor of anything it contains except possibly x itself. Let us call a class of 0 closed under successor, s, an "s-chain," or "chain" for short, and a class of 0 closed under s, except possibly for a unique element (unique element x), a "quasi-chain" ("quasi-chain omitting x"). (NB x is "omitted" as predecessor, not as successor.) That is,

> Z is a *quasi-chain* iff $Z(0) \;\&\; \exists!x\forall y(Z(y) \;\&\; y \neq x \to Z(s(y)))$, $\qquad$ (5.15)

and

> Z is a *quasi-chain (possibly) omitting* x iff $Z(0) \;\&\; \forall y(Z(y) \;\&\; y \neq x \to Z(s(y)))$. $\qquad$ (5.16)

Clearly DT defines the intersection of all quasi-chains omitting x, which is just the finite initial segment of 0 up to and including x. Suppose we now adopt as an axiom (or manage to derive),

$$\forall x \exists F[F(0) \;\&\; \forall y\{F(y) \;\&\; y \neq x \to F(s(y))\}] \qquad \text{(QC)} \qquad (5.17)$$
$$\text{(to be called (FQC))}.$$

"QC" stands for "quasi-chains." Observe that if we interpret the second-order quantifiers in QC and DT as ranging over only all *finite* subsets of the domain (i.e., as in *weak second-order logic*), DT still succeeds in defining "x is a natural number" in the right way, i.e., the class of all such x ordered by s forms an ω-sequence. Given this restriction, QC guarantees that every item belongs to a finite initial segment, and this suffices to rule out non-standard integers. Moreover, any two models of the usual successor axioms together with DT and QC, so interpreted, must be isomorphic. So far this is just a sketch of an alternate route to the well-known result that a categorical axiom system for the natural number system can be given in weak second-order logic. In fact, Monk (1976, 488–89) obtains this by extending Robinson arithmetic Q with the single axiom,

$$\forall x \exists F[F(x) \;\&\; \forall y(F(s(y)) \to F(y))], \qquad (5.18)$$

in which 'F' ranges over finite subsets of the domain of interpretation. However, the proof that this system is categorical would normally be given in set theory or in full second-order logic (with impredicative comprehension). Our question here is whether this is predicatively recoverable.

The point is that the usual impredicativity involved in defining "x is a natural number" is strictly avoided. When we say

> x is a natural number iff x belongs to every (i.e., the!)
> finite quasi-chain omitting x, (*) (5.19)

we introduce the (infinite) class of natural numbers without quantifying over classes of that type, and provably so (in second-order logic, using Dedekind's definition of X *is infinite*, viz., there is a 1–1 mapping of X onto a proper subset of X). Moreover, every finite quasi-chain is specifiable outright (by a disjunction of the form "$x = 0$ v $x = s(0)$ v ... v $x = s^k(0)$", for each k), so the definition meets predicativist demands, quantifying over only previously introduceable classes.

However, as (*) already suggests, predicativism must treat infinite as well as finite classes. Both the full statement of mathematical induction and the statement of categoricity of a system for arithmetic involve quantification over infinite classes. Somehow the language of weak second-order logic must be extended.

A natural way to do this is simply to add a second style of higher-order variables (in addition to F, G, etc., say, for finite sets, as above), X, Y, etc., ranging over arbitrary subsets of the domain, and R, S, etc., ranging over (binary) relations (which turns out to be sufficient). In addition to the usual second-order logical axioms other than comprehension, the following restricted principle of *finite comprehension* is added:

$$\exists X \forall y[X(y) \leftrightarrow \phi], \qquad (f\text{-Comp}) \qquad (5.20)$$

in which ϕ is an *f-formula* lacking free X, that is, a formula in which only quantifiers ranging over individuals and finite sets (and binary relations) occur. A similar comprehension principle governs introduction of binary relations. It turns out that this provides a suitable logic for realizing *predicative Dedekind arithmetic*: infinite as well as finite sets and relations may be introduced, but only as specified by *f-formulas*.

There are various ways to axiomatize arithmetic categorically and usefully — allowing a proof of mathematical induction — based on this background logic. (For example, see Feferman and Hellman 1995.) One way that remains close to the DT above is to add the axiom QC above to the first-order axioms of Robinson's Q, together with the associative law for addition (which turns out to be technically useful, as in the lemma below). Here, QC is to be written using a finite-set variable, F, thereby building in automatically restriction to finite initial segments. Let us rename this axiom "FQC," for "finite quasi-chains." Now we may also consider the corresponding axiom in which restriction to finite sets is achieved by explicitly writing out Dedekind's definition, $D\text{-}Fin(X) \equiv^{df}$ "every 1–1 map on X with values in X has X as range," thus:

$$\forall x \exists X[D\text{-}Fin(X) \ \& \ X(0) \ \& \ \forall y(X(y) \ \& \ y \neq x \rightarrow X(s(y)))]. \quad (\text{LFQC}) \qquad (5.21)$$

"LFQC" stands for "logicist finite quasi-chains." It is interesting to consider how far the predicativist can get with LFQC, as it makes use of the logicist analysis of "finite" rather than building it into the logical notation as given. As it turns out, LFQC, together with the first-order fragment of the system *PDA* (*predicative Dedekind arithmetic*) and the first of two further evident axioms to be given momentarily, suffices to derive the full principle of mathematical induction (for predicatively specified classes). (Cf. Feferman and Hellman 1995.) For this much, that is, the predicativist need not take "finite" as primitive. However, in order to recover the categoricity proof itself, weak second-order logical machinery must be invoked.

The system *PDA* is obtained by adding to the first-order axioms mentioned above the axiom FQC and the following two axioms: first, we require the assumption that *every* quasi-chain omitting x and terminating in x is finite:

$$\forall x \forall Y[\text{"}Y \text{ a quasi-chain omitting } x\text{"} \ \& \ \forall z(Y(z) \to z \le x) \to D\text{-}Fin(Y)] \qquad \text{(FTQC)} \qquad (5.22)$$

("finite terminating quasi-chains"), in which $z \le x \equiv^{df} \exists u(z + u = x)$; finally, we adopt the following link between notions of finitude:

$$\exists F(X = F) \to D\text{-}Fin(X), \qquad \text{(TFDF)} \qquad (5.23)$$

("truly finite implies Dedekind finite"), which obviously must hold in any (logically) *standard* model, in which the finite-set variables range over only all finite subsets of the domain.

Let us now sketch the proof that PDA or its first-order part plus (5.21) and (5.22) imply the second-order statement of mathematical induction. For this the following lemma is useful:

Lemma: Let $S_z \equiv^{df} \{y: y \le z\}$ (which clearly exists by f-Comp and is unique), and let Z be any quasi-chain omitting z; then $S_z \subseteq Z$.

Proof: Suppose there is a $u \in S_z - Z$. By definition of quasi-chain, if $v \in S_z - Z$, then $p(v) \in S_z - Z$, where p is the predecessor function. (Here the first-order axioms on addition are used.) Since $Z(0)$, p restricted to elements $\le u$, S_u, is a 1–1 function on $S_u - Z$, with values in $S_u - Z$, and lacking u in its range, showing that $S_u - Z$, hence S_u, and hence S_z, are Dedekind-infinite. But this contradicts the axiom FTQC, since S_z is obviously a terminating quasi-chain. This proves the lemma.

Now suppose induction fails, i.e., for some X, (i) $X(0)$, (ii) $\forall y(X(y) \rightarrow X(s(y)))$, but for some z, $\neg X(z)$. Observe that X is Dedekind-infinite, via the function s (by (i) and (ii)). Moreover, if $X(y)$ and $y < z$, then $s(y) < z$, for otherwise $s(y) = z$, whence $X(z)$ by (ii), contradicting the hypothesis $\neg X(z)$. Now let F be a Dedekind-finite quasi-chain omitting z (by the LFQC axiom). By the lemma, we have $S_z \subseteq F$ (putting F for Z), and we already must have that $F \subseteq S_z$ since if $F(y)$ and $y > z$, then F would be Dedekind-infinite (via s) by the assumption that F is a quasi-chain omitting z. Thus, in fact $F = S_z$ (showing, moreover, that F is the unique finite quasi-chain omitting z). Now, by the lemma again, $S_z \subseteq X$, and so $F \subseteq X$ (as X is *a fortiori* a quasi-chain omitting z). The argument is completed by noticing that if $F(y)$ and $y < z$, we must also have $s(y) < z$, since $X(y)$. (Recall our observations on X at the outset.) Thus, s shows that F must be Dedekind-infinite, which is the long-sought contradiction.

Thus, the predicativist can derive mathematical induction from elementary assumptions to the effect that "every initial segment is finite," and, moreover, this can be spelled out employing Dedekind's definition of "finite" rather than by taking this notion as an unanalyzed primitive. In fact, only arithmetical comprehension and not the full strength of f-Comp is required in this proof. So far, then, a serious tension between logicism and predicativism has not yet emerged.

However, when we turn to the proof of *categoricity* of our system, we find the machinery of weak second-order logic indispensable. However, that machinery does indeed suffice, and this is a noteworthy result on behalf of predicativism. The usual proof involves impredicative specification of an isomorphism (as noted above), and if this were unavoidable, the success of the above analysis would be only partial. However, the very same trick (DT) can be repeated at the level of specifying an isomorphism ψ between two given models $\mathcal{M}, \mathcal{N}$.

> $<u,v>$ belongs to ψ iff it belongs to every R such that
> "R is a function defined on $\mathcal{M}$ with values in $\mathcal{N}$ and R
> preserves zero and succession with the possible exception of $<u,v>$ itself,"   (**)   (5.23)

where it is clear how to write out the quoted conditions. Again the intersection so specified is a finite initial segment of the desired ψ, and we already know enough (induction!) to prove that each m in $\mathcal{M}$ is paired, by a finite R meeting the required conditions, uniquely with an n in $\mathcal{N}$ belonging to a finite initial segment, *and conversely*; so restricting the range of '$\forall R$' to *finite* functions (functions with finite domains) will not alter the specification of ψ. Again the specification so understood is not impredicative, provided we employ finite-set variables F, etc., to introduce finite functions. Here it emerges that the machinery of weak second-order logic is indeed essential, for if we attempt to dispense with it in favor of explicit restriction to finite functions employing Dedekind's definition (on the plan of LQFC above), we find ourselves appealing in ($**$) to an instance of *impredicative* comprehension, in virtue of the second-order quantifier over arbitrary functions in "Dedekind finite."

Thus we clearly see in the proof of categoricity of *PDA* a major trade-off between predicativism and logicism. Logicism provides a complete analysis of the notions "finite," "infinite," "cardinal number," etc., but at the price of impredicative comprehension and its "metaphysical commitments." Predicativism avoids the latter (at least to some extent), but must take the notion of "finite" as primitive in order to recover the categoricity of its system with arithmetic. Still, it can do this in a reasonable way, motivated, as suggested above, by the in principle specifiability of (hereditarily) finite sets and functions; *it need not begin by taking the natural number structure as given.*[9]

Adaptation of all of this to the modal-structural interpretation is straightforward. All that is required is that everything be rewritten with all quantifiers relativized to a hypothetical domain and with second-order function variables replacing the successor, addition, and multiplication constants throughout. The axioms of *PDA* (so rewritten) are then to be used in defining "ω-sequence."[10] With these changes, the interpretation and its justification go through, with the appropriate modifications in the proof of categoricity as indicated here. In this way, then, the entire modal-structural treatment of number theory is available for predicativist analysis.

What all this shows is that predicativist mathematics can be given an attractive start, taking even less for granted than is customary. Recent systems of Feferman (e.g., 1988), moreover, show that indeed a great many ordinary scientific applications can be accommodated within predicativist confines. Whether predicativism can indeed adequately serve the mathematical needs of the sciences and whether in any case it provides a philosophically acceptable framework for mathematics are major open questions with which we shall conclude — hereby assuring that they transcend the scope of this essay.

NOTES

This work was carried out with the support of the National Science Foundation Scholars Award DIR-8922435, which is gratefully acknowledged. I am also grateful to John Burgess, Solomon Feferman, Nicolas Goodman, and Stewart Shapiro for helpful correspondence.

1. To be sure, leading constructivists exceed a minimal agnosticism again and again, as, for instance, when Bishop writes, "Thus it is unlikely that there will ever exist a constructive proof that for every real $x \geq 0$ either $x > 0$ or $x = 0$" ([1967], 26), or when he writes, concerning Brouwer's continuity theorem, "It seems likely that we are in the tantalizing situation in which [the claim that every function from $\mathbb{R}$ to $\mathbb{R}$ is continuous] will never be proved and never be counterexampled" ([1967], 70).

2. Such a view is generally attributed to Brouwer; some of Bishop's remarks suggest that he espouses it; and it is articulated and defended by Dummett 1977.

3. The label is Stewart Shapiro's. See Shapiro 1994.

4. An important example is Bridges 1979.

5. For a constructive treatment of portions of the mathematical foundations of quantum mechanics, see Bridges 1981.

6. The content of this corollary was suggested to me by Stewart Shapiro.

7. A linear operator A is *closed*, just in case whenever both $\langle \phi_i \rangle \to \phi$ for a sequence $\langle \phi_i \rangle$ of vectors $\phi_i \in D_A$ (the domain of A) and $\langle A\phi_i \rangle \to \psi$, then $\phi \in D_A$ and $A\phi = \psi$. In this classical definition, $\to$ denotes convergence which may or may not be effective. Non-effective convergence is not recognizable as convergence by the constructivist. The Pour-El and Richards theorem produces cases of closed operators T and effectively convergent sequences of $\phi_i \in D_T$ such that, although each $T\phi_i$ is a constructive Banach space element, $\langle T\phi_i \rangle$ converges to a nonconstructive limit and the convergence is non-effective.

8. It is possible for the constructivist to recognize unbounded linear operators which remain undefined for certain constructive arguments in the respective domains of their classical counterparts, and a constructive theory of such truncated operators may be feasible. However, this course exacerbates the problem of leaving undefined physically meaningful quantities, such as the expectation values of the quantum magnitude represented by the unbounded operator, in question for many (even constructively recognizable) states in the domain of the classical operator but not the constructive counterpart. Similarly, temporal evolution out of such states, as given by the Schrödinger operator, will not be constructively recognized. Constructive quantum mechanics thus would be even more seriously "incomplete" in this sense than the classical Hilbert space formalism. For a valuable discussion of such problems in the classical setting, see Heathcote 1990.

9. It has been suggested that, despite this, this "weak second-order definition

does not fare significantly better on the score of avoiding impredicativity than the one based on full second-order logic" (Isaacson 1987, 156), since an exact representation of the natural number sequence (via finite initial segments) must indeed occur in the range of the second-order quantifiers. That is, of course, true and is the reason for the axiom FQC. But if that axiom is predicatively justifiable, as we have suggested it to be, why not adopt it? It is only by means of axioms on successor that we rule out too *small* a domain (forcing it to be infinite); any further condition that rules out too *large* a domain will thereby require an exact representation of the natural number sequence! There is still a vast difference between having to quantify over infinitely many finite sets (for a definition to work) and having to quantify over an infinite one (especially the very one you are trying to introduce)!

10. This is not necessarily the most economical way to proceed. By combining plural quantification (Boolos 1985) with mereology to introduce ordered pairs (Burgess, Hazen, Lewis 1991), the arithmetic basis can be reduced to just successor without transcending the bounds of either predicativism or nominalism (see Hellman 1994).

REFERENCES

Beeson, M. J. 1982. "Problematic Principles of Constructive Mathematics." In D. von Dalen et al., eds., *Logic Colloquium '80*. North Holland: 1982, 11–55.

———. 1985. *Foundations of Constructive Mathematics*. Berlin: Springer-Verlag.

Bishop, E. 1967. *Foundations of Constructive Analysis*. New York: McGraw-Hill.

Bishop, E., and H. Cheng. 1972. *Constructive Measure Theory*. Providence: American Mathematical Society.

Boolos, G. 1985. "Nominalist Platonism." *Philosophical Review* 94: 327–44.

Bridges, D. 1979. *Constructive Functional Analysis*. London: Pitman.

———. 1981. "Towards a Constructive Foundation for Quantum Mechanics." In F. Richman, ed., *Constructive Mathematics*. Springer Lecture Notes in Mathematics, no. 873, 260–73.

Burgess, J. P. 1984. "Synthetic Mechanics." *Journal of Philosophical Logic* 13: 379–95.

Burgess, J. P., A. P. Hazen, and D. Lewis. 1991. "Appendix on Pairing." In D. Lewis, ed., *Parts of Classes*. Oxford: Blackwell, 121–49.

Cooke, R., M. Keane, and W. Moran. 1985. "An Elementary Proof of Gleason's Theorem," in *Mathematical Proceedings of the Cambridge Philosophical Society 98*. Cambridge: Cambridge University Press, 117–28.

Dedekind, R. 1888. *Was sind und was sollen die Zahlen*. Brunswick: Vieweg, trans. as "The Nature and Meaning of Numbers." In *Essays on the Theory of Numbers*. New York: Dover, 1963, 31–115.

Dummett, M. A. E. 1977. *Elements of Intuitionism*. Oxford: Oxford University Press.

Feferman, S. 1984. "Between Constructive and Classical Mathematics." *Computa-*

tion and Proof Theory, Lecture Notes in Mathematics 1104. Berlin: Springer, 143–62.

——. 1988. "Weyl Vindicated: *Das Kontinuum* 70 Years Later." *Temi e prospettive della logica e della filosofia della scienza contemporanee 1* Bologna: Clueb: 59–93.

Feferman, S., and G. Hellman. 1995. "Predicative Foundations of Arithmetic." *Journal of Philosophical Logic* 24: 1–17.

Field, H. 1980. *Science without Numbers.* Princeton: Princeton University Press.

Gleason, A. M. 1957. "Measures on the Closed Subspaces of a Hilbert Space." *Journal of Mathematics and Mechanics* 6: 885–93.

Heathcote, A. 1990. "Unbounded Operators and the Incompleteness of Quantum Mechanics." *Philosophy of Science* 57: 523–34.

Hellman, G. 1989a. *Mathematics without Numbers.* New York: Oxford University Press.

——. 1989b. "Never Say 'Never'! On the Communication Problem between Intuitionism and Classicism." *Philosophical Topics* 17, no. 2: 47–67.

——. 1992. "The Boxer and His Fists: The Constructivist in the Arena of Quantum Physics." In *Proceedings of the Aristotelian Society, Supplement* 66: 61–77.

——. 1993a. "Gleason's Theorem Is Not Constructively Provable." *Journal of Philosophical Logic* 22, no. 2: 193–203.

——. 1993b. "Constructive Mathematics and Quantum Mechanics: Unbounded Operators and the Spectral Theorem." *Journal of Philosophical Logic* 22, no. 3: 221–48.

——. 1994. "*Real* Analysis without Classes." *Philosophia Mathematica (3)*, vol. 2, 228–50.

Isaacson, D. 1987. Arithmetical Truth and Hidden Higher-Order Concepts." *Logic Colloquium '85,* ed. Paris Logic Group. New York: Elsevier, North Holland, 1987, 147–69.

Jordan, T. F. 1969. *Linear Operators for Quantum Mechanics.* New York: Wiley.

Mackey, G. W. 1963. *The Mathematical Foundations of Quantum Mechanics.* Reading, Mass.: Addison-Wesley.

Monk, J. D. 1976. *Mathematical Logic.* (New York: Springer).

Quine, W. V. 1953. "Two Dogmas of Empiricism." In W. V. Quine, *From a Logical Point of View.* New York: Harper, 20–46.

Pour-El, M. B., and I. Richards. 1983. "Noncomputability in Analysis and Physics: A Complete Determination of the Class of Noncomputable Linear Operators." *Advances in Mathematics* 48: 44–74.

Putnam, H. 1967. "Mathematics without Foundations." *Journal of Philosophy* 64: 5–22.

Putnam, H. 1971. *Philosophy of Logic.* New York: Harper.

Richman, F. 1987. "The Frog Replies," and "The Last Croak." *Mathematical Intelligencer,* 9 3: 22–26.

Shapiro, S. 1994. "Mathematics and Philosophy of Mathematics." *Philosophia Mathematica (3)*, vol. 2, 148–60.

Troelstra, A. S. 1977. "Aspects of Constructive Mathematics." In J. Barwise, ed., *Handbook of Mathematical Logic.* Amsterdam: North Holland, 973–1052.

von Neumann, J. [1932] 1955. *Mathematical Foundations of Quantum Mechanics*. R. T. Beyer, trans. Princeton: Princeton University Press.

Wang, H. 1964. "Eighty Years of Foundational Studies," reprinted in H. Wang, *A Survey of Mathematical Logic*. Amsterdam: North-Holland, pp. 34–56.

Weinstein, S. 1983. "The Intended Interpretation of Intuitionistic Logic." *Journal of Philosophical Logic* 12: 261–70.

Weyl, H. 1949. *Philosophy of Mathematics and Natural Science*. Princeton: Princeton University Press.

6

Halfway Through the Woods:

Contemporary Research on Space and Time

Carlo Rovelli

Department of Physics, University of Pittsburgh

In this essay I describe the present state of affairs in *fundamental theoretical physics*, as I see it, with particular regard to the recent evolution of the notions of space and time. I believe that we are going through a period of profound confusion, in which we lack a general coherent picture of the physical world capable of embracing what, or at least most of what, we have learned about it. The "fundamental scientific view of the world" of the present time is characterized by an astonishing amount of perplexity, and disagreement, about what time, space, matter, and causality are.

To find a period previously accepted assumptions were questioned as deeply as they are now, one must go back four centuries. Between the publication of Copernicus's *De Revolutionibus*, which opened *the* scientific revolution *by antonomasia*, and the publication of Newton's *Principia*, which brought it to a spectacularly successful conclusion, 150 years elapsed. Over those 150 years, Western conceptions about time, space, cause, matter, and rationality were profoundly shaken, bitterly discussed, and finally reshaped into a new grand synthesis. The personæ of this intellectual adventure — Galileo, Descartes, Kepler, Huygens, Leibniz — had an acute awareness of being in the middle of a major conceptual change, but little clue as to where the change was leading. Each of them saw fragments of the final fresco. The fragments looked strange and sometimes inconsistent. And each struggled to combine those fragments, sometimes falling back to medieval views which in retrospect make us smile, sometimes with visionary intuitions. Newton saw the final picture, which then remained one of the wonders of Europe for over three centuries.

I have little doubt that we are in the middle of a similar process. At the beginning of this century, general relativity altered the classical understanding of the concepts of space and time in a way which, as I will argue below, is far from being already fully understood. A few years later, quantum mechanics challenged the classical account of matter and causality, to a degree that is still the subject of burning controversies. Since the discovery of general relativity we no longer are sure of what spacetime is, and since the discovery of quantum mechanics we no longer are sure of what matter is. The very distinction between spacetime and matter is likely to be ill-founded, as I will illustrate below. The picture of the physical world that had been available for three centuries had problematic aspects and was repeatedly updated, but it was fully consistent, and the core of its conceptual structure remained clear and stable from the formulation of the Cartesian program to the end of last century. I think it is fair to say that today we *do not have* a consistent picture of the physical world at all.

In addition to the uncertainty of the conceptual grasp within their respective domains, general relativity and quantum mechanics appear to be fully inconsistent with one another. This incongruity has taken us back to a pre-Newtonian situation in which physics of the sky and physics of the earth were distinct and the laws of motion of terrestrial objects were assumed to be different from the laws of motion of celestial objects. Today, microscopic objects — atoms and elementary particles — are assumed to be governed by quantum mechanics, while large-scale objects — stars and galaxies — by general relativity. Physics has been in a schizophrenic state of mind for more than half a century, which, after all, is roughly the same time Kepler's ellipses and Galileo's parabolas (quite incompatible, if taken as fundamental laws of motion) had to forcibly cohabit within physics. The pragmatism of the middle of this century, a pessimistic diffidence toward major synthetical efforts, and the enormous (sometimes terrifying) effectiveness of the modern *Scientia Nova* have restrained the majority of physicists from confronting such an untenable situation directly. But I do not think that this utterly pragmatic attitude can be satisfactory or productive in the long run.

I think that the responsibility for the search for the new synthesis is not for physicists alone. The conceptual notions on which the Newtonian synthesis was grounded emerged from a century of eminently *philosophical* debate on the extraordinary Copernican discovery that the solar system looks much simpler if seen from the point of view of the sun. Galileo asked: "If the earth really moved, could we perceive such a motion?" Descartes asked: "What is motion?" Kepler asked: "If planets revolve around the sun, isn't there perhaps an influence, a *force*, due to the sun, that drives the planets?" It is difficult to underestimate the role that a purely speculative and philosophical quest has had on the shaping of the classical scientific image of the world. Consider just a few examples: the idea that the world admits a description in terms of bits of solid matter moving through a background space and mechanically affecting each other; the idea that simple mathematical equations can describe any such motion and any such influence; determinism; the relativity of velocity; the distinction between primary and secondary qualities. Classical science would not have developed without these philosophical doctrines, which were elaborated, or revived, and integrated into the debate that followed Copernicus's discovery.

General relativity and quantum mechanics are discoveries as extraordinary as the Copernican discovery. I believe they are, like Kepler's ellipses and Descartes's principle of inertia, fragments of a future science. I think that it is time to take them seriously, to try to understand what we have actually learned about the world by discovering relativity and quantum theory, and to find the fruitful questions. Maybe the Newtonian age has been an accident and we will never again reach a synthesis. If so, a major project of natural philosophy has failed. But if a new synthesis is to be reached, I believe that philosophical thinking will be once more one of its ingredients. Due to the conceptual vastness of the problematic involved, the generality and accuracy of philosophical thinking and its capacity to clarify conceptual premises are probably necessary to help physics out of a situation in which we have learned so much about the world, but no longer know what matter, time, space, and causality are. As a physicist involved in this effort, I wish the philosophers who are interested in the scientific description of the world would not confine themselves to commenting and polishing the present fragmentary physical theories, but would take the risk of trying to look *ahead*.

I will begin this essay by describing the present reading of what general relativity asserts about space and time. General relativity is still the best physical theory we have about space and time, but its understanding has considerably matured in very recent years. Next, I will discuss quantum mechanics, considering the difficulties of regarding quantum mechanics as the fundamental theory of matter, and presenting a specific point of view on the problem of what precisely the theory is telling us about the world. Next, I shall summarize some recent technical advances and the numerous open problems connected with the construction of a quantum theory of gravity, which is my primary area of concern as a theoretical physicist. I will conclude with an exercise bound to fail: an attempt to look through the fog and to guess what a consistent general physical theory, taking into account what we have learned in this century about the physical structure of reality, may look like. If nothing else, this exercise is meant to emphasize how profoundly in the dark we are at the present time.

My perspective is necessarily personal and biased. More precisely, the ideas on general relativity that I describe are shared by many theoreticians of my generation, but they perhaps sound strange to several elder physicists. The ideas on quantum mechanics have been directly influenced by numerous sources, but are for the most part personal and controversial. A very personal view of the state of the art is the best we can offer, I think, from halfway through the woods.

General Relativity: A Contemporary Perspective

General relativity is a physical theory with a peculiar history: it was conceived almost in isolation, grounded in almost no experimental data and almost solely on the basis of pure speculation about the nature of space and time, and it has lately been spectacularly confirmed by a sequence of experimental results, which are still appearing — the last one having been recognized by the 1994 Nobel Prize in physics. For a few decades the scientific tribe of the general-relativists survived quite disconnected from mainstream physics, as a kind of respected but basically ignored cult: "those who study the shape of spacetime." Then, in the last decade, the theoretical physicists' community-at-large, having essentially solved the puzzle of atomic and subatomic physics[1] (perhaps also driven by the growing relevance of relativistic astrophysics and by the feeling that gravitational waves detection is now at hand) has begun to feel the discomfort of living with two incompatible "fundamental" theories, and is constantly devoting increasing attention to general relativity.

As the fortunes of the theory fluctuated, so did the way it was understood. Einstein's original emphasis on "general covariance" was re-peatedly challenged (see J. Norton 1993, for a comprehensive account of the controversy). Modern mathematics, and differential geometry in particular, allowed a coordinate-free formulation of the theory, with a substantial increase in clarity. The work of Dirac, Bergman, and Ko-mar first, and then Arnowit, Deser, and Misner, unraveled the dynami-cal structure of the theory, and the results of Penrose and Hawking demarcated its domain of validity. The reformulation of the theory proposed by Ashtekar revealed its closeness with the other physical field theories and a hidden level of simplicity.

In the fifties and sixties, much attention was focused on the concep-tual significance of the theory and on the precise relation between theory and "reality," as captured by experimental physics. The school of Peter Bergmann, in particular, addressed with clarity the central issue of *observability*: what precisely are the theoretical objects in the theory to which the measurements performed by experimenters correspond? During the seventies and eighties the attention shifted away from these general conceptual issues toward interesting, but more formal, devel-opments. In this regard, the work of Arnowit, Deser, and Misner, pre-cisely because of its unquestionable importance, has perhaps had a somewhat misleading effect, by focusing the attention of many re-searchers on nonphysical structures — such as the fictitious "ADM sur-face" — that make general relativity appear simpler than it really is.

Then, in recent years, the pressure for conceptual clarity, deriving from the efforts to reconcile the theory with quantum mechanics, has brought conceptual issues back onto the stage. In particular, the ob-servability issue — which ones are the "observables of the theory?" — is again a subject of debate. If quantum mechanics is correct, the true observables of the theory are "quantum variables," which namely are subject to Heisenberg's indeterminacy principle — not so the fields of the theory that are not "true observables." Thus, any attempt to study the quantum theory of gravity requires us to understand with precision which are the true observables of the theory.

The reader who is not very familiar with the technicalities involved in general relativity may wonder how scientists could disagree on "what quantities of the theory are to be compared with experiments," in a theory that has been used to make very successful predictions about nature. The answer (besides the obvious one that scientists can succeed in disagreeing on anything) lies in the concrete process through which information is extracted from the mathematics of general relativity: In any concrete calculation, concerning, say, the solar system, or a binary pulsar, or the deflection of light by the sun, an arbitrary coordinate system is picked out, all fields are written in terms of such a coordinate system, and, only at the very last stage, quantities independent of the chosen coordinate system are computed, to be compared with experimental data. These are, by definition, the "observables" of the theory, but the general relation between these observables and the original fields of the theory is far from transparent. In particular, it is not immediately obvious how to characterize these "observables" in general, as functions (functionals) of the basic fields of the theory.

The debate sparked by these problems has revived the conceptual investigations of the sixties, such as those by Peter Bergmann's group, and has led to a deeper understanding of general relativity. Philosophers and historians have contributed to this clarification process. The image of general relativity that emerges from these debates is far less simple than the one on which many contemporary textbooks are still based but is, I believe, more enchanting. General relativity discloses a deeply relational core,[2] which perhaps does justice to Einstein's early claims about it and to its specific philosophical lineage. At the same time, the clarification of this relational aspect of general relativity takes the ground away from under our feet and leaves us with a theory even harder to digest, as a fundamental theory, than before — and with difficulties surprisingly similar to the difficulties of quantum mechanics. In the rest of this section, I will summarize this modern perspective on general relativity, its justifications, and the problems it raises.

Motion with Respect to What?

According to popular accounts, the physical content of general relativity is captured by the two following tenets:

a. Spacetime is (best) described in terms of *4-dimensional curved geometry*.

b. The *geometry is dynamical*, that is, it is affected by the presence of matter and determined by differential equations (Einstein equations).

I contend that the physical content of general relativity reaches much further than tenets **a** and **b** imply.

Einstein understood the two points **a** and **b** with clarity already in 1912, at the time of his paper *Outline of a theory of gravitation*. Still, several years of intense labor on the theory had to lapse before general relativity was born. The 1912 paper lacked the third and major conceptual ingredient of the theory: that is, a profound change in the physical meaning of the spacetime points of the theory. In Einstein's (pre-coordinate-free differential geometry) language, he had to struggle to understand "the meaning of the coordinates."

This third and deeper novelty that general relativity contributes to theoretical physics is a full implementation of the late Cartesian and Leibnizian program of a completely relational definition of motion,[3] which *discards the very notion of space*:

c. Spatial, as well as temporal, *location* is defined solely in terms of contiguity between the interacting dynamical objects. *Motion* is defined solely as change of contiguity between interacting dynamical objects.

To a philosopher familiar with, say, Aristotle, the Descartes of the *Principles*, Leibniz, Berkeley, or Mach, this assertion may perhaps sound at first somewhat trivial. But modern theoretical physics was born from the distinction between space and matter and the idea that matter moves *in space*. Physics has emerged from the famous Newton's bucket, which proved that motion (more precisely, acceleration) is defined intrinsically, not with respect to other bodies. As Newton himself puts it, in polemic with Cartesian relationalism:[4]

So it is necessary that the definition of *places*, and hence of local motion, be referred to some motionless thing such as extension alone or *space*, in so far as space is seen to be truly distinct from moving bodies.

To a large extent, general relativity overturns this position. But the most radical aspect of this overturn — which, to my knowledge *is not* anticipated by the philosophical relational tradition, and which has been very seldom appreciated — is that this overturn does not refer just to space, but also to time. The time "along which" dynamics develops is discarded from general relativity, as well as the space "in which" dynamics takes place. I will return to this fundamental point in the last section.

I will try to clarify the path from the Newtonian absolutist definition of motion to the fully relational one in **c** in what follows. Let me begin by briefly reviewing the prerelativistic physical conceptualization of space and time. Newtonian and special-relativistic physics (which I collectively denote in this paper as prerelativistic physics, meaning pre-

general-relativistic), describe reality in terms of two classes of entities: spacetime and matter. *Spacetime* is described as a metric manifold, namely a differential manifold equipped with an additional structure: a metric, *i.e.*, a definition of the distance between any pairs of points. This metric is unique, fixed, and not subject to differential equations. *Matter* is whatever is described according to dynamical equations of motion as moving in spacetime.

The physical meaning of this pre-relativistic spacetime (how we individuate spacetime points, or how we know "where" a spacetime point is) is elucidated by the subtle and highly nontrivial notion of *reference system* introduced by von Neumann and Lange (who coined the expression) at the end of last century (Lange 1885), in light of relational attacks on Newton's position on absolute space and on the need, which is the core of prerelativistic mechanics, to combine the relational nature of position and velocity with the absolute nature of acceleration. As any good undergraduate textbook may testify, the pre-relativistic statements about spacetime can be detailed as follows.

We assert that there exist inertial reference systems, which move of uniform motion in relation to one another, and in which mechanics holds. These reference systems are conceptually thought of, as well as physically realized, in any laboratory or in any direct observation of nature, as collections of *physical objects*. The notion of localization is then grounded on the assumptions that:

i. A relation of "being in the same point, at the same time," which I will denote as *contiguity*, is physically meaningful and immediately evident (within the relevant approximation scale).
ii. The position of any dynamical material entity is determined *with respect to the physical objects of the reference system*, in terms of contiguity with these objects.

A key aspect of this machinery is the clear-cut distinction between physical objects that form the reference system (laboratory walls) and physical objects whose dynamics we are describing. Thus:

iii. The reference system objects are *not* part of the dynamical system studied, their motion (or better, *stasis*, by definition) is independent from the dynamics of the system studied.

This clear-cut distinction has the same content as the distinction between spacetime and dynamical matter. These first three assumptions are the "nonmetrical" ones. Then we have a metrical assumption:

iv. A notion of congruence exists, such that it is meaningful to say that the distance between a pair of objects A and B is the same as the distance between a second pair of objects C and D, and, that the time interval between a pair of events **A** and **B** is the same as the time interval between a pair of events **C** and **D**. In other words, we have sensible *rods* and *clocks* in our hands.

This allows the objects of the reference system to be *labeled* in terms of their physical distance x from an (arbitrary) origin and the events to be labeled in terms of the time t lapsed from an (arbitrary) initial time. Thus, thanks to von Neumann and Lange's work, we understand that if we are disturbed by the unobservability of the spacetime points as such, we can trade the x and t coordinates which label those points with x and t coordinates which label physical objects of a reference system. Conversely, we can read the reference system construct as the operational device that allows the properties of the background spacetime to be manifest.

It is irrelevant to the present discussion whether we like to ascribe the status of ontological reality to this spacetime, or whether we prefer a strictly empiricist position which relegates it to the limbo of the theoretical terms. The relevant point here[5] is that we may think of reality in terms of objects moving in space and time, and this works.

Given these assumptions, prerelativistic physics asserts the existence of classes of objects that are inertial reference systems, that is, in which the dynamical equations hold. Once more, the coordinates x and t express the contiguity (i) of given dynamical objects with reference-system objects (ii), which are dynamically independent from the dynamical object (iii), and are labeled in terms of physical distances and time intervals (iv).

Then came Einstein's struggle with "the meaning of the coordinates" in gravitational physics. As is well known, in the theory he built, x and t do not indicate physical distances and time intervals (these are given by formulae involving the metric field). An immediate interpretation of x and t is that x and t are *arbitrary* labels of arbitrary *physical reference system* objects. Assumptions (i), (ii), and (iii) still hold, but now *any* reference system, not just inertial ones, is acceptable. This interpretation of x and t is wrong, but unfortunately it still permeates many general relativity textbooks.

The important point is that assumption (iii) fails on physical grounds. This is a simple consequence of the fact that there is no physical object whose motion is not affected by the gravitational field,[6] but the consequences of this fact are far-reaching. The mathematical expression of the failure of (iii) is the invariance of Einstein's equations under active diffeomorphisms. In 1912, Einstein had no difficulty in dropping (iv). Dropping (iii) is a much more serious matter. In fact, if we drop (iii) there is no way we can cleanly separate spacetime from dynamical matter. The clear-cut prerelativistic conceptual distinction between spacetime and matter rests on the possibility of separating the reference system physical objects that determine spacetime from the dynamical objects to be studied. If spacetime points cannot be determined by using physical objects *external* to the dynamical system we are considering, what are the physical points?

According to the fascinating reconstruction provided by John Stachel (1989) and John Norton (1989), Einstein struggled with this problem through a very interesting sequence of twists. In his 1912 *Outline of a Theory of Gravity*, he claims, to the amazement of every modern reader, that the theory of gravity must be expressed in terms of the metric $g_{\mu\nu}(x, t)$, that the right-hand side of the equations of motion must depend on the energy momentum tensor, that the left-hand side must depend on $g_{\mu\nu}(x, t)$ and its derivatives up to second order (all this represents three quarters of general relativity), and that *these equations must* not *be generally covariant*! Einstein argues that general covariance implies either that the theory is nondeterministic, which we do not accept, or the failure of (iii), which is completely unreasonable, because if we accept it, we do not know what spacetime is. (For the details of this argument, which is the famous first version of the "hole argument," see Earman and Norton 1987; Stachel 1989; Norton 1989.)[7] Later, Einstein changed his mind. He accepted the failure of (iii), and characterized this step as a result "beyond [his] wildest expectations." The significance of the success of this step has not yet been fully absorbed by part of the scientific community, let alone by the wider intellectual community.

If we are forbidden to define position with respect to *external* objects, what is the physical meaning of the coordinates x and t? The answer is: there isn't one. In fact, it is well known by whoever has applied general relativity to concrete experimental contexts that the theory's quantities that one must compare with experiments are the quantities that are fully *independent* from x and t. Assume that we are studying the dynamics of various objects (particles, planets, stars, galaxies, fields, fluids). We describe this dynamics theoretically as motion with respect to x and t coordinates, but then we restrict our attention solely to the positions of these objects in relation to each other, and not to the position of the objects with respect to the coordinates x and t.

The position of any object with respect to x and t, which in pre-relativistic physics indicates their position with respect to external reference system objects, is reduced to a computational device, deprived of physical meaning, in general relativity.

But x and t, as well as their later conceptualization in terms of reference systems, represent *space* and *time*, in prerelativistic physics: the reference system construction was nothing but an accurate systematization, operationally motivated, of the general notion of matter moving *on* space and time. In general relativity, such a notion of space and time, or of spacetime, has evaporated: objects do not move with respect to spacetime, nor with respect to anything external; they move in relation to one another.

General relativity describes the relative motion of dynamical entities (fields, fluids, particles, planets, stars, galaxies) in relation to one another.

Assumption (i) above (*contiguity*) still holds in general relativity. Assumption (ii) (*need of a reference*) is still meaningful, since any position is determined with respect to a reference object. But the reference objects must be chosen *among the dynamical objects themselves that are part of the dynamical system* described by the theory.

Characteristic examples of this procedure follow. We may consider the two most recent experimental works that provided spectacular confirmations of general relativity predictions: solar system precise measurements, and measurement of the binary pulsar energy decrease via gravitational radiation. Let us consider *position* determination in the first example: The planets' positions are represented and computed in terms of completely arbitrary coordinates x and t. The statement that, say, Venus is in position x at time t is then completely meaningless, if taken alone. However, a statement about the relative position of Venus's and Mars's images on a photographic plate on the earth (also moving in the solar system), is a physical, testable statement of the theory. Similarly, let us consider a *time* measurement made in terms of the data from a binary pulsar system. The pulsar of the binary system sends a pulsing signal $f(t)$, where t is an arbitrary time coordinate. This signal is further modulated by the pulsar's revolutions around its companion; let $r(t)$ be such a modulation. The two functions $f(t)$ and $r(t)$ have no meaning by themselves, but if we invert $f(t) \to t(f)$ and we insert t in $r(t)$, we obtain $r(f) = r(t(f))$, which expresses the *physically meaningful* evolution of the modulation with respect to the physical time scanned by the pulses of the pulsar.

The *observables* of general relativity are the *relative* (spatial and temporal) positions (contiguity) of the various dynamical entities in the theory, in relation to one another. This is the relational core of general relativity; almost a homage to its Leibnizian lineage.

Contemporary Fundamental Physics is a Theory of Fields

The discussion above requires some clarifying comments. The reader may wonder what the relation is between the relational character of general relativity that I have emphasized, which implies some kind of disappearance of the very notion of space, and the statement about space contained in point A above.

In order to clarify this point, one must consider the historical evolution of the second horn of the spacetime-versus-matter dichotomy. *Matter* was understood in terms of massive particles until the profoundly innovative work of Faraday and Maxwell, which plays an important role in the present discussion. After Faraday and Maxwell, and particularly after the 1905 paper of Einstein himself, which definitively got rid of the hypothetical material substratum of the Maxwell field, *matter* is no longer described solely in terms of particles, but rather in terms of *fields*.

Two further developments have led to an almost complete displacement of a particle ontology in favor of a field ontology, in the physics of the late twentieth century. The first of these developments is relativistic quantum field theory, which admits a reading in terms of fields alone: electrons, quarks, and other fundamental "particles" are theoretically described as fields in the standard model Lagrangian.[8] For quarks in particular, which are the primary constituents of everything we mundanely view as the ponderable matter, their "particle" aspect has never been detected and is probably undetectable in principle. The second of these developments is general relativity itself, because the notion of physical particle is incompatible with general relativity. Indeed, a physical particle cannot be an *extended* rigid object, because rigid bodies are not admitted in the theory (they transmit information faster than light), nor can it be a *pointlike* massive object, because such objects too are incompatible with the theory (they disappear in their own black hole).[9] Thus, understanding the physical picture of reality offered by general relativity in terms of particles moving on a curved geometry is misleading.

In addition, I think it is important to emphasize the extent to which general relativity is a product, a main step, and one of the great successes in the process of understanding reality in terms of *fields*. The other steps are Maxwell electrodynamics, the $SU(3) \times SU(2) \times U(1)$ standard model, and Dirac's second quantized field theoretical description of the elementary particles, not minor steps, indeed. Einstein's program that led to general relativity, in fact, was to express Newtonian gravity in a *field theoretical* form, which could be compatible with special relativity. Einstein, we shouldn't forget, was profoundly fascinated by Maxwell's work.

Thus, in contemporary physics *matter* is understood also, and perhaps primarily, in terms of *fields*. The consequences of this fundamental fact on the understanding of general relativity are far reaching. The prerelativistic physical ontology is as follows:

Spacetime (fixed, nondynamical)		Matter (dynamical)	
Differential manifold	*Metric*	Fields: electromagnetic *gravitational* other interactions electron and quark other fields	Massive particles

Let us now see how this ontology is affected by general relativity. The technical core of general relativity is the *identification* of the gravitational field and the metric. Thus, general relativity describes the world in terms of the following entities:

Fixed, nondynamical		Dynamical	
Differential manifold	Field: metric = gravitational	Fields: electromagnetic other interactions electron and quark other fields	Massive particles

In general relativity, the metric and the gravitational field are the same entity: the metric/gravitational field. In this new situation, illustrated by the table above, what do we mean by *spacetime*, and what do we mean by *matter*? Clearly, if we want to maintain the spacetime-versus-matter distinction, we have a terminology problem.

In the physical, as well as philosophical, literature, it is customary to denote the differential manifold *as well as* the metric/gravitational field (see table above) as spacetime, and to denote all the other fields (and particles, if any) as matter.[10] But the table above shows that in the general relativistic world picture such a terminological distinction collapses. In general relativity, the metric/gravitational field has acquired most, if not all, the attributes that have characterized matter (as opposed to spacetime) from Descartes to Feynman: it satisfies differential equations, it carries energy and momentum, and, in Leibnizian terms, it *can act and also be acted upon*, and so on.

In the quote above, Newton requires motion to be referred to pure extension or *space*, "insofar as *space* is seen to be truly distinct from moving bodies," that is, insofar as space is pure nondynamical extension, which is certainly *not* the case of the gravitational/metric field. One cannot dispute terminology, but terminology can be very misleading: we may insist in denoting the metric/gravitational field as *space*, but the gravitational/metric field is much more similar to what Newton would have denoted as matter than to what he would have denoted as space.

Let me put it pictorially. A strong burst of gravitational waves could come from the sky and knock down the rock of Gibraltar, precisely as a strong burst of electromagnetic radiation could. Why is the first "matter" and the second "space"? Why should we regard the second burst as ontologically different from the second? Clearly the distinction can now be seen as ill-founded.

There is an alternative. Einstein's identification between gravitational field and geometry can be read in two alternative ways:

i. as the discovery that the gravitational field is nothing but a local distortion of spacetime geometry; or
ii. as the discovery that *spacetime geometry is nothing but a manifestation of a particular physical field*, the gravitational field.

The choice between these two points of view is a matter of taste, at least as long as we remain within the realm of nonquantistic and nonthermal general relativity. I believe, however, that the first view, which is perhaps more traditional, tends to obscure, rather than enlighten, the profound shift in the view of spacetime produced by general relativity.[11]

In the light of the second view, it is perhaps more appropriate to reserve the expression *spacetime* for the differential manifold, and to use the expression *matter* for everything which is dynamical, carries energy, and so on, namely all the fields *including the gravitational field*. (See the table above.) Again, one cannot dispute terminology. This is not to say that the gravitational field is *exactly* the same object as any other field. The very fact that it *admits* an interpretation in geometrical terms witnesses to its peculiarity. But this peculiarity can be understood as a result of the peculiar way it couples with the other fields.[12]

If, by *spacetime*, we indicate the differential manifold, then motion with respect to spacetime is nonobservable. Only motion with respect to dynamical matter is. Whatever terminology one chooses, the central point is that the prerelativistic theoretization of the structure of physical reality as *space* and *time* on which *dynamical matter* moves has collapsed in general relativistic physics. Physical reality is now described as a complex interacting ensemble of entities (fields), the location of which is only meaningful with respect to one another. The relation among dynamical entities of being *contiguous* (i) is the foundation of the spacetime structure. Among these various entities, there is one, the gravitational field, which interacts with every other one and thus determines the relative motion of the individual components of every object we want to use as rod or clock. Because of that, it admits a metrical interpretation.

I think it may be appropriate to close this section by quoting Descartes:[13]

If, however, we consider what should be understood by movement, according to the truth of the matter rather than in accordance with common usage, we can say that movement is the transference of one part of matter or of one body, from the vicinity of those bodies immediately *contiguous* to it, and *considered at rest*, into the vicinity of some others.

The quote by Newton I gave above was written directly in polemic with this relational Cartesian doctrine. The amount of energy Newton spent in articulating the details of the technical argument that supported his thesis against Descartes's relationalism, namely, the arguments that led to the conclusion quoted above, is proof of how seriously Newton thought about the issue. The core of Newton's arguments is that *one cannot construct a viable theory of motion out of a purely relational position*. In particular, Newton correctly understood that Descartes' own successful principle of inertia makes no sense if motion is purely relative.[14]

The well-known resistance of continental thinkers to Newtonian physics was grounded in the claimed manifest "absurdity" of Newton's definition of motion with respect to a nondynamical ("that cannot be acted upon") and undetectable space. The famous Leibniz-Clarke correspondence and Berkeley's criticisms are the prime examples of these polemics. From the present privileged point of view, everybody was right and wrong. Newton's arguments are correct if we add the qualification "in the seventeenth century" to the *one cannot construct* above. The construction of general relativity disproves the unqualified version of these arguments. As often happens in science, progress (Newtonian dynamics) needs some incorrect assumption, which later further progress (general relativity) is able to rectify.[15] I often wonder how Leibniz would have reacted by learning that by taking the fully relational nature of motion very seriously, a man living in Switzerland a few centuries down the road would have been able to correct 43 missing seconds-of-arc-per-century in Mercury's motion, which Newton's theory could not account for.

After three centuries of extraordinary success of the "absurd" Newtonian notion of "motion with respect to space," general relativity has taken us home, to a respectable but vertiginous understanding of the purely relational nature of space. In the game we have gained a strange extra bonus: the resurrected relationalism, combined with relativity, i.e., the discovery (by Einstein himself) of the intertwining of space with time, leads us to the thesis of the relational nature of time. This, I believe, is essentially unexplored land and is at the very core of the new physics. I will return to this point after a digression on quantum mechanics.

Quantum Mechanics: What Does It Say?

If two of the four conceptual pillars of classical physics — space and time — have been profoundly reshaped by general relativity, the other two — matter and causality — have been reshaped even more profoundly by quantum mechanics. There are various reasons for including a section on quantum mechanics in this work, which is mainly concerned with spacetime physics: first, the nonspatial way in which quantum mechanics deals with matter affects our understanding of spacetime; second, the physical degrees of freedom of the metric/gravitational field are quantum-mechanical, like any other degree of freedom, even if we do not have a consistent theory of their quantum mechanical behavior yet. This fact has a devastating effect on the view of spacetime as a metric manifold, as I will argue in the next section. Finally, I wish to mention here a perspective on quantum mechanics that addresses foundational issues relevant for the present discussion.

In spite of the success of quantum mechanics, it is difficult to overcome a sense of unease that quantum mechanics communicates, not so much when taken as a theory of microphysical objects, but certainly if we want to regard it as the universal foundation of our present view of the physical world. I believe that since quantum mechanics synthesizes most of what we have learned so far about motion, the issue is not to replace it, or to find new formulations of it, but to understand what it actually says about the world, or, equivalently, what precisely we have learned from experimental microphysics.

There is a historical precedent to this sense of unease and confusion with a successful physical formalism: special relativity. Special relativity is accredited to Einstein's celebrated 1905 paper, but the formal content of special relativity is coded in the Lorentz transformations, which were written by Lorentz, not by Einstein, and several years before 1905. Einstein's contribution was to understand the physical meaning of the Lorentz transformations. We could say, in a provocative manner, that Einstein's contribution to special relativity was the *interpretation* of the theory, not its formalism, since the formalism existed already.

The Lorentz transformations were perceived as unreasonable, even inconsistent, before 1905. Lorentz's tentative physical interpretation (a physical contraction of moving bodies, caused by complex electromagnetic interactions between the atoms of the bodies and the ether) was definitively unattractive and remarkably similar to certain "physical" interpretations of the wave function collapse presently investigated. Einstein discovered the reason for the unease: the implicit use of a concept (observer-independent time) inappropriate to describe reality, or, equivalently, a common assumption about reality (simultaneity is observer-independent) that was physically incorrect. The unease with the Lorentz transformations derived from a conceptual scheme in which an *incorrect notion*, absolute simultaneity, was assumed, yielding many sorts of paradoxical situations.

I suspect that the "paradoxical" situations associated with quantum mechanics (such as the famous and unfortunate half-dead Schrödinger cat [1935]) may derive from some analogous *incorrect notion* that we still employ in thinking about quantum mechanics (not in using quantum mechanics, since we seem to have learned to use it in a remarkably effective way). By finding this *incorrect notion*, we could perhaps free ourselves from the unease with our best theory of motion and fully understand what the theory is saying about the world.

Such a notion, I maintain, is the notion of absolute, or observer-independent, state of a system. (For an extensive and technical account of this bundle of ideas, see Rovelli 1996. Here I only summarize the essential considerations.)

Observers

I am going to argue that if we believe that quantum mechanics is not just a theory of the behavior of microscopic objects, but rather a general theory of motion, then we are necessarily led to conclude that:

a. In quantum mechanics different observers give different accounts of the same sequence of events.

Quantum mechanics deals with observations made by observers (expressed as measurement outcomes, or as histories, or in other ways, depending on the "interpretation" adopted). If

i. we insist that the observer is *also* a quantum mechanical entity, and, as such, can be, in turn, observed by a *second* observer, and if
ii. we compare the account of the *same* sequence of events, as provided by the first and the second observer,

then we can see that, whatever the interpretation of quantum mechanics one is using, the two observers provide different accounts of such a sequence of events.

Within the context of orthodox formalism, evidence for these assertions can be obtained by considering the following situation (for more details, see Rovelli 1996). An observer O (he) makes a measurement on a system S. We may think of O (for the moment) as a classical macroscopic measuring apparatus, including or not including a human being.

Assume that the quantity being measured, Q, may take two values, say A and B, and let $|A>$ and $|B>$ be the two states in which Q has value A and B, respectively. Let us assume that in a *given specific measurement*, which we denote as E, the outcome of the measurement is A. The system S is affected by the measurement, and at a time $t = t_2$ after the measurement, the state of the system is $|A>$.

Let us now consider this *same* sequence of events E, as theoretically described by a *second* observer (she), which we refer to as O', who describes a system formed by S and O. O' views both S and O as subsystems of the larger S-O system she is considering. We assume that O' does not perform any measurement on the S-O system during the t_1-t_2 interval, but that she knows the initial states of both S and O, and is therefore able to give a quantum mechanical account of the set of events E. The S-O system is described by the tensor product of the two Hilbert spaces of S and O. The physical process during which O measures the quantity Q of the system S is a physical interaction between O and S. In the process of this interaction, the state of O changes. If the initial state of S and O are not correlated, and if the interaction is appropriate for a measurement, then the coupled S-O system evolves into an *entangled* state, which is the superposition of a state in which S is $|A>$ and O has measured A (meaning, say, a pointer in an apparatus in O points toward the A mark), and a state in which S is in $|B>$ and O has measured B. This is the conventional description of a measurement as a physical process (Jauch 1968).

I stress the fact that we are describing an actual physical process E, taking place in a real laboratory. Standard quantum mechanics requires us to distinguish system from observer, but it allows us some freedom in drawing the line that distinguishes the two (von Neumann 1932). The peculiarity of the above analysis is only the fact that this freedom is exploited in order to describe the same sequence of physical events in terms of two different descriptions. In the first description, the line that distinguishes system from observer is set between S and O. In the second, between S-O and O'. We recall that we assume that O' is not making a measurement on the S-O system; there is no physical interaction between S-O and O' during the t_1-t_2 interval. (O' may perhaps make a measurement at a later time t_3: then, as shown by von Neumann (1932), the probabilities and the final quantum states that one computes assuming that the collapse happened between t_2 and t_2 or between t_2 and t_3 are in agreement.)

The description given by the observer O and the description given by the observer O' are two distinct *correct* descriptions of what happens to S during the sequence of events E. Consider time t_2. In the O description, the system S *is in the state* $|A>$ and the hand of the measuring apparatus indicates "A". While according to the O'-description, S *is not in the state* $|A>$ and the pointer of the measuring apparatus does not indicate "A."

According to a view on quantum mechanics often advocated, outcomes of measurements are the only physical content of the theory (Heisenberg 1927; van Fraassen 1991). Statement **a**, above, holds within such a view as well. The O' account, indeed, has nothing to say about reality at time t_2 (O' does not know anything about what happens between t_1 and some later time t_3 in which she performs a measurement: anything in between is like Heisenberg's "nonexisting" trajectory of the electron, of which we are not allowed to assert anything). Thus, O and O' give different accounts of S at $t = t_2$: O asserts that, at $t = t_2$, S *has the property A*; while O' claims that at $t = t_2$ *the quantity Q does not have any determined value*. (Or: the actualization of potentialities is observer dependent.) Thus, the two accounts of O and O' differ both as far as *states* are concerned as well as far as *measurement outcomes* are concerned. Different observers give different accounts of the same sequence of events.

Notice that the observer O' "knows" *that* O has made a measurement on S. She has a full account of the events in E at her disposal, and she knows that the state of the S-O system at t_2 is a correlated state. O' could measure an observable M that checks whether the pointer of O indicates the correct state of S. If she did so, then the outcome of this measurement would be "yes" with certainty. On the other side, O' does not know "*what* the value of the observable Q" is. (She could measure it, but the outcome of the measurement is not determined at $t = t_2$.) Thus, she does not know "what O knows about Q." We must distinguish the statement "O' knows *that* O knows about Q" from the statement "O' knows *what* O knows about Q." Of course, this is a consistent distinction, even common in everyday life (I know *that* you know the amount of your salary, but I do not know *what* you know about the amount of your salary.)

One can avoid conclusion **a** by denying that O, an observer, could be described as a quantum system. This is Bohr's premise (1949) (as far as I understand Bohr). If we accept it, then we have to separate reality into two categories of systems: quantum-mechanical systems on the one side, and classical observers (or a unique observer) on the other. Bohr claims explicitly that we must renounce giving a full quantum mechanical description of the classical world (Bohr 1949; Landau and Lifshitz 1977). Wigner (1961) pushes this view to the extreme consequences and distinguishes material systems (observed) from consciousness (observer). On the contrary, I wish to assume that:

b. All systems are equivalent: Nothing a priori distinguishes observer systems from quantum systems. If the observer O can give a description of the system S, then it is also legitimate for an observer O' to give a quantum description of the system formed by the observer O.

Let me now come to the main thesis, which follows from the premises **a** and **b**: If in quantum mechanics different observers may give different accounts of the same sequence of events, then *each quantum mechanical description has to be understood as relative to a particular observer*. Thus, a quantum-mechanical description of a certain system (state and/or value of measured quantities) should not be taken as an "absolute" (i.e., observer-independent) description of reality, but rather as a formalization, or codification, of the knowledge, or, more precisely, of the information, than an observer has about a system. Quantum mechanics can therefore be viewed as a theory about the relative information that systems have about each other. The various amounts of information are all related to each other. The fact that a system O has information about a system S is represented as a correlation between the state of S and the state of O, when viewed as information available to a third system O'.

If quantum mechanics describes relative information, the possibility is open that there is a deeper "underlying" theory that describes what happens "in reality." This is the thesis of the incompleteness of quantum mechanics (first suggested in Born 1926!). Examples of such hypothetical "underlying" theories are hidden variables theories (Bohm 1951; Belifante 1973). Alternately, the "wave-function-collapse-producing" systems can be "special" because of some not yet understood physics that becomes relevant due to a large number of degrees of freedom (Ghirardi, Rimini, and Weber 1986; Bell 1987), complexity (Hughes 1989), or quantum gravity (Penrose 1989). As is well known, there are no indications on *physical* grounds that quantum mechanics is incomplete. Indeed, physical praxis supports the view that quantum mechanics represents the best we can say about the world at the present state of experimentation and suggests that the structure of the world grasped by quantum mechanics is deeper, not more approximate, than the scheme of description of the world provided by classical mechanics.

On the other hand, one could consider motivations on *metaphysical* grounds, in support of the incompleteness of quantum mechanics: "Since reality has to be real and universal and the same for everybody, then a theory in which the description of reality is observer-dependent is certainly an incomplete theory. If such a theory were complete, our concept of reality would be disturbed."

However, if we are hunting for some incorrect assumption about reality, which we suspect to be the source of our unease with the theory, such kinds of arguments should make us suspicious and attentive. What we are looking for is *precisely* some apparently reasonable general assumption. Thus, I discard here the thesis of the incompleteness of quantum mechanics and assume

c. (Completeness) Quantum mechanics provides a complete and self-consistent scheme of description of the physical world, appropriate to our present level of experimental observations.

The conjunction of this assumption with **a,** and **b,** and the discussion above leads us to the following idea:

Quantum mechanics is a theory about the relative information that subsystems have about each other, and this is a complete description of the world.

I maintain that this conclusion is not self-contradictory. It simply implies that the "incorrect notion" at the source of our unease with quantum theory is the notion of "the *true,* or observer-independent, description of the world." If a complete description of the world is exhausted by the relevant information that systems have about each other, then there is no observer-independent description of the world: that is, there is neither an absolute "state of the system," nor an absolute "property that the system has at a certain time." The two accounts of the sequence of events E are both correct, even if distinct: any time we talk about a "state" or "property" of a system, we have to refer these notions to a specific observing system. I maintain that the apparent contradiction of the conclusion above, as well as the apparent paradoxes of quantum mechanics, can all be traced to the (incorrect) use of the concept of an observer-independent description of the world. They evaporate if we discard this concept, as the apparent paradoxes of the Lorentz transformations evaporate if we get rid of the notion of observer-independent simultaneity.

Thus, I suggest that quantum mechanics indicates that the notion of a "universal" description of the state of the world, shared by all observers, is a physically misleading concept. A *description* of the state of a system S exists only if somebody or something (namely, another system considered as an *observer*) is actually "describing" S, or, more precisely, has information about S. The state of a system is always a state of that system with respect to a certain observer (Crane 1995). A general physical theory is a theory about this *information* that systems may have about other systems.

Information

The relevance of the notion of information for understanding the nature of quantum physics has been advocated by John Wheeler (1988, 1989, 1992). The notion of information we employ here should not be confused with other notions of information employed in other contexts.

The technical notion of amount of information relevant in the present context is the one from information theory (Shannon 1949), which does not require us to distinguish between human and nonhuman observers, systems that understand meaning or that don't, very complicated or simple systems, and so on. In the technical information theory sense, information expresses the fact that a system is in a configuration, which is (has been) correlated to the configuration of another system (information source); the amount of information is the number of elements of a set of alternatives out of which the configuration is chosen. The relation between this notion of information and more elaborate notions of information is given by the fact that existence of the information theory information is a necessary condition for any "elaborate information" to exist. Thus, for a physical theory it is sufficient to deal with this information theory notion of information.

The fact that O has information about S means (when seen by O') that the states of O and S are correlated, or that the state of the S-O system (as described by O') is an eigenstate of the M operator. This notion of information is very weak and does not require us to consider information storage, thermodynamics, complex systems, meaning, or anything of the sort. Observers are not "physically special systems."

Axioms for Quantum Mechanics

It is possible to reconstruct quantum mechanics entirely, in terms of the concepts emerging from the discussion above (Rovelli 1996). Without entering into the technical details, I sketch here the main ideas. We assume the existence of an ensemble of systems, each of which can be equivalently considered as an *observing system* or as an *observed* system. A system (*observing system*) may have *information* about another system (*observed system*). Information is exchanged via physical interactions. We then have

Postulate 1 (Limited information): There is a maximum amount of *relevant information* that can be extracted from a system.

The physical meaning of Postulate 1 is that it is possible to exhaust, or "give a complete description of the system," in a finite time.

Postulate 2 (Unlimited questions): It is always possible to acquire *new* information about a system.

If, after having asked the N questions, the system O asks a further question of the observed system S, there are two extreme possibilities: either the answer to question Q is fully determined by the N questions already asked, or not. In the first case, no new information is gained. The second postulate states that there is always a way to acquire new information. Of course, previously relevant information now becomes irrelevant.

The motivation for this second postulate is fully experimental. In fact, we know that all quantum systems (and all systems are quantum systems) have the property that even if we know their quantum state $|\psi\rangle$ exactly, we can still "learn" something new about them by performing a measurement of a quantity O such that $|\psi\rangle$ is not an eigenstate of O. This is the essential *experimental* result about the world, which is coded in quantum mechanics. It is expressed in information theory terms in Postulate 2. Postulate 2 is true to the extent the Planck constant is different from zero: in other words, for a macroscopic system, getting to questions that increase our knowledge of the system after having reached the maximum of our information implies measurements with extremely high sensitivity.

It can be shown, using technology borrowed from quantum logic (Mackay 1963; Maczinski 1967; Jauch 1968; Finkelstein 1969; Piron 1972; Beltrametti and Cassinelli 1981), that out of these two postulates one can reconstruct almost completely the full formalism of quantum mechanics, including Hilbert spaces, projections, expectation values, and so on. The missing element (which is essentially the *complex* nature of the Hilbert space), can be added as a third ad hoc postulate (Rovelli 1996).

Notice that a statement on the information possessed by O is a statement about the physical state of O. Since the observer O is a regular system on the same ground as any other system, we may discuss his state in physical terms. However, since *there is no absolute meaning to the state of a system*, any statement regarding the state of O, *including any statement about its knowledge* is to be referred to some other system observing O. The notion of absolute *state* of a system, and thus *a fortiori* absolute state of an observer, is not defined. Therefore, the fact that an observer has information about a system is not an "absolute" fact: it is something that can be observed by a second observer O'. A second observer O' can have information about the fact that O has information about S. But bear in mind that any acquisition of information implies a physical interaction. O' can get new information about the information that O has about S only by physically interacting with the O-S system.

A common mistake in analyzing measurement problems in quantum mechanics is to forget that precisely as an observer can acquire information about a system only by means of physically interacting with it, in the same way two observers can compare their information only by physically interacting. This means that there is no way to compare the information possessed by O with the information possessed by O', without considering a physical interaction between the two, or between each of the two and a third system. Information, like any other property of a system, is a fully relational notion, and a question about the information possessed by O is in no way different from any other physical question. Thus the notion "A system O has information about a system S" is a physical notion that can be studied experimentally (by a third observer), in the same way as any other physical property of a system. In particular, the question "Do observers O and O' get the *same* answers out of a system S?" is a *meaningless* question. It is a question about the *absolute state* of O and O'. What is meaningful is to reformulate this question in terms of some observer. For instance, we could ask this question in terms of the information possessed by a further observer O', or, alternatively, by O' herself.

What has happened to the collapse of the wave function, namely, to the so-called nonunitary evolution during a measurement, from the present perspective? If O' knows the dynamics of the O-S system, she knows the two Hamiltonians of O and S *and* the interaction Hamiltonian. The interaction Hamiltonian cannot be vanishing because a measurement (O measuring S) implies an interaction: this is the only way in which a correlation can be dynamically established. From the point of view of O', the measurement is therefore a fully unitary evolution, which is determined by the interaction Hamiltonian between O and S. An interaction is a measurement that brings the states to a correlated configuration. On the other hand, O gives a dynamical description of S alone. Therefore he can only use the S Hamiltonian. Since between times t_1 and t_2 the evolution of S is affected by its interaction with O, the description of the unitary evolution of S given by O breaks down. *The unitary evolution does not break down for mysterious physical quantum jumps, due to unknown effects, but simply because O is not giving a full dynamical description of the interaction.* O cannot have a full description of the interaction of S with himself (O), because his information is correlation, and there is no meaning in being correlated with oneself.

Any Observation Requires an Observer

Let me summarize the point of view on quantum mechanics that I have discussed. I started from the distinction between observer and observed system. I assumed (**b**) that all systems are equivalent, so that any observer can be described by the same physics as any other system. In particular, I assumed that an observer that measures a system can be described by quantum mechanics. I have analyzed a fixed physical sequence of events E, from two different points of observation, the one of the observer and the one of a third system, external to the measurement. I concluded that two observers give different accounts of the same physical set of events (**a**).

Rather than backtracking before this observation and giving up the commitment to the belief that all observers are equivalent, I suggested taking this experimental fact at its face value and as a starting point for understanding the world. *If different observers give different descriptions of the state of the same system, this means that the notion of state is observer dependent.* I have taken this deduction seriously and considered a conceptual scheme in which the notion of absolute observer-independent state of a system is replaced by the notion of information about a system that an observer may possess.

In considered the postulates that this information must satisfy and that summarize present experimental evidence about the world. The first limits the amount of relevant information that a system can have; the second summarizes the novelty revealed by the experiments from which quantum mechanics has been deduced, by asserting that whatever the information we have about a system we can always get new information. Out of these postulates (plus a third one, not discussed here) the conventional Hilbert space formalism of quantum mechanics and the corresponding rules for calculating probabilities, can be derived (and therefore any other equivalent formalism).

The fact that a system O has information about a system S means that the states of S and O are correlated, meaning that a third observer O' has information about the coupled S-O system that allows her to predict correlated outcomes between questions to S and questions to O. Thus, correlation has no absolute meaning, because states have no absolute meaning and must be interpreted as the content of the information that a third system has about the S-O couple.

Finally, taking quantum mechanics as a complete description of the world at the present level of experimental knowledge (**c**), we are forced to accept the result that there is no "objective," or, more precisely, "observer-independent," meaning to the ascription of a property to a system. Thus, the properties of the systems are to be described by an interrelated net of observations and information collected from observations. Any complex situation can be described in toto by a further additional observer, and the interrelation is fully consistent. But there is no way to "exit" from the observer-observed global system. *Any observation requires an observer* (Maturana and Varela 1920). In other words, I suggest that it is a matter of natural science whether or not the descriptions that different observers give of the same ensemble of events is universal or not, and I maintain that

Quantum mechanics is the theoretical formalization of the experimental discovery that the descriptions that different observers give of the same ensemble of events are not universal.

The concept that quantum mechanics forces us to give up is the concept of a description of a system independent from the observer providing such a description, that is, the concept of absolute state of a system. The structure of the classical scientific description of the world in terms of *systems* that are in certain *states* is perhaps incorrect, and inappropriate to describe the world beyond the $\hbar \to 0$ limit. It is perhaps worthwhile to emphasize the fact that those considerations do not derive from the *theory* of quantum mechanics, but from a collection of *experiments* on the atomic world.

Combining General Relativity and Quantum Mechanics

If we wish to remain committed to the hypothesis, so far essentially undefeated, that the physical world admits a consistent theoretical description, we should expect that a quantum theory of the gravitational field is to be found.[16] There are three independent reasons for this expectation:

i. The gravitational field is a dynamical field, as is the electromagnetic field and the other fields: it should be a quantum mechanical quantity like those. The possibility that a nonquantistic *dynamical* gravitational field could interact with quantum fields has been repeatedly shown to be inconsistent, and is now considered nonviable by the large majority of physicists. Beyond the technical arguments that support this result, is the simple fact that quantum mechanics is consistent only if it is a fundamental theory, namely if quantum objects do not interact with classical objects. Bohr (1935, 1949) has shown that a single particle that violates the Heisenberg indeterminacy principle would allow all other physical quantities to violate the principle.

ii. The basic assumptions about physical reality that we employ in general relativity and in microphysics are in strident contradiction. As suggested in the introduction, we are back to the pre-Newtonian dichotomy between celestial physics and terrestrial physics. Our sole advantage with respect to pre-Newtonian physics is the existence of an intermediate area of phenomena that are described by an approximation of quantum mechanics (the classical limit, in which we may neglect the "small" Planck constant), as well as by an approximation of general relativity (the flat limit, in which we may neglect the "small" Newton constant). The two limits of the two theories are consistent with one another, and include the mundane classical nonrelativistic and special relativistic physics.

iii. We know for sure that there is a physical regime in which all present fundamental theories break down. This is the regime in which neither quantum mechanical effects nor the relativistic gravitational fields can be neglected, and corresponds, roughly, to the physical phenomena that happen at extremely short scale ($\sim 10^{-33}$ cm).

In spite of the combined efforts of some of the best scientific minds of the century, and in spite of very considerable partial results obtained, the quantum theory of gravity, which would combine general relativity and quantum mechanics, has not been found yet. Is the expectation that we can find such a theory at the present time reasonable? There are good reasons for doubting this. 10^{-33} cm is a dramatically short scale, which corresponds to energy scales immensely higher than anything we can hope to reach within the coming decades. Perhaps the physics "down there" is so different from what we know, that, in the lack of experimental evidence, any attempt to guess it is hubris.

But there are also good reasons for hope. We are indeed in one of those rare and happy situations in the history of physics in which we have two extremely well grounded theories contradicting each other. In the past, similar situations led to spectacular results: Remember what Newton was able to do by contemplating the contradiction between Galileo's terrestrial physics and Kepler's celestial one; what Einstein was able to do by contemplating the contradiction between Maxwell's electromagnetism and Galileo's transformations; or what Einstein was able to do by contemplating the contradiction between special relativity and Newton's gravitational law. All these powerful syntheses, and many others less spectacular, emerged precisely from the purely theoretical effort of combining two sets of *apparently* contradictory principles, both of which had been shown able to capture key aspects of physical reality.

A common aspect in these syntheses, which I believe is important for the sake of the present discussion, is that the new synthesis did not reject the key physical content of the apparently contradictory theories. Galileo discovered that the motion of free falling bodies is governed by their acceleration, not their velocity, and Kepler discovered the essential importance of the exact position of the sun, for the motion of the other planets. These two fundamental discoveries became the very ground of the Newtonian synthesis. In the twenties there still were physicists, such as Abraham, who claimed that *obviously* the Faraday-Maxwell discovery of the fields could not coexist with the relativity of velocity discovered by Galileo. Einstein found the way to understand theoretically that light propagates as a field-wave at a fixed speed, *and at the same time* velocity is relative. Such examples are innumerable. To understand quantum gravity, I maintain, we *do not* have to override quantum mechanics and/or override general relativity. We have to think hard, understand what *essential* discoveries about nature are coded into these two theories, and find the way to combine the two.

Of course, there is no guaranteed method for judging which are the essential aspects of a theory, the ones that survive the extension of the theory. This is where the speculative and intuitive aspect of the process of discovery is involved, and where we are required to have wide and far vision. This is precisely where the importance of a wide-angle approach, as opposed to a technical one, becomes crucial. The fact that modern science taught us the importance of technicalities shouldn't make us forget that with technicalities alone we get nowhere. Einstein could learn tensor calculus from Grossman and Levi-Civita, when he needed it; but without the profundity of his quest on the nature of spacetime, nurtured in Mach and Schopenhauer, we wouldn't have general relativity, and perhaps not even special relativity.

I believe that the essential discovery about nature coded into general relativity is precisely the one that Einstein himself considered such: namely, the general covariance of the theory. More precisely, it is the fact that position with respect to a "background" spacetime is a meaningless concept; only relative position of dynamical physical entities with respect to one another are observable. If I had to take a bet on twenty-second-century physics, this is the only idea I would bet will remain with us.

The core of quantum mechanics is still quite elusive. In the previous section, I presented an attempt to grasp it. Whatever it is, I believe that the replacement of classical determinism with quantum probabilities, and the unavoidability of the interference of the measuring apparatus on the observed system, are aspects of reality that we have understood and will remain with us.

Thus, what do we know about a quantum theory of gravity? In the physical regimes in which quantum mechanics as well as gravity are relevant, "something strange" happens to spacetime geometry. Quantum-mechanical quantities, such as the position of an electron, do not evolve along "trajectories," namely, do not have a precise value at every time; their evolution is given only in terms of probability distributions and expectation values for outcomes of measurements (or for suitable coarse-grained histories, according to the interpretation one likes better). The "trajectory" of the gravitational field, namely, the evolution of its configuration in time, is the four-dimensional (pseudo-) Riemanian geometry. Thus, to the level of accuracy required by general relativity and quantum mechanics together, *the very notion of space-time as a curved manifold disappears*, as does the notion of the trajectory of the electron. There is no curved spacetime manifold in a theory consistent with general relativity and with quantum mechanics. Spacetime geometry needs to be replaced by something as a "probability distributions of geometries."

This, however, is only part of the story. As I have argued, general relativity does not just replace a flat spacetime with a dynamical (and thus quantized) geometry. It also overrides the very notion of spacetime by reducing the geometry to an attribute of a particular field, the gravitational field, and by reducing the very concept of spacetime to a contiguity relation between the various dynamical entities. The observables, and the predictions of general relativity, concern the contiguity between physical objects. *These* observables acquire quantum properties, that is, are to be described in terms of their probability distribution in a quantum theory of gravity. Thus, the picture of reality that emerges by taking quantum mechanics and general relativity seriously is quite radical: spacetime as an "entity" has fully disappeared, and gets replaced by *probabilistic* contiguity relations between objects, which, by virtue of being probabilistic, will not harmonize within the reassuring picture of a curved pseudo-Riemanian geometry, as in the nonquantistic limit.[17] I will return to this point.

There are several lines of technical investigations towards quantum gravity which are presently being investigated (see Isham 1992; Rovelli 1991a). Two are of particular relevance. One is string theory, and the other is nonperturbative quantum gravity. Both lines of investigation have sketched a tentative general theory. The interest of string theory is that, even if the general (nonperturbative) theory does not yet exist, an approximation scheme is defined, which is well behaved, consistent, and connected to beautiful mathematics. Unfortunately, the approximation scheme breaks down precisely in the regime in which we are interested.

In general, scientists involved in string theory, most of whom come from the more pragmatically oriented tradition of high-energy physics, have so far avoided the issue of the physical meaning of the general theory, concentrating on technical aspects of internal consistency and recovery of low energy limits. If the world is described at the fundamental level by a nonperturbative string theory, then all the general conceptual structures in terms of which we think of the world (such as space, time, particles, and matter) are just approximations, recovered from the mathematics of the theory only within certain regimes. The very objects through which the theory makes contact with reality, namely, scattering amplitudes, do not exist beyond the regime in which spacetime is approximated by a background geometry. What the world is beyond these regimes, and how can we describe, measure, or just think about it, are issues still immersed in a profound mystery.

The other approach is nonperturbative quantum gravity, which cannot even claim a formal well-behaved approximation scheme. On the other side, several researchers in this area, most of whom come from the more speculatively oriented area of gravitational research, have devoted a certain attention to the interpretation of the theory and to the picture of reality around which the theory may be formed. Let me describe the results I consider most interesting.

To a first approximation, the picture that the theory offers is the following. Dynamical quantities, including the metric/gravitational field (see table above), are described as quantum entities.[18] Physically, this means two things: *i.* The state of the field is described in terms of a probability wave that assigns relative probabilities to alternative configurations of the field.[19] *ii.* Not all configurations are physically realizable, but only certain quantized ones. Similarly: *i.'* the gravitational/metric field is described in terms of a probability wave that assigns relative probabilities to alternative configurations of the metric field. Thus, we do not have a spacetime geometry, but a probability distribution of spacetime geometries. *ii.'* Not all configurations are physically realizable, but only certain quantized ones.

Among the most concrete results obtained so far in this direction is a set of solutions to the formal equations of the theory, which can be interpreted as descriptions of those particular quantized configurations of the metric field. In these configurations the continuous metric geometry breaks down at the scale of 10^{-33} cm. What remains at that scale is a fully discrete structure, described by a branch of mathematics denoted *knot theory*, which can be intuitively thought of as a tangle (or a "weave") of one-dimensional "loops of space" (Rovelli 1992; Rovelli and Smolin 1988, 1990; Ashtekar, Rovelli and Smolin 1992). These structures replace the metrical continuum, and emerge as the "quanta," or the quantized aspects of the gravitational/metric field, in the same sense in which a gas of photons is the quantum version of an electromagnetic wave, and represent a more accurate description of the microscopic manifestation of the wave itself. A characteristic possible experimental consequence of this discrete aspect of space is that the area of any physical surface might be quantized, namely, always be an integer multiple of an elementary *quantum* of area (Rovelli 1993a).

However, this quantized aspect of the metric/gravitational field is not the full story. The field operators do not live on a metric manifold, as in pregeneral relativistic quantum field theory, but on a differential manifold (see table), which is a much less structured entity, and, most crucially, is not observable as such. Because of that, the physical content of the theory is far more elusive than what an intuitive picture of a "discrete weave of space" may suggest.

As emphasized in the second section, general relativity is constructed in terms of an ensemble of variables largely redundant with respect to the quantities connected with observations. The nonphysical redundant (gauge)[20] part of the field, in gravity, is given by the dependence on the (x,t) coordinates themselves. The physical nonredundant component of the field is given by the coordinate-independent quantities, namely, the quantities that express the value of physical quantities *at the spacetime location* determined by other quantities. They are precisely those quantities that become quantum entities, are subject to quantum fluctuations, and are observable.

The content of the theory is in the distribution probabilities of the values of physical quantities at spacetime location determined by other dynamical entities.

I suspect that this rather ethereal structure is what we have to learn to master if we want to reach a unified picture of what we have learned on the physical world.

Relational Aspects

Notice that there are *two levels* at which a theory of this sort is relational. As quantum-mechanical observations, the observations described or predicted by the theory refer to the relation between an observing system and an observed system; as general relativistic observations they refer to the contiguity between an entity considered as dynamical and an entity considered as spatio-temporal reference.

Is there any connection between these two levels? If *locality* is to be included into the theory in some fundamental form, *contiguity* should also be an aspect of the observer/observed relation. This contiguity relation is indeed the only remnant of the spacetime structuring of the events of prerelativistic physics. I suspect a connection between the general relativistic relationalism and the quantum-mechanical one should exist, but no theoretical effort, to my knowledge, has yet succeeded in grasping such a connection (see, however, Crane 1994).

We are here at the very boundaries of the present understanding of the physical world.

Time

It is easy to talk about spacetime and think of it as a sophisticated version of our familiar space, but, of course, spacetime includes, for better or for worse, *time*. The notion of time is certainly among the most problematic. The difficulty of physics in dealing with the various layers of the notion of time has been repeatedly emphasized. For instance, the lack of any concept of "becoming" in physics (Reichenbach 1956; Grünbaum 1963), the loss of the notion of *present*, the strident friction between physical time and phenomenological time (see for instance the interesting and comprehensive discussion in Pauri 1994), are embarrassing facts that have made rivers of ink flow (see Macey 1991 for a collection of thousands references on this subject). The physics of this century, general relativity, quantum mechanics, and their hypothetical combination make the matter much more serious (see for instance Ashtekar and Stachel 1991).

If the spacetime continuum geometry breaks down at short scale (in time units, around 10^{-40} seconds), there is no way to think of the world as continuously "flowing through time." This is not going to affect in any way our description or understanding of temporal perception or mental processes, which are completely insensitive to such small time scales. But it is going to affect our basic description of the world. In a sense, "there is no time" at the fundamental level. This disappearance of time is dramatically evident in the mathematical formalism of the theory, both in nonperturbative canonical gravity and in string theory. In both theories, indeed, the fundamental equations essentially *do not contain a time variable*. In canonical quantum gravity, this is due to the fact that since the dependence on the coordinates x and t is a nonobservable gauge, the physical quantities in terms of which the quantum theory is constructed are independent from t. We still have clock times, which are phenomenologically connected to what we usually call time, but the different clock times are only probabilistically related at very short scale, and can no longer be thought of as approximating a universal absolute time variable. In nonperturbative string theory, the very spacetime manifold is born only as a particular vacuum solution of the full theory. In general, there isn't even a spacetime. In both cases, a reasonable evolution in time is hypothetically recovered only within particular regimes.

Can we deal with a fundamental physical theory radically without time? I believe we can. From the mathematical point of view, there are some technical issues to be solved, on which the debate is still open (see Rovelli 1991d, 1991e). But from the physical point of view, a profound rethinking of reality will go with it. This is not easy. Even the comparatively simple and extremely well-established alteration of the notion of time brought by special relativity is very difficult to digest. And many physicists, even many relativists, agree that the flow of time depends on the state of motion of the observer, but then cannot free themselves from thinking of the great big clock that ticks the universal time in which everything happens. Much more difficult is to fully familiarize oneself with the general relativistic relational notion of time; while it is easy to have a good intuition of physics on a given curved manifold, it is, on the other hand, rather difficult to have good intuition of the dynamics of the gravitational/metric field itself. The hypostatization of unphysical structures such as the ADM surface are good examples of the traps along the way, into which too many good physicists keep falling. But the radical step that quantum gravity seems to ask us to take, that is, learning how to think of the world in completely nontemporal terms, is definitely the hardest.

There are two speculative observations I would like to mention in this respect: the possibility of a deep relation between time and thermodynamics and the distinction between the observer's time and the system's time.

Time and Thermodynamics

In general relativity, the time coordinate t is unphysical, and we measure time by selecting a physical variable as a clock (the "internal time"). What is it that characterizes a physical variable as a good clock?

This is the general relativistic version of the classical difficulty of understanding how, from the physical formalism, one could recognize that the time variable is ontologically different from the other variables. Without pretending to address the full complexity of this issue, a hypothesis of a physical mechanism that would single out a particular variable as an internal time was put forward by the author of this work, and independently by the Fields Medal mathematician Alain Connes. The hypothesis can be summarized as follows:

i. That there is no time at the mechanical level. A time is determined only thermodynamically, that is, statistically.
ii. In different statistical states of the system, the time variable is different. Namely, the definition of time is state dependent. Since we can never measure the state of a field system exactly, we can have only statistical knowledge of the state of a system, and we are therefore forced to represent it as a (classical or quantum) probability distribution. It turns out that there is a very general way of computing a "time" flow out of any given statistical state. This is true in classical, as well quantum, mechanics. Explicit computations show that the flow computed out of "reasonable" statistical states are precisely what we would "reasonably" denote as time. (Rovelli 1993a, 1993b; Connes and Rovelli 1994)

This hypothesis is very speculative; to my knowledge it is the only attempt to study the emergence of a specific physical time, *including its physical attributes*, which in large measure are thermodynamical, from the timelessness of the fundamental theory.

Observer Time and System Time

Physics always deals with two distinct temporalities, which I may denote as the system's time and the observer's time. In classical non-relativistic physics, for instance, a physical system is described as evolving in an external parameter t, the system's time. However, the system is described *not just* as a single object that evolves in time. If it were, we would not have equations of motions and arbitrary initial data, but just a unique formula describing one actual motion. The theoretization of the dynamics of the system is done in terms of a space of arbitrary initial data (or a phase space), because the theory takes *explicitly* into account the fact that the description of reality begins at a certain moment (which is not supposed to be justified by the theory itself), and with a certain initial configuration. In other words, classical physics is *structured* in such a way to imply that there is something *external* to the system described, and that this something external determines the initial data. Notice that the possibility of setting up arbitrary initial data for a given system is, in principle, a necessary condition for the verifiability of a specific dynamical theory. Thus, the structure of classical mechanics assumes that there is "somebody" outside the system described that may "freely" set up the initial data, where the two words in quotes have been chosen provocatively.

In other words, if we had the deterministic universe of Laplace, in which everything that happens can be computed from the given position and velocities of all the universe's particles, *who* would compute the evolution? We may perhaps try to argue that such an initial configuration of positions and velocities could have included a small subset of these particles which happen to be in an initial configuration that included (in some sense) the full initial information about the others, plus the calculation capabilities of computing the positions of all other particles (and themselves), but it is clear that this leads us down a dangerous path. And in any case: what about all the other configurations, which, after all, most probably include our actual universe? Laplace, aware of the problem, introduced the extra-universe Demon, to make his picture meaningful. The Demon is a demon not so much because of its prodigious computational capacities, but essentially because it has to remain dynamically *external* to the universe. Classical physics, even if in a subtle way, makes sense only if we postulate that it describes a system that interacts with something external.

In the section on quantum mechanics, I argued that this explicit presence of the observer in the physical theory becomes more crucial in quantum mechanics. At the quantum-mechanical level of accuracy, the very realization of any configuration (state) of a system exists only in relation to an entangling, or an exchange of information, between the system and the observer. I have a best friend who, in learning quantum mechanics, wondered at the idea that all the atoms of his body were likely to be quantum-entangled with all other atoms of the universe. He was wrong. Because in the very moment we (observer) consider a system and start to ask it questions (make measurements over it), then, by definition, we are projecting the state of the system on the outcome of these observations, and therefore we are "disconnecting" it from whatever it was entangled with. In quantum mechanics, the very notion of a system's state depends on the fact that an observer is observing it.

Now the observer has his own time. This is the time *in which* one or the other of the different initial data is selected and the time *along which* the successive quantum measurements (questions, entanglements) are performed. If the observed system is described by a fundamental atemporal theory, but the general physical theory is formulated in terms of the observer/observed relation, can we still deal with a notion of observer time? This is, to my understanding, a cornerstone of the *conceptual* difficulties of constructing a synthesis of the present theories of the physical world.

If *time* is the order of the changes in the states of the systems, and if the state of the system is a relational notion, one that has meaning only if referred to an observer, can there be time outside the observer/observed relation? Is perhaps *time* precisely what emerges from this observer/observed relation? Is *time* perhaps precisely such a relation?

Conclusion

General relativity and quantum mechanics have dismantled the concepts on which the grand synthesis of the Cartesian-Newtonian physics rested for three centuries. We do not have a new synthesis. We have collected an impressive amount of fragmentary knowledge about the physical worlds, but we lack the general picture. We do not know how to think of space, time, matter, and causality in a way that is coherent and consistent with everything we have learned. I have tried in this essay to portray the present tangle of problems in which the physicists who struggle towards a new synthesis are immersed. I should stress once more that the perspective chosen, the emphasis given to the different issues are very personal and that several of the ideas I have illustrated are quite controversial.

I have described a modern view of general relativity that emphasizes the similarity between the gravitational and other physical "material" fields and the relational core of the theory. I have discussed the interpretational problems in quantum mechanics and described a view of the theory based on discarding the notion of an absolute (observer-independent) *state* of a system, in favor of a notion of relative observed-observer information. The relational description of the spatiotemporal structure of the world provided by general relativity reverberates strangely with the account of reality in terms of the observer-observed relation that is at the core of quantum mechanics. I have described the present state of the technical efforts to combine the two theories and construct a quantum theory of gravity and the conceptual jumps that these efforts seem to require us to take, in particular, the need to replace spacetime geometry with a complex probabilistic net of contiguity relations. Finally, I have indicated the problem of time as the area that I believe holds the major difficulties, as well as the key to future essential advances. We must probably learn how to think of reality in fully nontemporal terms, but we do not yet understand how.

My feeling is that the extraordinary physical discoveries of the twentieth century, general relativity and quantum mechanics in particular, point toward a profoundly new and deeper way of understanding physical reality. It is almost as if they are screaming something to us and we do not understand what. It has been claimed that we are at the end of physics. We are not close to the end of physics, nor to the final theory of everything. We are very much in the dark. We left the sunny grasses of Cartesian-Newtonian physics and are traveling through the woods, armed with everything we have learned and with our weak intuition, always wishing we were smarter. It would be a discouraging state of confusion, and we would feel lost, if it weren't that the trip is wonderful and the landscape so breathtaking.

NOTES

For ideas, discussions, friendship, and excitement, I am indebted to Abhay Ashtekar, José Balduz, Julian Barbour, Mauro Carfora, Paola Cesari, Alain Connes, Louis Crane, John Earman, Bob Geroch, Bob Griffits, Jonathan Halliwel, Jim Hartle, Gary Horowitz, Giorgio Immirzi, Chriss Isham, Junichi Iwasaki, Ted Jacobson, Al Janis, Karel Kuchar, Annalisa Marzuoli, Ted Newman, John Norton, Massimo Pauri, Roger Penrose, Lee Smolin, John Stachel, Bob Wald, John Wheeler.

1. The puzzle was to tidy up an immensity of experimental data. The solution is the $SU(3) \times SU(2) \times U(1)$ standard model. I believe that the common emphasis on the standard model's aesthetic weakness tends to obscure the fact that the standard model is one of the most successful theories ever invented. There is a curious reticence to admit that we are today in one of the rare moments of the history of physics in which we have a basic set of equations that could in principle describe and predict *everything* we are presently capable of observing.

2. As in Earman 1990, I denote as *relational* the philosophical position according to which it makes sense to talk of the motion of objects only in relation to one another, as opposed to the motion of objects with respect to space. The term *relational* has the advantage with respect to the term *relativistic* (used in this

context) in that it does not lead to confusion with Einstein's theories (of which special relativity is certainly fully nonrelational).

3. On the relational/absolute controversy, see Earman 1990 and Barbour 1989.

4. This is from Newton 1962. See Barbour 1989 for a fascinating account of the path that led to the discovery of dynamics.

5. I suspect everywhere.

6. The motion of every physical object is affected by the gravitational field in a substantial way, not as a perturbation eliminable in principle. Indeed, without knowledge of the gravitational field (that is, of the metric), we have absolutely no clue as to how an object will move.

7. Einstein was led to reject generally covariant equations by difficulties of recovering the weak field limit, about which he had incorrect assumptions that he corrected later. The general physical argument he provides against generally co-variant equations was probably prompted by those difficulties. But this does not dispense us from taking this argument seriously, particularly from the man who, out of general arguments of this sort, changed our view of the world drastically, and more than once.

8. These fields should not be confused, and have little to do, with the Schrö-dinger wave function; they are fields like the electromagnetic field, but with different spin-statistic properties.

9. Unless the approximation that they do not affect geometry is assumed: an approximation which is practically very useful, but conceptually is a denial of the very central assumption of the theory. One cannot found one's fundamental view of the world on an approximation that one knows is bound to fail.

10. To increase the confusion, the expression *matter* has a different meaning in particle physics and in relativistic physics. In particle physics, one usually distinguishes three entities: *spacetime*, *radiation*, and *matter*. Radiation includes fields such as the electromagnetic field, the gravitational field, and the gluon fields (bosons); matter includes the leptons and the quarks, which are also described as fields, but have more pronounced particlelike properties than photons or gluons (fermions). In relativistic physics terminology, which we follow here, one denotes collectively both *radiation* and *matter* (namely, both bosons and fermions), as *matter*. What I am arguing here is not that the metric has the same ontological status as the electrons (*matter-matter*), but that the metric/gravitational field has the same ontological status as the electromagnetic field (*radiation-matter*).

11. As Feynman put it, a good theoretical physicist keeps several alternative interpretations of the same phenomenon in his mind; these can be quite different but perfectly equivalent, as long as one remains within the ambit of that phenomenon. But no one can know *which one* of these interpretations will be the *good one* that generalizes to wider ambits. We are precisely in such a situation. The recent efforts to construct a quantum theory of gravity, or a unified theory of all interactions (for instance, via string theory), or a thermodynamics of the gravitational field, and other efforts, suggest that the second point of view, namely viewing the metric/gravitational field as a field like any other, is the fruitful one.

12. The gravitational field is characterized by the fact that it is the only field that couples with *everything*.

13. In the *Principia Philosophiae*, Sec. II-25, pg. 51.

14. To allow himself to make use of the notion of absolute space, Newton used the extreme resort of calling God to his aid: according to Newton, space has curious ontological properties because it is nothing but the *sensorium of God*.

15. This entire issue has strong resemblance with the polemics on the notion of *action-at-distance*, which is the other aspect of Newtonian mechanics that scandalized continental natural philosophers: pre-Newtonian physics disliked action-at-distance. Newton's genius introduced it, being aware of its difficulties (he calls it *repugnant*, at one point) but realizing its efficacy. Faraday, Maxwell and Einstein (again) were able to purify the prodigiously powerful Newtonian construct of this "repugnant" action-at-distance, without impairing the general scheme, but using the costly jewels of a much higher mathematical complexity and of the new idea of field (*tout se tien*).

16. The problem of finding a quantum theory of gravity must not be confused with the issue of "quantum cosmology," which, strictly speaking, is a subproblem of the problem of the interpretation of quantum mechanics. Quantum cosmology is the investigation of the possibility of providing a *quantum* description of the universe as a whole. One can see that this problem is logically disconnected from the problem of finding a quantum theory of the gravitational field, by considering the following: if the gravitational field *did not exist*, and the $SU(3) \times SU(2) \times U(1)$ standard model exhausted all known physics, we would *still* have a quantum cosmology problem, namely the issue of describing the quantum state of the whole universe; vice versa, if we did not care about describing the entire universe quantum mechanically, we would *still* have a quantum gravity problem in attempting to describe the behavior of the gravitational field in the small. It has been repeatedly suggested that the two problems (quantum gravity and quantum cosmology) can only be solved together. I still do not see why, but this is a quite unorthodox opinion. The history of physics is full of instances in which two apparently unrelated problems were solved together (such as combining electricity and magnetism, *and* understanding what light is), but what holds even greater hope is that unrelated problems would be connected (finding the theory of the strong interactions *and* getting rid of quantum field theory and its divergences is one of the latest examples).

17. More technically, neither the metric/gravitational fields $g_{\mu\nu}(x,t)$, nor any scalar (as the Ricci scalar $R(x,t)$) built from it, may become quantum observables, that is, give rise to independent Heisenberg relations. Because they depend explicitly on x, and t, while the *physical* quantities general relativity deals with (that is, the quantities that can be compared by means of experiments, univocally predicted, and so on) are *independent from x and t*. The quantum observables, whose probability distribution quantum gravity must provide, are quantities independent from x and t. They can, for instance, refer to the value of a certain field *in the place and at the time* determined by other physical objects dynamically interacting with the field. For a technical discussion of this important point, see Rovelli 1991b, c, d, e.

18. Quantum field operators.

19. The probability is given as the expectation value of relevant projection operators.

20. In a more technical language, only two out of the ten components of the metric field represent physical degrees of freedom. The others are unphysical (un-observable) *gauges*, whose role is just to permit a simpler, local, formulation of the dynamics. In electromagnetism, the Maxwell potential has four components, but only two of these correspond to physical degrees of freedom. Only these two are quantum variables. Namely, only these two are subject to Heisenberg indeterminacy principle.

REFERENCES

Ashtekar, A., C. Rovelli, and L. Smolin. 1992. "Weaving a Classical Metric with Quantum Threads." *Physical Review Letters* 69; 237–40.

Ashtekar, A., and J. Stachel, eds., 1991. *Conceptual Problems of Quantum Gravity.* Boston: Birkhäuser.

Barbour, J. 1989. *Absolute or Relative Motion? A Study from the Machian Point of View of the Discovery and the Structure of Dynamical Theories*, pt. 1. Cambridge: Cambridge University Press.

Belifante, F. J. 1973. *A Survey of Hidden Variable Theories.* Pergamon Press.

Bell, J. 1987. *Schrödinger: Century of a Polymath.* Cambridge: Cambridge University Press.

Beltrametti, E. G., and G. Cassinelli. 1981. *The Logic of Quantum Mechanics.* Reading, Mass.: Addison-Wesley.

Bohm, D. 1951. *Quantum Theory.* Englewood Cliffs, N.J.

Bohr, N. 1935. *Nature* 12.65.

———. 1949. Discussion with Einstein in *Albert Einstein: Philosopher-Scientist.* Open Court.

Connes, A., and C. Rovelli. 1994. "Von Neumann Algebra Automorphisms and the Time-Thermodynamics Relation in General Covariant Theories." Photocopy, University of Pittsburgh.

Crane, L. 1994. "Topological Field Theory as the Key to Quantum Gravity." In J. Baez, ed., *Knots and Quantum Gravity*, 121–34. Oxford: Clarendon Press.

———. *Journal Math. Phys.* 35: 16180.

Descartes, R. 1983 [1644]. *Principles of Philosophy.* Translated by V. R. Miller and R. P. Miller. Dordrecht: D. Reidel.

Earman, J. 1989. *World Enough and Spacetime: Absolute vs. Relational Theories of Space and Time.* Cambridge: MIT Press.

Earman, J., and J. Norton. 1987. "What Price Spacetime Substantivalism? The Hole Story." *British Journal for the Philosophy of Science* 38: 515–25.

Finkelstein, D. 1969. *Boston Studies in the Philosophy of Science*, vol. 5. Ed. R. S. Cohen and M. W. Wartofski. Dordrecht.

Ghirardi, G. C., A. Rimini, and T. Weber. 1986. *Phys Rev D* 34: 470.

Grünbaum, A. 1963. *Philosophical Problems of Space and Time.* New York: Knopf.

Hall, A. R., and M. B. Hall, eds. 1962. *Unpublished Scientific Papers of Isaac Newton.* Cambridge: Cambridge University Press.

Heisenberg, W. 1927. "Über den anschaulichen Inhalt der quantentheoretishen Kinematik und Mechanik." *Zeitschrift für Physik* 43: 172–98.

Hughes, R. I. G. 1989. *The Structure and Interpretation of Quantum Mechanics.* Cambridge: Harvard University Press.

Isham, C. J. 1992. "Quantum Gravity." Photocopy, Imperial College.

Jauch, J. 1968. *Foundations of Quantum Mechanics.* Reading, Mass.: Addison-Wesley.

Landau, L. D., and E. R. Lifshitz. 1962. *Quantum Mechanics: Non-Relativistic Theory.* Oxford: Pergamon.

Lange, L. 1885. "Über das Beharrungsgeset." *K. Sächsische Akademie der Wissenschaften zu Leipzig. Berichte über die Verhandlungen der Math.-phys. KL* 37: 333–51.

Macey, S. L. 1991. *Time: A Bibliographic Guide.* New York: Garland.

Mackey, G. W. 1963. *Mathematical Foundations of Quantum Mechanics.* New York: Benjamin.

Maczinski, H. 1967. *Bulletin de L'Académie Polonaise des Sciences* 15: 583.

Newton, I. 1962. "De Gravitatione et Aequipondio Fluidorum." Translation in A. R. Hall and M. B. Hall, eds., *Unpublished Papers of Isaac Newton,* 89–156. Cambridge: Cambridge University Press, 1962.

Norton, J. 1989. "Einstein's Search for Covariance 1912–1915." In D. Howard and J. Stachel, eds., *Einstein and the History of General Relativity, Einstein Studies,* vol. 1, 101–59. Boston: Birkhäuser.

———. 1993. "General Covariance and the Foundation of General Relativity: Eight Decades of Dispute." *Reports on Progress in Physics* 56: 791–858.

Pauri, M. 1994. *Dizionario delle Scienze Fisiche; Voce quadro: Spazio e Tempo.* Italy: Instituto Treccani.

Penrose, R. 1989. *The Emperor's New Mind.* Oxford: Oxford University Press.

Piron, C. 1972. "Survey of General Quantum Physics." *Foundations of Physics* 2: 287–314.

Reichenbach, H. 1956. *The Direction of Time.* Berkeley: University of California Press.

Rovelli, C. 1991a. "Ashtekar Formulation of General Relativity and Loop-space Non-perturbative Quantum Gravity: A Report." *Classical and Quantum Gravity* 8: 1613–75.

———. 1991b. "Quantum Mechanics Without Time: A Model." *Physical Review D* 42: 2638–46.

———. 1991c. "Quantum Reference Systems." *Classical and Quantum Gravity* 8: 317–31.

———. 1991d. "Time in Quantum Gravity: An Hypothesis." *Physical Review D* 42: 442–56.

———. 1991e. "What Is Observable in Classical and Quantum Gravity?" *Classical and Quantum Gravity* 8: 297–316.

———. 1992. "Knot Theory Spacetime." Feature article in Encyclopedia Britannica, *Year-Book of Science and Technology.*

———. 1993a. "A Generally Covariant Quantum Field Theory and a Prediction on Quantum Measurements of Geometry." *Nuclear Physics B* 405: 797–815.

———. 1993b. "Statistical Mechanics of Gravity and the Thermodynamical Origin of Time." *Classical and Quantum Gravity 10*: 1549–66.

———. 1993c. "The Statistical State of the Universe." *Classical and Quantum Gravity 10*: 1567–78.

———. 1996. "Relational Quantum Mechanics." *Int. Journal of Theoretical Physics 35*: 1637–78.

Rovelli, C., and L. Smolin. 1988. "Knot Theory and Quantum Gravity." *Physical Review Letters 61*: 1155–58.

———. 1990. "Loop-space Representation of Quantum General Relativity." *Nuclear Physics B 331*: 80–152.

Schrödinger, E. 1935. "Die gegenwärtige Situation in der Quantenmechanik." *Naturwissenshaften 23*: 807–12, 823–28, 844–49.

Shannon, C. E., and W. Weaver, 1949. *The Mathematical Theory of Communication*. Urbana: University of Illinois Press.

Stachel, J. 1989. "Einstein's Search for General Covariance, 1912–1915." In D. Howard and J. Stachel, eds., *Einstein Studies*, vol. 1, 63–100. Boston: Birkhäuser.

Van Fraassen, B. 1991. *Quantum Mechanics: An Empiricist View*. New York: Oxford University Press.

Von Neumann, J. 1932. *Mathematische Grundlagen der Quantenmechanik*. Berlin: J. Springer. Translated by R. T. Beyer as *Mathematical Foundation of Quantum Mechanics*, 1955. Princeton: Princeton University Press.

Wheeler, J. A. 1988. "World as System Self-synthesized by Quantum Networking." *IBM Journal of Research and Development 32*: 4–15.

———. 1989. "Information, Physics, Quantum: The Search for Links." In *Proceedings of the Third International Symposium on the Foundations of Quantum Mechanics*, 354–68. Tokyo.

———. 1992. "It from Bit and Quantum Gravity." Princeton University Report.

7

What Superpositions Feel Like

David Z Albert

Department of Philosophy, Columbia University

A number of famous stories in the physical literature start out by supposing the state of the world at any given instant can be completely described by means of a quantum-mechanical wave function, and that the linear quantum-mechanical equations of motion amount to a true and complete account of the evolutions of those wave-functions in time. These stories have absurd endings. That those sorts of stories end up that way has come to be called the measurement problem.

The sharpest and clearest of those stories, I think, is the one about Wigner's friend (1961).

Let me remind you of it.

It goes like this. Wigner has a friend who is a competent observer of the x-spins of electrons, and that friend is in possession of a device for measuring the x-spins of electrons, and that device is working properly.

What Wigner takes "competent" and "working properly" to mean is something like this: Suppose that Wigner's friend has resolved to measure the x-spin of a certain electron. If it's the case that the x-spin of the electron is with certainty "up," then an x-spin measuring device that is working properly will, by definition, once it has completed a measurement of the x-spin of that electron, indicate (by means of some sort of a *pointer*, for example) with certainty that the x-spin of that electron is "up"; and a competent observer, by definition, having looked at the measuring-device subsequent to that measurement, will with certainty come to *believe* that the x-spin of the electron is up. Of course, if the case is that the x-spin of the electron is with certainty "*down*," then a measuring device that is working properly will end up indicating *that*, and Wigner's competent friend will end up *believing* that, by definition, with certainty.

Mathematically, Wigner is supposing that the quantum-mechanical equations of motion entail that the composite physical system consisting of Wigner's friend and her measuring device and the measured electron will behave like this:

[resolved to measure the x-spin>$_F$ × [ready to measure the x-spin>$_M$ × [x-spin "up">$_e$ – – – – → [believes that x-spin is up>$_F$ × [indicates that x-spin is up>$_M$ × [x-spin "up">$_e$ (7.1)

and like this:

[resolved to measure the x-spin>$_F$ × [ready to measure the x-spin>$_M$ × [x-spin "down">$_e$ – – – – → [believes that x-spin is down>$_F$ × [indicates that x-spin is down>$_M$ × [x-spin "down">$_e$.   (7.2)

And these certainly seem like eminently reasonable necessary conditions for characterizing Wigner's friend and her measuring instrument as (respectively) "competent" and "working properly."

(It hardly needs saying, by the way, that there are immense oversimplifications in the way I've written down the states in (1) and (2). For example, the phrase "resolved to measure the x-spin" certainly doesn't completely describe the quantum state of Wigner's very complicated friend! The physical system called Wigner's friend obviously has many other physical properties too; properties like the positions of her toes, or the positions of those ions in her brain which determine, say, what sorts of ice cream are her favorites. Moreover, I am obviously adopting an absolutely naive account of how mental states supervene on physical states of the brains of sentient beings here. But I don't think that anything that's important is going to turn out to hinge on any of that. I think that all of what's about to happen would go through in much the same way in the context of considerably more *detailed* accounts of the physical structures of sentient observers and considerably less *naive* accounts of the supervenience of the mental on the physical too.)

OK. Now, suppose that as a matter of fact the initial state of the electron in question is neither the x-spin up state nor the x-spin down state but rather the y-*spin up* state; and suppose (as before) that Wigner's friend is resolved to measure the x-*spin* of this electron. So the state of the composite system that consists of the electron and the x-spin measuring device and Wigner's friend looks like this:

[resolved . . .> × [ready . . .> × [y-spin up> = [resolved . . .> × [ready . . .> × {½[x-spin up> + ½[x-spin down>}.   (7.3)

Well, it turns out that if Wigner's friend is indeed competent in accordance with Wigner's definition, and if her x-spin measuring device is working properly in accordance with Wigner's definition, and if both of those physical systems are taken to evolve strictly in accordance with the linear quantum-mechanical equations of motion, then it follows from the linearity of the quantum-mechanical equations of motion that

[the state in (3)> → ½{[believes that x-spin is up> × [indicates that x-spin is up> × [x-spin up> + [believes that x-spin is down> × [indicates that x-spin is down> × [x-spin is down>}.   (7.4)

The reason is that the initial state here — in (7.3) — can be expressed as a superposition of the two initial states in (7.1) and (7.2); and since that is so, the linearity of the equations of motion require that the *final* state here can necessarily be written as the same superposition of the two *final* states in (7.1) and (7.2).

And so what comes of supposing that the quantum-mechanical equations of motion give a true and complete account of the workings of the entire physical world is that we get forced to conclude that measuring processes like the one just described invariably and deterministically leave the world in a coherent superposition of *two* states, in one of which the x-spin of the electron is down and the pointer on the measuring device is *pointing* to the word "down" and the experimenter *sees* that that pointer is pointing to the word "down" and consequently *believes* that the x-spin of the electron is down, and in the other of which the x-spin of the electron is *up* and the pointer of the measuring device is pointing to the word "up" and the experimenter sees that the pointer is pointing to the word "up" and consequently believes that the x-spin of the electron is up.

And so what comes of supposing that quantum mechanics gives a true and complete account of the workings of the entire physical world is that we get forced to conclude that measuring processes like the one just described invariably and deterministically leave the world in a state in which (on the standard way of thinking about what it means to be in a superposition), there fails to be any determinate matter of fact about (among other things) where the pointer on the measuring device is pointing, and where the experimenter *takes* the pointer on the measuring device to be pointing!

And the trouble with *that*, of course, is that we are almost sure that we know that there *are* determinate matters of fact about where pointers on those sorts of devices are pointing at the ends of those sorts of measurements and that we are *absolutely* sure (so the story goes) that we know that there are determinate matters of fact about where we *take* those pointers to be pointing at the ends of those sorts of measurements.

So the conventional wisdom has always been that there must necessarily come a point in the course of this sort of measurement at which the purely quantum mechanical account of the workings of the world either fails to be true or at least fails to be complete; and that that point ideally ought not come any later than the moment at which the outcome of the measurement gets recorded in the position of the macroscopic pointer on the measuring device; and that that point absolutely cannot come any later than the moment at which the outcome of the measurement gets recorded in the *awareness* of the *experimenter*. And the problem of determining precisely how and where that failure occurs has emerged as the central problem at the foundations of quantum mechanics.

I

But (as everybody knows) there is a small underground tradition of what you might call revolutionary resistance to that conventional wisdom, a tradition that goes back to the late Hugh Everett III, who announced in 1957, in a heroic and astonishing paper, that he had discovered a way of coherently entertaining the possibility that the linear quantum-mechanical equations of motion are indeed (in spite of the arguments just described) the true and complete equations of motion of the whole world.

The trouble with Everett's paper is that it's terribly obscure. The trouble is that (notwithstanding the conviction of anybody who reads it that *there's something there*, that Everett is unmistakably *onto* something) it turns out to be extraordinarily difficult to dig any entirely explicit idea out of that paper; it turns out to be extraordinarily difficult to figure out exactly what Everett was trying to say there.

And so a number of rather different attempts at doing that, a number of *readings* of Everett's paper have emerged over the years. This essay addresses three of those.

II

The first, and the most famous, one is the *Many-Worlds* reading.

What Everett announced in his paper (to put things a little more concretely than I have) was that he had discovered some means of coherently entertaining the possibility that the states of things at the conclusions of measurements of the x-spins of y-spin up electrons really *are* (precisely as the linear equations of motion demand) superpositions like the one in (7.4). And the idea of the *many-worlds* reading of Everett's paper is that the means of coherently entertaining that possibility that Everett must have had in mind (or perhaps the one that he *ought* to have had in mind) is to take the two components of a state like the one in (7.4) to represent (literally!) *two physical worlds*. The idea is that in the course of an interaction like the one that leads from (7.3) to (7.4), *the number of physical worlds there are literally increases from one to two*, and that in each one of those worlds, x-spin measurement actually *has* an outcome and the observer actually has a determinate belief about that outcome, and that the linearity of the equations of motion will entail that those two worlds will for all subsequent time be absolutely unaware of one another.

But there turn out to be any number of reasons why that won't work. I'll just mention three of them.

To begin with, there's a mystery, on the many-worlds interpretation, about what it could mean to say, for example, that in the event that Wigner's friend carries out a measurement of the x-spin of an initially y-spin up electron, the "probability" that that measurement will come out "up" is 1/2. The trouble is that (as I just mentioned) that sort of a measurement, on *this* interpretation of quantum mechanics, will *with certainty* give rise to *two worlds*, in *one* of which there's a friend of Wigner's who sees that the outcome of the measurement is "up," and in the *other* of which there's a friend of Wigner's who sees that the out-come of the measurement is "down," and there *isn't* going to be any matter of fact about *which one* of those two worlds is the *real* one, or about which one of those two *friends of Wigner's* is the *original* friend.

And there's a sense in which the many-worlds interpretation, as it stands, isn't well defined. The trouble is that *what worlds there are*, at any particular instant, on this way of talking, depends on *what separate terms there are in the universal state-vector* at that instant; and what separate terms there *are* in that state-vector at that particular instant will depend on what *basis* we choose to *write that vector down in*; and of course nothing in the quantum-mechanical formalism *itself* will pick out any particular such basis as the *right* one to write things down in; and so, if there is going to be any *objective matter of fact* about what worlds there are, at any given instant, on the many-worlds way of talking, then some new general principle has to be *added* to the formalism which *does* pick out some particular basis as the right one to write things down in.

And there's a sense in which the many-worlds interpretation is just *incoherent*.[1]

Here's how that goes:

Consider the state in (7.4). That state is necessarily an eigenstate of some complete observable of the composite system which consists of Wigner's friend and her *x*-spin measuring device and the electron *e* (since *any* quantum state of *any* physical system is necessarily an eigen-state of some complete observable of that system). Call that observable *A*, and call the associated *eigenvalue a*.

OK. Now, imagine a universe that initially consists of Wigner, his friend, and of a measuring device called $m1$ for measuring *x*-spins, of a measuring device called $m2$ for measuring *A*, and of a *z*-spin up elec-tron. Suppose that Wigner and his friend agree that the following will take place: first, Wigner's friend will measure the *x*-spin of the electron, and when that's done, Wigner himself will carry out a measurement of *A*.

Once the *x*-spin measurement is done (but when the *A*-measure-ment is, as yet, undone) the universal state-vector is going to be:

$1/\sqrt{2}\{[r>_w[r>_{m2}[\text{believes that } x\text{-spin is up}>_F[\text{indicates that } x\text{-spin is up}>_{m1}[x\text{-spin up}>_e + [r>_w[r>_{m2}[\text{believes that } x\text{-spin is down}>_F[\text{indi-cates that } x\text{-spin is down}>_{m1}[x\text{-spin down}>_e\}.$ (7.5)

On the many-worlds interpretation, (7.5) represents a world in which there is a Wigner who is resolved to measure A and there is a friend who believes that the x-spin of e is up and so on, and *another* world in which there is *another* Wigner who is *also* resolved to measure A and another *friend* who believes that the x-spin of e is *down*, and so on. And of course there will be no determinate matter of fact, in *either one* of those worlds, about what the value of A is. But, since the universal state-vector in (7.5) is an eigenstate of A with eigenvalue a, the quantum mechanical equations of motion entail, with certainty, that the universal state-vector once the A-measurement is done is going to be:

$1/\sqrt{2}\{$[believes that $A = a>_w$[indicates that $A = a>_{m2}$[believes that x-spin is up$>_F$[indicates that x-spin is up$>_{m1}$[x-spin up$>_e$ + [believes that $A = a>_w$[indicates that $A = a>_{m2}$[believes that x-spin is down$>_F$[indicates that x-spin is down$>_{m1}$[x-spin down$>_e\}$.   (7.6)

And so the worlds which these two Wigners *see*, with *certainty*, when they go and look, turn out *not* to be the ones which the many-worlds interpretation describes them as *inhabiting*! And in the light of *that*, it becomes unclear what it could possibly mean to construe the two terms in (7.5) as descriptions of two separate physical worlds *at all*.

III

Let's switch to *another* reading (let's switch, that is, to *another* attempt to entertain the possibility that the linear quantum-mechanical equations of motion are the true and complete equations of motion of the whole world) which has been around for quite a long time, and which has lately been re-invented in a number of papers by Murray Gell-Mann and Jim Hartle (1990) and Wojciech Zurek.

The idea is that there's a certain feature of the interactions between macroscopic physical systems and their environments that makes it the case that superpositions of macroscopically different states (unlike superpositions of *microscopically* different ones) *can* be regarded, and *ought* to be regarded, as situations in which either *one* or *another* of those macroscopically different states actually *obtains*.

Let me spell that out with some care. Let's start with the basics.

Remember why it is that the proposition that a certain physical system is in a coherent superposition of the states [$A>$ and [$B>$ has always been thought to be incompatible with the proposition that that system is in *either* the state [A> *or* the state [B>, but we do not know which: It's because there are necessarily real, physical, measurable properties of the coherent superposition of [A> and [B> (*whatever* states [A> and [B> are) which are properties of *neither* [A> nor [B> *separately*, and which (consequently) also cannot be properties of a situation which is *either* [A> or [B>, but we don't know which. That's what's taught us to say, of such superpositions that they represent situations not in which we are *ignorant* of whether it is [A> or [B> that obtains, but rather situations in which there is simply not any *fact* to the matter as to whether [A> or [B> obtains.

Consider, for example, an electron which is in a coherent superposition of a z-spin "up" and a z-spin "down" state, with equal coefficients. That superposition has the property that, with certainty, the value of the x-*spin* is "up," and neither the z-spin "up" state nor the z-spin "down" state nor any situation in which either one or the other of those states (but we do not know which) obtains has that property.

And precisely the same sort of thing is true of, say, the z-spin of an electron in an atom of hydrogen in its ground state, in which the spins of the electron and the proton are quantum-mechanically *entangled* with one another. That state (which is a coherent superposition of one state in which the z-spin of the electron is "up" and another in which the z-spin of the electron is "down") is associated with a definite value not of any spin-observable of the electron by itself nor with any spin-observable of the *proton* by itself, but rather with a definite value (namely, zero) of the total spin angular momentum of the *two*-particle system. It happens that there is no state of that two-particle system in which the z-spin of the electron is "up" and the total spin angular momentum is zero, and that there is no state in which the z-spin of the electron is "down" and the total spin angular momentum is zero, and so the total spin angular momentum's being zero is simply *incompatible* with the hypothesis that there is any fact of the matter about the value of the electron's z-spin.

And so it goes for superpositions in general.

Now, what Gell-Mann and Hartle and Zurek and their many predecessors have pointed out is that it's in the *nature* of macroscopic observables (in virtue of their *being* macroscopic observables) that their *values* almost instantaneously get *recorded* in extraordinarily complicated properties of their *environments*.

What all those guys have pointed out (to be a bit more specific) is that it's in the nature of macroscopic observables that their values almost instantaneously get recorded in certain properties of the *air molecules* in their vicinities, and of the *photons* in their vicinities, and of the *cosmic rays* in their vicinities, and so on.

So states like the one in (7.4), if the two terms in (7.4) genuinely represent *macroscopically different* circumstances, will almost instantaneously evolve into states as follows:

$1/\sqrt{2}${[C>[believes that x-spin is up> × [indicates that x-spin is up> × [x-spin up> +[C'>[believes that x-spin is down> × [indicates that x-spin is down> × [x-spin is down>}   (7.7)

where [C> and [C'> are orthogonal states of the *environment* of Wigner's friend and her measuring apparatus. And so the observable which distinguishes the linear superposition of the two macroscopically different states in (7.4) from either one of those two states *separately* (that is, the analogue of the total spin angular momentum of the hydrogen atom in the previous example) is necessarily going to be an observable of (among other things) a *huge* collection of air molecules and photons and cosmic rays; and consequently that observable is necessarily going to be an *extraordinarily difficult* one to *measure*!

And that's all right.

But all that certainly does *not* entail what Gell-Mann and Hartle and Zurek and their predecessors seem to *think* it does (that is, it does not entail that superpositions of macroscopically different states can be regarded as situations in which either one or another of those states — but we do not know which — actually *obtains*). The question of whether superpositions can be regarded as either/or situations is, alas, precisely the same in the macroscopic level as on the *microscopic* one. The fact of the matter, according to quantum theory, is that the world has certain definite physical properties, when such superpositions obtain, which it would *not* have in the event that one or another of the *superposed* states obtained. The *practical measurability* of those properties is beside the point; all that is relevant to the question at issue here is their *reality*.

And (so long as what's being entertained is that the quantum-mechanical equations of motion are the true and complete equations of motion of the entire physical world) the reality of those properties cannot be in dispute.

And so this reading doesn't pan out either.

Let's try one more.

This one is due to me, as far as I know.

IV

The idea here is to read Everett as being skeptical that, when you come right down to it, the problem of measurement has any *empirical basis*.

It goes like this:

Suppose (contrary to the first reading) that there is only one world; and suppose (contrary to the *second* reading) that the *standard* way of thinking about what it means to be in a superposition is invariably the *right* way, both for *microscopic and macroscopic* systems.

Let's see if we can figure out what it would *feel* like, if all that's true, to be the experimenter in a state like the one in (7.4).

What's the right way to figure something like that out?

Here's a way to get started:

Suppose (as we've been supposing throughout this discussion) that the linear quantum-mechanical equations of motion were invariably true, and (consequently) that observers like the one described above frequently did end up, at the conclusions of x-spin-measurements, in states like the one in (7.4).

Let's see if we can figure out what those equations would entail about how an observer like that, in a state like the one in (7.4), would *respond* to *questions* about how she feels (that is: about what her mental state is). Maybe that will tell us something.

The most obvious question to ask is: 'What is your present belief about the x-spin of the electron?' That question, though, turns out not to be of much use here. Here's why: Suppose that the observer in question (the one that's now in the state in (7.4)) gives honest responses to such questions; suppose, that is, that when her brain-state is [believes that x-spin up> she invariably responds to such a question by saying the word "up," and that when her brain state is [believes that x-spin down> she invariably responds to such a question by saying the word "down." The problem is that precisely the same linearity of the equations of motion which brought about the superposition of different brain-states in the state in (7.4) in the first place will now entail that if we were to address this sort of a question to this sort of an observer, when (7.4) obtains, then the state of the world after she *responds* to the question will be a superposition of one in which she says "up" and another in which she says "down"; and it will not be any easier to interpret a "response" like that than it was to interpret the superposition of brain-states in (7.4) of which that response was intended to describe!

But there are other sorts of questions that turn out to be more informative.

Let me remind you, to begin with, of something about quantum-mechanical superpositions. It's something extremely trivial, but what it entails about the measurement problem has somehow escaped wide notice so far.

It's just this: It follows from the linearity of the operators which represent observables of quantum mechanical systems that *any* measurable physical property which happens to be shared by all of the individual mathematical terms of some particular superposition (written down in any particular basis) will necessarily also be shared by the full superposition, considered as a single quantum-mechanical state, as well.

Consider, for example, an electron which is in a superposition of being in one corner of this room and being in another one; and suppose that I have a measuring device that is designed to tell me *not* precisely where the electron is, but only whether it is *somewhere in the room*. Then what follows from the linearity of the operator which represents the "somewhere-in-this-room" observable is that when the state I just described obtains, that measuring device will necessarily end up indicating that that electron *is* in the room.

O.K. Let's apply that to the superposition of states in (7.4).

Suppose that we were to say to Wigner's friend, "Don't tell me whether you believe the x-spin of the electron to be up or if you believe it to be down, but tell me merely whether or not *one* of those two is the case; tell me, in other words, merely whether you now *have* any particular definite belief (not uncertain and not confused and not vague and not superposed) about the value of the x-spin of this electron."

Well, if we were to ask Wigner's friend that when the state [believes that x-spin up>$_F$[indicates that x-spin up>$_m$[x-spin up>$_e$ obtains, and if Wigner's friend is indeed an honest and competent reporter of her mental states, then she would presumably answer, "Yes, I *do* have some determinate belief about that at present; one of those two *is* the case"; and of course she would answer in precisely the same way in the event that [believes that x-spin down$>_F$[indicates that x-spin down$>_m$[x-spin down$>_e$ obtains. So responding to this particular question in this particular way (by saying yes) is an observable property of Wigner's friend in both of those states, and consequently (and this is the punch line here) it will also be an observable property of her in any superposition of those two brain states, and consequently (in particular) it will be an observable property of her in (7.4).

That's really odd. Look what we've found out: On the one hand, the dynamical equations of motion predict that Wigner's friend is going to end up, at the conclusion of a measurement like the one we've been talking about, in the state in (7.4), in which (on the standard way of thinking about what it means to be in a superposition) Wigner's friend has *no particular belief whatever* about the x-spin of the electron; and on the other hand we have just now discovered that those same equations also predict that when a state like (7.4) obtains, Wigner's friend is necessarily going to be convinced (or at any rate she is necessarily going to *report*) that she *does* have some particular belief about the x-spin of the electron.

Let's go on. Suppose that Wigner's friend carries out a measurement of the x-spin of an electron whose y-spin is initially definite with an x-spin measuring device called $m1$, and suppose that when that's done (that is: when a state like (7.4) obtains), h carries out a *second* measurement of the x-spin of that electron with a *second* x-spin measuring device called $m2$. When *that* is done, the state of Wigner's friend, the two measuring devices, and the electron (if the measuring devices are good, and if Wigner's friend is competent, and if everything evolves in accordance with the linear dynamical equations of motion) is going to look like the following:

$1/\sqrt{2}${[believes that the outcome of the first measurement is "up" and believes that the outcome of the second measurement is "up"$>_F$[indicates that up$>_{m1}$[indicates that up$>_{m2}$[x-spin up$>_e$ + [believes that the outcome of the first measurement is "down" and believes that the outcome of the second measurement is "down"$>_F$[indicates that down$>_{m1}$[indicates that down$>_{m2}$[x-spin down$>e$}.   (7.8)

And suppose that at that point (when (7.8) obtains) we were to say to Wigner's friend, "Don't tell me what the outcomes of either of those two x-spin-measurements were; just tell me whether or not you now believe that those two measurements both had definite outcomes, and whether or not those two outcomes were *the same.*

It will follow from the same sorts of arguments as we gave above that Wigner's friend's response to a question like that (even though, as a matter of fact, on the standard way of thinking, *neither* of those experiments had any definite outcome) will necessarily by "Yes: they both had definite outcomes, and both of those outcomes were the same."

And it will also follow from the same sorts of arguments that if two observers were both to carry out measurements of the x-spin of some particular electron whose y-spin was initially definite, and if they were subsequently to talk to one another about the outcomes of their respective experiments, then both of those observers will report, falsely, that the *other* observer has reported some definite particular outcome of *her* measurement, and both of them will report that that reported outcome is completely in agreement with their own.

And all of these results are obviously only special cases of a much more general and fundamental principle (which the reader will now have no trouble in proving for herself, by means of precisely the same sorts of arguments as we've just gone through), which is that the linearity of the quantum-mechanical equations of motion entails that there cannot possibly be any such thing as a physical system (measuring device, automaton, sentient observer or whatever) which reliably reports whether there is any determinate matter of fact about the value of any particular observable property of the world or not.

And so it turns out not to be altogether impossible (even if there is only one world, and even if the standard way of thinking about what it means to be in a superposition is invariably the right way of thinking about it) that the state we end up in at the conclusion of a measurement of the x-spin of a y-spin up electron is precisely the one in (7.4).

It turns out, that is, that the quantum-mechanical equations of motion can function as a scientific instrument for being radically skeptical even about the determinateness of the most mundane and everyday features of the external macroscopic physical world, and even about the most mundane and everyday features of *our own mental lives.*

And so the conventional formulation of the measurement problem (that is: the one that I rehearsed at the outset of this talk; the one according to which there is some particular point in the course of the sort of measurement we've been talking about here at which the hypothesis that the quantum-mechanical equations of motion are the true and complete equations of motion of the whole world somehow flatly contradicts something that we know, with certainty, by pure introspection, about our own thoughts) is all wrong.

V

Here's something else that's interesting; something that's relevant to the question of *probabilities*. Suppose that Wigner's friend is confronted with an infinite collection of electrons, all of which initially y-spin up, and that Wigner's friend undertakes to measure the x-spin of *each one* of those electrons.

Before those measurements start, the state of Wigner's friend and of those electrons (whose names are 1, 2, ...) and of Wigner's friend's x-spin measuring devices (whose names are, respectively, $m1, m2, \ldots$.) is:

$$[\text{ready}>_F[\text{ready}>_{m1}[y\text{-spin up}>_1[\text{ready}>_{m2}[y\text{-spin up}>_2[\text{ready}>_{m3}[y\text{-spin up}>_3. \ldots \quad (7.9)$$

Once the measurement of the x-spin of electron 1 is done, the state is:

$$1/\sqrt{2}\{[\text{believes that 1 up}>_F[\text{indicates that 1 up}>_{m1}[x\text{-spin up}>_1 + [\text{believes that 1 down}>_F[\text{indicates that 1 down}>_{m1}[x\text{-spin down}>_1\} \times [\text{ready}>_{m2}[y\text{-spin up}>_2[\text{ready}>_{m3}[x\text{-spin up}>_3 \ldots ; \quad (7.10)$$

and once the measurement of the x-spin of electron 2 is done, the state is:

$$1/\sqrt{4}\{([\text{believes that 1 up and 2 up}>_F[\text{indicates that 1 up}>_{m1} \times [\text{indicates that 2 up}>_{m2}[x\text{-spin up}>_1[x\text{-spin up}>_2) + ([\text{believes that 1 up and 2 down}>_F[\text{indicates that 1 up}>_{m1}[\text{indicates that 2 down}>_{m2}[x\text{-spin up}>_1[x\text{-spin down}>_2) + ([\text{believes that 1 down and 2 up}>_F[\text{indicates that 1 down}>_{m1}[\text{indicates that 2 up}>_{m2} \times [x\text{-spin down}>_1[x\text{-spin up}>_2) + ([\text{believes that 1 down and 2 down}>_F[\text{indicates that 1 down}>_{m1}[\text{indicates that 2 down}>_{m2}[x\text{-spin down}>_1[x\text{-spin down}>_2)\} \times [\text{ready}>_{m3}[y\text{-spin up}>_3 \ldots \quad (7.11)$$

and so on. The number of superposed conditions in the overall state of the world will increase geometrically (like the numbers of the branches in the diagram in figure 1, as you work your way up) as the number of y-spin-measurements increases.

Suppose that once the first N of those measurements are complete we say to F, "Don't tell me what the x-spin of electron 1 or electron 2 or any particular one of the first N electrons turned out to be; tell me merely whether or not you believe that each one of those electrons now has a definite x-spin, and, if so, tell me also what *fraction* of those first N electrons turned out to be x-spin up.

That won't tell us much, as it stands. The answer to the first question (as we've already seen) is going to be yes, but of course F will not produce any coherent answer to the *second* question; once F has responded to *that* question, the state of the world will be a superposition of states in which F answers that question in various different ways.

But here's something curious: It happens that in the limit as N goes to infinity (that is: in the limit as the number of x-spin-measurements which h has so far performed goes to infinity) the state of the world will, with certainty, *approach* a state in which F *will* answer that question in a perfectly *determinate* way, and in which the answer F gives will with certainty be "½" (which is, of course, precisely what *ordinary* quantum mechanics will predict, with certainty, about that response in that limit).

And that turns out to be an instance of something a good deal more general, which runs as follows: Suppose that an observer is confronted with an infinite ensemble of identical systems in identical states, and that she carries out a certain identical measurement on each of them. Then, even though there will actually be no matter of fact about what h takes the outcomes of *any* of those measurements to be, nonetheless, as the number of those measurements which have already been carried out goes to infinity, the state of the world will approach (not as a merely *probabilistic* limit, *but as a well-defined mathematical epsilon-and-delta-type limit*) a state in which the reports of h about the *statistical frequency* of any particular outcome of those measurements will be perfectly definite, and also perfectly in accord with the standard *quantum mechanical* predictions about what that frequency ought to be.

VI

Now, what I once hoped was that all this might amount to an argument that there just isn't any such *thing* as a measurement problem.

What I hoped was that that might make it entertainable that (even if the standard way of thinking about what it means to be in a superposition is the *right* way of thinking about what it means to be in a superposition) the linear dynamical laws are nonetheless the *complete* laws of the evolution of the *entire world*, and all of the appearances to the contrary (like the appearance that experiments have outcomes, and the appearance that the world does not evolve deterministically) turn out to be just the sorts of *delusions* which *those laws themselves* can be shown to *bring on*!

I call that hypothesis (or rather, that *reading* of *Everett*'s hypothesis) "the bare theory."

But there turn out to be all sorts of reasons why the bare theory can't be quite right either.

Note, for example, that if the bare theory were true, then there would be matters of fact about what we think about (say) the frequencies of "*x*-spin up" outcomes of measurements of the *x*-spins of *y*-spin up electrons *only* (if at all) in the limit as the number of those measurements goes to infinity. And so, if the bare theory were true (and since only a finite number of such measurements has ever actually been carried out by any one of us, or even in the entire history of the world), then there could not now be any matter of fact (notwithstanding our delusion that there is one) about what we take those frequencies to *be*. And so, if the bare theory is true, then there can be no matter of fact (notwithstanding our delusion that there is one) about whether we take those frequencies to be in accordance with the standard quantum-mechanical *predictions* about them. And so, if the bare theory is true, then it is unclear what sorts of reasons we can possibly have for *believing* in anything like quantum mechanics (which is what the bare theory is supposed to be a way of making *sense* of) in the first place.

And as a matter of fact, if the bare theory is true, then it turns out to be extraordinarily unlikely that the present quantum state of the world is one of those on which there is even a matter of fact about whether any sentient experimenters exist at all. And of course in the event that there *isn't* any matter of fact about whether or not any sentient experimenters exist, then it becomes unintelligible even to inquire (as we've been doing here) about what sorts of things such experimenters will *report*.

And then (as far as I can tell) all bets are off.

VII

And so it now seems to me *not* to be entertainable (in any of the ways I've mentioned, or in any *other* way I've heard of) that the linear quantum-mechanical equations of motion are the true and complete equations of motion of the whole world.

But I think, nonetheless, that there are interesting lessons in this last stuff about what superpositions *feel like*.

What that stuff *does* show, I think, is that precisely that feature of those equations which make it *clear* that they cannot possibly be the true and complete equations of motion of the whole world (that is, their *linearity*) *also* makes it radically *un*clear *how much* of the world and *which parts* of the world those equations possibly *can* be the true and complete equations of motion.

What I think it shows (to put it another way) is that there can be *no such thing* as a definitive list of what there have absolutely got to be matters of fact about which is scientifically fit to serve as an "observational basis" from which all attempts at fixing quantum mechanics up must *start out*.

What I think it shows is that what there are and what there aren't determinate matters of fact about, even in connection with the most mundane and everyday macroscopic features of the external physical world, and even in connection with the most mundane and everyday features of *our own mental lives*, is something which we will ultimately have to *learn* (in some part) from whatever turns out to be the *best* way of fixing quantum mechanics up.

VIII

But of course all of this raises a number of difficult questions about (among other things) the very business of *seeking out* the best way of fixing quantum mechanics up!

Let me say something, just by way of finishing up (and which will need to be absurdly brief and absurdly inconclusive) about that.

To begin with, there are all sorts of nonobservational criteria of theory choice (various sorts of *simplicity*, various sorts of *economy*; stuff like that) which have been much discussed in the philosophy of science, and which will of course come into play in choosing the best way of fixing quantum mechanics up. We will of course require that any acceptable way of fixing quantum mechanics up make the *correct* predictions about the values of those observable properties of the world whose values it entails there are *matters of fact* about! And we shall want to adopt some generally *conservative* principle on the question of *what there are* matters of fact about; we shall want (that is) to adhere as closely as we *can* (all other things being equal) to the common-sense position that there are invariably determinate matters of fact at least about the most mundane and everyday features of our own mental lives, and of the macroscopic external physical world.

The lesson of the stuff about what superpositions feel like, I think, is just that the linearity of the quantum-mechanical equations of motion can function as a scientific instrument for injecting some *slack* into that conservative principle, more or less at any point we like, if we find we need some.

And it happens that a good deal of my own work over the past few years has gone into showing that every proposal for fixing up quantum mechanics that we presently know about (collapse theories and hidden-variable theories and so on) *does* need some; every one of those proposals (it turns out) fails to accommodate our intuitions, at one point or another, either about the determinateness of certain mundane and everyday features of the macroscopic external physical world, or else about the determinateness of certain mundane and everyday features of the mental lives of sentient observers, or else about both.

And the lesson of what I've been talking about, I think, is that the linearity of the quantum-mechanical equations of motion can function as an instrument for learning how to live with some of that.

NOTES

1. The following argument owes much to my work with Jeff Barrett. A more detailed account is about to be submitted for publication.

REFERENCES

Albert, D. Z. *Quantum Mechanics and Experience.* Cambridge: Harvard University Press.

Gell-Mann, M., and J. B. Hartle. 1990. "Quantum Mechanics in the Light of Quantum Cosmology." In W. Zurek, ed., *Complexity, Entropy, and the Physics of Information.* Santa Fe Institute Studies in the Sciences of Complexity, vol. 8. Reading, Mass.: Addison-Wesley.

Wigner, E. P. 1961. "Remarks on the Mind-Body Question." In I. J. Good, ed., *The Scientist Speculates*, 284–302. New York: Basic Books.

8

The Preparation Problem in Quantum Mechanics

Linda Wessels

Department of History and Philosophy of Science, Indiana University

Similarities between preparations and measurements make it easy to suppose that something like the measurement problem also plagues preparation procedures. Both measurements and preparations are typically modeled as interactions: A system is measured when it interacts with a measuring device M and as a result M exhibits the outcome of the measurement; a system O is prepared when it interacts with a preparation device and as a result ends up in the state to be prepared. Both are interactions that are supposed to leave one of the interacting systems in a particular state: Because M is supposed to exhibit the measurement result, it should come out of the interaction in a state that assigns probability one to the result it exhibits (the eigenstate of the value exhibited); because the preparation interaction is supposed to prepare O in a quantum mechanical state, O should come out of the interaction in that state. The measurement problem stems from the fact that the standard quantum mechanical treatment of interactions does not assign the appropriate state to the measuring device at the end of the interaction. Given the above similarities between measurement and preparation, it appears on the face of it that an analogous problem will arise for preparations. In fact, it is often assumed that the two problems are fundamentally the same, and hence can both be solved in the same stroke by an appropriate solution to the measurement problem.

Little work has been done on analyzing preparation processes themselves. Henry Margenau (e.g., 1963) clearly distinguished between preparation and measurement, but he did not examine preparation in detail. His primary aim was not to shift attention to preparation, but to clear the way for his own challenge to some of the standard claims about measurement. Most quantum mechanics textbooks include at least some discussion of measurement and the difficulties associated with analyzing measurement processes, but few even have an index entry for preparation. If preparations are mentioned at all (and in some they are not, e.g., Cohen-Tannoudji et al. 1977), it is simply assumed that successful preparation procedures exist, and in sufficient variety that at least in principle, virtually any mathematically possible quantum mechanical state can be prepared (e.g., Fano and Fano 1972, 134–36; French and Taylor 1978, 251–52). Philosophers and physicists concerned with foundations of quantum mechanics have tended to pay more careful attention to preparation, but generally focus on the experimental and foundational roles of preparation (e.g., Hughes 1989, 61–62, 85, 88; Jauch 1968, 93–94), not on the nature of preparations themselves or on difficulties that arise in quantum mechanical descriptions of preparation interactions.

The most extended discussion of preparation yet available appears in Healey's (1989) recent book on the "interactive interpretation" of quantum mechanics, in which preparations play a unique and central role. But when it comes time to spell out his theory/interpretation of preparation in detail, Healey considers only preparations that also are or involve measurement (1989, 104–15). When he discusses what he calls "the preparation problem," he clearly takes this to be a difficulty for preparations that is strictly analogous to the measurement problem (ibid., 186–87). Thus even in Healey's treatment of preparation, it is his concern with measurement and the measurement problem that guides his analysis and interpretation of preparation. Healey recognizes this; at the end of the book, he points out the need for a more thorough and general analysis of preparation and preparation reactions than he has provided.

This essay is a first step toward taking a more careful look at preparation processes, both in the abstract and in the laboratory. Even the initial look provided here shows that the apparent similarities between measurement and preparation have been misleading. A simple analysis provided in the next section reveals a fundamental difference between preparation and measurement. The measurement problem is a foundational problem posed when one attempts to apply quantum mechanics to describe measurement interactions. No analogue of this foundational problem arises when quantum mechanics is applied to preparations. The notion that something similar to the measurement problem also exists for preparations is not completely misguided, however. In the section after that, we will see that difficulties which echo certain features of the measurement problem do arise when one attempts a quantum-mechanical description of procedures typically thought to be preparations. Thus, it will be argued, a preparation problem does exist that is in some ways similar to the measurement problem. But it is not a foundational problem, as is the measurement problem. Rather, it is a problem that arises when one tries to use quantum mechanics to explain laboratory practice. In the subsequent section, it will become clear that the similarities between the two problems are not enough to justify the common assumption that a solution to the measurement problem will also solve the preparation problem. The solution to the two problems may be linked for some preparatory procedures — those that are also measurements. But there are commonly used preparatory procedures that are not measurements. An alternative approach for such nonmeasuring procedures will be suggested, and an example of how this works will be sketched.

What the Preparation Problem Is Not

The reason why no analogue to the measurement problem arises for preparations becomes apparent when we trace the similarity between measurement and preparation interactions; at a certain point the similarity will fail and the crucial difference will emerge.

A measurement of a system S for observable A can be modeled as an interaction between S and a measuring device D, as on the left in figure 8.1. For the sake of simplicity I will assume throughout that observable A as well as any other observables discussed have only discrete spectra. Because it is a measuring device, M is assumed to have an observable R (a "reading" observable) that plays a crucial role in the measurement. Let $\{\rho_i\}$ and $\{r_i\}$ be the eigenstates and eigenvalues of R, and $\{\alpha_i\}$ and $\{a_i\}$ the eigenstates and eigenvalues of A, respectively. In the usual treatment of a measurement process, it is assumed that for M to be a device appropriate for measurements of A, there must be for each eigenvalue $a_k \in \{a_i\}$ a corresponding eigenvalue $r_k \in \{r_i\}$, such that r_k is the reading exhibited by M to indicate that a_k is the result of the measurement for

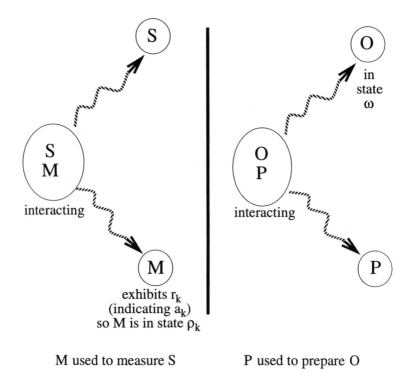

Figure 8.1 Analogous features of measurement and preparation.

A. It is further assumed that when M exhibits reading r_k, this is macroscopically observable, so that when M exhibits reading r_k, M "has" value r_k for observable R. Thus at the end of the measurement, M must be in state ρ_k, the eigenstate corresponding to the value r_k it exhibits.[1]

Similarly, the preparation process can be modeled as an interaction between two systems that leaves one of the interacting systems in a specific state. If the object of the preparation process O is to be prepared in quantum mechanical state ω, the process is modeled as an interaction between O and a preparation device P which at its finish leaves O in ω, as on the right in figure 8.1.

Quantum mechanics describes the result of an interaction between a pair of systems by tracing the time evolution of the state of the compound system composed of the interacting systems; applying the time-dependent Schrödinger equation to the state of the compound just prior to the interaction, the state of the compound at any time during or after the interaction can be determined. In the case of our measurement interaction, the compound is composed of M, and S; call it $M + S$. Its state at the end of the interaction, ψ_{M+S}, can be written most generally in terms of the members of $\{\rho_i\}$, the eigenstates of the reading observable R of M and the members of a set $\{\phi_j\}$ that is a basis for H_s. Here and following, for any system X, H_x represents the Hilbert space of states available to X. Thus in general,

$$\psi_{M+S} = \Sigma c_{ij}\rho_i \otimes \phi_j, \tag{8.1}$$

where $\rho_i \otimes \phi_j$ is the tensor product of ρ_i and ϕ_j. In the case of our preparation interaction, the compound is composed of P and O; call it $P + O$. Similarly, its state at the end of the interaction, ψ_{P+O}, can be written most generally in terms of the members of a set $\{\pi_i\}$ that is a basis for H_P, and the members of a set $\{\omega_j\}$ that is a basis for H_O:

$$\psi_{P+O} = \Sigma c_{ij}\pi_i \otimes \omega_j. \tag{8.2}$$

But given the characteristic features of measurement and preparation interactions noted, we expect the state of the compound system at the end of a measurement or preparation interaction to be only a special case of the general form indicated. At the end of a measurement, the measuring device should be in one of the ρ_i, hence the state of the compound $M + S$ should take the special form

$$\psi_{M+S} = \rho_i \otimes \phi, \tag{8.3}$$

where ϕ is some state in H_S. Similarly, at the end of a preparation, the system that is the object of preparation should be in the prepared state ω; hence the state of the compound $P + O$ should take the special form

$$\psi_{P+O} = \pi_i \otimes \omega. \tag{8.4}$$

What kinds of quantum-mechanical interactions will yield these special forms for measurement or preparation? The answer to this question for measurements is: None. Once an additional characteristic feature of measurement interactions is added to those already noted, it can be proven that no interaction is describable by the Schrödinger equation which for arbitrary initial states of S will yield a state of the form

(8.3) for $M + S$. The characteristic feature is this: If just prior to the measurement S is in an eigenstate of A, the reading exhibited at the end of the measurement must be the eigenvalue corresponding to that eigenstate. Thus if S is in the eigenstate α_k just prior to the measurement, M must exhibit the reading r_k. Clearly, this feature ensures that the interaction will satisfy the constraints set by quantum mechanics itself on a measurement: If S is in an eigenstate of A prior to measurement, then quantum mechanics assigns probability 1 to getting the eigenvalue of A associated with this eigenstate as the result of the measurement. The following argument highlights the role this feature plays in ruling out the possibility that measurements are interactions that quantum mechanics can handle in its usual way.

Consider again a measurement on system S for observable A where A has a discrete spectrum $\{a_i\}$, and a set of eigenstates $\{\alpha_i\}$ that form a basis for H_S.

a. Suppose S is in an eigenstate of A, say α_k, just prior to the measurement. Then according to quantum mechanics, the measurement interaction between M and S must result in M exhibiting value r_k of its reading observable, indicating that the value a_k is the result of the measurement. Since M must have value r_k of observable R, it must be in the eigenstate ρ_k at the end of the measurement interaction. If we assume that the quantum-mechanical treatment of this interaction assigns the appropriate state at the end of this measurement, the time-dependent Schrödinger equation will describe the state transition due to the interaction as:

$$\rho_o \otimes \alpha_k \rightarrow \rho_k \otimes \phi_k, \qquad (8.5)$$

where ρ_o is the initial state of M, prior to the measurement, and ϕ_k is whatever state the interaction puts S into when it begins in state α_k.

b. Now consider what happens when S begins in a state $\Sigma c_i \alpha_i$, in which more than one c_i is nonzero. If the evolution of state is again governed by the same Hamiltonian, then since the Schrödinger equation is linear, (8.5) implies the following state transition for this general case:

$$\rho_o \Sigma c_i \otimes \alpha_i \rightarrow \Sigma c_i \rho_i \otimes \phi_i. \qquad (8.6)$$

The final state of $M + S$ is indeed a special case of the general expression (8.1), but it is not the special case required by the characteristic features of measurement depicted in figure 8.1 — it is not represented by (8.3). Thus, for any interaction I, if quantum mechanics assigns the appropriate state to $M + S$ at the end of I for the cases in which S starts out in an eigenstate of A, then for cases in which S does not start out in an eigenstate of A, quantum mechanics will not assign an appropriate state to $M + S$ at the end of I. Hence no interaction can both have the characteristic features of a measurement interaction and satisfy the Schrödinger equation. The measurement problem can be formulated in a variety of ways, but it is basically the problem of reconciling this impossibility with the undeniable fact that measurements are made for quantum-mechanical observables by setting up interactions between measuring devices and the systems to be measured, definite results are exhibited by the measuring devices used, and these results satisfy the predictions of quantum mechanics.

There is no such impossibility in the case of preparation. At least in principle there may exist interactions which both have the characteristic features of a preparation interaction and satisfy the Schrödinger equation. This is because the characteristic features of preparations do not include an analogue to the additional feature of measurement highlighted above. In the case of preparation there is no requirement that the interaction between P and O correlate the end state of one with the initial state of the other. Let π_o be the state of P at the beginning of the preparation interaction. Suppose that the quantum-mechanical description of the preparation interaction between P and O assigns ω to O at the end of the interaction whenever O starts out in the kth eigenstate of A, α_k. Then the state transition due to the interaction has the form:

$$\pi_o \otimes \alpha_k \rightarrow \pi_k \otimes \omega, \tag{8.7}$$

where π_k is the state in which P is left when O begins in α_k. It is consistent with quantum mechanics to suppose that also, for any other eigenstate α_i as initial state for O, the interaction will also evolve according to:

$$\pi_o \otimes \alpha_i \rightarrow \pi_i \otimes \omega, \tag{8.8}$$

where again, π_i is the state to which P evolves when O begins in α_i.

Now what does the linearity of the Schrödinger equation imply for this interaction when O begins in some arbitrary state $\Sigma c_i \alpha_i$? It follows from (8.7) and (8.8) that in this general case:

$$\pi_o \otimes \Sigma \alpha_i \rightarrow (\Sigma \pi_i) \otimes \omega. \tag{8.9}$$

Thus for *any* initial state of O, O is assigned ω at the end of the interaction.

Not all preparations need proceed in this uniform way, of course. The point here is that in principle there are preparation interactions that satisfy the Schrödinger equation, and so can be treated quantum-mechanically in the same way as any ordinary interaction. The measurement problem arises out of an in-principle impossibility. There is no quantum-mechanically allowed interaction between a measuring device M and a system S such that for any initial state of S the quantum-mechanical description of the interaction will assign to M at the end of the interaction an eigenstate ρ_k corresponding to the eigenvalue r_k displayed by M as a result of the measurement.[2] There is no in-principle conflict between the way that quantum mechanics describes interactions and the standard assumptions about what characterizes a preparation interaction, however. There is no analogue of the measurement problem for preparation.

A Different Problem for Preparations

Nevertheless, there are difficulties associated with giving a quantum-mechanical analysis of preparations. This becomes apparent when we look at procedures that are actually used in the laboratory for preparation or are assumed to achieve preparation in standard discussions of how quantum mechanics works — what I call "supposed-preparations." We will see that according to quantum mechanics many, perhaps even most, supposed-preparations do not succeed in preparing.

First, however, note that some supposed-preparations are successful according to quantum mechanics. It would be nice to have an example of a procedure that is uniform, one in which system O is prepared in some state ω no matter what the state of O prior to preparation, but I do not know of any.[3]

There are successful nonuniform supposed-preparations, however. A simple example is one that uses a Stern-Gerlach apparatus oriented

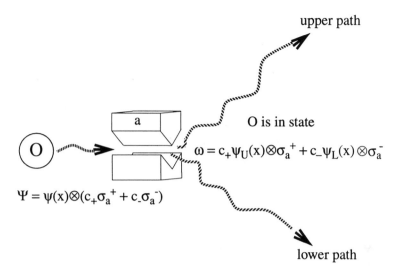

Figure 8.2 Successful preparation using a Stern-Gerlach apparatus.

in some direction *a*. Spin-½ particles enter the Stern-Gerlach apparatus from the left in state ψ, as in figure 8.2. If a scintillation screen is placed to the far right behind the exit region of the apparatus, the position of a particle after it has exited can be measured. As is well known, two regions will show a positive result for position, one higher than the exit point of the apparatus, the other lower. A particle found to have position in the upper region is, by virtue of that result, also found to have a value of + for the observable S_a, spin in direction *a*; a particle found to have position in the lower region is thereby found to have value − for S_a. Let the eigenstates for S_a be $\{\sigma_a^+, \sigma_a^-\}$, where σ_a^+ gives probability one for value + of S_a and σ_a^- probability one for value − of S_a. Since $\{\sigma_a^+, \sigma_a^-\}$ forms a basis for the spin states of any spin-½ particle, the state of the incoming beam can be expressed as $\psi(x) \otimes (c_+ \sigma_a^+ + c_- \sigma_a^-)$. Due to the influence of the magnetic field in the Stern-Gerlach apparatus, such a particle exits on the right in state $c_+ \psi_U(x) \otimes \sigma_a^+ + c_- \psi_L(x) \otimes \sigma_a^-$, where $|\psi_U(x)|^2 \approx 0$ except along the path leading from the exit point of the Stern-Gerlach apparatus to the upper region on the screen, and $|\psi_L(x)|^2 \approx 0$ except along the path leading from the exit point of the Stern-Gerlach apparatus to the lower region. Thus, according to quantum mechanics, a particle exiting the Stern-Gerlach apparatus has been prepared in state $\omega = c_+ \psi_U(x) \otimes \sigma_a^+ + c_- \psi_L(x) \otimes \sigma_a^-$. This state will not be the same for all possible initial states of O; initial states with different values of c_+ and c_- will yield different prepared states ω. Hence this is not a uniform preparation. But it is a process that according to quantum mechanics does prepare O in some quantum mechanical state. This quantum-mechanical description of the preparation depends, of course, on the assumption that the state of a particle entering the Stern-Gerlach apparatus from the left is not itself entangled in some superposition involving another system with which it has previously interacted, that is, that it is correctly assigned a quantum-mechanical state ψ. We will examine that assumption later.[4]

Pseudo-Preparations

Many supposed-preparations do not succeed in their preparation task, however — at least according to quantum mechanics. I call supposed-preparations of this sort "pseudo-preparations." In many cases, the type of failure involved is familiar. Consider the common form of pseudo-preparation depicted in figure 8.3: System O interacts with a preparation device P in order to prepare O in state ω. The quantum-mechanical description of such interactions, using the Schrödinger equation to give the evolution of state during and after the interaction, assigns to the compound system $P + O$ at the end of the interaction a superposition of tensor products of states of P and states of O, a state of the form $\psi_{P+O} = \Sigma b_{ij}\pi_i \otimes \psi_j$, where $\{\pi_i\}$ and $\{\psi_j\}$ are bases for H_P and H_O respectively. Since ω can always be expressed in terms of the $\{\psi_j\}$, there exist a $\pi \in H_P$ and coefficients c_ω and $\{c_{ij}\}$ such that $\psi_{P+O} = c_\omega \pi \otimes \omega + \Sigma c_{ij}\pi_i \otimes \psi_j$. Those processes for which some of the $c_{ij} \neq 0$ will be pseudo-processes since quantum mechanics fails to assign ω to O at the end of the interaction. Generally in such pseudo-processes it is also the case that ψ_{P+O} cannot be factored into a simple tensor product of a state for P and a state for O. In such cases, then, not only is O not assigned ω, but no quantum-mechanical states are assigned at all to P and O separately. Such supposed-preparations suffer from what we can call "the problem of entangled states." A rough quantum-mechanical analysis of two procedures usually thought to prepare particular quantum-mechanical states shows that they actually suffer from this problem of entangled states.

Consider first a procedure commonly assumed to prepare a beam of spin-½ particles in σ_a^+, the $+$ eigenstate for S_a. It is standardly claimed that this can be done using a Stern-Gerlach apparatus oriented in direction a plus a blocking screen P located behind the Stern-Gerlach appa-

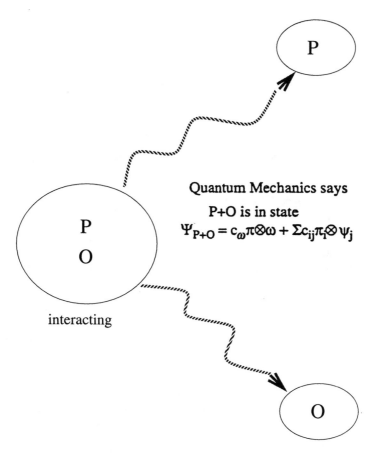

Figure 8.3 Common pseudo-preparation scheme.

ratus along the lower path, as in figure 8.4. The rationale for supposing this is a preparation for σ_a^+ seems to be that spin-½ particles exiting the Stern-Gerlach apparatus are in a superposition $\psi' = c_+\psi_U(x)\otimes\sigma_a^+ + c_-\psi_L(x)\otimes\sigma_a^-$. The first part of this superposition is the state for a beam of spin-½ particles located along the upper path and having value + for S_a; the second is the state for a beam along the lower path, with value − for S_a. Now if the lower path is blocked by a screen P, only the upper beam is left, that is, only a beam of particles with spin state σ_a^+.

But, of course, this rationale presupposes that an exiting beam of particles in the superposition $c_+\psi_U(x)\otimes\sigma_a^+ + c_-\psi_L(x)\otimes\sigma_a^-$ really consists of two sub-beams, one moving along the upper path with spin

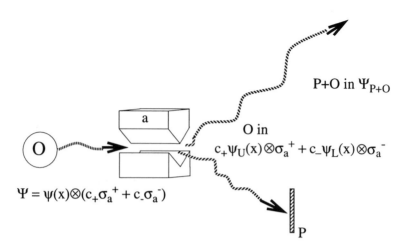

$$\Psi = \psi(x)\otimes(c_+\sigma_a{}^+ + c_-\sigma_a{}^-)$$

Figure 8.4 Pseudo-preparation of O in σ_a^+ using a Stern-Gerlach apparatus.

state σ_a^+, the other moving along the lower path with spin state σ_a^-. This is simply to misinterpret the significance of assigning a superposition.

Look at how quantum mechanics describes the interaction between a particle O exiting the Stern-Gerlach apparatus and the blocking screen P. Let π_o be the state of P prior to the interaction. Initially the composite system $P + O$ is in state $\pi_o\otimes(c_+\psi_U(x)\otimes\sigma_a^+ + c_-\psi_L(x)\otimes\sigma_a^-)$. To find the state of $P + O$ after this interaction is over, we can examine what would happen in two hypothetical cases. If upon exiting the Stern-Gerlach apparatus O were in $\psi_U(x)\otimes\sigma_a^+$, then since $|\psi_U(x)|^2 \approx 0$ in the region of P, the state of O would be almost unaffected by the presence of P, and likewise, the state of P would be almost unaffected by the presence of O. Let π_e represent the state to which P will evolve in the absence of any interaction with O. Then the evolution of the state of $P + O$ is (approximately) given by $\pi_o\otimes\psi_U(x)\otimes\sigma_a^+ \to \pi_e\otimes\psi_U(x)\otimes\sigma_a^+$. If, on the other hand, O were in state $\psi_L(x)\otimes\sigma_a^-$ upon exiting the apparatus, the interaction between P and O would affect the states of both; according to quantum mechanics, the states of P and O would end up entangled in some superposition state for the compound $P + O$. The evolution of the state of $P + O$ under this circumstance will have (approximately) the form $\pi_o\otimes\psi_L(x)\otimes\sigma_a^- \to \Sigma c_{ij}\pi_i\otimes\psi_j$ where $\{\pi_i\}$ and $\{\psi_j\}$ are bases for H_P and H_O, respectively. Since in the general case, the state of O prior to interaction with P is a weighted sum of the states in the two hypothetical cases, we can use the linearity of quantum-mechanical state evolution to determine the evolution of the state of $P + O$: $\pi_o \otimes (c_+ \psi_U(x) \otimes \sigma_a^+ + c_- \psi_L(x) \otimes \sigma_a^- \rightarrow c_+ \pi_e \otimes \psi_U(x) \otimes \sigma_a^+ + c_- \Sigma c_{ij} \pi_i \otimes \psi_j$.

According to quantum mechanics, then, this interaction does not yield spin-½ particles along the upper path with spin state σ_a^+. Rather, it produces compound systems $P + O$ in a superposition state that cannot in general be expressed as a simple tensor product of a state for P and a state for O. That is, this supposed procedure for the preparation of systems in state σ_a^+ suffers from the problem of entangled states.

A second example is the supposed-preparation assumed in the double-slit thought experiment. A source S yields a beam of particles in state ψ that approach screen P from the left, as in figure 8.5. In screen P are two slits, A and B. It is usually assumed that a portion of the beam will emerge on the right of screen P, in a state of the form $\omega = c_A \psi_A(x) + c_B \psi_B(x)$, where $|\psi_A(x)|^2 \approx 0$ except just behind slit A, and $|\psi_B(x)|^2 \approx 0$ except just behind slit B. This state then evolves as the particles travel to the detecting screen D. Of course, the interest in this thought experiment derives from the fact that the pattern of scintillations on screen D will be similar to the pattern produced by interfering waves even though each scintillation is a small spot indicative of an impinging particle. Our concern with what goes on in preparation leads us to focus

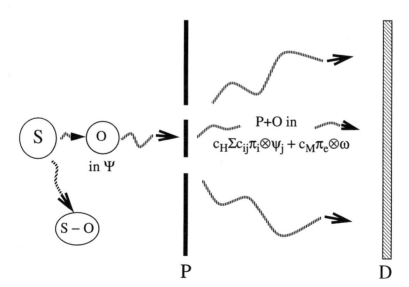

$P+O$ in

$c_H \Sigma c_{ij} \pi_i \otimes \psi_j + c_M \pi_e \otimes \omega$

in Ψ

$S - O$

P   D

Figure 8.5 Pseudo-preparation using a particle behind a screen with two slits.

not on what happens at the detecting screen D, however, but on what happens in the interaction between the particles and the slitted screen P. How does quantum mechanics describe this interaction?

We can use the fact that ψ, the state of a particle O coming from the left, can be decomposed into a sum $c_H \psi_H(x) + c_M \psi_M(x)$ where $|\psi_H(x)|^2 \approx 0$ just to the left of slits A and B (i.e., it gives probability one for finding O in a position just in front of one of the slits), while $|\psi_M(x)|^2 \approx 0$ for all positions on P that are not along slits A and B (i.e., it gives probability one for finding O in a position that corresponds to hitting P). If O were in state $\psi_M(x)$ just prior to the interaction, the subsequent state of P would be almost unaffected by the presence of O, and vice versa. Thus for such an O, the state evolution of $P + O$ due to such an interaction can be (approximately) represented as $\pi_o \otimes \psi_M \rightarrow \pi_e \otimes (c_A \psi_A(x) + c_B \psi_B(x))$ where π_o is the state of P just prior to the interaction, and π_e is the state to which π_o evolves in the absence of any interaction between P and O. If, on the other hand, O were in state $\psi_H(x)$ just prior to the interaction, then the probability would be one that O interacts with P. The evolution of the state of $P + O$ due to the interaction would then be represented by $\pi_o \otimes \psi_H \rightarrow \Sigma c_{ij} \pi_i \otimes \psi_j$ where $\{\pi_i\}$ and $\{\psi_j\}$ are bases for H_P and H_O, respectively. Using again the linearity of state evolution, the transition due to the interaction in the general case will then be (approximately) $c_H \psi_H(x) + c_M \psi_M(x) \rightarrow c_H \Sigma c_{ij} \pi_i \otimes \psi_j + c_M \pi_e \otimes (c_A \psi_A(x) + c_B \psi_B(x))$. Again, according to quantum mechanics, O is not prepared in the state this procedure is assumed to produce, $c_A \psi_A(x) + c_B \psi_B(x)$; indeed, quantum mechanics assigns no state at all to O at the end of the interaction. This supposed-procedure also suffers from the problem of entangled states.

The Preparation Problem

The pattern of difficulty is clear. There are processes standardly assumed to prepare a system O in state ω which quantum mechanics says are not really doing this job. Rather, a quantum-mechanical analysis of the process assigns to a compound system composed of O and other objects a state in which ω appears as only part of a superposition. The fact that such pseudo-preparations are nonetheless commonly assumed to prepare O in ω raises a question: Why do we think that pseudo-preparations are successfully preparing? If quantum mechanics does not pick them out as preparation procedures, what does? The answer seems to be: empirical evidence. Certain processes are used as preparation processes because there is empirical reason to think they do this job. Indeed, there seem to be two ways that empirical evidence gets used to determine what processes prepare a system O in a state ω.

First, the state of O can be directly determined by doing measurements on an ensemble of systems all (supposedly) prepared in the same way as O. Quantum textbooks contain various rules, developed on the basis of quantum mechanics itself, for determining the state of a system on the basis of measurements on such an ensemble. There is, for example, a standard rule for detecting whether a beam of systems is in a spin eigenstate. An unusually careful statement of that rule is found in Blum's *Density Matrix Theory and Its Applications*:

> If it is possible to find an orientation of the Stern-Gerlach apparatus for which a given beam is *completely* transmitted, then we will say that the beam is in a pure spin state . . . [corresponding to spin in the] unique direction which is parallel to the direction of the alignment of the Stern-Gerlach apparatus. (1981, 2)

What happens when we apply this rule to check the state of particles along the upper path exiting a Stern-Gerlach apparatus when the lower beam is blocked—the set-up depicted in figure 8.4? A measurement of the intensity of the upper beam will give the same result whether or not another Stern-Gerlach apparatus, oriented in the same direction as the first, is placed in its path. According to the rule, then, we should conclude that particles in the upper beam are in σ_a^+. One might be tempted to fault the rule in this case. The question of whether in this circumstance an "upper beam" even exists might legitimately be raised. But consider a second, more general method for determining the state of a system.

It is standardly said that under the appropriate circumstances the state of a system O can be determined (up to phase) by doing measurements for some observable A on an ensemble of systems similar to O (i.e., supposedly prepared in the same way as O), and gathering statistics on the frequency with which each of the eigenvalues of A is found (e.g., Blum 1981, 2–11, 14–18, 24–25, 35, 38; Fano and Fano 1972, 134–36; Fine 1969; French and Taylor 1978, 235–36, 252; Park and Band 1971). These statistics can be used to calculate (up to phase) the coefficients c_i in a formulation of the state of O in terms of the eigenstates of A, $\{\alpha_i\}$: $\Sigma c_i \alpha_i$. (A further series of measurements can be used to determine the appropriate phase factors also.) Basically this is the rule (in the more general form appropriate for observables represented by unbounded operators) that is assumed, for example, when we are told that we can verify the state assigned in the two-slit experiment depicted in figure 8.4 by doing measurements for position to the right of P on an ensemble of systems (pseudo-) prepared in the same way as O. A scintillation screen D might be placed just to the right of P, for example, and the intensity of response at the various points on D recorded. The spatial distribution of intensity on D gives $|\psi(x)|^2$, from which $\psi(x)$ can be calculated (up to phase). The result will be (approximately) a state of the form $c_A\psi_A(x) + c_B\psi_B(x)$—even though, as we have seen, the quantum-mechanical description of the interaction between O and the screen P assigns no quantum-mechanical state at all to O on the right of P.

A second sort of evidence that convinces us that a particular pseudo-preparation does prepare systems in a state ω is hypothetico-deductive: If we assume the process prepares a system O in ω, then predictions about the subsequent behavior of O and systems that subsequently interact with O are correct. In the two-slit experiment of figure 8.5, for example, if we assume that a system O emerging to the right of screen P is in state $\omega = c_A\psi_A + c_B\psi_B$, then as the myriad discussions of this experiment affirm, the correct predictions are obtained for subsequent measurements on O and systems that interact with O. Likewise, if we assume that a system O emerging on the upper path from a Stern-Gerlach apparatus oriented in direction a, with the lower path blocked (figure 8.4), is prepared in the state σ_a^+, correct predictions are obtained for the results of subsequent measurements on O or systems that subsequently interact with O.

Thus, pseudo-preparations are thought to prepare a system O in state ω because the empirical evidence used to determine the state in which O has been prepared indicates a preparation has taken place. This empirical evidence may be the sort that is used to calculate ω directly, or it may be indirect confirmation of the assignment of ω—a match between the results of experiments involving O and the predictions made for the outcomes of those experiments on the basis of assuming ω for O. Of course, to completely determine the state of O using either method would often require an infinity of measurements, so in practice only reasonable inference to ω can be made. In principle, however, each of these empirical approaches yields a state assignment to O that gives correct predictions for the outcomes of a variety of subsequent experiments involving O.

But why does the state ω determined empirically for O give correct predictions when quantum mechanics assigns Ψ_{P+O} to $P + O$ instead? Generally the predictions made on the basis of assigning Ψ_{P+O} to $P + O$ will be different from those given when ω is assigned to O. This is the puzzle associated with preparation that might well be called "the preparation problem." It is similar to the measurement problem in that it arises because the Schrödinger equation assigns an entangled state to a compound system while empirical evidence indicates a well-defined state for one of the compound's components. But unlike the measurement problem, it emerges upon examination of particular quantum mechanics experiments, rather than from an in-principle difficulty encountered in developing a general theory of preparation. Thus this preparation problem may not seem especially pressing — particularly if the actual use of pseudo-preparations as preparation procedures is infrequent, or perhaps occurs only in the philosopher's thought experiments. Even a preliminary look at laboratory procedures suggests that the use of pseudo-preparations is pervasive, however. To see why, consider one more example of a pseudo-preparation.

Source Preparations

A typical procedure for obtaining a beam of particles in a particular state ω is to use a natural source — a chunk of radioactive material, for example, or a filament heated to a high temperature (see Dunlap 1988, 285–95, for other, more detailed examples). Systems O are (supposedly) prepared in state ω simply by virtue of the fact that they are emitted in ω from a source S. The emission may be spontaneous, as when the source is a radioactive sample, or induced, as when a thin film is bombarded with an appropriate beam or a fine wire subjected to an electric current. I will call such procedures "source preparations."

Consider the quantum-mechanical description of a simple kind of source preparation in which system O is (supposedly) emitted in state ω. Prior to the (supposed) emission, the state of S can be written as the state of a compound system $P + O$, where P is the portion of S that is left behind when S emits O — a state of the form $\Sigma c_{ij}\pi_i\psi_j$, where $\{\pi_i\}$ and $\{\psi_j\}$ are bases for H_P and H_O, respectively. Over the course of time, the quantum-mechanical evolution of this state will yield a state for $P + O$ that can be written

$$\Psi_{P+O} = c_E\pi_E\otimes\omega + \Sigma c'_{ij}\pi_i\otimes\psi_j, \tag{8.10}$$

where $|c_E|^2$ is the probability that O will be found emitted in ω, the state-to-be prepared, and π_E ϵ element of H_P is the state of P when O is emitted. Typically, the coefficients c'_{ij} only go to zero in the limit of infinite time. Thus, for any time at which one might actually do experiments using S, the quantum-mechanical treatment of the evolution of ψ_{P+O} assigns to $P+O$ a superposition of a state corresponding to the emission of O in ω and a state corresponding to no emission. Thus, even this simple case of source preparation suffers from the problem of entangled states. More complex cases of source preparation are possible. For example, O may be omitted in one of several different possible states, or the emission of O and P may be only one of several possible modes of decay. In these cases, S will be assigned an even more complicated superposition of states, representing the various emission and nonemission modes. Such as superposition is, of course, a telltale sign that these complex source preparations also suffer from the problem of entangled states.

That source preparations suffer from the problem of entangled states is important because a preliminary examination of the various supposed-preparation procedures actually used in laboratory experiments indicates that most, maybe even all, actual experiments involve source preparations. A supposed-preparation procedure P "involves" a source preparation when either: (i) P is a source preparation; or (ii) P is a multistage procedure where the first stage is a source preparation; or (iii) P functions by acting on the state of incoming systems, and these input systems have themselves been prepared using a procedure involving a source preparation. There are familiar instances of each of these three. Typically the experiments discussed in connection with the Bell inequalities are supposed-preparations of type (i), for example. The pairs of particles or photons analyzed for spin or polarization in many of these experiments are obtained by waiting for appropriately chosen atoms to "give off" two particles in the desired correlated state. The supposed-preparation in the two-slit experiment depicted in figure 8.5 provides an example of type (ii): systems are first obtained from a source, and then allowed to impinge on the screen P. The supposed-preparation depicted in figure 8.4, using a Stern-Gerlach apparatus plus blocking screen, can be of type (iii). In that supposed-preparation, it is assumed that the system to be prepared enters the Stern-Gerlach apparatus in some specified state ψ, which is then altered by the magnetic field of the apparatus. If the procedure by which the system is supposedly prepared in ψ is itself a source preparation or a multistage procedure that begins with a source preparation, then the supposed-preparation achieved by allowing the system to pass through the Stern-Gerlach apparatus "involves" a source preparation. Indeed, a Stern-Gerlach preparation "involves" a source preparation even when the procedure that supposedly prepares the input system is not of type (i) or (ii), but itself only involves a source preparation by virtue of some prior use of a source.[5]

Since source preparations suffer from the problem of entangled states, it follows (by induction) that all supposed-preparations that involve a source preparation also suffer from this problem. This means that even supposed-preparation procedures that *in principle* do not suffer from the problem of entangled states may nonetheless *as a matter of fact* suffer from the problem when actually carried out in the laboratory. Consider again, for example, the Stern-Gerlach preparation that does not in principle suffer from the problem of entangled states, depicted in figure 8.2. If O enters the Stern-Gerlach apparatus in a state $\Psi = \psi(x) \otimes (c_+ \sigma_a^+ + c_- \sigma_a^-)$, quantum mechanics says O will emerge in state $c_+ \psi_U \otimes \sigma_a^+ + c_- \psi_L \otimes \sigma_a^-$. But this quantum-mechanical analysis depends crucially on supposing that as O enters the Stern-Gerlach apparatus, it can be assigned Ψ. If the procedure that previously prepared O itself suffered from the problem of entangled states, however, then the state of the incoming O will be entangled with that of some other system, say P. The state of the compound P + O will be of the form $c_\Psi \pi \otimes \Psi + \Sigma c_{ij} \pi_i \otimes \Psi_j$, where π and the members of $\{\pi_i\}$ are states in H_P, and $\{\Psi_j\}$ is contained in H_O. The linearity of the Schrödinger equation implies that the state of P + O after being influenced by the Stern-Gerlach apparatus will be $c_\Psi \pi' \otimes (c_+ \psi_U(x) \otimes \sigma_a^+ + c_- \psi_L(x) \otimes \sigma_a^-) + \Sigma c'_{ij} \pi_i \otimes \Psi_j$, where π' is the state to which π evolves, and the c'_{ij} are the values the c_{ij} take on at this later time. Hence the "prepared system" O is not assigned a quantum state, but only the compound system P + O. Thus in such circumstances, the Stern-Gerlach preparation procedure also suffers from the problem of entangled states — not because of the way the preparation procedure works, but because the state of the input system was entangled — as a result of the way it was (pseudo-) prepared.

The apparent pervasiveness of source preparations in the laboratory supports the following conjecture:

(C) Most, perhaps even all, supposed-preparation procedures used in actual quantum mechanical experiments involve source preparations.

And this in turn implies that most, perhaps even all, supposed-preparation procedures used in actual quantum mechanical experiments suffer from the problem of entangled states. Hence if (C) is correct, most, perhaps even all, supposed-preparation procedures actually used in the laboratory are pseudo-preparations. The preparation problem is, then, a pervasive problem. Empirical evidence gives good reason to believe that the supposed-preparation procedures used in the laboratory are successful, but quantum mechanics itself seems to say that almost none, perhaps even none at all, genuinely succeed. How can these two facts be reconciled?

Solving the Preparation Problem

A Piecemeal Approach

One might try to solve the preparation problem by denying that the Schrödinger equation even applies to pseudo-preparations. On this approach, the disparity between what our empirical evidence says is the state prepared and what the Schrödinger equation says is the outcome of a pseudo-preparation would be resolved by rejecting the latter. This approach is similar to that traditionally taken to solve the measurement problem. Most formulations of quantum mechanics include a special postulate that has been added to the original theory, the Projection Postulate, which says that at the end of a measurement for observable A such as depicted in figure 8.1, the compound system $M + S$ is not in the state assigned by the Schrödinger equation, but in a tensor product of the eigenstates associated with r_k, the value exhibited by M, and a_k, the corresponding value of A, respectively. (This "solution" just shifts the measurement problem to the questions of precisely when and why the Projection Postulate should be applied. It also puts more restriction on the end state of S than is needed to "solve" the measurement problem. See Healey 1989.) Similarly, one might try adding a "Preparation Postulate" which says that "at the end of a supposed-preparation process, the supposedly prepared system O does not have its state entangled in the way the Schrödinger equation says, but," then specify the state that should be assigned to S. But how can this specification be made? No objective feature of pseudo-preparation processes *in general* seems to pick out the "right" state to assign the prepared system. In the case of measurement, the result of the measurement does this job. In the case of preparation, however, actual practice seems to depend on empirically testing each particular type of supposed-preparation procedure to determine the supposedly prepared state. Or often, in lieu of actual empirical testing, one simply uses a semiclassical analysis to figure out what the results of an empirical test would be. The analysis is based on characteristic features of the particular type of procedure involved, rather than on some a priori notion of what characterizes supposed-preparations in general. Perhaps a closer analysis of what goes on in pseudo-preparations would reveal a determining feature common to all such processes, but at this point neither actual practice nor theory gives a clue as to what that feature might be.

This plus the diversity of procedures that are used as preparations suggests a piecemeal approach to solving the preparation problem: Instead of searching for a single solution that works for all pseudo-preparations, develop different solutions for different types of pseudo-preparations. On this approach, the primary task will be to identify the relevant types, then characterize precisely the class of pseudo-preparations corresponding to each type. One such class, for example, is the class of pseudo-preparations that are also measurement processes satisfying the Projection Postulate. In such pseudo-preparations the system supposedly prepared S is also the system measured. Hence, according to the Projection Postulate S is left at the end of the measurement in the eigenstate corresponding to the eigenvalue obtained as the result of the measurement. Therefore, the Projection Postulate guarantees that the interaction prepares S in that eigenstate. Thus it appears that the approach rejected as unpromising for pseudo-preparations in general — deny that the Schrödinger equation applies — will work for interactions in this particular class. A solution to the measurement problem will also provide a solution to the preparation problem for pseudo-preparations in this class.[6] But not all supposed-preparations are measurements. Other classes of pseudo-preparations will require other solutions.

There may be other classes of pseudo-preparations that could be handled in a similar way, that is, that could be construed as interactions that do successfully prepare by violating the Schrödinger equation. The task would be to identify the objective features of the interactions in such a class that determine the state they succeed in preparing, thus providing a basis for formulating a principle of evolution for interactions in the class analogous to the Projection Postulate for measurement interactions. This might be called the "replacement approach" to solving the preparation problem since it depends on replacing the Schrödinger equation by a new principle of evolution for the preparation interactions in such classes.

But for at least some classes of pseudo-preparations, the solution to the preparation problem may lie not in replacing the Schrödinger equation, but in explaining why these interactions can be used as preparations even though they do satisfy the Schrödinger equation, by showing that the state assigned by the Schrödinger equation to the compound $P + O$ at the end of the interaction gives the same predictions for subsequent experiments involving O as the state ω determined for O by empirical evidence. To carry out this "explanation approach," one would identify various classes of pseudo-preparations for which one could (1) specify a condition C that is satisfied by all members of the class; and (2) prove that if a pseudo-preparation satisfies C, then assigning ω to O gives the same predictions on subsequent experiments involving O as assigning Ψ_{P+O} to $P + O$. The following section provides an analysis which suggests that there is at least one well-defined class of interactions containing a number of common pseudo-preparations for which this approach yields a solution to the preparation problem.

Spatially Concentrated Pseudo-Preparations

Consider first the two-slit experiment as depicted in figure 8.5, where a scintillation screen D, placed some distance to the right of the two-slit screen P, is used to measure the position of O. A rough quantum-mechanical analysis shows that the same predictions for the results of this measurement are obtained whether ω is assigned to O or Ψ_{P+O} to $P + O$.
Let:

$t =$ time at which O is measured for position at D;
$\psi' =$ state to which ψ has evolved at t, for any ψ;
$\mathbf{P}(\Delta r) =$ probability that a system O emitted from S will be found at position in region Δr of D;
$P_X^r =$ projector onto the Δr subspace of H_X, for any system X; and
$Tr(A\psi) =$ trace of $A\psi$, for any operator A and state ψ.

We want to compare the two expressions for $P(\Delta r)$ obtained when the two different states are assigned, ω to O and ψ_{P+O} to $P + O$.

(a) Suppose O is in ω just to the right of screen P. When the state of a particle emitted from S is expressed as $\psi = c_H\psi_H(x) + c_M\psi_M(x)$, the probability that a particle emitted from S will be found to the right of screen P is $|c_M|^2$. Since ω' is the state to which ω has evolved when O is measured for position at screen D, the conditional probability that if O is in ω just to the right of P, then O will be found in region Δr at D, is $Tr(P_o^r\omega')$. Thus,

$$P(\Delta r) = |c_M|^2 Tr(P_o^r\omega').$$

(b) Suppose $P + O$ is in Ψ_{P+O}. Then,

$$P(\Delta r) = Tr(P_{P+O}^r\Psi'_{P+O}) = Tr(c_H P_{P+O}^r\Sigma c_{ij}\pi'_i\otimes\Psi'_j + c_M P_{P+O}^r(\pi'_e\otimes\omega')).$$

But $P_{P+O}^r = I\otimes P_O^r + P_P^r\otimes I - P_P^r\otimes P_O^r$, so

$$P(\Delta r) = Tr(c_H P_{P+O}^r\Sigma c_{ij}\pi'_i\otimes\psi'_j) + c_M\pi'_e\otimes P_O^r\omega' + c_M P_P^r\pi'_e\otimes\omega'$$
$$- c_M P_P^r\pi'_e\otimes P_O^r\omega'.$$

Since P is a macroscopic object at some distance from D, $|\pi_e|^2 \approx 0$ in the region of screen D, hence $P_P^r\pi'_e \approx 0$. Likewise, since the system formed when a particle hits P is also macroscopic and occupies the same region as P, $P_{P+O}^r\Sigma c_{ij}(\pi'_i\otimes\psi'_j) \approx 0$. Hence,

$$P(\Delta r) \approx Tr(c_M\pi'_e\otimes P_O^r\omega') = |c_M|^2 Tr(P_O^r\omega'). \text{ QED.}$$

A similar analysis can be given for a system O subjected to the pseudo-preparation depicted in figure 8.4 where O is (allegedly) prepared in the spin eigenstate σ_a^+ using a Stern-Gerlach apparatus plus blocking screen. Suppose O is then measured for S_b using a second Stern-Gerlach apparatus, this one oriented in direction b, placed some distance along the upper path behind the first, plus a scintillation screen D placed behind the second apparatus, as in figure 8.6. Again, a rough quantum mechanical argument shows that assigning the state σ_a^+ to systems emerging along the upper path from the first Stern-Gerlach apparatus will give the same predictions for the results at screen D as assigning Ψ_{P+O} to $P + O$. This argument depends on asserting that

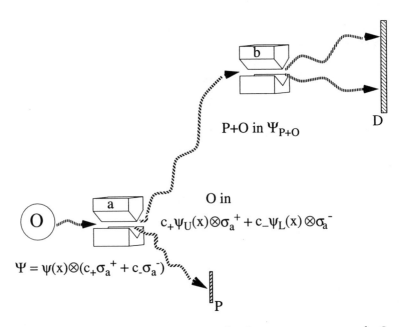

Figure 8.6 Pseudo-preparation of O in σ_a^+ with subsequent measurement for S_b.

there is virtually zero probability of getting a positive position reading at screen D for either P or a system composed of P and a particle that has collided with it. Hence the probability that $P + O$ will contribute to any positive result for the position measurement at D will depend only on c_M and $\psi_U \otimes \sigma_a^+$ in Ψ_{P+O}. Since this argument hinges on considerations so similar to that given above for the two-slit experiment, I will not go through the details here. I will turn instead to another example that does not suppose that the measurement involved is for position.

Consider the situation depicted in figure 8.7 involving a simple source (pseudo-) preparation: A system O (allegedly) emerges from a source $S = P + O$ where P is the source S with O removed, then O is measured for observable A using a measuring device M some distance to the right of S.

Let:

t_o = time at which measurement for A on O with M begins;
t_e = time at which measurement for A on O with M ends;
ψ' = state to which ψ has evolved at t_o, for any ψ;
ψ'' = state to which ψ has evolved at t_e, for any ψ;

$\{a_n\}$ & $\{\alpha_n\}$ = eigenvalues and eigenstates of observable A of O;
$\{r_i\}$ & $\{\rho_i\}$ = eigenvalues and eigenstates of observable R of M;
$\quad\quad\mathbf{P}(r_k)$ = probability that M will display reading r_k at t_e, indicating result a_k for a measurement of O on S.

We want to compare the expression for $\mathbf{P}(r_k)$ obtained when ω is assigned to O with the expression obtained when ψ_S is assigned to $P + O$. A rough quantum-mechanical analysis indicates that both assignments give the same probability.

a. Suppose O is emitted at time t from S in ω. Using the notation of equation (8.10), S is in $\psi_S = c_E \pi_E \otimes \omega + \Sigma c'_{ij} \pi_i \otimes \psi_j$ at t. Thus the probability that a system originally in S will be emitted in ω at t is $|c_E|^2$. The state of O at t_o, ω', can be written using the $\{\alpha_n\}$: $\omega' = \Sigma b_n \alpha_n$. Then at t_o the state of $M + O$ is $\rho_o \otimes \Sigma b_n \alpha_n$. Due to the measurement interaction between M and O this has evolved at t_e to $\Sigma b_n \rho_n \otimes \mu_n$ where for each n, μ_n is the state to which α_n evolves as a result of measurement. Hence, the conditional probability that if O is in ω when it is emitted, M will exhibit r_k, is $|b_k|^2$. Thus,

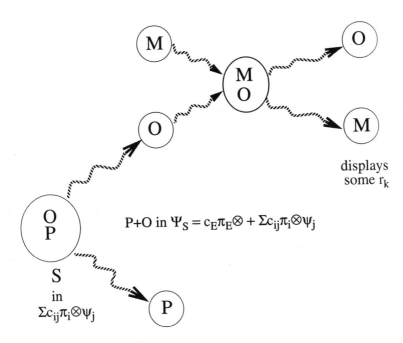

displays
some r_k

P+O in $\Psi_S = c_E \pi_E \otimes + \Sigma c_{ij} \pi_i \otimes \psi_j$

S
in
$\Sigma c_{ij} \pi_i \otimes \psi_j$

Figure 8.7 Source pseudo-preparation of O with subsequent measurement.

$$\mathbf{P}(r_k) = |c_E|^2 |b_k|^2.$$

b. Suppose $P + O$ is in state ψ_S. Then at t_o the state of $M + O$ will be

$$\rho_o \otimes \Psi_S' = c_E \rho_o \otimes \pi' \otimes \omega' + \Sigma c_{ij}' \rho_o \otimes \pi_i' \otimes \psi_j'$$

$$= c_E \pi' \otimes \rho_o \otimes \Sigma b_n \alpha_n + \Sigma c_{ij}' \rho_o \otimes \pi_i' \otimes \psi_j'.$$

Due to the measurement interaction between M and O, this has evolved at t_e to:

$$c_E \pi' \otimes \Sigma b_n \rho_n \otimes \mu_n + \Sigma c_{ij}' \rho_o'' \otimes \pi_i'' \otimes \psi_j''.$$

But a source with zero probability of emitting a system O, that is, a source in the second summand of this superposition, $\Sigma c_{ij}' \rho_o'' \otimes \pi_i'' \otimes \psi_j''$, has negligible probability of producing a positive reading in M. Hence $\rho_o' \in \{\rho_o\} \cup \{\rho_m\}$ where $\{\rho_m\}$ are those states in H_M (if any) that have eigenvalues which signify that no result for A has been found. Thus the second summand in the state of $M + O$ at t_e contributes nothing to the probability that r_k is exhibited by M at t_e. Therefore,

$$\mathbf{P}(r_k) = \{c_E\}^2 \{b_k\}^2. \text{ QED.}$$

These quantum-mechanical arguments are very sketchy and depend on assuming that the nonzero portions of certain quantum-mechanical states (in spatial representation) will be well concentrated and their "tails" so small that they can be neglected. Nonetheless they suggest that there may be a class of common pseudo-preparations, what we might call "spatially concentrated pseudo-preparations," for which the preparation problem can be solved using the explanation approach outlined above. Intuitively, surrounding a spatially concentrated pseudo-preparation will be regions $\{R_i\}$ far enough away from the influence of the preparation device P that there will be virtually no interaction between a measuring device placed in one of the R_i and P, or between such a measuring device and any other system "located" at the site of P. In more precise quantum-mechanical terms, the condition concerns the position probabilities given by the quantum-mechanical states of P, O, and $P + O$ for the regions $\{R_i\}$. Let $\mathbf{P}_X(R_i / \psi)$ be the conditional probability of finding system X in region R_i, given the assignment of ψ to X, for any state ψ and system X, and let $>_s$ stand for "significantly greater than." The above examples suggest the following definition: The class of spatially concentrated pseudo-preparations consists of pseudo-preparations of the form depicted in figure 8.3 such that:

(SC) There is a time t after the time for which O is assigned ω, and there are regions $\{R_i\}$ in the vicinity of P such that $P_O(R_i/\omega') >_s 0$, but also $P_P(R_i/\pi') \approx 0$ and $P_{P+O}(R_i/\Sigma c_{ij}\pi'_i \otimes \psi'_j) \approx 0$ where for any state ψ, ψ' is the state to which ψ has evolved at t.

The Explanation Approach

The above sketchy analyses are only a first step, at best simply suggesting the viability of the explanation approach. More accurate and detailed quantum mechanical analyses of the two-slit and Stern-Gerlach pseudo-preparations are needed to show that both really do satisfy (SC) to some reasonable approximation. In addition, it remains to be proven in detail that if a pseudo-preparation satisfies (SC), then the assignments of ω to O and of Ψ_{P+O} to $P + O$ give the same predictions for a variety of subsequent experiments involving O. Further work is also needed to find out what other pseudo-preparations satisfy (SC), in order to determine how wide this class is. Finally, one would like to identify other classes of pseudo-preparations that might be treated similarly, in order to see how far the explanation approach will reach.

One important feature of the explanation approach should not be overlooked, however. The strategy of this approach is to explain why certain types of pseudo-preparations can be used in laboratory and thought experiments. But it should not be supposed that this will be achieved by proving that the state supposedly prepared by such pseudo-preparation gives the same predictions as the entangled state assigned by the Schrödinger equation for *all possible* subsequent measurements. In the case of spatially concentrated pseudo-preparations, for example, it is simply not true that the assignments of ω to O and of Ψ_{P+O} to $P + O$ at the end of a pseudo-preparation give the same predictions for all possible subsequent experiments involving O. For, in general, the assignments of ω to O and of Ψ_{P+O} to $P + O$ will give different predictions if correlations between measurements on O and P are examined, or correlations between measurements on any systems that have interacted with O and P, respectively, subsequent to the pseudo-preparation; different predictions may also be given for any measurements on O itself if O and P have interacted a second time — after the pseudo-preparation but before the measurement in question. Thus, this approach to solving the preparation problem will only explain why certain sorts of pseudo-preparations can be used experimentally by showing that the state supposedly prepared by such pseudo-preparations will yield the same predictions as the state assigned by quantum mechanics *for the usual sorts of experiments* — that is, *for all practical purposes.* There is little doubt that "FAPP solutions" fail to address the foundational difficulty raised by the measurement problem. When it comes to the preparation problem, however, the situation is different. The preparation problem is not a foundational problem. Quantum mechanics has no trouble providing a theory of how preparation procedures might work in principle. It is only when quantum mechanics is applied to actual laboratory procedures or to thought experiments that the question arises: Why can the supposed-preparation be used on these particular occasions as though it did successfully prepare? The question raises a pragmatic issue and calls for an answer of the form: Because in this type of situation this particular tool used in this particular way achieves the desired end. The explanation approach answers: Because, on these particular occasions, supposing the procedure does prepare gives the same predictions as quantum mechanics. Since the issue is not whether quantum mechanics can yield appropriate state assignments for a theory of preparation proper, but only whether one can explain why a state other than that assigned by quantum mechanics can be substituted in particular situations, an explanation specific to the particular situations in question is sufficient to solve the problem.

Conclusion

There is a significant difference between the preparation problem and the measurement problem. The measurement problem arises from an "in-principle" difficulty — it is impossible for any interaction to satisfy both the assumptions that characterize measurement interactions and the standard quantum-mechanical description of interactions. The preparation problem, on the other hand, arises from a *de facto* difficulty. According to quantum mechanics, many (and perhaps all) of the processes we do, as a matter of fact, take to be preparation processes and actually use as preparations in the laboratory, do not succeed in preparing. While the in-principle difficulty behind the measurement problem clearly calls for the sort of general and perhaps radical solution proposed by many recent attempts to solve it, the preparation problem invites and perhaps requires a more piecemeal approach. For pseudo-preparations that are also measurements, the preparation problem might be taken care of by a solution to the measurement problem, using what I have called a replacement approach. For other types of pseudo-preparations, however, a different approach has been suggested: Identify classes of pseudo-preparations that can be assumed to evolve according to the Schrödinger equation and nonetheless be shown to yield end states that make them usable — at least for all practical purposes — as preparation procedures. This will depend on being able to show that given the characteristics that define the class, the standard quantum-mechanical analysis of interactions in the class will explain why using ω as the state for O will give correct predictions on all of the typically measured behavior of O and systems that have interacted with O subsequent to the pseudo-preparation. It remains to be seen whether these two approaches — replacement and explanation — can solve the preparation problem for the wide variety of pseudo-preparations used in the laboratory and appealed to in discussions of thought experiments.

NOTES

Earlier versions of this essay were presented to the Center for Philosophy of Science, University of Pittsburgh, and the History and Philosophy of Science Colloquium, Indiana University; helpful comments from those present are gratefully acknowledged. I would also like to thank Jeremy Butterfield, Arthur Fine, Roger Newton, Osvaldo Pessoa, Michael Redhead, Don Robinson, Laura Ruetsche, and Abner Shimony for helpful comments and discussion. The research for this paper was funded in part by the National Science Foundation under grant DIR-9122666.

1. At this point, my treatment is both narrower and wider than the reader familiar with literature on the measurement problem might expect. Strictly speaking, the characterization of measurement and the arguments concerning the measurement problem should employ the notion of a mixed state, and measurements should be characterized as an interaction for which the end state of M is a mixture of the ρ_i. For the sake of simplicity, I have formulated the characterization and arguments of this essay in terms of pure states only. The reader is invited to recast them in terms of mixtures to verify that nothing significant is lost or gained by this simplification.

Usually, expositions of the measurement problem presuppose von Neumann's theory of quantum-mechanical measurement. That theory assumes that when M exhibits r_k, S must end up in α_k at the end of the measurement, since von Neumann supposed that the value found for A in a particular measurement would certainly also be found if the measurement were immediately repeated. But the measurement problem does not depend on making this particular assumption about the state of S at the end of a measurement, so no such assumption is made here.

2. The measurement problem is often said to arise because in general the state of M at the end of a measurement is entangled in a superposition state for $M + S$. Strictly speaking, the difficulty is that in general the state of M at the end of a measurement is not one of the eigenstates $\{\rho_i\}$. This difficulty occurs not simply when there is entanglement. If a uniform preparation of the sort illustrated in the preceding paragraph is also a measurement, for example, we identify M with P and S with O and see that the state transition due to the preparation/measurement for an arbitrary initial state of S is $\rho_o \otimes \Sigma c_i \alpha_i \rightarrow (\Sigma c_i \rho_i) \otimes \omega$. System S is put into state ω, and M is left in a pure state, not entangled in a superposition state for $M + S$. The difficulty is that this pure state is not one of the eigenstates $\{\rho_i\}$ but a superposition of these eigenstates.

3. Several people have suggested that the supposed-preparation procedure used in the two-slit experiment is an example of uniform preparation: Let a beam impinge on a screen with two slits in it so that the particles that emerge on the other side of the screen are in the usual two-slit state. Others have suggested the simple procedure of waiting for a collection of atoms of some particular element, which are initially in a variety of excited states, to decay, until all end up in the same ground state. But as later arguments in this essay show, the first procedure is not uniform — beams in different initial states yield different final states for the particles emerging from the slits. And neither procedure is even a successful preparation, according to standard quantum mechanics; both are what will later be called pseudo-preparations. A variety of more complicated examples have been offered also, but on closer inspection they also turn out to be only pseudo-preparations.

4. From the perspective of quantum field theory, we can see that the reason quantum mechanics does not say that the state of O is entangled is that quantum mechanics treats the magnetic field of the Stern-Gerlach device semiclassically. The effect of the magnetic field on O is captured by using an operator counterpart of the classical expression for the magnetic field and inserting this in the Hamiltonian governing the evolution of the state of O, rather than (as in quantum field theory) treating the field as another quantum system and tracing the interaction between this system and O. On the latter approach, the state of O would be entangled with that of the field at the end of the interaction, giving a theoretical approach that would not describe this as a preparation of O in state ω. Thus while quantum mechanics does take this procedure to be a successful preparation (at least in principle, so long as it can be assumed that O enters the Stern-Gerlach device in a quantum-mechanical state ψ), the more accurate treatment by quantum field theory does not. The issue here, however, is how quantum mechanics describes preparations.

5. Of course, exactly which systems involved in a physical process count as parts of the preparation device, which are parts of a previous (pseudo-)preparation, and

which are included as parts of a measuring device, is to some extent relative to the way the physical situation is described. A complete analysis of preparation (and measurement), and any solutions to the difficulties associated with supposed-preparations (and measurements), will have to take this descriptive relativity into account. The analyses and solutions suggested below are flexible enough to accommodate this descriptive relativity, but an explicit proof to that effect is still needed.

6. It would seem that a solution to the measurement problem would also provide a similar solution for pseudo-preparations that are measurements which do not satisfy the Projection Postulate, except that the state in which S is prepared would not necessarily be an eigenstate of the observable measured. However, one can imagine that particularly ornate solutions to the measurement problem might not yield such a tidy consequence. A more careful examination of how this plays out for particular proposed solutions to the measurement problem will be needed to determine the implications for this broader class of pseudo-preparations.

REFERENCES

Blum, K. 1981. *Density Matrix Theory and Its Applications.* New York: Plenum Press.

Cohen-Tannoudji, C., B. Diu, and F. Laloë. 1977. *Quantum Mechanics*, volumes I and II. Translated by S. R. Hemley, N. Ostrowsky, and D. Ostrowsky. New York: Wiley & Sons.

Dunlap, R. A. 1988. *Experimental Physics.* New York: Oxford University Press.

Fano, U., and L. Fano. 1972. *Physics of Atoms and Molecules.* Chicago: University of Chicago Press.

Fine, A. 1969. "On the General Theory of Measurement." *Proceedings of the Cambridge Philosophical Society* 65: 111–22.

French, A. P., and E. F. Taylor. 1978. *An Introduction to Quantum Mechanics.* New York: Norton.

Healey, R. 1989. *The Philosophy of Quantum Mechanics: An Interactive Interpretation.* Cambridge, England: Cambridge University Press.

Hughes, R. I. G. 1989. *The Structure and Interpretation of Quantum Mechanics.* Cambridge: Harvard University Press.

Jauch, J. M. 1968. *Foundations of Quantum Mechanics.* New York: Addison-Wesley.

Margenau, H. 1963. "Measurements and Quantum States." *Philosophy of Science* 30: 1–16, 138–57.

Park, J. L., and W. Band. 1971. "A General Theory of Empirical State Determination in Quantum Physics." *Foundations of Physics* 1: 211–26, 339–57.

9

Schrödinger's Cat and Other Entanglements of Quantum Mechanics

Jeffrey Bub
Department of Philosophy, University of Maryland, College Park

In this essay I show that there is essentially only one way to solve the measurement problem of quantum mechanics. I show how to construct all possible modal interpretations of quantum mechanics that solve the measurement problem formally in a certain sense. One such interpretation is Bohm's (1952) hidden-variable theory, and all other modal interpretations are hidden-variable theories that preserve the dynamics of quantum mechanics. So a modal interpretation in the sense I propose is a type of Bohmian theory.

The essay is divided into three subsections. In the first I discuss the relation between classical and quantum mechanics by way of motivating the inquiry in the next section, in which I consider how to construct all possible maximal determinate sublattices of quantum propositions that we can take as having definite (but perhaps unknown) truth values, subject to certain constraints that provide the basis for a solution to the measurement problem in terms of a modal interpretation. In the third section I show how to exploit a proposal by Vink (1993) to develop a generalized Bohmian dynamics for any modal interpretation constructed according to the prescription in the second section. Such a modal interpretation supplied with a dynamics solves the measurement problem by allowing the description of model quantum-mechanical universes that evolve dynamically so as to correlate determinate values of "measured" dynamical variables with determinate values of "pointer" variables.

The Transition from Classical to Quantum Mechanics

Classical mechanics describes model universes in which the state of the universe is specified by an assignment of values to all (generalized) position and momentum variables, hence to all dynamical variables as functions of positions and momenta. The equations of motion yield rates of change for positions and momenta in terms of partial derivatives of the Hamiltonian, a function of positions and momenta, and generate a possible history of the universe from the position and momentum values specified at some particular time.

The states of subsystems of a classical universe are similarly specified by assignments of values to the generalized positions and momenta of the subsystems. So the subsystems in a classical universe are separable as classical systems with local states that are partial states of the global state of the universe, and we can understand a classical universe as consisting of separable classical systems interacting under the influence of the forces encoded in the Hamiltonian. The evolution of the universe over time — a history of the universe — is given by a particular dynamical evolution of the state.

The state of a classical universe plays two distinct roles: a *diachronic* role as the carrier of the dynamics over time, and a *synchronic* role as the specification of a "possible world" at a particular time, that is, the selection of a set of determinate values for the dynamical variables as one of the possible sets of values for these variables, equivalently the selection of a set of properties for the systems in the universe as one of the possible sets of properties. In its diachronic role, a classical state is a *dynamical state*: it carries the dynamics of the universe. In its synchronic role, a classical state is a *property state*: it specifies the properties of the universe as one of the possible sets of properties.

Measurement in a model classical universe can be understood as the evolution of the universe into a global state in which the local state of some subsystem M, regarded as a measuring instrument, becomes correlated with the local state of another subsystem S, regarded as the measured system, so that the local state of S is indicated by the local state of M. More generally, a measurement will involve the correlation of some dynamical variable R of M, the "pointer" or indicator variable, with some dynamical variable A of S, so that the value of R indicates the value of S. Such dynamical evolutions exist unproblematically in a classical universe.

Quantum mechanics is derived as a generalization of classical mechanics in which commutation relations are imposed on canonically conjugate classical dynamical variables. We obtain a noncommutative algebra of dynamical variables, equivalently a non-Boolean algebra of properties, that is representable as an operator algebra on a Hilbert space, a linear vector space over the complex numbers. This is the import of Schrödinger's proof of the equivalence of his wave mechanics formulation of quantum theory and Heisenberg's matrix mechanics formulation of the theory, and von Neumann's Hilbert space representation of matrix and wave mechanics as a general theory of quantum mechanics.

In a classical mechanical universe, the algebra of dynamical variables is a commutative algebra of real-valued functions on the classical phase space coordinatized by positions and momenta. The algebra of properties is a subalgebra of this algebra: the algebra of idempotent functions, that is, two-valued functions with values 1 and 0, corresponding to the truth values "true" and "false," respectively, for the associated propositions (e.g., the proposition asserting that the value of a dynamical variable A lies in a certain range). This algebra of properties or propositions is a Boolean algebra, ideally isomorphic to the power set of phase space, but more properly, to a subset of the power set forming a σ-field (usually the Borel sets). So properties or propositions in classical mechanics are represented by subsets of phase space. The proposition asserting that the value of A lies in a certain range is represented by the subset of classical states — phase space points — for which this proposition is true, via the functional relationship between A and positions and momenta. Classical states correspond to atoms in the Boolean algebra of properties, that is, minimal nonzero elements of the algebra (singleton subsets of phase space). Each state defines a two-valued homomorphism on the Boolean algebra of properties that selects a maximal filter or ultrafilter of properties as the collection of properties belonging to the state in the synchronic sense, the properties represented by subsets of phase space containing the state as phase point.

In quantum mechanics, the dynamical variables in the noncommutative algebra are represented by operators that can be represented as linear sums of projection operators onto subspaces of Hilbert space, with real coefficients. So a dynamical variable A might be represented as:

$$A = \Sigma a_i P_{a_i}$$

where the P_{a_i} are projection operators onto the eigenspaces of A, and the a_i are the distinct possible nonzero eigenvalues of A. (I use the same symbol for a dynamical variable and the operator representing the dynamical variable. That each dynamical variable can be represented in this way or, more generally, by an integral instead of a sum, is the content of the spectral representation theorem.) The projection operators represent idempotent dynamical variables with two possible values, 1 and 0; that is, they represent properties or propositions in quantum mechanics. The subalgebra of projection operators is isomorphic to the lattice of subspaces of Hilbert space, an atomic, orthocomplemented lattice. The proposition asserting that the value of A lies in a certain range is represented by the subspace of Hilbert space corresponding to the associated projection operator.

In the non-Boolean lattice of projection operators or subspaces of Hilbert space representing the properties of a model quantum-mechanical universe, the one-dimensional subspaces or rays are atoms, minimal elements in the lattice. The rays in Hilbert space represent states in the diachronic sense; that is, they represent dynamical states that evolve via unitary transformations defined by Schrödinger's time-dependent equation of motion. On the *orthodox* interpretation of quantum mechanics, the rays are also taken to represent states in the synchronic sense; that is, they are taken to represent property states. The collection of properties determined by a ray in Hilbert space is taken as the collection of subspaces containing the ray,[1] just as the collection of properties belonging to a classical state is the collection of properties represented by subsets of phase space containing the state as a point in phase space. But while the classical property state corresponds to a two-valued homomorphism on the Boolean algebra or Boolean lattice of classical properties, the orthodox quantum-mechanical property state does not correspond to a two-valued map on the lattice of subspaces representing quantum properties that reduces to a two-valued homomorphism on each Boolean sublattice of properties. (Each Boolean sublattice represents a family of propositions of the form "the value of A lies in the range [ ]" for all possible [ ].) I refer to such maps as Boolean homomorphisms. By the Kochen and Specker (1967) theorem, there are no Boolean homomorphisms on the lattice of subspaces of a Hilbert space except in the case of a two-dimensional Hilbert space, and even in this case the collection of subspaces containing a ray — the property state according to the orthodox interpretation — does not correspond to a collection of subspaces selected by a Boolean homomorphism. (The orthodox property state in the case of a two-dimensional Hilbert space H is the four-element Boolean algebra generated by the ray representing the state and the orthogonal ray as atoms. A Boolean homomorphism on the lattice of subspaces selects one ray in each orthogonal pair of rays in H.)

The orthodox decision to take the ray representing a dynamical state in quantum mechanics as defining a property state in the sense just described has the consequence that we can no longer apply quantum mechanics to model quantum-mechanical universes in the same way in which classical mechanics is applied to model classical universes. According to the orthodox interpretation, at any particular time the ray representing the dynamical state specifies the property state as the collection of properties represented by the subspaces containing the ray. But since this property state does not correspond to a Boolean homomorphism on the lattice of properties, we cannot understand the properties in this collection as obtaining in the model universe and the properties not in this collection as not obtaining; that is, we cannot take the corresponding propositions as true and false respectively, for this will involve a contradiction. (There will be infinitely many sets of propositions of the form "the value of A lies in the range [ ]," for all possible [ ], that would all have to be taken as false, because no proposition asserting that A lies in some range belongs to the property state.) We could take the propositions represented by subspaces orthogonal to the ray as false, but this leaves the propositions represented by subspaces that are neither orthogonal to the ray nor contain the ray indeterminate neither true nor false.

On the orthodox interpretation, the ray representing the quantum state assigns *probabilities* to *all* the propositions represented by subspaces of the Hilbert space: probability 1 to propositions in the property state defined by the ray, probability 0 to propositions represented by subspaces orthogonal to the ray, and nonzero probabilities to all other propositions. It follows that these probabilities cannot be represented on a classical Kolmogorov probability space as measures over different possible property states or "possible worlds." So a nonzero probability assigned to a proposition asserting that the value of a dynamical variable A lies in a certain range is said to be "the probability of finding the proposition to be true on measurement" or "the probability of finding the value of the dynamical variable in the range on measurement." This means that quantum mechanics can only be applied to model quantum-mechanical universes in the sense of providing probabilities for the results of measurements on these universes by some agent or device *external* to these universes. To mark this distinction between classical and quantum mechanics, dynamical variables in quantum mechanics are referred to as "observables," which are understood to have no determinate value unless the ray representing the quantum state lies in one of the eigenspaces of the observable.

Clearly this notion of "measurement" is completely undefined dynamically, as is the notion of an observable having no determinate value at one time and coming to have a determinate value at some other time as the outcome of a "measurement" of the observable in question (or a measurement of some observable that commutes with the observable in question). One would like, as Bell (1987) put it, to have an interpretation of quantum mechanics, our most fundamental theory of motion, in terms of "beables" instead of "observables."

The impossibility of an *internal* account of measurement in quantum mechanics (along the lines of the account in a classical mechanical universe), entailed by the orthodox interpretation of the dynamical state as representing a property state, is what is known as the *measurement problem* of quantum mechanics.

To see what goes wrong with an internal account of measurement on the orthodox interpretation, consider a model quantum-mechanical universe consisting of two systems: a system designated as the measuring instrument M and a system designated as the measured system S. Suppose that the quantum state of $S + M$ — a ray in Hilbert space — takes an initial product form represented by the unit vector $\alpha_i \otimes \rho_0$ in the ray, where α_i lies in an eigenspace of some observable A of S (so that A initially has some value a_i according to the orthodox interpretation) and ρ_0 lies in an eigenspace of some "pointer" or indicator observable R of M (so R initially has some zero value according to the orthodox interpretation). There exists a unitary evolution:

$$\alpha_i \otimes \rho_0 \rightarrow \alpha_i \otimes \rho_i, \text{ for each } i,$$

that correlates the pointer vector with the eigenvector of the measured observable A, and hence can be taken as representing a measurement of A in this model universe. In this case, each sybsystem in the universe is associated with its own quantum state, both before and after the measurement interaction represented by the unitary evolution of the composite state, as in a classical mechanical universe. According to the orthodox interpretation, the property state after the measurement interaction is defined by the collection of properties represented by subspaces that contain the final product state $\alpha_i \otimes \rho_i$, and this is a set of A-properties and correlated R-properties, indexed by i. But now, since the dynamics is linear, it follows that

$$(\Sigma c_i \alpha_i) \otimes \rho_0 \rightarrow \Sigma c_i \alpha_i \otimes \rho_i$$

for an initial state in which the subsystem S has no determinate A-value and is represented by a state in the factor space of S that is a linear superposition of A-eigenstates. In this case, each subsystem is no longer associated with its own state after the measurement interaction represented by the unitary evolution of the composite state: the final state is the "entangled" state $\Sigma c_i \alpha_i \otimes \rho_i$, a linear superposition of product states. According to the orthodox interpretation, the property state is defined by the collection of properties represented by subspaces that contain the final entangled state $\Sigma c_i \alpha_i \otimes \rho_i$. But no A-property belongs to this collection, and no R-property belongs to this collection. So the unitary evolution cannot be taken as representing a measurement.

As Schrödinger (1935) emphasized, if M represents a cat and R takes two possible values, associated with the cat being alive and dead, and the cat interacts with a microsystem S, such as an atom that can either decay or not decay in a certain time (where these events are associated with the two possible values of A), the decay event triggering a device that kills the cat, then the cat will be neither alive nor dead after the measurement interaction, according to the orthodox interpretation.

The upshot of this analysis is that "measurement," in the sense required by the orthodox interpretation of the dynamical state as yielding probabilities of measurement outcomes, is completely undefined dynamically. It is also clear that an observable that has no determinate value cannot come to have a determinate value as the result of a measurement construed dynamically, and so the requirement of the orthodox interpretation that observables come to have determinate values when measured has no dynamical justification.

The orthodox notion of the property state of a quantum-mechanical universe is by no means forced on us by the transition from classical to quantum mechanics. The proposal to take the quantum state, represented by a ray in Hilbert space, as a dynamical state specifying possibilities and probabilities in a quantum-mechanical universe only, together with a characterization of the property state differing from the orthodox interpretation, was first put forward by Bas van Fraassen (1979, 1981, 1991) as the *modal interpretation* of quantum mechanics. (Van Fraassen introduces a distinction between dynamical states and value states. His value states are property states in my sense.) The aim of a modal interpretation is to make possible a purely internal account of measurement, as in classical mechanics.

A variety of modal interpretations are now in the literature, with different proposals for the definition of a property state in quantum mechanics. (Modal interpretations have been proposed by Kochen 1985, Krips 1987, Healey 1989, 1993, Dieks 1989, 1994, and Bub 1991, 1992a, 1992b, 1993, 1994. The interpretations proposed by Kochen, Healey, and Dieks exploit the polar decomposition theorem, and this type of modal interpretation is perhaps the most visible in the literature.) They share the assumptions (i) that the quantum state represented by a ray in Hilbert space is to be understood as a dynamical state specifying possibilities and probabilities only, and that an *interpretation* of quantum mechanics requires the definition of an associated property state; and (ii) that the orthodox notion of property state is unsatisfactory because it leads to the measurement problem. What we want is a concept of property state that will support a purely internal account of measurement.

Determinate Sublattices

There are no Boolean homomorphisms on the lattice L of subspaces representing the properties of a model quantum-mechanical universe — maps from L to $\{0,1\}$ that reduce to two-valued homomorphisms on each Boolean sublattice of L — if the Hilbert space is more than two-dimensional. This is the content of the Kochen and Specker theorem (1967). The number of quantum-mechanical observables required to generate a sublattice of propositions over which such maps cannot be defined is surprisingly small, especially if the Hilbert space is more than three-dimensional. Mermin (1990) generates a contradiction from a map on the propositions generated by nine observables in an eight-dimensional Hilbert space.

The problem of how small we can make the set of observables and still generate a Kochen-Specker contradiction is interesting mathematically, but of no foundational significance for the interpretation of quantum mechanics. To provide an interpretation of quantum mechanics that supports an internal account of measurement, we want to know how large we can take the set of observables *without* generating a Kochen-Specker contradiction, that is, we are interested in maximal sets of "observables" that can be taken as "beables."

Specifically, according to the orthodox interpretation the quantum state assigns probabilities to the propositions represented by all subspaces of the Hilbert space H of a model quantum-mechanical universe; that is, to all propositions in a lattice L isomorphic to the lattice of subspaces of H. We know we cannot assign truth values to *all* these propositions by a two-valued map on L that reduces to the usual Boolean assignment on each Boolean sublattice of properties (that is, a two-valued homomorphism on each Boolean sublattice), so we know that we cannot interpret the probabilities defined by the quantum state epistemically as measures over the different possible truth-value assignments to all the propositions in L; that is, as measures over property states. We also know that any single observable can be taken as determinate, that is, as a beable, for any quantum state, so we may suppose that fixing a quantum state represented by a ray e in H and an arbitrary observable R places restrictions on what propositions can be taken as determinate for e in addition to R-propositions. To support property states in quantum mechanics, the sublattices of interest are the maximal sublattices $D(e,R)$ of L that satisfy at least the following two conditions:

(i) Truth condition: Truth values can be assigned to all the propositions of $D(e,R)$, where each assignment of truth values is defined by a Boolean homomorphism on L—a two-valued map on $D(e,R)$ that reduces to a two-valued homomorphism on each Boolean sublattice of $D(e,R)$.

(ii) Probability condition: The probabilities defined by e on the propositions of $D(e,R)$ can be represented as measures over the different possible truth-value assignments to $D(e,R)$, i.e., as measures on a Kolmogorov probability space (X, F, μ), where X is the set of Boolean homomorphisms on $D(e,R)$ and $\mathrm{tr}(ea) = \mu(\{h:h(a) = 1\})$ for any proposition $a \in D(e,R)$. Here $\mathrm{tr}(ea)$ is the probability assigned by the quantum state e to the proposition represented by the subspace a, and

$\mu(\{h:h(a) = 1\})$ is the measure of the set of Boolean homomorphisms assigning the value 1 (i.e., true) to a.

A sublattice $D(e,R)$ will be decomposable into two disjoint sets of propositions: the set $I(e,R)$ of impossible propositions for e and R, and the set $P(e,R)$ of possible propositions for e and R. The impossible propositions are the propositions in $D(e,R)$ assigned zero probability by the measure μ corresponding to e, and the possible propositions are the propositions in $D(e,R)$ assigned nonzero probabilities by μ. Notice that if any proposition $a \in I(e,R)$, then any proposition $b \in D(e,R)$ such that $b \leq a$ will belong to $I(e,R)$. In particular, any atom in a belonging to $D(e,R)$ will also belong to $I(e,R)$. This is not the case for propositions in $P(e,R)$. Rather, if $a \in P(e,R)$, then any proposition $b \in D(e,R)$ such that $a \leq b$ will belong to $P(e,R)$. It follows that a sublattice $D(e,R)$ partitions the Hilbert space H into two orthogonal subspaces that span H: the subspace $i(e,R)$ corresponding to the supremum of the propositions in $I(e,R)$, and the subspace $i(e,R)^{\perp}$ orthogonal to $i(e)$. (The supremum of the propositions in $P(e,R)$ corresponds to H.) Note that the subspace $i(e,R)$ represents an impossible proposition for e and R because e must be orthogonal to $i(e,R)$, and hence $i(e,R)$ is assigned zero probability by the measure μ corresponding to e.

Call the sublattices of propositions satisfying conditions (i) and (ii), together with the following four conditions, the *determinate sublattices* for e and R:

(iii) Eigenstate condition: If e is an eigenstate of R, then $D(e,R)$ contains the proposition e (the proposition represented by the one-dimensional subspace e).

(iv) Impossibility condition: If $a \in D(e,R)$ and $a \leq e^{\perp}$, then $b \in D(e,R)$ if $b \leq a$.

(v) Refinement condition: If R is a refinement of $R^{\#}$ (so that the eigenspaces of $R^{\#}$ are either the same as the eigenspaces of R or spans — suprema — of some of the eigenspaces of R), and the possible atoms of $D(e^{\#},R^{\#})$, for any states $e^{\#}$, are a proper subset of the possible atoms of $D(e,R)$, then $D(e,R) \subset D(e^{\#},R^{\#})$, for all such states $e^{\#}$.

(vi) Measurement condition: If the unit vector in the ray representing the quantum state takes the form of a polar decomposition $\Sigma c_i \alpha_i \otimes \rho_i$ with respect to eigenvectors α_i of some observable A and eigenvectors ρ_i of the observable R, so that A-propositions become correlated with R-propositions in the quantum state, then $D(e,R)$ includes the Boolean algebra of A-propositions (i.e., all propositions represented by the eigenspaces of A).[2]

The impossibility condition requires that if a proposition b implies any proposition $a \in I(e,R)$ that is impossible for a quantum state e and observable R, then b must be determinate and hence impossible for e and R. In particular, any atom in the subspace spanned by the impossible propositions for e and R must also belong to $D(e,R)$ and be impossible for e and R.

It now follows from the eigenstate condition and the impossibility condition that if $R = I$, the unit observable, then $D(e,I) \supset L_e$, where L_e is the sublattice generated by the atom e and the atoms represented by all the rays in the subspace orthogonal to e. Since L_e is maximal, $D(e,I) = L_e$.

To see that L_e is maximal, first note that if the Hilbert space H is more than two-dimensional and we add a multidimensional subspace $b \notin L_e$ to L_e and generate a lattice L_{eb} by closure, then every ray in b belongs to L_{eb}. For any ray $f \leq b$ that is not orthogonal to e, $e \lor f \in L_{eb}$ (because $e \lor f = e \lor g$, for some ray g orthogonal to e, and $e \lor g \in L_{eb}$ because $e \in L$ and $g \in L$). So $(e \lor f) \land b = f \in L_{eb}$.

Suppose we add a subspace $b \notin L_e$ to L_e and generate a lattice L_{eb} by closure. If b is a ray, then $b^{\perp} \in L_{eb}$ is multidimensional. So it suffices to consider adding a multidimensional b. It follows that every ray in b belongs to L_{eb}. Consider a specific ray $f \leq b$. Since $f \in L_{eb}$, $f^{\perp} \in L_{eb}$, and so again every ray in $f^{\perp}$ belongs to L_{eb}.

Now consider any ray h in H not already in L_{eb}, and not in the plane spanned by f and e. The plane $e \lor h$ belongs to L_{eb}, because $e \lor h = e \lor r$, for some ray r orthogonal to e. Also the plane $f \lor h$ belongs to L_{eb}, because $f \lor h = f \lor s$, for some ray s orthogonal to f. So $(e \lor h) \land (f \lor h) = h \in L_{eb}$, which means that every ray in H that is not in the plane spanned by f and e belongs to L_{eb}. But if k is a ray in the plane spanned by f and e, then k is not in the plane spanned by e and some other ray $f^{\#}$ in the subspace b. So a similar argument with $f^{\#}$ instead of f will show that $k \in L_{eb}$.

It follows that if we add $b \notin L_e$ to L_e and obtain L_{eb} by lattice closure, then L_{eb} must contain every ray in H, and so $L_{eb} = L$, the lattice of all subspaces of H. Since H is assumed to be at least four-dimensional (to allow representations of the quantum state that can be interpreted as measurements), there are no Boolean homomorphisms on L by the Kochen and Specker theorem, and so L_e is maximal.

The refinement condition is motivated by the requirement that if R is a refinement of $R^\#$, then the transition from $D(e,R)$ to $D(e^\#,R^\#)$ should simply involve the transformation of some of the possible atomic propositions in $D(e,R)$ to impossible atomic propositions in $D(e^\#,R^\#)$, with the appropriate modification to the subspace of impossible propositions to conform to the impossibility condition. The latter condition requires that every ray in the subspace of impossible propositions in $D(e^\#,R^\#)$ belongs to $D(e^\#,R^\#)$, which means that $D(e^\#,R^\#)$ contains some impossible propositions that do not belong to $D(e,R)$, and so $D(e,R) \subset D(e^\#,R^\#)$.

Since every observable can be regarded as a refinement of the unit observable I, the refinement condition entails the special refinement condition: $D(e,R) \subset D(e^\#,I)$ if the possible atoms of $D(e^\#,I)$, for any states $e^\#$, are a proper subset of the possible atoms of $D(e,R)$. Since $D(e^\#,I) = L_{e^\#}$, for any $e^\#$, and contains only one possible atom represented by the ray $e^\#$, the special refinement condition requires that $D(e,R) \subset D(e^\#,I)$ for any possible atom represented by a ray $e^\#$ in $D(e,R)$.

In the case that the Hilbert space H is a tensor product space with the observable R associated with a factor space, and the vector representative of the quantum state e takes a polar form $\Sigma c_i \alpha_i \otimes \rho_i$, where the ρ_i are eigenvectors of R, the measurement condition together with the impossibility condition entails that $D(e,R)$ contains elements corresponding to all the eigenspaces of any observable with k eigenvectors in the directions e_{r_i}, where the e_{r_i} are the nonzero projections of e onto the m eigenspaces of R. This follows because the measurement condition requires that $D(e,R)$ contains the propositions associated with the eigenspaces of any observable A with k eigenvectors α_i, as well as R, and hence the propositions asssociated with the eigenspaces of an observable with k eigenvectors $\alpha_i \otimes \rho_i$ (the nonzero projections of $\Sigma c_i \alpha_i \otimes \rho_i$ onto the eigenspaces of R) and $n\text{-}k$ eigenvectors $\alpha_j \otimes \rho_k$, $j \neq k$, corresponding to tensor products of A-eigenvectors and R-eigenvectors assigned zero probability by e (where n is the dimensionality of H). The impossibility condition requires in addition that all atomic propositions represented by rays in the subspace spanned by the $n\text{-}k$ vectors assigned zero probability by e are in $D(e,R)$, which means that the propositions associated with all observables with eigenvectors coinciding on the k tensor products $\alpha_i \otimes \rho_i$ of A-eigenvectors and R-eigenvectors assigned nonzero probability by the quantum state belong to $D(e,R)$.

It can now be shown that the *unique* choice for $D(e,R)$ is $L_{e_{r_1}e_{r_2}...e_{r_k}} = \{e_{r_i}, i = 1, \ldots, k\}'$, the commutant in L of $\{e_{r_i}, i = 1, \ldots, k\}$, where the k rays $e_{r_i} = (e \vee r_i^\perp)/\wedge r_i$ are the nonzero projections of the quantum state e onto the eigenspaces r_i of R, in the case that the Hilbert space H is a tensor product space with R associated with a factor space, and the vector representative of the quantum state e takes the polar form $\Sigma c_i \alpha_i \otimes \rho_i$.[3]

The determinate sublattice $L_{e_{r_1}e_{r_2}...e_{r_k}}$ is generated by the atoms e_{r_i}, $i = 1, \ldots, k$, and the atoms represented by all the rays in the subspace orthogonal to the subspace $e_{r_1} \vee e_{r_2} \vee \ldots e_{r_k}$ spanned by the e_{r_i}. There will be k Boolean homomorphisms on $L_{e_{r_1}e_{r_2}...e_{r_k}}$ if the subspace $(e_{r_1} \vee e_{r_2} \vee \ldots e_{r_k})^\perp$ is more than two-dimensional, where the ith homomorphism maps the proposition e_{r_i} onto 1. If the subspace $(e_{r_1} \vee e_{r_2} \vee \ldots e_{r_k})^\perp$ is less than three-dimensional, there will also be Boolean homomorphisms that map this subspace onto 1 and each of the rays e_{r_i}, $i = 1, \ldots, k$, onto 0, but these Boolean homomorphisms will all be assigned zero measure by the measure μ corresponding to e on the Kolmogorov probability space of Boolean homomorphisms on $L_{e_{r_1}e_{r_2}...e_{r_k}}$ (since e is orthogonal to $(e_{r_1} \vee e_{r_2} \vee \ldots e_{r_k})^\perp$). To generate the probabilities defined by e for the propositions in $L_{e_{r_1}e_{r_2}...e_{r_k}}$ on the Kolmogorov probability space, the Boolean homomorphism (more precisely, the corresponding singleton subset), that maps e_{r_i} onto 1 is assigned measure $\mathrm{tr}(er_i)$, for $i = 1, \ldots, k$. The set of determinate observables associated with $L_{e_{r_1}e_{r_2}...e_{r_k}}$ includes any observable with k eigenvectors in the directions e_{r_i}.[4]

For a proof of uniqueness, see Bub (1994). The proof proceeds by first showing that $D(e,I) = L_e$ is maximal in the sense that the truth condition is violated if we add any element to L_e and require lattice closure. (In Bub 1994 I characterized this sublattice as $D(e,e)$, where the second occurence of e is understood to represent the projection operator corresponding to the range e. Clearly, $D(e,e) = D(e,I)$.) Second, $D(e,R) = L_{e_{r_1}e_{r_2}}$ is shown to be maximal in the case that R has only two distinct eigenvalues. In this case, the determinate sublattice is generated by the atoms e_{r_1}, e_{r_2} and the atoms represented by all the rays in the subspace $(e_{r_1} \vee e_{r_2})^\perp$. This lattice is maximal in the sense that no element not in $L_{e_{r_1}e_{r_2}}$ can be added to $L_{e_{r_1}e_{r_2}}$ without generating a lattice by closure that is either $L_{e_{r_1}}$ or $L_{e_{r_2}}$, or the lattice $L_{e_{r_1} \vee e_{r_2}}$, generated by all the rays in the plane $e_{r_1} \vee e_{r_2}$ and all the rays in the subspace $(e_{r_1} \vee e_{r_2})^\perp$. Since H is assumed to be at least four-dimensional, the subspaces $(e_{r_1})^\perp$ and $(e_{r_2})^\perp$ are at least three-dimensional. This means that there are no Boolean homomorphisms on $L_{e_{r_1}}$ that assign 1 to the proposition e_{r_2}, and no Boolean homomorphisms on $L_{e_{r_2}}$ that assign 1 to the proposition e_{r_1}, by the Kochen and Specker theorem. (Since any Boolean homomorphism that maps any ray in $(e_{r_1})^\perp$ onto 1 must map e_{r_1} onto 0, it follows that any such homomorphism must map one of each orthogonal n-tuple of rays in $(e_{r_1})^\perp$ onto 1 and the remaining members of the n-tuple onto 0. But there are no such Boolean homomorphisms on $L_{e_{r_1}}$ if H is at least four-dimensional.) Hence there will be insufficiently many Boolean homomorphisms on $L_{e_{r_1}}$ or $L_{e_{r_2}}$ to construct a Kolmogorov probability space on which the probabilities defined by e for all the propositions in $L_{e_{r_1} e_{r_2}}$ can be generated by measures over the two-valued homomorphisms. So the probability condition is violated.

The lattice $L_{e_{r_1} \vee e_{r_2}}$ violates both the impossibility condition and the refinement condition. The lattice $L_{e_{r_1} \vee e_{r_2}}$ violates the impossibility condition because e lies in the plane $e_{r_1} \vee e_{r_2}$ and so the ray orthogonal to e in this plane represents an impossible proposition for e. It follows that every ray in the subspace spanned by this ray and the subspace $(e_{r_1} \vee e_{r_2})^\perp$ must belong to the determinate sublattice, and this is not the case for $L_{e_{r_1} \vee e_{r_2}}$. (If we were to add all these rays to $L_{e_{r_1} \vee e_{r_2}}$, we would generate the lattice L by closure.) To see that $L_{e_{r_1} \vee e_{r_2}}$ violates the (special) refinement condition, notice that the condition requires that $L_{e_{r_1} \vee e_{r_2}} \subset D(e^\#, I)$, for $e^\# = e_{r_1}$ or $e^\# = e_{r_2}$. This requires that $L_{e_{r_1} \vee e_{r_2}} \subset L_{e_{r_1}}$ and $L_{e_{r_1} \vee e_{r_2}} \subset L_{e_{r_2}}$, and this is not the case.

Finally, $D(e, R) = L_{e_{r_1} e_{r_2} \dots e_{r_k}}$ is shown to be maximal in the general case where R has k distinct eigenvalues, with $k > 2$. In this case, adding any element to $L_{e_{r_1} e_{r_2} \dots e_{r_k}}$ generates a sublattice by closure that violates either the probability condition (because some of the rays e_{r_i} are assigned zero by every Boolean homomorphism) or the (special) refinement condition.

Uniqueness follows when e takes the polar form for R, because $D(e, R)$ must include the propositions associated with the eigenspaces of any observable with k eigenvectors in the directions e_{r_i}, by the measurement condition and the impossibility condition, which means that $D(e,R) \supset L_{e_{r_1} e_{r_2} \dots e_{r_k}}$. Since $L_{e_{r_1} e_{r_2} \dots e_{r_k}}$ is maximal, it follows that $D(e,R) = L_{e_{r_1} e_{r_2} \dots e_{r_k}}$ when e take the polar form for R.

When e does not take the polar form for R, the special refinement condition requires that $D(e,R)$ is included in the intersection of the sublattices $L_{e^\#}$, where $e^\#$ is any possible one-dimensional atom or ray in $D(e,R)$. Since $D(e,R)$ includes the Boolean algebra of R-propositions, that is, all the eigenspaces of R, the rays in $D(e,R)$ must all lie in these eigenspaces (for each ray will correspond to a Boolean homomorphism on $D(e,R)$ that selects one of the eigenspaces of R as the eigenspace containing the ray). At most one ray $e_i \in D(e,R)$ can lie in each eigenspace r_i of R because these rays will generate a lattice $L_{e_1 e_2 \dots e_m}$, where the e_i, $i = 1, 2, \dots, m$, define the projections of some ray e° onto the m eigenspaces of R (i.e., $e_i = e^\circ_{r_i}$), and such lattices are maximal. So, the maximal determinate sublattices $D(e,R)$, when e does not take the polar form of R, must be the sublattices $L_{e^\circ_{r_1} e^\circ_{r_2} \dots e^\circ_{r_k}}$, where the $e^\circ_{r_i}$ are the nonzero projections of some ray e° onto the eigenspaces of R, for all rays e° in H. Since we assume that the evolution of $D(e,R)$ over time is determined by the (continuous) evolution of e, it follows that $e^\circ = e$, and so $D(e,R) = L_{e_{r_1} e_{r_2} \dots e_{r_k}}$ whether or not e takes the polar form for R.[5]

Dynamics

The analysis in the previous section solves a problem left open by the Kochen and Specker theorem. Kochen and Specker set out to investigate the question of whether quantum mechanics can support a hidden variable interpretation, where the hidden variables define property states satisfying the truth condition and the probability condition of the previous section. They reduced this question to the question of whether the lattice L of quantum propositions is embeddable into a Boolean algebra. An embedding requires the existence of sufficiently many Boolean homomorphisms to separate any pair of distinct elements in L. As it turns out, there are no Boolean homomorphisms on L in the general case. This raises the question of determining the sublattices of L that support hidden variable interpretations of quantum mechanics in the required sense. I have characterized such interpretations as modal interpretations satisfying certain conditions motivated by the requirement that the interpretations should support a purely internal account of measurement.

What the analysis in the last section shows is that a hidden variable or modal interpretation of quantum mechanics that preserves the dynamics of the theory can be constructed, in principle, by stipulating any observable R as always determinate. This privileged "pointer" observable, together with the quantum state represented by a ray e in H, then determines a unique sublattice, $D(e,R) = L_{e_{r_1}e_{r_2}\dots e_{r_k}}$, on which property states can be defined as Boolean homomorphisms. When e takes the polar form associated with a measurement by the "pointer" R, the determinate sublattice $D(e,R)$ represents the sublattice of propositions that we can take as measured via R-propositions, in the sense that there exist suffficiently many Boolean homomorphisms or property states on $D(e,R)$ to generate a Kolmogorov probability space on which the probabilities defined by e for the propositions in $D(e,R)$ can be represented as measures over the different possible property states on $D(e,R)$. (Note that the quantum state e, on this interpretation, defines probabilities for the propositions in $D(e,R)$ only.)

To complete this interpretation, we need to formulate a dynamics for these property states consistent with the dynamical evolution of e. There is a 1-1 correspondence between property states and distinct R-values in the determinate sublattices $L_{e_{r_1}e_{r_2}\dots e_{r_k}}$; that is, there are at most m Boolean homomorphisms on $L_{e_{r_1}e_{r_2}\dots e_{r_k}}$ that are assigned nonzero measure by the measure μ corresponding to e on the Kolmogorov probability space of Boolean homomorphisms, where each of the m homomorphisms selects one of the m distinct R-values. So the dynamical evolution of property states defined by Boolean homomorphisms on $D(e,R)$ is completely determined by the dynamical evoluton of determinate R-values.

Vink (1993) has shown how to formulate a dynamics for any discrete observable by generalizing a proposal by Bell (1987) for constructing stochastic Bohm-type trajectories for fermion number density as a "beable" for quantum field theory. (Vink proposes to take *all* observables as determinate, recognizing that functional relationships between observables will not be preserved in general. From the perspective developed here, this is unnecessarily redundant.)

Bohm (1952) takes the Schrödinger time-dependent equation of motion for the wave function in configuration space

$$i\hbar \frac{\partial \psi}{\partial t} = -\frac{\hbar^2}{2m} \nabla^2 \psi + V\psi$$

and extracts two real equations from this complex equation by substituting $\psi = Re^{iS/\hbar}$ (note that R is not related to the privileged "pointer" observable defined previously):

$$\frac{\partial S}{\partial t} + \frac{(\nabla S)^2}{2m} + V - \frac{\hbar^2}{2m} \frac{\nabla^2 R}{R} = 0$$

$$\frac{\partial R^2}{\partial t} + \nabla \cdot \left(R^2 \frac{\nabla S}{m}\right) = 0$$

The first equation (derived from the real part of the Schrödinger equation) can be interpreted as a Hamilton-Jacobi equation for the motion of particles under the influence of a potential function V and an additional "quantum potential" $-(\hbar^2/2m)(\nabla^2 R/R)$. The trajectories of these particles are given by the solutions to the equation:

$$\frac{dx}{dt} = \frac{j}{\rho}\left(= \frac{\nabla S}{m} = \frac{\hbar}{m} \frac{\text{Im}(\psi^* \nabla \psi)}{|\psi|^2}\right).$$

So the trajectories $x(t)$ depend on the initial configuration ψ.

The second equation (derived from the imaginary part of the Schrödinger equation) can be written as a continuity equation for a density ρ and current j:

$$\frac{\partial \rho}{\partial t} + \nabla \cdot j = 0$$

where $\rho = R^2 = |\psi|^2$ and $j = R^2(\nabla S/m) = (\hbar/m)\text{Im}(\psi^* \nabla \psi)$. The continuity equation guarantees that if $\rho = |\psi|^2$ initially, ρ will remain equal to $|\psi|^2$ at all times.

Vink considers an arbitrary maximal set of commuting observables $R^i (i = 1, \ldots, I)$, with eigenvectors $|r_{n^1}^1, r_{n^2}^2, \ldots, r_{n^I}^I\rangle$, where the $n^i = 1, \ldots,$ N^i label the finite and discrete eigenvalues of R^i. Suppressing the index i, these are written as $|r_n\rangle$. The time evolution of the state vector in an n-dimensional Hilbert space H_n is given by the equation of motion:

$$i\hbar \frac{d}{dt}|\psi(t)\rangle = H|\psi(t)\rangle$$

or

$$i\hbar \frac{d}{dt}\langle r_n|\psi\rangle = \langle r_n|H\psi\rangle = \sum_m \langle r_n|H|r_m\rangle\langle r_m|\psi\rangle$$

in the R-representation.

The imaginary part of this equation yields the continuity equation:

$$\frac{d}{dt}P_n = \frac{1}{\hbar}\sum_m J_{nm}$$

where the density P_n and the current matrix J_{nm} are defined by:

$$P_n(t) = |\langle r_n|\psi(t)\rangle|^2;$$

$$J_{nm}(t) = 2\mathrm{Im}\langle\psi(t)|r_n\rangle\langle r_n|H|r_m\rangle\langle r_m|\psi(t)\rangle.$$

For the nonmaximal (degenerate) observables R^i, the density and current matrices are defined by summing over the remaining indices, for example,

$$P^i_n = \sum_q |\langle r^i_{n^i},q|\psi\rangle|^2$$

where q denotes $r^j_{m^j}, j \neq i$, and similarly for J^i_{nm}.

We want a stochastic dynamics for the discrete observable R consistent with the continuity equation. Suppose the jumps in R-values are governed by transition probabilities $T_{mn}dt$, where $T_{mn}dt$ denotes the probability of a jump from value r_n to value r_m in time dt.

The transition matrix gives rise to time-dependent probability distributions of R-values, $P_n(t)$, which must satisfy the master equation:

$$\frac{d}{dt}P_n(t) = \sum_m T_{nm}P_m - T_{mn}P_n$$

and this equation must be consistent with the continuity equation

$$\frac{d}{dt} P_n(t) = \frac{1}{\hbar} \sum_m J_{nm}$$

that is, we require

$$\frac{J_{nm}}{\hbar} = T_{nm} P_m - T_{mn} P_n$$

We want solutions for T, given P and J, with $T_{mn} \geq 0$. Since $J_{mn} = -J_{nm}$ (hence $J_{nn} = 0$), the previous equation yields $n(n-1)/2$ equations for the n^2 elements of T. So there are many solutions. Bell's (1987) choice was:

For $n \neq m$:

$$T_{nm} = \frac{J_{nm}}{\hbar P_m}, J_{nm} \geq 0$$

$$T_{nm} = 0, J_{nm} \leq 0$$

For $n = m$, T_{nn} is fixed by the normalization $\sum_m T_{mn} dt = 1$.

Vink shows that Bell's choice leads to Bohm's theory in the continuous limit, when R is position in configuration space. For example, consider a single particle on a one-dimensional lattice. Let $x = an$, with $n = 1, 2, \ldots, N$ and a the lattice distance. Writing $\psi = Re^{iS/\hbar}$, Vink shows that

$$J_{mn} = \frac{\hbar}{ma} [S'(an) P_n \delta_{n,m-1} - S'(an) P_n \delta_{n,m+1}]$$

and

$$T_{mn} = \left[\frac{S'(an)}{ma} \right] \delta_{n,m-1}, S'(an) \geq 0;$$

$$T_{mn} = \left[\frac{S'(an)}{ma} \right] \delta_{n,m+1}, S'(an) \leq 0.$$

For positive $S'(an)$ the particle can jump from site n to site $n + 1$ with probability $|S'(an) dt|/ma$, and for negative $S'(an)$ the particle can jump from site n to $n - 1$ with the same probability. Since each jump is over a distance a, the average displacement in a time interval dt is:

$$dx = \frac{S'(x)dt}{m},$$

that is,

$$\frac{dx}{dt} = \frac{S'(x)}{m}.$$

As $a \to 0$, $S' \to \partial_x S$, and so in the continuous limit:

$$\frac{dx}{dt} = \frac{\partial_x S}{m}$$

as for the continuous trajectories in Bohm's theory. Vink shows that the dispersion vanishes in the limit as $a \to 0$, and so the trajectories become smooth and identical to the trajectories in Bohm's theory as $a \to 0$.

The choice of a privileged always determinate "pointer" observable R, together with a dynamics for the property states on the determinate sublattice $D(e,R)$ via the dynamics for R, provides an interpretation of quantum mechanics that supports a purely internal dynamical account of measurement. The sublattice $D(e,R)$ evolves in time as e evolves in time dynamically. At all times, the "pointer" observable R is determinate, with a value defined by one of the Boolean homomorphisms on $D(e,R)$, and no other observables are determinate at all times, apart from R and the unit observable I. At any particular time, other observables will be determinate, depending on e (these will generally be global observables that cannot be associated with observables of subsystems represented as factor spaces of H) and at certain times, when e takes the polar form for R, some of these other observables can be associated with subsystems that are measured by R. The determinate values of all observables other than R supervene on the values of R and have no independent ontological status, for the dynamical evolution of property states is completely determined by the evolution of R. The ontology of a quantum-mechanical universe is therefore fully specified by dynamical systems whose property states are characterized by the dynamical variable R via a global quantum state e.

Note that the dynamical evolution of a quantum-mechanical universe cannot be characterized by the dynamical evolution of R alone without e. The evolution of R reflects only part of the unitary evolution of the quantum state, in effect the evolution characterized by the real part of Schrödinger's time-dependent equation in the R-representation. This has to be consistent with the continuity equation derived from the imaginary part of Schrödinger's equation. So the ontology requires both R and e, but no other dynamical variables. The whole dynamical story can be told for all physical processes, including all processes that can be represented as measurements with respect to the "pointer" R in a model quantum-mechanical universe.

Of course, everything depends on the choice of R. We can only give an internal dynamical account of measurement processes that can be represented as dynamical evolutions that result in quantum states that take the polar form with respect to the "pointer" R. With an inappropriate choice for R, very few physical processes that we now consider to be measurements will be representable in this way. If we take the orthodox choice of R as the unit observable I, then no measurement processes will be representable dynamically.

With every choice of R, there will be a class of measurement processes that can be represented in terms of a dynamical account internal to a model quantum-mechanical universe. Call a class of measurement processes that includes all the interactions we propose to count as yielding measurements a *complete* class of measurement processes. This class of measurement processes will be represented dynamically in model quantum-mechanical universes by unitary transformations that produce quantum states in the polar form $\Sigma c_i \alpha_i \otimes \beta_i$, the dynamical representations of measurements in such universes. By an *appropriate* "pointer" observable for a class of measurements, I mean an observable R with eigenstates ρ_i, satisfying at least the condition that all measurements in the class can be represented by unitary transformations that produce quantum states in the polar form $\Sigma c_i \alpha_i \otimes \rho_i$ for R, as well as certain other conditions if we want the dynamical evolution of R to be well-behaved in ways characteristic of a good measurement "pointer."

Bohm (1952) argues, in effect, that the appropriate pointer observable R for a complete class of measurement processes should be taken as position in configuration space. Bell (1987, 175) suggests fermion number density, "The distribution of fermion number in the world certainly includes the positions of instruments, instrument pointers, ink on paper . . . and much much more." If the choice of R is regarded as unsuitable at some point, perhaps as we change our idea of what counts as a measurement, then one might want to consider some other grand "pointer" observable defined by the individual "pointers" of all the physical systems we propose to count as potential measuring instruments for this new complete class of measurement interactions. Or we might conceivably want to consider certain observables associated with the brains of sentient beings as defining the observable R.

The choice of R will also be constrained in ways that depend on contingent features of our universe. While Bohm's choice of position in configuration space might be appropriate, it would not be appropriate to define R in terms of momentum rather than position. Such a "pointer" would behave rather strangely, largely because of the nature of the potentials we encounter in our universe. Even with an appropriate choice of R, there will always be some probability of apparently anomalous behavior; for example, stochastic transitions between values of R associated with different values of a measured observable immediately after a measurement. Such probabilities can be shown to be ignorably small by decoherence arguments, as in Bohm's theory. The application of decoherence here is, of course, unobjectionable because the events in question are always determinate and do not depend for their determinateness on the smallness of certain probabilities. Rather, decoherence explains why certain events that we would regard as anomalous will occur very rarely.

Whatever we take as the privileged "pointer" observable R, the previous analysis fixes a unique choice for the sublattice of propositions $D(e,R)$ that we can take as determinate together with R propositions, when the state e takes the polar form for R associated with a measurement. What the determinate sublattice $D(e,R)$ represents is the sublattice of propositions we can take as measured by R-propositions, in the sense that there exist sufficiently many Boolean homomorphisms on $D(e,R)$ to generate a Kolmogorov probability space on which the probabilities defined by e for the propositions in $D(e,R)$ can be represented as measures over the different possible property states or truth-value assignments to the propositions in $D(e,R)$. Each such property state correlates a determinate R-value with a determinate value for any measured observable.

So the picture of a quantum-mechanical universe is this: The quantum state e evolves dynamically in time and can be understood as representing an objectively real field in R-space that influences the evolution of R-values. The always determinate values of R evolve stochastically under the influence of a transition matrix T_{mn} that depends on e. The determinate sublattice $D(e,R)$ defines the propositions assigning ranges of values to observables that we can take as determinate together with R — it defines what we can talk about as measured. But this "measurement talk" about the values of observables other than the "pointer" R is really redundant. The only real change in a quantum-mechanical universe is the change in e and R, and this suffices to account for all physical processes, including the processes that we interpret as measurements.

What this analysis shows is that Bohm-type theories — modal interpretations of quantum mechanics with a privileged "pointer" observable — are really the only game in town if we want a dynamical solution of the measurement problem that preserves the linear dynamics of the theory.

NOTES

This essay was written during a stay at the University of California, Irvine. I am indebted to Jeff Barrett, Peter Woodruff, and Jeroen Vink for informative discussions.

1. In terms of observables, on the orthodox interpretation, the quantum state defined by a ray in Hilbert space selects the set of determinate observables — the set of observables that have determinate values — as those observables for which the state is an eigenstate. The determinate value of the observable is the eigenvalue corresponding to the quantum state as eigenstate of the observable. This is sometimes referred to as the "eigenvalue-eigenstate link." There is a clear statement of this interpretation in Einstein, Podolsky, and Rosen 1935. Indeed, the paper is formulated as a *reductio* for this interpretation: Einstein, Podolsky, and Rosen argue that it follows from the orthodox interpretation, together with certain realist assumptions, that quantum mechanics is incomplete.

2. Modal interpretations that exploit the polar decomposition theorem appeal to the existence of a unique decomposition of the form $\Sigma d_i \beta_i \otimes \sigma_i$, in terms of the eigenvectors of some observables B and T, when the coefficients $|d_i|$ are all distinct. The measurement condition here requires only that observables correlated with R, when the state takes the polar form for R, are determinate.

3. I use the same symbol, r_i, here for eigenvalues of R and the associated eigenspaces. The symbol $\{\ \}'$ indicates the commutant in L of $\{\ \}$, the set of all operators that commute with the projectors in $\{\ \}$. The symbol "$\perp$" represents the

orthocomplement. My previous proposal in Bub 1991, 1992a, 1992b, 1993, 1994 was, in effect, to take $D(e,R)$ as $\cap \{e_{r_i}, r - e_{r_i}\}'$, where the intersection is taken over the k nonzero projections of e onto the m eigenspaces of R. (See Cassinelli and Lahti 1995 for a formal characterization of different versions of the modal interpretation.) Evidently, $\cap \{e_{r_i}, r - e_{r_i}\}' \subseteq \{e_{r_i}, e = 1, \ldots, k\}'$, because $\cap \{e_{r_i}, r - e_{r_i}\}' \subseteq \cap \{e_{r_i}, H - e_{r_i}\}'$ and $\cap \{e_{r_i}, H - e_{r_i}\}' = \{e_{r_i}, e = 1, \ldots, k\}'$.

4. Note that $L_{e_{r_1}e_{r_2}\ldots e_{r_k}}$ includes all the propositions in $D(e,I)$ that are compatible with R, i.e., all propositions in $D(e,I)$ that, taken together with the propositions in the Boolean sublattice of R-propositions, generate a Boolean sublattice of L under lattice closure (equivalently, all propositions that commute with R). In fact, $L_{e_{r_1}e_{r_2}\ldots e_{r_k}}$ is generated by the Boolean sublattice of R-propositions and the propositions in $D(e,I)$ that are compatible with all R-propositions. The sublattice $L_{e_{r_i}}$ generated by the atom $e_{r_i} = (e \vee r_i^\perp) \wedge r_i$ includes all the propositions in L_e compatible with r_i, and $L_{e_{r_1}e_{r_2}\ldots e_{r_k}}$ is just the intersection of the k sublattices $L_{e_{r_i}}$, for the k nonzero projections e_{r_i} of e onto r_i.

5. In Bub 1994 I proved uniqueness on the basis that $D(e,r)$ should include all the propositions associated with any observable equivalent to R, with respect to ideal measurements on a system in the state e. An ideal measurement on any observable with k eigenvectors in the directions e_{r_i}, the k nonzero projections of e onto the eigenspaces of R, will yield the same set of outcomes with the same probabilities as a measurement of R on a system in the state e. So, I argued, these observables are indistinguishable from R if the system is in the state e. This now seems to me irrelevant to the modal interpretation proposed, in which R plays the role of a privileged always determinate "pointer" observable in a closed quantum-mechanical universe. See also Bub and Clifton 1996 and Bub 1997 (both written after this essay) for a different and better-motivated proof of the uniqueness theorem.

REFERENCES

Bell, J. S. 1987. "Beables for Quantum Field Theory." In *Speakable and Unspeakable in Quantum Mechanics*. Cambridge: Cambridge University Press, 167–80.

Bohm, D. 1952. "A Suggested Interpretation of the Quantum Theory in Terms of Hidden Variables, I and II." *Physical Review* 85: 166–79, 180–93.

Bub, J. 1991. "Measurement and 'Beables' in Quantum Mechanics." *Foundations of Physics* 21: 25–42.

———. 1992a. "Quantum Mechanics as a Theory of 'Beables.'" In A. van der Merwe, F. Selleri, and G. Tarozzi, eds., *Bell's Theorem and the Foundations of Modern Physics*. Singapore: World Scientific, pp. 117–24.

———. 1992b. "Quantum Mechanics Without the Projection Postulate." *Foundations of Physics* 22: 737–54.

———. 1993. "Measurement: It Ain't Over Till It's Over." *Foundations of Physics Letters* 6: 21–35.

———. 1994. "On the Structure of Quantal Proposition Systems." *Foundations of Physics* 24: 1261–80.

———. 1997. *Interpreting the Quantum World.* Cambridge: Cambridge University Press.

Bub, J., and R. Clifton. 1996. "A Uniqueness Theorem for Interpretations of Quantum Mechanics." *Studies in the History and Philosophy of Modern Physics* 26: 181–219.

Cassinelli, G., and P. J. Lahti. 1995. "Quantum Theory of Measurement and the Modal Interpretation of Quantum Mechanics." *International Journal of Theoretical Physics* 34: 1271–81.

Dieks, D. 1989. "Resolution of the Measurement Problem Through Decoherence of the Quantum State." *Physics Letters A* 142: 439–46.

———. 1994. "Measurements, Modal Interpretation of Quantum Mechanics, and Macroscopic Behaviour." *Physics Review A,* 49: 2290–2300.

Einstein, A., B. Podolsky, and N. Rosen. 1935. "Can Quantum Mechanical Description of Reality Be Considered Complete?" *Physical Review* 47: 777–80.

Healey, R. 1989. *The Philosophy of Quantum Mechanics: An Interactive Interpretation.* Cambridge: Cambridge University Press.

———. 1993. "Measurement and Quantum Indeterminateness." *Foundations of Physics Letters* 6: 307–16.

Kochen, S. 1985. "A New Interpretation of Quantum Mechanics." In P. Lahti and P. Mittelstaedt, eds., *Symposium on the Foundations of Modern Physics.* Singapore: World Scientific, 151–70.

Kochen, S., and E. P. Specker. 1967. "On the Problem of Hidden Variables in Quantum Mechanics." *Journal of Mathematics and Mechanics* 17: 59–87.

Krips, H. 1987. *The Metaphysics of Quantum Theory.* Oxford: Clarendon Press.

Mermin, N. D. 1990. "Simple Unified Form for the Major No-Hidden-Variables Theorems." *Physical Review Letters* 65: 3373–76.

Schrödinger, E. 1935. "Die Gegenwärtige Situation in der Quantenmechanik." *Natuurwissenschaften* 23: 807–18, 823–28, 844–49. Trans. by J. D. Trimmer in J. A. Wheeler and W. H. Zurek, eds., *Quantum Theory and Measurement.* Princeton: Princeton University Press, 1983.

Van Fraassen, B. C. 1979. "Hidden Variables and the Modal Interpretation of Quantum Statistics." *Synthese* 42: 155–65.

———. 1981. "A Modal Interpretation of Quantum Mechanics." In E. Beltrametti and B. C. van Fraassen, eds., *Current Issues in Quantum Logic.* New York: Plenum, pp. 229–58.

———. 1991. *Quantum Mechanics: An Empiricist View.* Oxford: Clarendon Press.

Vink, J. C. 1993. "Quantum Mechanics in Terms of Discrete Beables." *Physical Review A* 48: 1808–18.

10

Deterministic Chaos and the Nature of Chance

John A. Winnie
Department of History and Philosophy of Science, Indiana University

There was a time when it seemed that we understood determinism. If a system of differential equations had a unique solution for a given set of boundary and initial conditions, then the entire future evolution of the system in question was thereby determined. In a deterministic world, despite the complexities and vagaries of large scale observation, predictability was guaranteed, at least in principle. Beneath these surface generalizations there was a world of order and clarity whose workings could be gleaned by controlled experiment and observation. In such a world, chance lay only on the surface of things, the result of ignored forces and imprecise reckonings.

Chaos theory has changed this picture considerably — some would say so drastically that determinism is no longer recognizable. Even within the rigid confines of determinism, chance has been elevated from a by-product of human ignorance to a fundamental ontological feature of reality. In this essay, I will examine some of the results in chaos theory that have led to this view, and critically examine the claim that chaos theory has not only legitimized chance ontologically, but has ended in making it indistinguishable from determinism itself.

Determinism and Chance

The relation between determinism and chance has never been a simple matter. First of all, there is chance on the level of those apparently random occurrences called "coincidences." The psychologist Carl Jung tells of how, while a patient was relating her dream involving an Egyptian beetle-shaped amulet called a scarab, he heard a scratching at the window. Opening the window, he discovered, to his amazement, a beetle outside — a scarab beetle! Impressive and unusual as such events are, they are no threat to a deterministic world view; coincidences, as Aristotle pointed out, may simply be causal chains in rare (or rarely noticed) collision.

More important to us here is the apparent randomness or chance that appears when we either deliberately or unavoidably examine a deterministic physical system in broad detail. At this level, there seem to be two ways in which chance may appear. The first is that the macro variables or factors under study are only related by statistical or probabilistic laws. Heat exchange, diffusion, and suicide rates may all be treated statistically at the macro level, compatible with the existence (or even derived from the existence) of a deterministic model of the underlying microprocess. The simplest of such models are those whose state at time $t + 1$ depends only upon its state at time t. These are called "Markov processes," and when we imagine dependencies which extend even farther into the past, the processes are said to be "n-step Markov processes." While such processes at the macro level are compatible with determinism at the micro level, the dependence of future on past states in such models allows the probabilities involved to be interpreted as the result of overlooking or ignoring additional factors which, had we taken them into account at the outset, would allow exact prediction.

There is, however, another possibility. There are some probabilistic models in which the next state does not depend on the current or past states. In a sense, these models depict systems that have no history: what happens next not only fails to depend on where the system is, but also fails to depend on where it has been. Such systems are called *Bernoulli systems* (or *Bernoulli processes*), familiar to those with a nodding acquaintance with probability theory.[1] Roughly, Bernoulli processes are finite (or at most countably infinite) state processes where the state at time $t + 1$ is given by throwing a (possibly loaded) multisided die. Those Bernoulli processes in which each state is equally likely, such as spinning a roulette wheel or flipping a coin, are often used as examples of the most random of processes, while Bernoulli processes that have unequal probability distributions tend to behave in a more orderly, but still history-independent, manner.[2]

One of the main questions posed by this essay can be put as follows. Let us suppose that a system viewed at the macro level can be accurately described as an equidistributed Bernoulli process. Can that same system, viewed more finely, be described by a deterministic model? In other words, can a deterministic model generate a random, or equidistributed, Bernoulli process? Or must the deterministic underpinnings of a physical system inevitably seep up to the macro level and reveal themselves by the presence of short- or long-term patterns of statistical dependence of future states on past states?

The view that determinism and randomness are incompatible appears frequently in discussions of randomness in the literature of computer science. Consider von Neumann's proposed method of generating a sequence of random (say, ten-digit) numbers. Starting with an arbitrary ten-digit number n, square n, and then extract the middle ten digits to get the next number in the sequence. Apply the same method to the result to get the next number, and so on. Now notice what D. Knuth says about this method:

There is a fairly obvious objection to this technique: how can a sequence generated in such a way be random, since each number is completely determined by its predecessor? The answer is that this sequence *isn't* random, but it *appears* to be. (1981, 3)

Notice that, according to Knuth, it is not the fact that such sequences are ultimately cyclic that prevents them from being random, but simply that they are *determined*. In the literature, such sequences are called *pseudo-random*, and most often they are so-called because they are generated by a deterministic rule, a function.[3]

Since the claim that microdeterminism inevitably invades the macro world to produce pattern will be one of the central concerns of this essay, let us give it a name and call it the principle of *deterministic macro ordering*, or simply the *macro ordering* principle.

Deterministic macro ordering. *If a dynamical system is deterministic, then any coarse classification of its states will exhibit behavior that fails to be a Bernoulli process.*

The idea here is that, unlike Markov processes, Bernoulli processes are independent of their history: the next state's probability does not depend on the currently occupied state.

At this point, I have not provided good reasons for believing in either the truth or falsity of this principle. In order to pursue the matter further, it will first be necessary to become clearer about some of the fundamental concepts involved, especially the concept of 'randomness' and its connection with Bernoulli processes.

Randomness, Stochastic Processes, and Dynamical Systems

There are (at least) two notions of "randomness" that will be used throughout the following. One sense is the randomness or patternlessness of a sequence of symbols taken from some finite alphabet. I shall call this *sequence randomness*. The other sense of randomness comes from the theory of stochastic processes and, while related to the first sense, is more general. (As we shall see, Bernoulli processes generate sequence randomness most of the time.) Both notions will be used throughout, so we shall need some idea of what is involved in both cases, although I shall try to keep technical details to a minimum.[4]

Given a finite set of symbols, and a finite sequence of such, what are we to mean when we say that the sequence is "random," in the sense of disordered, irregular, or patternless? Kolmogorov's approach to this question was to define the degree of irregularity of a (finite) sequence in terms of the length of the shortest method of describing it. Methods of description, in turn, are specified as algorithms or programs to be executed by a Universal Turing Machine or computer of the appropriate kind; hence, the resulting measure of complexity becomes the length of the shortest program that will generate the sequence in question. Roughly, and for sequences that are sufficiently long, the complexity $Kol\{s_i\}$ of a sequence $\{s_i\}$ will be high (and the sequence said to be "random") when the length of its shortest generator is close to the length of the sequence itself. Using a simple counting argument, it is easy to show that most sequences of length N (when N is sufficiently large) are Kolmogorov random, although another, more difficult, argument shows that the property of Kolmogorov randomness is not computable. (See Li and Vitanyi 1993, 2.2, 2.7.)

The Kolmogorov complexity of *infinite* sequences, for our purposes, will be taken to be the limiting frequency of the Kolmogorov complexity of $\{s_i\}$ in the long run, that is:

$$Kol(s) = \lim_{N \to \infty} \frac{Kol(s_N)}{N}, \tag{10.1}$$

where s_N is the initial segment of $s = \{s_i\}$ of length N. Notice that if there is an algorithm that computes the infinite sequence s, then regardless of that algorithm's (finite) length, the Kolmogorov complexity of that infinite sequence will be 0. Thus any reasonable definition of algorithmic randomness (say $Kol(s) > 0$) for infinite sequences implies the nonexistence of such an algorithm.

Another approach to randomness is by way of the study of *stochastic processes*. We begin with a set of states $S = \{S_1, S_2, \ldots, S_k\}$, a probability space $<\Omega, F, p>$, and a set of functions $\{X_0, X_1, X_2, \ldots\}$ (called *random variables*) each from Ω to S.[5] The intuition here is that each member ω of the event space Ω consists of an infinite sequence of happenings, such as an infinite sequence of tosses of a (possibly unfair) die. The first toss occurs at time t_0, the second at time t_1, etc. Each of the individual tosses has exactly one of the attributes in S. (In this example, a natural choice for S would be the set $\{1,2,3,4,5,6\}$.) To find out which attribute or state ω is in at time t, apply X_i. Thus $X_0(\omega)$ yields the state of ω at time $t = 0$, and the sequence $X_0(\omega), X_1(\omega), X_2(\omega), \ldots$ describes the infinite sequence of states that ω passes through indefinitely into the future.

When the probability of a set $\{\omega: X_n(\omega) = S_{i_n}, \ldots, X_{n+k}(\omega) = S_{i_{n+k}}\}$, for all states S_i and all integers $k, n \geq 0$, is independent of n (in other words, the start time n doesn't matter), then the process is said to be *stationary*. When the random variables are independent, the process is said to be *independent*, and when each variable has the same probability distribution ($p\{\omega: X_i(\omega) = S_j\} = p\{\omega: X_k(\omega) = S_j\}$), for all states S_j and integers $i, k \geq 0$, then the variables are said to be *identically distributed*. Processes such as infinite sequences of tosses of a (possibly unfair) coin are then stationary, independent, and identically distributed processes. This is the more precise characterization of what I have earlier called *Bernoulli processes*. When the stochastic process has just two states, say $S = \{0, 1\}$ and the random variables are equidistributed ($p(X_i = 0) = \frac{1}{2}$), then we have the model for the tossing of a fair coin. Equidistributed Bernoulli processes might thus be said to be "most random," with other Bernoulli processes, depending upon the unevenness of their probability distributions, coming in somewhere behind.[6] Notice that, equidistributed or not, the next state of a Bernoulli process is statistically independent of its previous state.

An equidistributed Bernoulli process, as one might expect, will generate Kolmogorov random sequences most of the time. To see this in the case of reasonably long finite sequences of length k, notice that if the process is equidistributed Bernoulli, then every sequence of length k is equally likely. But, by the result mentioned earlier, almost every sequence of length k is Kolmogorov random; hence, most likely, a sequence generated by an equidistributed Bernoulli process will be one of these.

Still stochastic, but more regular and history-dependent, are the *Markov processes*. These are stationary identically distributed processes whose random variables need not be independent, but in which X_i depends at most on X_{i-1}. The chances of being in a given state at time t_i may depend on the state at the previous time. Since Bernoulli processes are, strictly speaking, special cases of Markov processes, let us call Markov processes in which independence fails to hold, *proper* Markov processes. Along the same lines as the above definition, we have *n-step* proper Markov processes, depending on how far back into the past the dependencies of the random variables reach.

From the stochastic systems viewpoint, we thus have various types of random processes. The most random are the equidistributed Bernoulli processes — sequences of tosses of a fair coin or spins of a roulette wheel. More regular, but still independent of their past, come the Bernoulli processes that are not equidistributed — such as tosses of an unfair coin. And finally — at least for our purposes here — there are the proper Markov processes. In these processes, the influence of the past upon the future is felt, but probabilistically, not decisively. Intuitively and roughly, proper Markov processes would be expected to generate sequences that are highly patterned and thus come in low on anyone's measure of sequence complexity.

When we now come to consider dynamical systems, we seem to be at the opposite pole from chance. Here we are dealing with a state space X, almost always supplied with a topology (and often much in addition, such as a metric and an invariant measure), and a function f from the state space to itself that provides the dynamics. Simple dynamical systems such as these are called *maps*; when the state space is provided with a set of such functions $\{f_t\}$ (indexed by the reals or the positive reals) that form a (one-parameter) group, then the resulting system is called a *flow*. Roughly, maps depict discrete time dynamical systems, while flows describe continuous time systems. If p is a point in the state space X, then the (future) orbit of p is, in the case of maps, the sequence $(p, f(p), f^2(p), \ldots)$, where f^n is the nth iterate of the function f. (In the case of flows, the future orbit is a curve from the non-negative reals to X.) When the orbits of nearby points separate rapidly and behave quite differently in the long run, the system is said to be *unstable* or *chaotic*. The study of such deterministic but unstable systems, nowadays called "*chaos theory*," has led to a number of results which have greatly affected our conceptions of what deterministic systems can and cannot do, especially when it comes to their abilities to mimic random behavior. Before turning to these results, however, let us go back for a moment to stochastic systems and explore a very general, but important, connection between such systems and deterministic dynamical systems.

Earlier, we defined a stochastic process in terms of a (finite) state space $S = \{S_1, \ldots, S_k\}$, an (infinite) set of random variables $\{X_0, X_1, \ldots\}$, and a probability space (Ω, F, p). When ω is in Ω then the sequence $(X_0(\omega), X_1(\omega), \ldots)$ depicted the successive states that ω goes through as time advances. We may now simply replace each ω in the probability space Ω by the corresponding sequence $(X_0(\omega), X_1(\omega), \ldots)$ to obtain a new probability space (Ω^*, F^*, p^*), where F^* and p^* are just like F and p, except that each state ω is replaced by the corresponding ω^*. Clearly, the two probability spaces are isomorphic.

Next, notice that the old random variables are still here, now in the form of the slot functions u_i.[7] In other words, $X_i(\omega) = u_i(\omega^*)$. Earlier, we used these random variables to make time advance, but now there is another way: define the *shift function T* on Ω^* to be the function that simply strips off the first item in ω^*, that is,

$$T((\omega^*)(0), \omega^*(1), \omega^*(2), \ldots)) = (\omega^*(1), \omega^*(2), \ldots). \quad (10.2)$$

Now consider the deterministic dynamical system (Ω^*, T). Start with a state ω^* in Ω^* and iterate the shift function T to get its orbit: $Orbit(\omega^*) = (\omega^*, T(\omega^*), T^2(\omega^*), \ldots)$.

Next, partition the state space Ω^* into k subsets $S^* = \{S_1^*, S_2^*, \ldots, S_k^*\}$, where a member of Ω^* is in *subset S_i^** just in case its first ($t = 0$) slot contains state S_i. Now track the members of the orbit of ω^* in terms of their membership in one of the above partitions. The result will be what is called the *itinerary* of ω^* (with respect to this partition) and may be represented by a sequence of states: $(S_{i0}, S_{i1}, S_{i2}, \ldots)$, where $T^n(\omega^*)$ is in S_{in}. Notice that this sequence of states is *exactly the same sequence* as that given, in the context of stochastic system theory, by $(X_0(\omega), X_1(\omega), X_2(\omega), \ldots)$. Furthermore, let A^* be a (measurable) subset of Ω^*. Let $B^* = T^{-1}[A^*]$ be the set of states in Ω^* that enter A^*

in a single iteration of T (B^* might be called the *immediate past* of A^*).
It is now easy to show that if the random process is stationary, then
$p(B^*) = p(A^*)$, that is, the mapping T is what is called a *measure-preserving* map.[8] What we have come up with is thus a dynamical
system of the form $(\Omega^*, F^*, p^*, T, S^*)$, where T is a measure-preserving
map of Ω^* and S^* is a partition of Ω^*. Thus the itineraries of a deterministic dynamical system (Ω^*, T) with respect to the partitioning S^*
exhibit the same statistical behavior as the sample realizations of the
corresponding stationary stochastic process.

This approach (according to Ornstein 1974, 4) goes back to Kolmogorov, and thus antedates the results which I shall be discussing later.
Nevertheless, its paradigm example — a Bernoulli shift — is crucial for
the issue of the compatibility of determinism and randomness. The
simplest case is that of the sequence of tosses of a fair coin. We begin
with a probability space $(X, p) = (\{0, 1\}, p^*)$, where $p^*(0) = p^*(1) =$
½. For the universe of the state space of our dynamical system, we
take $\{0, 1\}^\infty$, the set of all (one-sided) infinite binary sequences (such as
$(0, 1, 1, 0, \ldots)$). For our map we take the shift map, for our probability
measure m, the product measure obtained from (X, p),[9] and partition
the space into two sets $S = \{S_0, S_1\}$, the members of S_0 having a 0 as their
first element, the members of S_1 having a 1 as their first element. The
result (Ω, m, T, S), is thus a partitioned dynamical system, with T (the
shift map) an m-measure-preserving map. Using the partition S, the
itineraries of this system generate the elements of the probability space
of the corresponding stochastic process, and the fact that the mapping
T is measure-preserving implies (when T is ergodic) that the partitions
are occupied with the appropriate long-run relative frequencies.

Since most binary sequences are random in the Kolmogorov sense, it
follows at once that the partitioned dynamical system just described
generates Kolmogorov random — not merely apparently random —
sequences. Thus, without further ado we are safe in asserting that the
principle of Macro Ordering is false in the following sense:

*(A) There are abstract deterministic dynamical systems that, when appropriately coarse grained, have, for the most part, Kolmogorov random itineraries
and for which the collection of these itineraries constitutes (with the appropriate probability measure) a (p = ½) Bernoulli process.*

(A simple example of this is the tent function, which we shall examine
in more detail later.)

There are two things to keep in mind at this point, however. The first is that the above result applies to abstract dynamical systems of a very special kind. There may be good reasons, however, for believing that such systems simply do not occur in nature. The second is that, while the result shows that deterministic underpinnings do not always reveal themselves in a given coarse graining, it may nevertheless be the case that there is always *some* coarse graining (perhaps not too fine at that) where the deterministic basis of the stochastic behavior becomes evident. Before considering these matters in more detail, however, let us briefly expand on the above results in the wider context of dynamical systems that need not be Bernoulli shifts.

Brudno's Theorem and Random Trajectories

From this point on, we shall be considering only dynamical systems of a certain type, called *K-systems*. Without going into the details of the definition, these systems are defined in such a way that their entropy with respect to any nontrivial partition is positive. Consequently, K-systems all have positive entropy and, hence, at least one positive Lyapunov exponent.[10] They are, as a result, systems that anyone would call "chaotic," so much so that it would not be going too far to define chaotic systems as simply K-systems. Now, there are partitionings of such systems, it has been shown, that have mostly Kolmogorov random itineraries. This result, first proved by Brudno 1983, would seem to provide good reason to believe that the principle of Deterministic Macro Ordering is false — and so it does. It is, however, easy to misconstrue Brudno's theorem as showing more than this. Let us see why.

In order to understand exactly what Brudno's theorem shows, we first need to be clear about what it means to talk about the complexity of an individual orbit of a dynamical system, in particular, a map.[11] The first step is to partition the state space into a finite number of "bins" $B = \{B_1, B_2, \ldots, B_k\}$. We obtain the itinerary of a given state x_0 by tracking its orbit as it successively enters the bins of our partition B. Thus, corresponding to the orbit $\{x_0, x_1 = f(x_0), x_2 = f^2(x_0), \ldots\}$, we obtain the sequence of bins $\{B_{i_0}, B_{i_1}, B_{i_2}, \ldots\}$, where x_j is in B_{i_j}. Now replace each partition B_i by a symbol for that partition B_i^* to obtain the symbol sequence $B_{i_0}^*, B_{i_1}^*, B_{i_2}^*, \ldots$ and we have an infinite sequence of symbols taken from a finite pool. The Kolmogorov complexity of this infinite sequence can now be taken as the measure of the complexity of the orbit of x_0 with respect to the partition B. More precisely, the complexity of x_0 with respect to the partition B is given by:

$$K(x_0, f, B) = \lim_{N \to \infty} \frac{1}{N} Kol(B_{i_0}, B_{i_1}, \ldots, B_{i_N}). \qquad (10.3)$$

Here, *Kol* is the Kolmogorov complexity of the finite sequence which is an initial segment of the itinerary of the state x_0.[12] The complexity of the trajectory of x_0, $K(x_0, f)$, is now defined as the supremum of (10.3) taken over all partitions (actually all open covers) of the state space. Brudno's theorem now states that for almost all states x_0 (all but a set of measure zero) the complexity of the trajectory of x_0 will be equal to the entropy of the system (X, f). Putting it simply, systems with positive entropy will have orbits whose complexity will be equal to their entropy.

A simple example of a positive entropy system is the tent map (see figure 10.1). A standard result is that its entropy is *log* 2 (see Schuster 1988). In fact, if we take as our partition of the unit interval the sets $AB = \{A = [0, \frac{1}{2}), B = [\frac{1}{2}, 1]\}$, then for almost all itineraries, $K(x_0, tent, AB) = 1$, resulting in random itineraries. If, on the other hand, we take the partition $ABC = \{A = \{[0, \frac{1}{4}), B = [\frac{1}{4}, \frac{3}{4}]\}, C = [\frac{3}{4}, 1]\}$, then although most orbits spend about a third of their time in each partition, the itineraries thus generated are neither Kolmogorov random

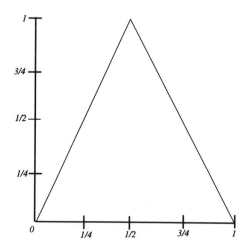

Figure 10.1 The tent function: If $x \leq \frac{1}{2}$, tent$(x) = 2x$; otherwise, tent$(x) = 2(1 - x)$.

nor Bernoulli random. For example, an A^* in an itinerary is never followed by a C^*, a B^* is never followed by an A^*, and a C^* is never followed by a C^*. If we look as this system as one composed of three states $\{A, B, C\}$, then the system is a Markov process with transition table given by

$$P_{i,j} = \begin{bmatrix} 0.5 & 0.5 & 0.0 \\ 0.0 & 0.5 & 0.5 \\ 0.5 & 0.5 & 0.0 \end{bmatrix} \qquad (10.4)$$

The orbits have not changed, of course, but due to the change in partitioning, a deterministic system first seen as a "random" Bernoulli process in now seen as a "nonrandom," but stochastic, Markov process. Let me put this less metaphorically. Whether or not a deterministic system represents a Bernoulli process or an (n-stage) Markov process is relative to a partitioning of the state space. The very same set of trajectories or orbits may in one case yield itineraries of high Kolmogorov complexity and model a Bernoulli process, and in another partitioning may result in itineraries of much lower complexity and model much more regular, history-dependent, stochastic processes.

Next, let us return to Brudno's theorem. Notice that the above definition of the complexity of an orbit takes this to be the maximum (actually, the *supremum*) complexity obtained over the set of *all possible partitions*. Typically, the partitions of maximum complexity will be fairly coarse, and simple refinements will yield itineraries of lesser Kolmogorov complexity. Thus, while Brudno's theorem guarantees the existence of partitions that yield complex itineraries (when the system has positive entropy), these partitions are neither typical nor especially fine grained. The moral, then, should be clear. Brudno's theorem tells us that *if our observations are sufficiently coarse and aptly chosen (to yield the appropriate partitioning)*, then a deterministic system can generate a sequence of such observations that are random, both in the sense of long-run Kolmogorov complexity and in the sense that the resulting collapsed system is a Bernoulli process. It does not follow from this result, however, that all — or even most — coarse grainings of the state space will yield a similar result. Indeed, it turns out that other coarse grainings may be highly regular stochastic processes and, as we shall see, processes that may expose the deterministic underpinnings of the Bernoulli processes themselves.

Chaotic Determinism and Stochastic Systems

Patrick Suppes has argued that when a deterministic system is chaotic, the fact that it may be represented as a stochastic system makes the application of the terms "deterministic" or "stochastic" to such systems a matter of choice, rather than an empirical issue.

Deterministic metaphysicians can comfortably hold to their view knowing they cannot be empirically refuted, but so can indeterministic ones as well. (Suppes 1993, 254)

In support of this view, Suppes cites a theorem of Ornstein on a certain class of *K*-systems.

The modern research on dynamical systems, whose lineage extends back to the deep analysis of Poincaré of motion in celestial mechanics in the nineteenth century, has produced a variety of philosophically significant results, but none more so than that expressed in the following theorem.

Theorem 4. (Ornstein) There are processes which can equally well be analyzed as deterministic systems of classical mechanics or as indeterministic semi-Markov processes, no matter how many observations are made. (Suppes 1993, 254)

Needless to say, the above statement is not the theorem proved by Ornstein, but a paraphrase with a philosophical gloss. In this section, I shall begin by discussing a similar interpretation of Brudno's theorem and argue that it is misleading. Then we shall turn to Ornstein's theorem proper, where I shall argue that Suppes' interpretation is similarly misleading.

As we have seen, Brudno's theorem seems to guarantee that chaotic systems have, for the most part, random trajectories. Thus, it might seem that deterministic chaos theory leads to a blurring of the distinction between deterministic and probabilistic, even random, dynamical systems. Let us consider once again the tent function:

$$tent(x) = 2x, \text{ if } (x \le 0.5), \text{ else } 2(1 - x). \tag{10.5}$$

As I mentioned earlier, this is a classical example of a chaotic dynamical system (defined on the unit interval [0, 1]) with a Lyapunov exponent equal to *log* 2. It is also an easy matter to reveal its dynamics by representing each state as a binary decimal such as .011010001 . . . and showing that the succeeding state is obtained by shifting the decimal to the right and dropping the first digit. (In this example, the next state is thus .11010001. . . .) When the binary decimal begins with the digit 1, however, the shift must be followed by complementing each of the remaining digits. Since binary decimals are between 0 and ½ when they begin with a 0, and they are between ½ and 1 when they begin with a 1, as we follow the trajectory of a point by the above process of shifting and (possibly) complementing, the first digit in the representation tracks the state's orbit as it moves from the lower half of the state space (A) to the upper half (B). Thus, corresponding to the initial state in the example above, we obtain, first, the *trajectory*: .011010001 . . . , .11010001 . . . , .0101110 . . . , .101110 . . . , .10001 . . . , . . . and the corresponding *itinerary*: $A^*, B^*, A^*, B^*, B^*, \ldots$, which symbolically depicts the successive partitions (the A or low partition, the B or high partition) into which we have coarse grained the space. Notice that the itinerary of a state depends upon how the space is partitioned. In the above example, the state space has been divided into two parts; hence, the itinerary is a binary string. In general, a state space may be partitioned or coarse grained into any finite number n of parts, yielding itineraries that are infinite n-ary strings. The fact that the same trajectory will generate many different itineraries, depending on the coarse graining, will be extremely important later. For now, let us return to the tent function and its behavior.

As we have seen, the evolution of a given state of our system can be represented by a "shift and complement" operation, sometimes referred to as a "Bernoulli shift." One more piece to this so far deterministic picture and randomness emerges. It turns out that, for all but a set of measure zero, the successive digits in the binary expansions of numbers in the unit interval are a random sequence. This implies that, for almost any initial state of the system, its AB itinerary is represented by a sequence of A^*'s and B^*'s that cannot be generated by any computation. It also implies that, no matter how extensive our knowledge of the history of that itinerary, we are unable on that basis to confidently predict its next outcome. This feature of such systems — and the tent function is only one of many such with this property — has led some writers to wax poetically on the downfall of classical determinism:

These subsystems may exhibit randomness, even though the total system is deterministic. This is the lesson the Bernoulli shift teaches us. Like the queen of England, determinism reigns but does not govern. (Ekeland 1988, 63)

The idea here seems to be that we are always, in a sense, coarse graining the state space with our observations. In our example, we might imagine having an *AB* device; applied to the physical system, it duly records whether or not the system is in the lower half of the state space (0 to ½) or in the upper half (½ to 1). If the former, an *A* * is returned; if the latter, a *B* *. Examining our deterministic system with such a device, we obtain a random sequence of readings. Admittedly, the reason for this is that by taking such a measurement we do not learn all there is to know about the system. But this, the argument continues, is always our situation: we never are in a position to assert that our measurement has succeeded in extracting enough information to dispel the spectre of randomness. For that, we would need perfect accuracy, which is, of course, impossible.

It is at this point, I believe, that the argument goes philosophically astray. We are given an example of how a deterministic system *can* generate a random series of observations, and then led to believe that this is the typical situation. To see exactly why this is not the case, let us return to the tent function example once more. This time, imagine that we have, not just an *AB* instrument, but an αβγμ device. If the system is in a state between 0 and ¼, an α is returned; between ¼ and ½ a β is returned, etc. The result of such a string of observations will, as before, be written as a string of digits, this time from the symbol set {α*, β*, γ*, μ*} . . . Now let us imagine that we observe the evolution of the system using this device, obtaining an itinerary that looks something like: γ*, γ*, β*, α*, μ*, β*, α*, ...

First, we ask: will this sequence, like its *AB* counterpart, be Bernoulli random? The answer to this question is a definite "No!" To see why, consult figure 10.2 below. Notice that if the system is in one of the α states, then its next state *must* be either an α or a β state. Similarly, if the system is in a β state, then its next state must be a γ or a μ state. A γ state, likewise, must be followed by a γ or a μ state, and a μ state must be succeeded by an α or a β state. These results may be summarized as follows:

$$\alpha \rightarrow \alpha \text{ or } \beta$$

$$\beta \rightarrow \gamma \text{ or } \mu$$

$$\gamma \rightarrow \gamma \text{ or } \mu$$

$$\mu \rightarrow \alpha \text{ or } \beta$$

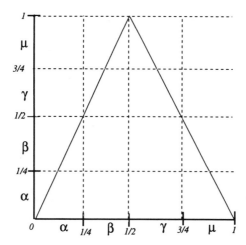

Figure 10.2 The tent function with its state space goarse-grained into four compartments of equal measure.

Clearly, with such a set of constraints, the set of observations recorded by an αβγμ device are anything but random: the chances of the system entering one of the states α, β, γ, or μ clearly is a function of its current state.[13] So the first conclusion to be drawn from this example is that a deterministic process that is random on one level of coarse graining may not be so when the level of coarse graining is only slightly refined using the above procedure. Infinite precision is by no means necessary, in a deterministic system, to have randomness disappear and be replaced by more regular forms of probabilistic behavior.

But this is not all. Given our αβγμ device, we are now in a position to predict *exactly* the next reading of the *AB* device. Consulting the above table, we see that if the system is in state α, then it must next be in either state α or in state β. In either case, however, the system is then in a condition which will register as condition *A** on the *AB* device. Hence we see that α → *A*: state α will be followed by state *A*. Similarly, if the system is in state β, then it will next be in γ or μ, that is, in state *B*. Continuing in this way we obtain a new table relating the results of a measurement taken at time *t* by the $\alpha\beta\gamma\mu$ device to a measurement at time $t + 1$ taken by the AB device:

$$\alpha \rightarrow A$$

$$\beta \rightarrow B$$

$$\gamma \rightarrow B$$

$$\mu \rightarrow A$$

Determinism has now reappeared in (almost) all its glory, and the previously "random" sequence has become totally predictable!

Notice that we have not redeemed determinism—if that is what we have done—by assuming that the new coarse graining gives us a perfectly accurate and complete account of the current state of the system. On the contrary, *the graining remains coarse*, and yet the AB itinerary becomes predictable. Hence, although it is possible that data corresponding to a sequence of coarse measurements on a deterministic physical system will be a random sequence, it does *not* follow that the same will be true when only a slightly less crude instrument is employed. Such an instrument will permit the perfect predictability of its slightly coarser companion.

Here is the general proof of this last claim. Suppose that we have a deterministic system, $S = (S, f)$, where f is a continuous function from S to S. Let $P = \{P_1, P_2, \ldots, P_n\}$ be an open partition of S.[14] This partition allows us to define a set of symbols $P^* = \{P_1^*, P_2^*, \ldots, P_n^*\}$ which we may then use to describe the itineraries of the various points in S. Now suppose that these itineraries are for the most part random, or more generally, stochastic. Nevertheless, we may now move to the somewhat finer partition, given by

$$P+ = \{P_1 \cap f^{-1}[P_1], P_1 \cap f^{-1}[P_2], \ldots, P_1 \cap f^{-1}[P_n], \ldots,$$

$$P_n \cap f^{-1}[P_1], \ldots, P_n \cap f^{-1}[P_n]\}.$$

In other words, the members of the new partition are all the subsets of the form $P_{ij} = P_i \cap f^{-1}[P_j]$, where $i, j = 1, 2, \ldots, n$. Since f is continuous, these sets are also open, and the result is a new, finer, open partition of the state space S. But notice now that if a state is in, say, $P_2 \cap f^{-1}[P_3]$, then it *must* go into partition P_3 on its next iteration. In general, if a state is in $P_{ij} = P_i \cap f^{-1}[P_j]$, then its successor must be in partition P_j (and in some partition P_{kj}, although k cannot, in general, be predicted exactly). Hence the new, finer partitioning permits the exact prediction of the next entry of the more coarsely grained itinerary.

Thus, the fact that a chaotic deterministic system, by Brudno's theorem, has *some* partitioning that yields a set of random or stochastic observations in no way undermines the distinction between deterministic and stochastic behavior for such systems. Only by steadfastly refusing to look at other partitionings does the systems appear to be irreducibly random or Bernoulli. As successive partitionings are examined (say, by using the method above) the determinism underlying the preceding, coarser observations emerges. To be sure, at any stage of the above process, the system may be modeled stochastically, but the successive stages of that modelling process provide ample — inductive — reason for believing that the deterministic model is correct.

As I stated above, Suppes has argued for the empirical equivalence of chaotic deterministic and stochastic models on the basis of a theorem of Ornstein. Since this theorem applies to a subclass of K-systems called *Bernoulli flows*, the setting is somewhat different than that of the tent function; hence, I shall begin by outlining the basic framework.

I have already talked about maps that are Bernoulli shifts. A *Bernoulli flow* is a flow $\{f_i\}$ such that f_1 is isomorphic to some Bernoulli shift. In other words, a Bernoullli flow is a flow that looks like a Bernoulli shift when you confine your attention to times that are integers. The intuition here is that, since (not necessarily equiprobable) Bernoulli shifts are among the most random maps, the flows that generate them should be regarded as among the most random possible flows.

Next, an appparently quite different kind of dynamical system, what Ornstein calls a *semi-Markov process*. The general idea is to embed a finite-state Markov process in a manifold. Imagine a state space M in which a finite number of points P of the manifold M have been singled out. Now imagine that our process behaves as follows. At any time t, the state of the process is at one of the singled out points in P, where it remains for a while. It then jumps to one of the other points of the finite set, as determined by a Markov process transition table on the points in set P. In this way, the process proceeds to occupy (and reoccupy) the states in P, remaining in each state for a definite, but arbitrarily selected, period of time. Clearly, such a system is not deterministic, since the same state upon separate occasions may be followed by different states.

What Ornstein has shown is that *any Bernoulli system is an arbitrarily close approximation to some semi-Markov process.* To illustrate, Ornstein introduces the notion of an ϵ-viewer. The idea is that any time we view the Bernoulli flow through the viewer, we find the system at one of a finite number of points in *P*. When this occurs, however, the actual state of the system is farther than ϵ from that point at most ϵ of the time. (So, when ϵ is small, the system is almost always within ϵ of the point viewed.) Given a particular Bernoulli flow, choose your value of ϵ. Then, Ornstein's theorem states, there will exist an ϵ-viewer such that the resulting system is a semi-Markov process.[15] Since certain classical systems, such as a billiards on a square table containing a convex obstacle, are Bernoulli systems (Gallavotti and Ornstein 1974), Ornstein comments:

Our theorem also tells us that certain semi-Markov systems could be thought of as being produced by Newton's laws (billiards seen through a deterministic viewer) or by coin flipping. This may mean that there is no philosophical distinction between processes governed by roulette wheels and processes governed by Newton's laws. (Ornstein and Weiss 1991, 37–8)

Presumably, from the passage I quoted earlier, this is also Suppes' view, but put somewhat more cautiously.

Notice that the mere existence of a deterministic model for a stationary stochastic process is not the main point here. That, we saw earlier, follows from the shift map approach to such stationary processes. As Ornstein adds:

. . . we should note that our model for a stationary process (1.2) means that random processes have a deterministic model. This model, however, is abstract, and there is no reason to believe that it can be endowed with any special additional structure. Our point is that we are comparing, in a strong sense, Newton's laws and coin flipping. (Ibid., 40)

Similarly, Suppes is also at pains to point out that it is the applicability of Ornstein's theorem to classical systems that is of primary importance.

The existence of physically realistic models of natural phenomena for which such a theorem holds is the basis for skepticism about the empirical nature of any general claims for determinism. The simplest concrete models for which the theorem holds are those with a single billiard ball moving on a table on which is placed a convex-shaped obstacle. As the theorem indicates, the motion of the ball as it hits the obstacle from the various angles is not predictable in detail, but only in a stochastic fashion. (Suppes 1993, 254)

Let us now examine these claims more closely.

First of all, notice that Ornstein's "viewers" are, in effect, partitions or coarse grainings of the state space. States that are mapped onto the same point in the (finite) set P by the viewer are in the same partition. As the earlier discussion of Brudno's theorem showed, however, the fact that deterministic systems generate stochastic behavior under a given partitioning in no way undermines the fundamental determinism of such systems. Slightly finer partitionings reveal the predictability of their coarser predecessors, while exhibiting at the same time their own brand of history-dependent stochastic behavior. Furthermore, the stochastic behavior (the probability distribution) of the successive partitionings also follows from the deterministic model.

While Ornstein's theorem applies to certain types of flows, not maps, the conceptual situation is similar. The fact that a Bernoulli flow can be partitioned in such a way as to yield a (semi-) Markov process merely illustrates what has been acknowledged all along: Some deterministic systems, when partitioned, generate stochastic processes. No one of these stochastic processes can, however, generate the deterministic flow. The deterministic flow is, if you like, a recipe for generating stochastic processes, none of which can, in return, generate its parent flow.

Suppose, for example, we are observing a Bernoulli flow using a viewer of some specific accuracy ϵ_0. Then we have, in effect, partitioned the state space so that only finitely many states P are distinguishable, and we may now suppose that Ornstein's theorem applies, so that we "see" a system jumping from state to state in P and behaving like a semi-Markov process with a certain transition table p_{ij}. Let us now suppose that we decide to model that system as simply a Markov process with that transition table. While this model, by hypothesis, adequately describes the behavior of the system when seen through the ϵ_0 viewer, it does not tell us what to expect when we change to a somewhat finer viewer. In fact, the semi-Markov model does not even imply that the more finely viewed process will be a simple Markov process, rather than, say, a two-step Markov process. The deterministic model thus outstrips any single Markov model in its conceptual and predictive power.

Consider, however, the following possibility. Suppose that our viewer is the finest or most accurate viewer currently available. Then, do not predictions about what *would* happen *if* we had a finer viewer become empirically empty? And since Ornstein's theorem applies to viewers of any accuracy, are we not always in the empirical position described by the philosophical commentaries of Ornstein and Suppes?

If the level of accuracy described above is merely a limitation of our current observational technology, then any strong claim about the empirical equivalence of the deterministic and stochastic models would clearly be incorrect. The deterministic model would then be in the position of making predictions about potentially observable situations about which the stochastic model is mute. If there were, however, some *in principle* limitation on viewer accuracy and we were currently at those limits, then the case would be apparently much stronger. The difficulty with this construal of the equivalence of deterministic chaotic and stochastic systems is that the systems in question are classical, that is, Newtonian systems. And in these systems there is no intrinsic limitation on the accuracy of measurement or viewers!

In short, some deterministic chaotic systems, when suitably partitioned, can exhibit genuinely stochastic, Bernoulli, or even random, behavior. This much follows simply from Kolmogorov's shift mapping approach to stationary stochastic processes. Recent results by Brudno and Ornstein have shown that such stochastic behavior results from any system with a positive Lyapunov exponent or any Bernoulli flow. Furthermore, as Ornstein has also shown, some classical systems are Bernoulli flows and thus support stochastic behavior upon being suitably partitioned. While these results are mathematically and philosophically of great interest in showing the extent to which Newtonian determinism at the fine level can generate stochastic or even random behavior upon coarse graining, they do not, I have argued, support the further claim that the chaotic deterministic and stochastic models are empirically indistinguishable.

Randomness and the Continuum

Up to this point, I have emphasized that deterministic systems can generate random sequences without thereby undermining the fundamentally deterministic nature of these systems. The moral to be drawn from deterministic chaos is that it allows genuine chance at some level of coarse graining or observation, yet at the same time, the behavior of the system at ever slightly finer levels of coarse-grained observation, while remaining stochastic, reveals the underlying deterministic nature of the system.

There is, however, an important caveat that must be added to the first of these two claims, the view that determinism allows randomness at some level of coarse graining. In a nutshell, the problem is that these results strongly depend upon the assumption that the state space of the model is a continuum, rather than either a countably infinite or a (presumably) large finite set.

J. L. McCauley (1993) argues for a general account of deterministic chaos that uses only algorithmic objects and functions. For example, let us assume that we restrict the tent function to the computable numbers in the unit interval.[16] The result is a model of a dynamical system with a countable infinity of states, yet as I have shown elsewhere (Winnie 1992), the resulting function has all the usual characteristics of chaos: transitivity, sensitive dependence upon initial conditions, and a dense set of periodic points. As a result, it is difficult to see how such a model could be empirically distinguished from its ontologically richer continuum cousin.

Nevertheless, there is no room for genuine long-term randomness in such a model. To see this, consider the standard AB partitioning discussed earlier, where $A = [0, \frac{1}{2}]$ and $B = [\frac{1}{2}, 1]$. As we saw there, in the standard continuum model, for most (all but a set of measure zero) initial points, the resulting AB itinerary is a Kolmogorov random sequence. In the computable model, however, each itinerary is computable exactly. For example, starting with $p = \sqrt{2}/2$, the first n steps of p's itinerary may be computed by using your favorite algorithm to compute its binary expansion to the first $n + 1$ places. Another algorithm (complement and shift) can now be used to extract the first n steps of the itinerary. Since the computation of the itinerary by this method is exact, we have thus shown that there exists an algorithm for computing the tent map AB itinerary of any point in the state space as far out in the sequence as we care to go. Clearly, it follows at once that this itinerary is not random in the Kolmogorov sense. (This procedure generalizes easily, even when the map involved is not a shift map. Begin with an algorithm for the initial state p which allows its computation to be an arbitrary degree of precision. This, in turn, allows us to compute $f^n(p)$ to arbitrarily high precision (assuming f is computable), and so to locate $f^n(p)$ in its appropriate partition.)[17]

Staying with the tent map for the moment, notice that the algorithm for generating the AB itinerary of a starting point is fairly brief. Hence, while the above argument shows that the infinite trajectory of any starting point is not random, the brevity of the algorithm for computing *any* trajectory would seem to show that all reasonably lengthy finite trajectories must also fail to be random. This would indeed be so, except for one hitch — the algorithm for the starting point itself. To see the difficulty here, suppose that we consider an initial point that is a rational number p with an initial binary expansion of length k (k large) that is Kolmogorov random. (After this initial segment, the number simply repeats the initial segment *ad infinitum.*) Since the algorithm for computing the AB itinerary of p requires an algorithm for generating p itself, and (because the initial segment of length k is Kolmogorov random) the latter algorithm must be of a length comparable to k, any algorithm for computing the first k steps of the AB itinerary of p must be of a length greater than k. Thus, although there is an algorithm that generates the entire itinerary of p, there is no brief algorithm that generates its first k steps. As a result, while computable chaos is incompatible with the randomness of an *infinite* trajectory, it remains compatible with, and — as the above example shows — sometimes generates *finite* sequences that are Kolmogorov random. Nevertheless, determinism eventually triumphs, and, as the itinerary advances, the result is an itinerary that is not Kolmogorov random, and not just in the (infinite) long run, but in the finite long run. As the itinerary's length grows beyond that of the algorithm that generates it, its complexity decreases, and determinism generates pattern.

The situation is similar for large finite state spaces. In these cases, the dynamical map is always computable, since a finite list will do the job. Of course, in some cases, much shorter algorithms will be available. For example, in the case of the tent function, we might define its restriction to a state space of $k + 1$ states (rational numbers of the form i/k, $= 0,1, \ldots, k$) by defining $tent^*(i/k)$ to be the greatest rational in the state space less or equal to $tent(i/k)$. When such a technique is used to define finite state ($k = 100{,}001$) counterparts to the quadratic function $ax(1 - x)$, for values of a between 2.5 and 4, we obtain the bifurcation diagram shown in figure 10.3.

As this familiar figure shows, the (finite) long-term behavior of the finite state quadratic map does not appear to differ appreciably from that of the standard continuum model. (Of course, the computational renderings of the standard models are also finite state (due to roundoff), but the number of states involved is typically much higher than the one hundred thousand figure of this example.) Thus, there are good reasons for holding that continuum models and large finite state models (or even relatively small finite state models with noise) are empirically

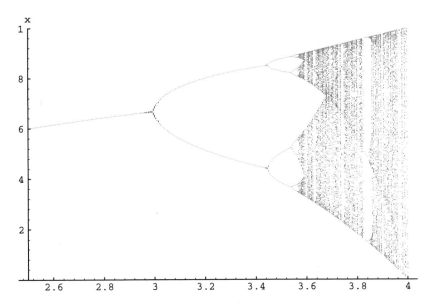

Figure 10.3 Bifurcation diagram for the finite state ($k = 100,001$) version of the tent function. Initial points are chosen at random.

indistinguishable.[18] The orbits of finite state deterministic models are, however, always eventually periodic: eventually, the orbit must return to some previously occupied point. Thus, while in some cases there may be some relatively short Kolmogorov random itineraries generated by such orbits, in the finite long run periodicity emerges, and determinism has its day.

These considerations show that, while chaotic determinism is compatible with randomness in models wherein the state space is a continuum, the same need not be so when the state space is countably infinite or finite. When the state space is countably infinite, the computable chaotic models show that pattern always emerges in the finite long run; when the state space is finite, periodicity is the eventual fate of every orbit and, hence, every itinerary. Deterministic randomness in an actual physical system is thus a possibility, but *only if we take the continuum postulated by our model seriously.* When we discount continuum models on grounds of their empirical indistinguishability from large or noisy finite models, then random itineraries are merely artifacts of a convenient mode of representation. Similarly, and more directly, when we take continuum models to be merely smooth idealizations of an underlying discrete set of physical states (as in some population and thermodynamical models), then once again the random itineraries of such models are merely conceptual artifacts. So long as a realistic interpretation of the continuum assumed by a chaotic dynamical model remains open, so too for the ontological status of randomness in the physical system modeled.

NOTES

I am greatly indebted to Zeno Swijtink, Diana Coffa, and Linda Wessels for a number of helpful conversations. Eric Bedford (Mathematics, Indiana University) was especially helpful with much of the relevant mathematics.

1. Actually, this is not quite true. There are systems (not all of them Bernoulli) called "K-systems," whose future events are also independent of their past. (This will be discussed in greater detail later.)

2. But here care must be taken. Roulette wheels are spun, and coins are flipped, by processes that are not themselves wheels or coins. It is the entire system, wheel + spinner or coin + flipper, that behaves (if indeed it does) like a Bernoulli process. Once this is seen, it becomes clear that if these systems are deterministic on some finer level of analysis, we currently have no idea of how to model it. We may suspect that small variations in the pressure of hand or thumb may produce "random" results, but analyses of the physics of coin-flipping do not bear this out. Thus according to analyses such as Ford 1983, a weak pattern in the thumb pressure would result in a pattern in the output string of "heads" and "tails."

3. Thus, Harel writes: "The implicit assumption that it [the computer simulation of tossing a fair coin] can be done is unjustified, since a real digital computer is a totally deterministic entity, and hence, in principle, all of its actions can be predicted in advance. Therefore [notice!], a computer cannot generate truly random numbers, and hence cannot simulate the truly random tossing of fair coins" (1992, 341).

4. What I shall here call "sequence" randomness, John Earman calls "performance" randomness. We both seem to use "process randomness" in the same sense. (For more details on these two notions, see Earman 1986, chaps. 8 and 9.) Kolmogorov complexity is treated at length in Li and Vitanyi 1993, and a thorough foundational account is provided in van Lambalgen 1987. The contemporary approach to the theory of stochastic processes is nicely presented in Bhattacharya and Waymire 1990. Classical results on the ergodic theory of dynamical systems can be found in Arnold and Avez 1989 (first published in 1968), Alekseev and Yakobson 1981, and Walters 1981. For a more recent survey, see Sinai 1989. The account that follows draws heavily on these sources, while at the same time trying to skirt the technical difficulties that would follow from detailed references to definitions and theorems.

5. The mathematically knowledgeable reader will recognize that the above

account is restrictive in many ways. State spaces need not be finite, and the random variables may be indexed by a continuous or multidimensional set. (See Bhattacharya and Waymire 1990, chap. 1, for examples and a more general definition.)

6. The entropy of a distribution is sometimes used as such a measure, where $E = \Sigma p_i \log p_i$. See Ekeland 1991, 42–47 for an informal account. For a defense of the randomness of biased coins, see Earman 1986, 143–44.

7. If $\{ s_n \}$ is a sequence, then the slot function u_i applied to that sequence is simply s_i, the occupant of the ith "slot."

8. See Billingsley 1965, 3–4.

9. For details, see ibid., 5–6.

10. For the relevant definitions and theorems, see Walters 1981, chap. 4.

11. Orbit complexity can be defined for a flow f_t in terms of the orbit complexity of the corresponding map f_1.

12. For simplicity's sake, I have neglected some of the technical complications resulting from the fact that Brudno worked with open coverings which might overlap. For a more detailed account, see Alekseev and Yakobson 1981, 294–96.

13. As it turns out, the resulting system becomes a Markov process, a stochastic system where the probability of the succeeding state depends only upon the current state.

14. This means that the P_i are disjoint, open, and their closure is S.

15. Viewers are what Ornstein calls "α-congruences," a kind of approximate isomorphism of dynamical systems. For the details, see Ornstein and Weiss 1991, sec. 2.

16. A *computable real number* is, roughly, one for which there exists an algorithm or computer program that allows its computation to any desired degree of accuracy. Since there are at most a countable infinity of programs, the same is true of the set of computable real numbers.

17. This method of computing the itinerary of a system is due to Palmore, more thoroughly discussed and explored in McCauley 1993.

18. For more details on this (complex) matter, see Winnie 1994. This figure, and the others in this paper, were drawn using *Mathematica*, Wolfram 1994. Wimsatt and Schank 1992 is a computing package that includes algorithms for rendering finite state systems such as the above.

REFERENCES

Alekseev, V. M., and M. V. Yakobson. 1981. "Symbolic Dynamics and Hyperbolic Dynamical Systems." *Physics Reports* 75 5:287–325.

Arnold, V. I., and A. Avez. 1989. *Ergodic Problems of Classical Mechanics*. Redwood City, Calif.: Addison-Wesley.

Bhattacharya, R., and E. Waymire. 1990. *Stochastic Processes With Applications*. New York: John Wiley.

Billingsley, P. 1965. *Ergodic Theory and Information*. New York: John Wiley & Sons Inc.

Brittan, G. G., Jr., ed. 1991. *Causality, Method, and Modality.* The Netherlands: Kluwer Academic Publishers.

Brudno, A. A. 1983. "Entropy and the Complexity of the Trajectories of a Dynamical System." *Trans. Moscow Math. Soc.* 2:127–51.

Dennis, J. 1992. *It's Raining Frogs and Fishes.* New York: HarperCollins.

Earman, J. 1986. *A Primer on Determinism.* Dordrecht: D. Reidel.

Ekeland, I. 1988. *Mathematics and the Unexpected.* Chicago: University of Chicago Press.

———. 1991. *The Broken Dice, and Other Tales of Mathematical Chance.* Chicago: University of Chicago Press.

Ford, J. 1983. "How Random Is a Coin Toss?" *Physics Today.* April, 40–47.

Gallavotti, G., and D. S. Ornstein. 1974. "Billiards and Bernoulli Schemes." *Comm. Math. Phys.* 38:83–101.

Harel, D. 1992. *Algorithmics: The Spirit of Computing.* Reading, Mass.: Addison-Wesley.

Knuth, D. 1981. *The Art of Computer Programming,* Vol. 2: *Semi-Numerical Algorithms.* Reading, Mass.: Addison-Wesley.

Li, M., and P. Vitanyi. 1993. *An Introduction to Kolmogorov Complexity and Its Applications.* New York: Springer-Verlag.

McCauley, J. L. 1993. *Chaos, Dynamics and Fractals.* Cambridge: Cambridge University Press.

Ornstein, D. S. 1974. *Ergodic Theory, Randomness, and Dynamical Systems.* New Haven: Yale University Press.

Ornstein, D. S., and B. Weiss. 1991. "Statistical Properties of Chaotic Systems." *Bulletin (New Series) of the American Mathematical Society* 24 (Jan.):11–116.

Schuster, H. G. 1988. *Deterministic Chaos.* 2nd. rev. ed. New York: VCH Publishers.

Sinai, Ya. G., ed. 1989. *Dynamical Systems II.* New York: Springer-Verlag.

Suppes, P. 1991. "Indeterminism or Instability, Does It Matter?" In *Brittan 1991,* 5–22.

———. 1993. "The Transcendental Character of Determinism." *Midwest Studies in Philosophy* 18:242–57.

Van Lambalgen, M. 1987. *Random Sequences.* Ph.D. diss., University of Amsterdam.

Walters, P. 1981. *An Introduction to Ergodic Theory.* New York: Springer-Verlag.

Wimsatt, W., and J. Schank. 1992. *Modelling — A Primer.* Dept. of Philosophy, University of Chicago.

Winnie, J. 1992. "Computable Chaos." *Philosophy of Science* 59:263–75.

———. 1994. "Sparse Models of Chaos." Manuscript; current version available on request.

Wolfram, S. 1991. *Mathematica,* 2nd ed. Reading, Mass.: Addison-Wesley.

11

Models, the Brownian Motion, and the Disunities of Physics

R. I. G. Hughes
Department of Philosophy, University of South Carolina

> Today our theories of physics, the laws of physics, are a multitude of different parts and pieces that do not fit together very well.
>
> **Richard Feynman (1967, 30)**

Feynman made this comment in 1964, at a time when the prospects for a grand unified theory of absolutely everything looked particularly bleak. But even if such a theory is vouchsafed to us, I will suggest, physics will still exhibit disunity. Or rather, it will exhibit disunities, for disunity comes in various forms. In this essay I will examine three kinds of disunity, without claiming that my list is exhaustive. At each stage of the discussion, the topic of unity and disunity will be intertwined with another: the use of models in physics. In the central part of the essay I will examine in some detail Einstein's use of models in his analysis of Brownian motion, and at the end I shall draw some general conclusions about the way we should regard physical theories.

Disunities of the First Kind

In Part Six of his *Discourse on the Method*, Descartes acknowledges that the enterprise of deducing all the regularities of nature from a small number of principles faces a problem. He writes,

I must admit that the power of nature is so ample and so vast, and these principles so simple and so general, that I notice barely any particular effect of which I do not know at once that it can be deduced from the principles in many different ways; and my greatest difficulty is usually to discover in which of these ways it depends on them. ([1637] 1985, 1:144)

His solution is "to progress to the causes by way of the effects," (ibid., 150), that is, to put forward and then to test "suppositions," as he calls them here, or "hypotheses," as they came to be known.[1] In the *Principles of Philosophy*, he tells us that the method of hypothesis is particularly valuable in the investigation of "imperceptible things."

Just as when those who are accustomed to considering automata know the use of some machine and see some of its parts, they easily conjecture from this how the other parts which they do not see are made; so, from the perceptible effects and parts of natural bodies, I have attempted to investigate the nature of their causes and of their imperceptible parts. ([1644] 1983, pt. 4, sec. 203, 286)

Notice that the direction of the investigation is again from effects to causes.

Descartes's mention of automata links his procedures to our own. These mechanical devices are analogous to things which we know more abstractly as *models* of the phenomena, and which find employment in the twentieth century version of the method of hypothesis. Consider, for example, the "two fluid" model of superconductivity.

Superconductivity is now a well-known and widely applied phenomenon. At very low temperatures the electrical resistance of some metals vanishes, so that no voltage is needed to keep a current flowing through them. It was first observed in 1911 by Kammerlingh Onnes, who found that at a transition temperature about 4° above absolute zero the resistance of mercury suddenly fell off by a factor of at least 10^5 (see Kubo and Nagamiya 1969, 186). The phenomenon has now been observed in many metals and alloys, and at considerably higher temperatures.

In the two decades after its discovery, much experimental information became available about this phenomenon, but the first promising theoretical treatment of it was the two-fluid model proposed by Fritz and Heinz London in 1935 (ibid., 186). This postulates a distinction between two kinds of electric current, a "normal current" and a "superconducting current." The "London equations" describing the superconducting current are exactly those required to account for the Meissner effect — the observed effect that no magnetic field resides in a superconductor (ibid., 188–89). Thus the Londons offer us a modern counterpart of a Cartesian hypothesis: a model whose characteristics are determined by arguing from effects to causes.

Oddly, both the normal and the superconducting currents are thought of as composed of electrons. Hence, the model leaves a major question unanswered: how can the same entities form two distinguishable fluids? To answer this, another model is needed. It was provided by Bardeen, Cooper, and Schrieffer in the mid-1950s. Whereas the only force between free electrons is the repulsion due to the Coulomb interaction between them, on the BCS model a pair of electrons in a crystal lattice can also "use the lattice" to exert an attractive force on one another. The number of these "Cooper pairs" suddenly increases at the transition temperature, and they comprise the superconducting current (Christmas 1988, chap. 13.2).

For twenty years prior to the appearance of the BCS model, however, physicists had perforce to be content with a "local" model for the phenomenon of superconductivity, a model whose behavior conforms to fundamental theory (in this case electromagnetic theory), but which contains elements whose specification is independent of the equations of the theory. The practice of using such a model gives rise to what I will call a *disunity of the first kind*.

This is a purely formal or, if you like, structural disunity. The use of these models means that physics is not set out as a theory in the logician's sense; it cannot be presented as the deductive closure of a set of axioms. To echo Descartes, at any stage of physics there is a logical gap between the principles of our fundamental theories, "so simple and so general," and many of the "particular effects" to which these principles apply.

Disunities of the Second Kind

The principles explicitly appealed to by Descartes, and which served as a constraint on his hypotheses, were indeed wholly general. In the *Principles of Philosophy* he summarizes them as follows:

I do not accept or desire in Physics any other principles than in Geometry or abstract Mathematics; because all the phenomena of nature are explained thereby, and certain demonstrations of them can be given.[2]

By 1700, however, the principles of physics had become more specific, and physics was becoming divided into subdisciplines. Regarded as physical theories, the mechanics presented in Newton's *Principia* and the theory of light set out in his *Opticks* are effectively independent of each other. Two hundred and fifty years later, the divisions were firmly established within academic syllabi; the books I learned from as a schoolboy, all by the estimable G. R. Noakes, were entitled *A Textbook of Heat, A Textbook of Light, A Textbook of Electricity and Magnetism*, and so on. In other words, within physics we find *disunity of the second kind*.

I am not here drawing attention to divisions within the social organization of physics, like the distinction between solid-state physicists and particle physicists, or between low-temperature physicists and astrophysicists. These divisions may, but need not, correspond to the differences I am concerned with, namely, the different headings under which phenomena are brought and the different, complementary sets of principles used to explain them. Electromagnetic induction, for example, is explained by reference to Maxwell's laws of the electromagnetic field, the Carnot cycle by reference to the principles of thermodynamics.

The use of models in seemingly disparate areas of research, far from creating disunities of the second kind, may in fact serve to reduce them. For the models used to explain phenomena from different fields may *converge*; that is to say, they may share a significant common feature. It was a mark in favor of the Londons' two-fluid model of superconductivity, for example, that a similar model was used by Gorter and Casimir to explain the observed changes in specific heat of a superconductor around the transition temperature. In this instance the success of a single model for the two phenomena is not very surprising, since we may assume that we are dealing with different aspects of the same physical change. More striking are the cases in which the phenomena come from very different research programs.

The classic case of convergent modeling is furnished by the atomic hypothesis. In his 1913 book *Les Atomes*, Perrin listed thirteen instances of phenomena successfully explained by a molecular model, and he could have added others. Furthermore, in each case the success of the model hinged on postulating the same number of molecules per mole of the substance involved. The number in question is, of course, Avogadro's number. As this selection shows, the range of phenomena is striking:

(a) Brownian motion;
(b) alpha decay from radioactive substances;

(c) X-ray diffraction by crystals;
(d) black body radiation;
(e) electrochemistry.[3]

The papers on Brownian motion that Einstein published between 1905 and 1908 were a major contribution to this chapter in the history of science, and at this point I turn aside — or appear to turn aside — from my main argument to examine the first of them in some detail. I have two reasons for doing so: the first is that the 1905 paper offers a remarkable case study of the use of models in physics; the second is that it will lead us toward disunities of the third (and final) kind.

Einstein and Brownian Motion

"Brownian motion" is the name given to the random movements of little particles suspended in a liquid. Though these movements are small, they are large enough to be observed through a microscope.[4] The molecular explanation of the motion is that in any short period of time a great number of molecules of the liquid will bounce off the suspended particle and tend to propel it in some direction or other. The movements we observe are due, not to individual collisions between the suspended particle and particularly energetic molecules of the liquid, but to statistical variations in the momentum imparted to the particle in different directions as time goes on; the particle may receive more blows pushing it from left to right than from right to left, for example, and, a bit later, a greater number pushing it up than down.

It is only with hindsight that Einstein's 1905 paper can be unequivocally described as concerned with Brownian motion. As he announces in the title, the motion he sets out to analyze is "the movement of small particles . . . demanded by the molecular-kinetic theory of heat." In other words, his aim is not to explain a motion that has already been observed, but to predict and describe a motion that might prove observable. Whether the observed and the predicted motions will turn out to be one and the same remains to be seen; Einstein writes,

It is possible that the movements to be discussed here are identical with the so-called "Brownian molecular motion"; however, the information available to me regarding the latter is so lacking in precision, that I can form no judgment in the matter. (1905, 1)

Neither here nor elsewhere in the paper does Einstein point out how important it is for his project that the two motions be identified. If they were independent of each other, then the phenomena Einstein predicts would always be masked by Brownian movement, and so could never provide evidential support for the molecular-kinetic theory of heat he favors.[5]

As it is, the evidence they provide is remarkably indirect. Where I have described the cause of Brownian motion rather vaguely as being due to "statistical variations in the momentum imparted to the particle in different directions," Einstein is even vaguer. "The irregular movement of the particles," he says, is "produced by the thermal molecular movement" (1905, 11). A page later, he says again that the "irregular movements" of the suspended particles "arise from thermal molecular movement" (ibid, 12). These two references to "thermal molecular movements" are the nearest Einstein comes to a causal account of the random motion of the suspended particles.

Instead, his strategy is to apply to suspensions the account of solutions he put forward in his doctoral dissertation.[6] On this account, when a solute is taken into solution, the molecules of the solute behave like small particles moving around within the solvent. They effectively constitute a "gas" of molecules, in that their behavior can be described in terms of the models provided by the kinetic theory of gases. These models represent a gas as a collection of molecules in random motion; the (absolute) temperature T of the gas is seen as proportional to the main kinetic energy of the molecules, and the pressure P that it exerts on the walls of a container as the force per unit area exerted by these molecules in bouncing off them. According to Einstein's theory of solutions, it is no coincidence that van't Hoff's law for dilute solutions has exactly the same form as the phenomenological gas law:[7]

$$PV = RTz \qquad (11.1)$$

When this law is applied to solutions, P becomes the osmotic pressure resulting from the presence of a mass z of solute in the solution.

On this analysis the molecules of the solute behave like particles suspended in the surrounding solvent. Conversely, Einstein suggests, a suspension of particles in a liquid may be expected to behave like a solution. In particular, the suspended particles may be expected to exert an "osmotic pressure" due to thermal motions:

According to this theory a dissolved molecule is differentiated from a suspended body *solely* by its dimensions, and it is not apparent why a number of suspended particles should not produce the same osmotic pressure as the same number of molecules. (Einstein 1905, 3)

Observe what is happening here: (a) a dissolved substance is modeled as a suspension of molecules, and (b) a gas is modeled as a collection of molecules in random motion. Both these subject-model relationships are then inverted. By coupling (a) with the inversion of (b), Einstein models the hypothetical collection of molecules in the solution as a gas, and so obtains an expression for the osmotic pressure P. Then, by an inversion of (a), he models a suspension as a solution, and interprets P as an "effective pressure" exerted by the suspended particles.

In an important respect, however, the suspended particles and the postulated solute molecules both differ from the molecules of an ideal gas. They move within a viscous fluid, whereas the molecules of an ideal gas do not. Einstein uses this fact in deriving an expression for the diffusion coefficient D of the particles (or solute molecules), which specifies the rate at which particles diffuse from regions of high concentration to regions of low concentration.[8]

He obtains,

$$D = \frac{RT}{N} \bullet \frac{1}{6\pi\eta a} \qquad (11.2)$$

Here η is the coefficient of viscosity of the fluid, and a the radius of the particles (assumed spherical).

This equation is very revealing. R and T appear because the particles are being modeled as a gas obeying equation (11.1); Avogadro's number N appears because, in accordance with the kinetic theory of gases, the gas is being modeled as a Newtonian system of particles; and η and a appear because these particles are represented as moving through a viscous fluid. Its mission accomplished, the hypothetical pressure P exerted by the suspended particles has disappeared from view.

Einstein now considers the thermal motions which, on the kinetic theory, bring about this diffusion. As I mentioned earlier, there is no analysis of what causes these motions. All Einstein assumes is (i) that there is a specific probability function $\Phi(\Delta)d$ that the particles will move a distance between Δ and $\Delta + d$ in a certain time interval τ, (ii) that this function is nonzero only when Δ is small, and (iii) that it is symmetric about $\Delta = 0$.

Einstein confines himself to motion in one dimension. From assumptions (i)–(iii) he obtains the root mean square value λ of the distance moved by a particle in time t:

$$\lambda = \sqrt{(2Dt)} \tag{11.3}$$

where D is defined in terms of Φ and shown to be equal to the diffusion coefficient. Hence (11.2) and (11.3) together give

$$\lambda = \sqrt{\left[2t \cdot \frac{RT}{N} \cdot \frac{1}{6\pi\eta a} \right]} \tag{11.4}$$

from which,

$$N = \frac{RTt}{3\pi\eta a \lambda^2} \tag{11.5}$$

Assuming a value for N of 6×10^{23} molecules per mole, Einstein calculates that, if $t = 1$ minute, then $\lambda \sim 6\mu$, a distance large enough to be observed through a microscope.

In this way, Einstein's inversion of the relation between a solution and his model of a solution effects a reversal of the phenomenological/theoretical distinction. In the case of a solution, the osmotic pressure is measurable (a 1% sugar solution typically exerts an osmotic pressure of about $\frac{2}{3}$ atm.; see Pais 1982, 87), but the molecules and their motions are not. In contrast the "effective pressure" attributed to particles in a suspension is too small to measure. In equation (11.1) z is measured in moles (or gm molecular wt.). On the molecular theory, therefore, $z = n/N$, where n is the number of molecules in the volume V. Equation (11.1) then becomes:

$$P = \frac{RT}{V} \cdot \frac{n}{N} \tag{11.1*}$$

When this equation is applied to the suspension, n/N is tiny, and so is P. Thus, no measurement of P can provide a direct challenge to classical thermodynamics, according to which no such pressure exists.[9] On the other hand, unlike the molecules in a solution, the particles in a suspension are themselves observable, and so are their motions.

Indeed, as we now know, not only are they observable, but Perrin and his co-workers found ways to measure the relevant quantities in equation (11.5) and obtained thereby a value for Avogadro's number. These results, along with others, established the reality of molecules in the eyes of everyone except Ernst Mach. Exactly what form of argument leads from the predictive success of the molecular hypothesis to the conclusion that molecules are real has been the subject of recent philosophical debate (see, e.g., Fine 1986, chaps. 7–8). I will not enter the debate here, except to suggest that Einstein's use, not only of molecular models but also of their inverses, offers the possibility of a stronger form of argument than that of inference to the best explanation.[10]

Einstein's analysis, however, contains a paradoxical feature, to which I have so far drawn no attention. Throughout his discussion he uses a model which is almost invisible to the reader; it remains literally in the background. Recall that the molecules of a solute and the particles of a suspension are both treated as systems of particles moving through a viscous fluid. Equation (11.2), for example, contains the factor $6\pi\eta a$ because, by Stokes's Law, which is used in its derivation, the drag F on a sphere moving with speed v through a fluid is given by $F = 6\pi\eta a v$. But ordinary hydrodynamics, within which this law appears, models a fluid as a continuous and homogeneous medium. And in section 2 of Einstein's 1905 paper, in which he shows that "the existence of an osmotic pressure can be deduced from the molecular-kinetic theory of Heat" (1905, 9), he assumes that "the liquid is homogeneous *and exerts no force on the particles*" (10, emphasis added).

This, on the face of it, is bizarre. On the one hand Einstein announces that an experimental confirmation of the result he predicts will be critical for the molecular-kinetic theory of heat, on which matter is treated as particulate. On the other, in making these predictions he treats matter (at least in the fluid state) as continuous and homogeneous. Furthermore, he is quite explicit about what he is doing. In a 1908 review of his work on solutions and suspensions he writes,

But when the dissolved molecule can be looked on as approximately a sphere, which is large compared with a molecule of the solvent, we may ascertain the frictional resistance of the solute molecule according to the methods of ordinary hydrodynamics, which do not take account of the molecular constitution of the liquid. (1908, 73)

If you find this peculiar, then you are probably not a physicist. A good physicist has a finely tuned sense of when to use one model and when another. Like Einstein, she or he is untroubled by a disunity of the third kind.

Disunities of the Third Kind

A disunity of the second kind arises when physicists employ different, but complementary, sets of principles; a disunity of the third kind when these principles are mutually at odds.

Consider, for example, the models of plasmas proposed by David Pines (1987) in the volume *Quantum Implications*.[11] He uses the term "plasma" rather generally, to cover those states of matter in which electrons are not attached to specific parent atoms. This can occur in a high-temperature gas, or in a conducting solid. Pines suggests that solid-state plasmas should be regarded as "quantum plasmas," and their behavior modeled in terms of quantum theory. Gaseous plasmas, however, can be modeled as "classical plasmas," and their behavior described in terms of Newtonian mechanics, augmented by Coulomb's Law. In one and the same paragraph, Pines tells us that essentially similar systems can profitably be thought of as governed by different, in fact incompatible, sets of principles.

I should emphasize again that I am not criticizing this practice. But I do criticize philosophical accounts of scientific theorizing that cannot accommodate it, or that ignore the fact that it goes on. In the course of just three pages of *The Structure of Scientific Revolutions*, for instance, Thomas Kuhn tells us that a scientific revolution involves "the rejection of an older paradigm" and "the destruction of a prior paradigm," and that in a revolution "the second [theory] must displace the first" (1970, 95–97). Within this picture there is no room for the fact that textbooks of physics and course listings of university physics departments are replete with presentations of theories which lost out. Feynman's (1965) lectures on physics, for example, include two lectures on ray optics, which lost out to the wave theory sometime between 1810 and 1840; likewise, the calendar of the University of Oregon, where I am writing this, contains a graduate course listing for Physics 611–612: *Theoretical Mechanics*, and the course description makes it clear that its subject matter is the classical mechanics allegedly rejected, destroyed and displaced by the revolutions of twentieth-century physics.

Courses like this are not rear-guard actions by theoretical conservatives, nor are they given to acquaint graduate students with the history of their subject. One reason for giving them is that they are pedagogically useful. Thus, in the preface to his well-known book on classical mechanics, Goldstein writes,

[Classical mechanics] has a twofold role in preparing the student for the study of modern physics. First, classical mechanics, in one or other of its advanced formulations, serves as the springboard for the various branches of modern physics. . . . Secondly, classical mechanics affords the student an opportunity to master many of the mathematical techniques necessary for quantum mechanics while still working in terms of the familiar concepts of classical physics. (1950, xi)

Clearly the University of Oregon's course in classical mechanics is organized with this end in view; the course outline reads, "Lagrangian and Hamiltonian mechanics, small oscillations, rigid bodies"; Lagrangian and Hamiltonian mechanics are precisely the "advanced formulations" of classical mechanics that Goldstein had in mind. Similarly, at the end of his second lecture on ray optics, we find Feynman (ibid.) discussing Fermat's least time principle in a way that anticipates his own path-integral approach to quantum theory.

But pedagogical utility cannot be the whole story. Classical mechanics is still used by practicing physicists, as the example of Pines's classical model of a plasma shows. Nor is this kind of disunity simply a matter of choosing theoretical models of one type for one phenomenon and of a different type for another. It can appear even within an apparently unified approach. Orthodox quantum mechanics, for instance, makes extensive use of "semiclassical" models, within which the quantum states of systems are represented by ψ-functions, while the Hamiltonian operators that determine the evolution of these states contain terms from classical electromagnetic theory for the fields within which these particles move. One might think that the use of these models serves to link two theories and so to unify physics. But even as a disunity of the second kind is bridged, a disunity of the third kind is created. Orthodox quantum theory uses Schrödinger's equation to describe how the ψ-function changes with time. Implicit in this equation is the background of a neo-Newtonian account of space and time. In contrast, the symmetries of classical electromagnetic theory demand the space-time postulated by the special theory of relativity. Thus, underlying the two theories in question we find two different, and mutually incompatible, accounts of space-time.

This conflict between orthodox quantum theory and the special theory of relativity has been known — and lamented — since quantum theory was articulated in the 1920s. Yet the existence of this anomaly, while it spurred efforts to find alternatives to the Schrödinger equation,[12] did not — and does not — inhibit physicists from using semiclassical models.

Two points are worth making about this practice. Both echo remarks I made earlier about Einstein's use of the classical theory of continuous fluids. The first, Feyerabendian, point is that semiclassical models were theoretically useful; they enabled physicists to make advances that would have been impossible had they been too fastidious to tolerate a disunity of the third kind. The second is that physicists were perfectly aware of the limits to what could be tolerated. To quote Landau and Lifshitz:

In non-relativistic quantum mechanics, the magnetic field may be regarded as an external field only. The magnetic interaction between the particles is a relativistic effect, and a consistent relativistic theory is needed to take it into account.[13]

The Structure of Theories: Two Views

These matters carry implications for our views on the structure of physical theory. Before considering these implications, however, I will sketch the two views that have been dominant in post-Galilean physics. I begin by quoting Alfred Tarski.

Every scientific theory is a system of sentences which are accepted as true and which may be called Laws or Asserted Statements, or, for short, simply Statements. (1965, 3)

Tarski here expresses a view characteristic of the logical empiricists, and dominant in the philosophy of science in the middle decades of this century. If we add to his specification the requirement that the statements of the theory are to be laid out as a formal deductive system, we get the following thesis, also characteristic of that period:

The sciences are properly expounded in formal axiomatized systems. . . . The sciences are to be axiomatized: that is to say, the body of truth that each defines is to be exhibited as a sequence of theorems inferred from a few basic postulates or axioms. And the axiomatization is to be formalized: that is to say, its sentences are to be formulated within a well-defined language, and its arguments are to proceed according to a precisely and explicitly specified set of logical rules. (Barnes 1975, xi)

I will use the terms "statement view" and "axiomatic view" synonymously, my choice reflecting the aspect of the view that I wish to emphasize.

The view has a long and distinguished history. In fact, the passage above is a summary by a recent translator, not of the views of the logical empiricists, but of "the essential thesis of Book A" of Aristotle's *Posterior Analytics*. His choice of phrase is only slightly mischievous. Later ancestors of the logical empiricists' view can be found in the methodological writings of Descartes and Newton. Descartes — at least, on some occasions and on one interpretation of his ambiguous writings — sought a science deducible from simple and general principles. At the end of Book II of the *Principles of Philosophy*, for example, he announces that he "will accept as true only what has been deduced from indubitable common notions so evidently that it is fit to be considered as a mathematical demonstration" ([1644] 1983, 2:64).[14] Newton, in Question 31 in the third edition of the *Opticks*, proposes that "analysis" — the sifting of regularities from phenomena — should be followed by "synthesis." The latter involves setting out a theory in the geometrical manner. Newton describes the procedure in terms of causes:

The synthesis consists in assuming the causes discover'd, and establish'd as principles, and by them explaining the phaenomena proceeding from them, and proving the explanations. ([1730] 1952, 404–05)

For Newton, as for Descartes, "causes" are not just efficient causes, but the general explanatory principles under which phenomena can be subsumed.[15]

Two hundred years after the publication of the *Opticks*, Duhem eschewed Newton's causal vocabulary but gave an account which, while more abstract than Newton's, is structurally similar to it.

A physical theory . . . is a system of mathematical propositions deduced from a small number of principles which aims to represent as simply, as completely, and as exactly as possible a set of experimental laws. ([1906] 1954, 20)

From "a small number of propositions which will serve as first principles in our deductions," consequences are drawn; these consequences "may be translated into judgements bearing on the physical properties of . . . bodies," and "these judgements are compared with the experimental laws which the theory is intended to represent" (ibid.). Duhem's talk of "translation" anticipates, in turn, the logical empiricists' view of a theory as a linguistic entity. The empiricists were also heirs to the investigations into formal logic and the foundations of mathematics pursued by Frege, Hilbert, and Russell. The result was a prescription for physical theory analogous to Hilbert's presentation of geometry: to be properly understood it was to be set out as a formal calculus comprising a set of statements, and these (in the case of physical theory) were to be "partially interpreted" in terms of an observational language.[16]

For the last two decades, however, the "statement view" of theories has been under attack;[17] it has come to resemble an aged heavyweight, summoned yet again out of retirement to act as a punching bag for a younger, nimbler opponent. I will take a few swings at it myself, while giving qualified endorsement to the new orthodoxy, the so-called "semantic view" of theories. On this view, to put forward a theory is (1) to provide a *theoretical definition*, that is, to specify a class of models; and (2) to propose a *theoretical hypothesis*, that is, to claim that a model from that class can adequately represent part of the world.[18] Equivalently, we may say that the theoretical definition defines a particular kind of system (e.g., a Newtonian system or a quantum system) and its behavior, while the theoretical hypothesis asserts that a part of the world (e.g., a plasma) can be regarded as a system of that kind.

In a particular case, we may choose to set out the theoretical definition axiomatically, but there is no requirement that we do so, let alone that the axioms be expressible in a particular formal language (e.g., the language of first-order logic with equality). Furthermore, axiom systems which differ in their choice of axioms or primitive terms but still define the same class of models are regarded as characterizing the same theory.

Among the virtues claimed—I think correctly—for this approach are that it steers philosophers of science away from a preoccupation with questions of language,[19] and that it offers a perspicuous way to display and compare the structures of foundational theories like classical mechanics and quantum theory.[20] Given the first virtue, it may be surprising to find it called the "semantic view." The name derives from the fact that, in speaking of models, the early proponents of the view had in mind the model-theoretic approach to semantics pioneered by Tarski.

Disunities and the Structure of Theories

In the final section of this essay, I return to the disunities of physics and argue that their existence not only provides strong evidence against the statement view of theories, but also points to an oversimplification in standard presentations of the semantic view.

Following Giere, I provided a capsule summary of the latter view in terms of a theoretical definition and a theoretical hypothesis. According to this view we have, on the one hand, a "part of the world" and on the other, a theoretically defined system, and we hypothesize that the one can be represented by the other.[21] What this simple duality leaves out, however, is that between the phenomenal world and the abstract models supplied by our foundational theories, physics interposes intermediate models, as we may call them. What is more, these intermediate models play an essential role in our theorizing.

I will distinguish three types of intermediate models, though the distinctions between them are not cut and dried. The first type, discussed in some detail by Nancy Cartwright, is used at a level very close to foundational theory. Frequently, when a mathematical theory is applied to a particular phenomenon or physical system, the system or phenomenon is modeled in a schematic and abstract way; the resulting "specially prepared, usually fictional description of the system" Cartwright (1983, 158) calls a *simulacrum*. The foundational theory provides a mathematical model of the behavior, not of the original system, but of the simulacrum. Cartwright describes a two-level, rather than a two-stage process, one that is dialectical rather than sequential. We do not first model the original system and then apply the foundational theory to the resulting model. Rather, the art of this kind of theorizing is to find a description of the system under scrutiny in terms that are amenable to theoretical treatment. To use Cartwright's example (ibid., 131), the helium-neon laser can be treated as "a collection of three-level atoms in interaction with a single damped mode of a quantum field, coupled to a pumping and damping reservoir." Spelling out the behavior of these three interacting components — the atoms, the field, and the reservoir — in terms of quantum field theory may require considerable theoretical ingenuity, but it can be done.

Intermediate models of the second type are exemplified in Pines's treatment of plasmas (see "Disunities of the Third Kind" above). The plasmas he considers are already theoretically described; they are states of matter in which electrons have been stripped from their parent atoms. As in the case of simulacra, foundational theory here models the behavior of models, but the latter, unlike simulacra, are not tailored to a particular theory; indeed, as Pines tells us, they can be modeled either as classical or as quantum systems. The great majority of applications of theory are of one of these two kinds. Very rarely is it possible to talk of empirical consequences of a foundational theory *tout court.*

The experiments by Aspect on coupled quantum systems, the experiments which showed that observed statistical correlations violate Bell's inequalities and corroborate quantum theory, offer a possible exception. Even this example, however, involves modeling at the third intermediate level, that of experiment. To start with the obvious: implicit in the use of any scientific instrument is the invocation of a theory; Newton would not have known what to make of so basic a tool as a moving-coil galvanometer. And, on the semantic account, any theory is specified in terms of models. In addition, (1) the winnowing of statistical data from the output of experiments typically involves a complex interplay of instrumental and theoretical methodology, and (2) these data need to be associated with the ostensible topic of the experiment.

I will here follow the example of many philosophers of science and most physics textbooks by setting the models used at the experimental level to one side. I do so at risk of distorting the practices of physics (and also, no doubt, of displaying an ideological propensity to render the material processes of science invisible).[22] Be that as it may, in this essay I will focus on intermediate models of the first and second types.

These are the models used in the twentieth century version of the method of hypothesis, the models whose use gives rise to disunities of the first kind. But, as I observed earlier, the method of hypothesis appears as a problem only to someone who assumes that the ideal scientific theory should be set out in the geometrical manner. Although I introduced intermediate models by pointing out that the semantic view, as often presented, paid no heed to their use, that omission can easily be remedied; we merely have to rewrite the theoretical hypothesis as the claim that a part of the world *as we describe it* can be regarded as a system of the kind specified by the theoretical definition. Thus amended, the semantic view has the added virtue that it gives an account of the application of theories to specific phenomena, which phenomena can themselves be described in theoretical terms. In contrast, the statement view deals easily with only a limited class of applications: those that involve specifying the boundary conditions that apply in a particular case, and which need to be entered at appropriate places in the equations of the theory.

Turning to disunities of the second kind, we again find that the semantic account can capture a crucial component of scientific theorizing. This is the practice, noted in "Disunities of the Second Kind" above, of convergent modeling, the use, that is, of the same elements in models appearing in more than one theory. Recall that the atomic hypothesis is put to work when a perfect gas is modeled as a Newtonian system, when the transfer of charge through an electrolyte is modeled as a movement of ions, and so on. In each of these cases the method of hypothesis is employed, within which the atomic hypothesis makes its appearance in an intermediate model of the second kind. To the extent that the axiomatic approach to physical theory finds difficulty in accommodating disunities of the first kind, it will also fail to give due weight to the practice of convergent modeling.

But perhaps I have overstated that extent. My assumption that the method of hypothesis poses problems for the axiomatic account could be challenged. This account can accommodate any particular disunity of the first kind, one might argue, by adding additional terms to the theoretical vocabulary, together with such axioms as are necessary to specify the relations between these terms and the terms of the foundational theory. The apparent disunity would then melt away.

As it did so, however, the proponent of the axiomatic approach would be left with a dilemma: the more successfully the disunity was assimilated within the theory, the less credible would be the claim that the approach articulated the structure of actual scientific theories.

One response to the dilemma would be to grasp its second horn, and to deny that the articulation of actual scientific theories was any part of the program's aim; that aim is simply to provide a "canonical linguistic formulation [that] will display the essential epistemic features of scientific theories."[23] But the horn, grasped in this way, instantly metamorphoses into a nettle. As every student of seventeenth-century science knows, the use of the method of hypothesis then raised crucial epistemic questions. Descartes claimed certainty for the general principles underlying his science, but only "moral certainty" for the results obtained using the method of hypothesis, while Newton, in his later years, disallowed its use altogether.[24] There is no reason to think that these issues have been conclusively resolved. A formulation of physical theory in which disunities of the first kind melted away would thus forfeit its claim to display the theory's essential epistemic features.

More generally, if the claim is made that essential features of scientific theories are being displayed in a presentation that differs structurally from any that scientists actually use, it needs to be argued for with some care. It presupposes a prior examination of those theories to isolate the features in question. There is, however, scant evidence that the logical empiricists undertook one. Even so subtle and informed a philosopher as Ernest Nagel (1961, 90), when he set out to "look more closely at the articulation of theories," declared without preamble that they could be usefully analysed in the way the axiomatic program prescribes, in terms of a formal calculus with rules for its interpretation.[25]

In contrast, the semantic account often gives a literal description of theoretical practice. The abstract of a paper in a 1994 issue of the *Journal of Physics*, for example, opens with these words:

Slow motion of a Newtonian fluid past a porous spherical shell has been examined. The flow in the free fluid region (inside the core and outside the shell) is governed by the Namier-Stokes equations, whereas the fluid in the porous region (shell region) is governed by the Brinkman model. (Bhatt and Sacheti 1974, 37)

The abstract is set out in precise conformity with (the amended version of) Giere's (1985) schematic account of a theory. Two different mathematical models are to be applied to the flow of that fictitious substance, a "Newtonian fluid."

For a more interesting example of how an emphasis on models and modeling allows us to portray the structure of scientific theorizing in a perspicuous way, consider my analysis of Einstein's first paper on Brownian motion. His argument, as I read it, involves the modeling of one phenomenon, Brownian motion, in terms of another, the behavior of solutions; solutions, in turn, are modeled as ideal gases; and, where convenient, the solvent and the liquid in which the Brownian motion occurs are both modeled as homogeneous fluids. Though it would be tedious to insist on it, each of these moves could be represented by the Giere schema of theoretical definition coupled with theoretical hypothesis. In contrast, no conceivable axiomatization of the theories that Einstein employs would bring out his argumentative strategy. Reference to models and the representations they provide is crucial to an understanding and appraisal of what he is doing. This may not be true of all theories; for example, we need not appeal to models to understand the best known of Einstein's 1905 papers, in which he sets out the special theory of relativity. But even there we do not distort his theory if we regard it, as did Minkowski, as specifying a class of models for space-time.[26]

I turn finally to disunities of the third kind. Faced with these, the statement view of theories runs into an immediate problem. For, as Tarski (1965) suggests, the primary virtue of a statement is that it is true. When two theories are incompatible, however, at most one of them can be captured by a set of true statements. How then, on the statement view, are we justified in using the resources both of classical mechanics and of quantum theory, or of ray optics and of the wave theory of light? Indeed, since we have good inductive evidence that even our best theories will turn out to be provisional, the statements that — on the statement view — those theories comprise are all likely to be false.

To meet this problem the devotee of the statement view can adopt one of two strategies. The first, and less promising, is to appeal to a notion of "approximate truth." But when two theories are incommensurable, the terms of at most one of them can refer. To describe the statements of the other as "approximately true" would involve, therefore, not just a modification of our theory of truth, but a wholesale recasting of it.

The second strategy is to adopt an instrumentalist position, to say, that is, that theoretical statements are neither true nor false, properly speaking; rather, the theory as a whole operates as a set of principles used for drawing conclusions about observable facts (see Nagel 1961, 129–40). Two objections may be raised to this strategy. The first is that it relies on an unworkable distinction between observational and theoretical terms. (For a critique of this distinction, see Suppe 1972.) The second is that it makes no distinctions within the class of theoretical terms. No theoretical term whatsoever is to be granted a reference; no statement about theoretical entities is to carry a truth-value. The position thus commits us to a thoroughgoing antirealism. Yet we may well want to adopt different ontological attitudes toward different theoretical entities. We may believe firmly in the existence of atoms, yet be skeptical about the anomalously heavy quark. These are distinctions that the instrumentalist position does not permit us to make.

The decisive argument against this approach, however, is surely this. The instrumentalist position is offered as a refuge from a major problem for the statement view of theories. In providing one, it suggests that what purport to be statements about theoretical entities should be regarded otherwise. In other words, it defends the statement view by telling us that the view is seriously misleading.

With that the proponent of the semantic view can concur. For the amended version of the semantic view does not suffer from these problems. It allows us to be discriminating in our ascriptions of existence. If we model Xs as a collection of Ys, we may take any ontological attitude we wish toward the Ys, or for that matter toward the Xs.[27] And adopting the semantic view does not preclude our making theoretical statements. These statements may be appraised as true or false if we are realists about the theoretical entities referred to in them; if we are not, they will be appraised as true-in-the-model or false-in-the-model. Models themselves are neither true nor false; they are more or less adequate representations of the world. That is why we feel free to help ourselves to models from mutually incompatible theories. True, we are then left with a difficult question: what exactly is the relation of the model to its subject? Differently put: in what sense, or senses, does a physical theory *represent* the world?

But in asking this we shall at least be asking the right question. To my mind, it is one of the key questions that philosophers of physics need to address.[28]

NOTES

1. For a general discussion of the "Method of Hypothesis" and its subsequent history up to the nineteenth century, see Laudan 1981.

2. Descartes [1644] 1983, pt. 2, sec. 264, 76. Note that, in talking of "certain demonstrations," Descartes has conveniently forgotten the need for hypothetical reasoning; he remembers it some two hundred pages later, by which time certainty has been diluted to "moral certainty."

3. Salmon 1984, 217–19. As Wesley Salmon pointed out to me, item (e) is not on Perrin's list.

4. A brief history of work on Brownian motion prior to 1905 is given by Furth in note 1 to Einstein 1956, 86–8.

5. This may explain why in 1906 Einstein accepted so readily Siedentopf's assurance that "he and other physicists . . . had been convinced by direct observation [*sic*] that the so-called Brownian motion is caused by the irregular thermal movements of the molecules of the liquid" (Einstein 1906b, 19).

6. The dissertation (Einstein 1906a) was completed only eleven days before the 1905 paper was submitted for publication, and was not itself published until 1906.

7. In fact van't Hoff defined *ideal solutions* as "solutions which are diluted to such an extent that they are comparable to ideal gases" (Pais 1982, 87).

8. The derivation is wonderfully cunning; for an analysis of Einstein's argument, see Pais 1982, 90–98.

9. Einstein (1905, 2–3) notes the conflict between the two theories, but does not mention that the difference is too small to detect.

10. For a critique of inference to the best explanation, see van Fraassen 1989, 142–49; for an argument in terms of a common cause, see Salmon 1984, 213–27.

11. I discuss this essay in more detail in Hughes 1993.

12. See Dirac 1928, and Messiah 1960, ch. 20: "L'équation de Dirac."

13. Landau and Lifshitz 1977, 454; see also the remainder of chap. 15: "Motion in a Magnetic Field."

14. Note that Descartes does not demand that the deduction be a mathematical deduction, just that it carry the epistemic force of one.

15. The term *principle* too is ambiguous; in 1700 it could be used not only to denote a law (as in the "principle of the parallelogram of forces"), but also to describe a tendency in a particular sort of body (as in "vital principle"). See OED. Note also that, for Newton, *prove* means "verify by experiment."

16. In fact, Carnap's (1939) account of scientific theory appears at the end (secs. 23–25) of his monograph, *Foundations of Logic and Mathematics*.

17. See Suppe 1974, especially the editorial introduction, "The Search for Philosophic Understanding of Scientific Theories."

18. Here I paraphrase Giere 1985, 76–80.

19. See in particular van Fraassen 1980, 53–56. Some advocates of the view, however, still displayed this preoccupation; see, e.g., Suppe 1974, 228.

20. See, e.g., Hughes 1989, chap. 2, where these two theories are compared.

21. My terminology is slightly different from the formulation in (e.g.) Giere 1985, 80 and van Fraassen 1989, 222 (see note 27 below), but this dichotomy appears in both.

22. I plead in my defense that I am at work on a case study of a particular experiment in solid-state physics.

23. I quote Suppe 1972, 3; he too thinks that the program fails in that aim.

24. Newton had mechanical hypotheses in mind, but the methodological prescription of Question 31 in the *Opticks* ([1730] 1952) also leaves no room for the method of hypothesis abstractly conceived. Despite this, it is hard to consider Newton's talk of "least parts" of light (ibid., 1) or of the "fits" into which these least parts must be thrown (ibid., 281), as other than hypotheses, to be "proved" by experiment.

25. Note also that, at the beginning of the same chapter (Nagel 1961, 80), he identifies "theories" with "theoretical laws."

26. It may be objected that here I am disregarding important distinctions between the types of models that physicists use, between, for example, constitutive models like Bohr's model of the atom, and abstract, structural models like those proposed in space-time theories. I agree that there are differences between these types of models; however, I deny that they are important. For present purposes, they can be ignored; witness the fact that I am ignoring them.

27. In contrast, Giere and van Fraassen's versions of the theoretical hypothesis both stipulate that the subject of the model (the *X*s) must be a "real system" (Giere 1985, 80; van Fraassen 1989, 222).

28. In March 1993 I presented versions of the first four sections of this paper in a seminar at the University of Oregon and then at the Center for History and Philosophy of Science at the University of Pittsburgh. I would like to thank those who were present on those occasions, and Wesley Salmon in particular, for valuable comments.

REFERENCES

Aristotle [4th C., B.C.] 1975. *Posterior Analytics.* Trans. J. Barnes. Oxford: Clarendon Press.

Barnes, J. 1975. "Introduction." In Aristotle, 1975.

Bhatt, B. S., and N. C. Sacheti. 1994. "Flow Past a Porous Spherical Shell Using the Brinkman Model." *Journal of Physics D* 27:37–41.

Carnap, R. 1939. *Foundations of Logic and Mathematics.* Chicago: University of Chicago Press.

Cartwright, N. 1983. *How the Laws of Physics Lie.* Oxford: Clarendon Press.

Christmas, J. R. 1988. *Fundamentals of Solid State Physics.* New York: John Wiley.

Churchland, P. M., and C. A. Hooker, eds. 1985. *Images of Science.* Chicago: University of Chicago Press.

Descartes, R. [1644] 1983. *The Principles of Philosophy.* Trans. V. R. Miller and R. P. Miller. Boston: Reidel.

———. [1619–49] 1985. *The Philosophical Writings of Descartes.* 2 vols. Trans. J. Cottingham, R. Stoothoff, and D. Murdoch. Cambridge: Cambridge University Press.

Dirac, P. A. M. 1928. "The Quantum Theory of the Electron." *Proceedings of the Royal Society A* 117:610–24.

Duhem, P. [1906] 1954. *The Aim and Structure of Physical Theory.* Trans. P. P. Wiener. Princeton: Princeton University Press.

Einstein, A. 1905. "On the Movement of Small Particles Suspended in a Stationary Liquid Demanded by the Molecular-Kinetic Theory of Heat." In *Einstein* 1956, 1–18; orig. pub. in *Annalen der Physik* 17:549–59.

———. 1906a. "A New Determination of Molecular Dimensions." In *Einstein* 1956, 36–62; orig. pub. in *Annalen der Physik* 19:289–306.

———. 1906b. "On the Theory of the Brownian Movement." In *Einstein* 1956, 19–35; orig. pub. in *Annalen der Physik* 19:371–81.

———. 1908. "The Elementary Theory of the Brownian Motion." In *Einstein* 1956, 68–85; orig. pub. in *Zeitschrift für Elektrochemie* 14:235–39.

———. 1956. *Investigations on the Theory of the Brownian Movement.* Ed. R. Furth and trans. A. D. Cowper. New York: Dover Books.

Feynman, R. P., R. B. Leighton, and M. Sands. 1965. *The Feynman Lectures on Physics.* 3 vols. Reading, Mass.: Addison-Wesley.

Feynman, R. 1967. *The Character of Physical Law.* Cambridge, Mass.: MIT Press.

Fine, A. 1986. *The Shaky Game.* Chicago: University of Chicago Press.

French, P. A., T. E. Uehling Jr., and H. K. Wettstein, eds. 1993. *Midwest Studies in Philosophy XVIII: Philosophy of Science.* Notre Dame: University of Notre Dame Press.

Giere, R. N. 1985. "Constructive Realism." In Churchland and Hooker 1985, 75–98.

Goldstein, H. 1950. *Classical Mechanics.* Reading, Mass.: Addison-Wesley.

Hiley, B. J., and F. D. Peat, eds. 1987. *Quantum Implications: Essays in Honour of David Bohm.* London: Routledge & Kegan Paul.

Hughes, R. I. G. 1989. *The Structure and Interpretation of Quantum Mechanics.* Cambridge: Harvard University Press.

———. 1993. "Theoretical Explanation." In French et al. 1993, 132–53.

Kubo, R., and T. Nagamiya, eds. 1969. *Solid State Physics.* Trans. from 2nd ed. New York: McGraw-Hill.

Kuhn, T. S. 1970. *The Structure of Scientific Revolutions.* 2nd ed. Chicago: University of Chicago Press.

Landau, L. D., and E. M. Lifshitz. 1977. *Quantum Mechanics, Non-Relativistic Theory.* 3rd ed. rev., with the assistance of L. P. Pitaevskii. Trans. J. B. Sykes and J. S. Bell. Oxford: Pergamon Press.

Laudan, L. 1981. *Science and Hypothesis.* Boston: Reidel.

Messiah, A. 1960. *Mécanique Quantique.* 2 vols. Paris: Dunod.

Nagel, E. 1961. *The Structure of Science.* New York: Harcourt, Brace & World.

Newton, I. [1730] 1952. *Opticks,* based on 4th ed. New York: Dover Books.

Pais, A. 1982. *"Subtle Is the Lord": The Science and Life of Albert Einstein.* New York: Oxford University Press.

Perrin, J. 1913. *Les Atomes.* Paris: Alcan.

Pines, D. 1987. "The Collective Description of Particle Interactions from Plasmas to the Helium Liquids." In Hiley and Peat 1987, 66–84.

Salmon, W. C. 1984. *Scientific Explanation and the Causal Structure of the World.* Princeton: Princeton University Press.

Suppe, F. 1972. "What's Wrong with the Received View of the Structure of Scientific Theories." *Philosophy of Science,* 39, 1–19.

Suppe, F., ed., 1974. *The Structure of Scientific Theories.* Urbana: University of Illinois Press.

Tarski, A. 1965. *Introduction to Logic and to the Methodology of the Deductive Sciences,* 3rd ed. New York: Oxford University Press.

Van Fraassen, B. C. 1980. *The Scientific Image.* Oxford: Clarendon Press.

———. 1989. *Laws and Symmetries.* Oxford: Clarendon Press.

Induction, Scientific Methodology, and the Philosophy of Science

12

The Continuum of Inductive Methods Revisited

Sandy L. Zabell

Departments of Mathematics and Statistics, Northwestern University

Let $X_1, X_2, X_3, \ldots$ denote a sequence of observations of a phenomenon (for example, the successive letters in an encrypted text, the successive species observed in a previously unexplored terrain, the success or failure of an experimental surgical procedure). In the classical Johnson-Carnap continuum of inductive methods (Johnson 1932, Carnap 1952), the outcomes that can occur are assumed to be of $T < \infty$ possible types or species that are known and equiprobable prior to observation. If, in a sample of n, there are n_1 outcomes of the first type, n_2 of the second, and so on, then (under appropriate conditions) the Johnson-Carnap continuum gives as the conditional epistemic probability of observing an outcome of the ith type on the $(n + 1)$-st trial the value

$$f(n_i, n) =: \frac{n_i + \alpha}{n + T\alpha} \ (\alpha > 0).$$

Note an important consequence of this: if $t < T$ species have been observed during the first n trials, then the probability of observing a new species on the next trial is a function of t and n,

$$g(t, n) =: 1 - \sum_{i=1}^{t} \frac{n_i + \alpha}{n + T\alpha} = \frac{T\alpha - t\alpha}{n + T\alpha};$$

by assumption, of course, $g(t, n) = 0$ for all $t \geq T$.

From its inception, the Johnson-Carnap continuum has been the

351

subject of considerable controversy, and a number of its limitations have been pointed out by its critics; see, for example, the discussions in Howson and Urbach 1989 and Earman 1992. Among the most important of these is the failure of the continuum to permit the confirmation of universal generalizations and its assumption that the possible types that can arise are known in advance.[1]

In this essay I discuss a new three-parameter continuum of inductive methods, discovered by the statistician Pitman, which has a number of attractive features. First, it not only permits the confirmation of universal generalizations, but its mathematical derivation reveals this to be one of its essential elements. Second, it does not assume that the species to be observed are either known in advance or limited in number. Third, it interweaves two distinct continua: the observation of a new species both establishes a new category for confirmation *and* increases the likelihood of observing further new species down the road.

Of course, it is possible to achieve these desiderata in an ad hoc fashion: the confirmation of universal generalizations can certainly be achieved by assigning point masses to the initial probabilities in the de Finetti representation (Wrinch and Jeffreys 1919); an unlimited number of categories can be accommodated if one abandons the requirement that they be epistemically symmetric (see, e.g., Zabell 1982); and there have been several proposals in the literature regarding the use of rules of succession to predict the occurrence of new species (De Morgan 1845, 414–15; Kuipers 1973; Zabell 1992).

The compelling aspect of the new system discussed here is that all three of these features emerge as a natural consequence of a new postulate: as before, it is assumed that if a species has been observed n_i times out of n in the past, then the probability of observing that species again on the next trial is a function $f(n_i, n)$ of n_i and n alone, and that the probability of observing a new species is a function $g(t, n)$ of t (the number of species observed thus far) and n, but it is *not* assumed that $g(t, n) = 0$ for t greater than some prespecified value.

This essay is divided into four parts: in the first and second, the new continuum is explained, and some of its philosophical consequences explored; in the third, some prior literature is discussed; and in the fourth, the mathematical derivation of the continuum is given.

1. Exchangeable Random Partitions

The Johnson-Carnap continuum gives probabilities for exchangeable random *sequences*; the continuum discussed here gives probabilities for exchangeable random *partitions*.[2] In brief, if the different possible species are known in advance, it is possible to state the probability of seeing a particular sequence of individuals; if the different species are not known in advance, then it is only possible to state probabilities — *prior to observation* — for events framed in terms of the first species to be encountered, the second species to be encountered, and so on. (That is, one can state before the event "the species that occurs on the first trial will also occur on the third and fourth trials," but one cannot state the event, "a giant panda will occur on the first, third, and fourth trials," unless one already knows that giant pandas exist.)

Thus, if a total of t different species are represented in a sequential sample of n individuals (observed at times $1, 2, \ldots, n$), and A_j is the set of times at which the jth species encountered is observed, then the sets $A_1, A_2, \ldots, A_t$ form a partition of the time set $\{1, 2, \ldots, n\}$; and it is to such partitions that probabilities are assigned. For example, suppose one encounters the sequence of transmitted symbols

QUOUSQUETANDEMABUTERECATALINAPATIENTIANOSTRA

There are $t = 26$ different possible letters that can occur in the sequence; there are a total of $n = 44$ letters observed in the text; the observed frequencies are $n_a = 7, n_t = 6, n_e = 5, n_i = n_n = n_u = 4, n_o = n_q = n_r = n_s = 2, n_b = n_c = n_d = n_l = n_m = n_p = 1$; and all other frequencies are zero.

Suppose, however, that a Romulan having no prior knowledge of our civilization encountered this sequence of symbols. It would not know in advance the 26 symbols in our alphabet. Thus, it notes that a total of $t = 16$ different symbols occur in the initial segment of length 44; that the first symbol encountered (the symbol "Q") occurred at positions 1 and 6 (that is, $A_1 = \{1, 6\}$); that the second symbol encountered (the symbol "U") occurred at positions 2, 4, 7, and 17 (that is, $A_2 = \{2, 4, 7, 17\}$), and so on. The sets $A_1, A_2, \ldots, A_{16}$ generate a partition $<A_1, A_2, \ldots, A_{16}>$ of the set $\{1, 2, \ldots, 44\}$. (See Table 1). The point is that — not knowing beforehand of the existence of our alphabet — a Romulan can hardly be expected to describe beforehand, let alone assign probabilities to, events such as the above 44-symbol sequence, but it is certainly possible for such a being both to describe and assign probabilities to the possible partitions (such as $<A_1, A_2, \ldots, A_{16}>$) that might arise from such a sequence.

In strictly mathematical terms, a *random partition* Π_n is a random object whose values are partitions π_n of a set $\{1, 2, \ldots, n\}$, and in the sampling of species problem it is precisely such random entities that one must consider (see Zabell 1992 for further discussion). In order to derive a continuum of inductive methods for such random partitions, some assumptions naturally have to be made concerning the underlying random structure governing their behavior. The four assumptions made here fall naturally into two classes or categories: one assumption is of a general nature, parallel to Johnson's *permutation postulate* for sequences (Johnson 1924); the other three assumptions are much more restrictive, parallel to Johnson's *sufficientness postulate* (Johnson 1932), and limit the possible random partitions that can arise to a three-parameter family. The general assumption that the random partitions Π_n be *exchangeable* is discussed in the remainder of this section; the other three assumptions are discussed in the next.

Thus, let us consider the definition of an exchangeable random partition. Consider a partition $\boldsymbol{\pi} = <A_1, A_2, \ldots, A_t>$ of the set $\{1, 2, \ldots, n\}$, and let $n_i =: n(A_i)$ denote the number of elements in the set A_i. Corresponding to $\boldsymbol{\pi}$ is the *frequency vector*

$$\mathbf{n} = \mathbf{n}(\boldsymbol{\pi}) =: <n_1, n_2, \ldots, n_t>.$$

In turn, let a_j denote the number of frequencies n_i equal to j; then corresponding to the frequency vector is the *partition vector* (or "allelic partition")

$$\mathbf{a} = \mathbf{a}(\boldsymbol{\pi}) =: <a_1, a_2, \ldots, a_n>.$$

(Because $n_j \leq n$, the number of components of the partition vector never can exceed n.) In table 12.1 the various processes of "delabeling" a sequence, partitioning the time set $\{1, 2, \ldots, n\}$ into subsets A_j, and computing the species frequencies n_i and the components a_j of the partition vector a are illustrated for the example discussed above.

The random partition Π_n is said to be an *exchangeable random partition* (the concept is due to J. F. C. Kingman) if all partitions $<A_1$,

TABLE 12.1 Delabeling

In the first column, the S_i indicates the ith symbol or species to be observed in the example; the second column records this symbol; the third column, the subset of times when this species occurs; the fourth column, the size n_i of this subset (the number of times the ith species occurs); the fifth column, a_j, the number of species that occur j times (that is, the number of times the number j occurs in the preceding column). The resulting (unordered) partition of n is conveniently summarized as $1^6 2^4 3^1 4^2 5^1 6^1 7^0 8^1$.

S_1	Q	$\{1, 6\}$	$n_1 = 2$	$a_1 = 6$
S_2	U	$\{2, 4, 7, 17\}$	$n_2 = 4$	$a_2 = 4$
S_3	O	$\{3, 40\}$	$n_3 = 2$	$a_3 = 1$
S_4	S	$\{5, 41\}$	$n_4 = 2$	$a_4 = 2$
S_5	E	$\{8, 13, 19, 21, 34\}$	$n_5 = 5$	$a_5 = 1$
S_6	T	$\{9, 18, 24, 32, 36, 42\}$	$n_6 = 6$	$a_6 = 1$
S_7	A	$\{10, 15, 23, 25, 29, 31, 38, 44\}$	$n_7 = 8$	$a_7 = 0$
S_8	N	$\{11, 28, 35, 39\}$	$n_8 = 4$	$a_8 = 1$
S_9	D	$\{12\}$	$n_9 = 1$	—
S_{10}	M	$\{14\}$	$n_{10} = 1$	—
S_{11}	B	$\{16\}$	$n_{11} = 1$	—
S_{12}	R	$\{20, 43\}$	$n_{12} = 2$	—
S_{13}	C	$\{22\}$	$n_{13} = 1$	—
S_{14}	L	$\{26\}$	$n_{14} = 1$	—
S_{15}	I	$\{27, 33, 37\}$	$n_{15} = 3$	—
S_{16}	P	$\{30\}$	$n_{16} = 1$	—

CHECK: $44 = \sum_{j=1}^{44} ja_j = 1 \times 6 + 2 \times 4 + 3 \times 1 + 4 \times 2 + 5 \times 1 + 6 \times 1 + 8 \times 1.$

$A_2, \ldots, A_t>$ having the same partition vector have the same probability; that is, if π_1 and π_2 are partitions of $(1, 2, \ldots, n\}$, then

$$\mathbf{a}(\pi_1) = \mathbf{a}(\pi_2) \Rightarrow P[\Pi_n = \pi_1] = P[\Pi_n = \pi_2].$$

In brief, the partition vector is a set of "sufficient statistics" for the random partition.[3] There is a sense in which this definition is natural; if one takes an exchangeable *sequence*, requires that its probabilities be category symmetric, and passes to the random partition it generates, then such partitions are exchangeable in the above sense. Note that the partition vector specifies an "unordered partition" of the number n; thus, an exchangeable random partition assigns the same probability to all partitions of the set $\{1, 2, \ldots, n\}$ that give rise to the same unordered partition of the number n.

Thus far we have considered one exchangeable random partition Π_n. Suppose we have an infinite sequence of them: $\Pi_1, \Pi_2, \Pi_3, \ldots,$

such that Π_n is a partition of $\{1, 2, \ldots, n\}$ for each $n \geq 1$. There is a natural sense in which such a sequence of partitions $\{\Pi_n : n \geq 1\}$ is *consistent*: if $m < n$, then the random partition Π_n of $\{1, 2, \ldots, n\}$ induces a random partition of $\{1, 2, \ldots, m\}$; and one requires that for all $m < n$ that this induced random partition (denote it $\Pi_{m,n}$) coincides with the random partition Π_m (that is, one has $\Pi_{m,n} = \Pi_m$ for all $m < n$, $1 \leq m < n < \infty$). (To be precise, given a set T and a subset S of T, the partition $<A_1, A_2, \ldots, A_t>$ of T induces the partition $<A_1 \cap S, A_2 \cap S, \ldots, A_t \cap S>$ of S. The probability of the induced partition π of $\{1, 2, \ldots, m\}$ is then the probability of all partitions π^* of $\{1, 2, \ldots, n\}$ that induce π.) If $\{\Pi_n : n \geq 1\}$ is an infinite consistent sequence of exchangeable random partitions, then the random partition Π of the integers that $\{\Pi_n : n \geq 1\}$ gives rise to is also said to be exchangeable.

The simplest examples of infinite consistent sequences of exchangeable random partitions (or *partition structures*) are Kingman's *paintbox processes* (see, e.g., Aldous 1985, 87): one runs an infinite sequence of independent and identically distributed random variables $Z_1, Z_2, \ldots, Z_n, \ldots$, and then "delabels" the sequence (that is, passes to the partitions $\Pi_1, \Pi_2, \ldots, \Pi_n, \ldots$ of the time set that such a sequence gives rise to).[4]

There is a rich theory here: Just as the general infinite exchangeable sequence is formed from a mixture of independent and identically distributed sequences of random variables (this is the celebrated *de Finetti representation theorem*), the general infinite exchangeable random partition is a mixture of paintbox processes (this is the *Kingman representation theorem*). Indeed, every element of the classical theory of inductive inference for exchangeable random sequences has a counterpart in the theory of exchangeable random partitions (see Zabell 1992). But this powerful mathematical machinery is not needed here; just as in the classical Johnson-Carnap approach, it is possible in special cases to deduce directly from simple postulates the predictive probabilities of the partition structure characterized by those postulates; and this is the approach taken below.

This approach, although it has the twin merits of expository simplicity and philosophical clarity, reverses the actual process of historical discovery. The two-parameter family of partitions structures $\Pi_{\alpha,\theta}$ discussed in the next section were not discovered via their predictive probabilities. The Berkeley statistician Jim Pitman, working instead from the perspective of the Kingman representation, originally discovered them via their "residual allocation model" (RAM) characterization, derived their predictive probabilities from the RAM, and then suggested to me that it might be possible to characterize them by the form of their predictive probabilities (along the lines discussed in Zabell 1982). This paper states and proves a sharply formulated version of Pitman's conjecture.

It turns out that in the simplest such characterization it is necessary to add in a component corresponding to the confirmation of universal generalizations. The result is, therefore, a pleasant surprise: a simple three-parameter continuum of inductive methods that meets a fundamental objection to Carnap's original continuum for random sequences. The corresponding three-parameter family of partition structures already appears in Pitman's work: Corollary 3 in Pitman (1992b) characterizes the distribution of the "size-biased permutation" of the atoms in the Kingman representation of exactly such structures. The fact that this three-parameter family admits of two very different characterizations, both natural in their own right, is perhaps not without mathematical and philosophical significance.

In sum, the new continuum proposed here does *not* assume that the possible categories or species are known in advance. Successive observations are made, from time to time new species are encountered, and at each stage the number of outcomes so far noted in each category is recorded. At any given time n a total of t species have been encountered, and the number of outcomes that fall into each category is summarized by the partition vector $\mathbf{a} =: <a_1, a_2, \ldots, a_n>$, where a_1 is the number of species that appear once in the sample, a_2 is the number of species that appear twice in the sample, and so on.

In technical terms, it is assumed in the new continuum that *the probabilities that describe such a process give rise to an infinite exchangeable random partition.*

2. The New Continuum

Consider the problem of predicting the next outcome, given a sample of size n, for an infinite exchangeable random partition. The first assumption (1) made is:

(1) $P[\Pi_n = \pi_n] > 0$ for all partitions π_n of $\{1, 2, \ldots, n\}$;

that is, it no particular species scenario is ruled out or deemed, a priori, to be impossible.

Let $Z_{n+1} \in S_i$ denote the event that the $(n + 1)$st individual to be observed turns out to be a member of the i-th species to have already occurred. The second assumption (2) made is:

(2) $P[Z_{n+1} \in S_i \mid <n_1, n_2, \ldots, n_t>] = f(n_i, n),\qquad 1 \leq i \leq t.$

That is, the *predictive probability* of observing the ith species on the next trial depends only on the number n_i of that species thus far observed, and the total sample size n. (Note the function $f(n_i, n)$ does not depend on the species i.)[5]

The third and final assumption (3) made is:

(3) $P[Z_{n+1} \in S_{t+1} \mid <n_1, n_2, \ldots, n_t>] = g(t, n).$

That is, the probability of observing a new species (since t species have been observed to date, this is necessarily the $(t + 1)$st to be observed), is a function of the number of species thus far observed and the total sample size. It is a remarkable fact (proved in section 4) that if just these three conditions are imposed for all $n \geq 1$, then the functions $f(n_i, n)$ and $g(t, n)$ must be members of a three-dimensional continuum having parameters θ, α, and γ:

The Continuum of Inductive Methods for the Sampling of Species

Case 1: Suppose $n_i < n$ (and therefore $t > 1$; the universal generalization is *disconfirmed*). Then

$$f(n_i, n) = \frac{n_i - \alpha}{n + \theta}; \qquad g(t, n) = \frac{t\alpha + \theta}{n + \theta}.$$

Case 2: Suppose $n_i = n$ (and therefore $t = 1$; the universal generalization is confirmed). Then

$$f(n_i, n) = \boxed{\frac{n - \alpha}{n + \theta}} + \boxed{c_n(\gamma)} \qquad g(t, n) = \boxed{\frac{\alpha + \theta}{n + \theta}} - \boxed{c_n(\gamma)}$$

Increment due to confirmation of universal generalization:

Here $c_n(\gamma) = \dfrac{\gamma(\alpha + \theta)}{(n + \theta)\left[\gamma + (\alpha + \theta - \gamma)\prod\limits_{j=1}^{n-1}\left(\dfrac{j - \alpha}{j + \theta}\right)\right]}$

The predictive probabilities in Case 1 are precisely the ones that Pitman (1992c) derived from the $\Pi_{\alpha,\theta}$ process. The numbers $c_n(\gamma)$ in Case 2 represent adjustments to these probabilities that arise when only one species is observed; in the present language, of seeing the partition $<A_1>$ consisting of the single set $A_1 = \{1, 2, 3, 4, \ldots, n\}$. The parameter γ is related to the prior probability ϵ that only one species is observed in an infinite sequence of trials; it is shown in the last section, "Derivation of the Continuum," that $\gamma = (\alpha + \theta)\epsilon$.[6]

If an infinite exchangeable random partition $\Pi = \{\Pi_1, \Pi_2, \Pi_3, \ldots\}$ satisfies Assumptions (1)–(3), then its predictive probabilities must have the above form for *some* α, θ, and γ. Not all values of α, θ, and γ are possible however. The question therefore arises as to which values of the parameters α, θ, and γ can actually be realized by an exchangeable random partition. (Of course, if such a random partition exists, it is necessarily unique because it can be generated by the given predictive probabilities). It is not difficult to prove that such exchangeable random partitions exist in precisely the following cases:

Range of possible parameter values in the continuum:

$$0 \leq \alpha < 1; \theta > -\alpha; \text{ and } 0 \leq \gamma < \alpha + \theta.$$

This corresponds to the new inductive continuum, discussed here.[7]

The subfamily of partitions $\Pi_{\alpha,\theta}$ that arise in the special case when $\gamma = 0$ are, as mentioned at the end of the previous section, the discovery of the Berkeley statistician Jim Pitman (1992a–d), who has extensively studied their properties; for this reason, it is referred to below as the *Pitman family* of infinite exchangeable random partitions.[8] (Other values of α and θ are possible if one either relaxes Assumption (1) or does not require the exchangeable random partition to be infinite; for simplicity, these cases are not discussed here.) The important class of partition structures that arise in the further special case $\alpha = 0$ had been discovered earlier by Warren Ewens (see Zabell 1992).

The two parameters α and θ can be interpreted as follows. The parameter θ is related to the a priori probability of observing a new species; the parameter α is the effect of subsequent observation on this likelihood. Given a new species, the information corresponding to the *first* observation of this species plays two roles and is divided into two parts. First, the observation is of a new species; thus, it gives us some reason to think that subsequent observations will continue to give rise to new species, and it contributes the term $\dfrac{\alpha}{n + \theta}$ to $g(t, n)$. Second, the observation is of a particular species; thus the observation also gives us some reason to think that the species in question will continue to be observed in the future, and it contributes the term $\dfrac{1 - \alpha}{n + \theta}$ to $f(n_i, n)$. In contrast, the parameter θ is related to the a priori likelihood of observing a new species, because no matter how many species have been observed, there is always a contribution of $\dfrac{\theta}{n + \theta}$ to $g(t, n)$.[9]

There is a simple urn model that describes the new continuum in the case $\gamma = 0$ and $\theta > 0$. Imagine an urn containing both colored and black balls, from which balls are both drawn and then replaced. Each time a colored ball is selected, it is put back into the urn, together with a new ball of the same color having a unit weight; each time a black ball is selected, it is put back into the urn together with two new balls, one black and having weight α ($0 \leq \alpha < 1$), and one of a new color, having weight $1 - \alpha$.

Initially the urn contains a single black ball (the *mutator*) having a weight of θ. Balls are then successively drawn from the urn; the probability of a particular ball being drawn at any stage is proportional to its selection weight. (The mutator is always selected on the first trial.) It is not difficult to see that the predictive probabilities at a given stage n are those of the Pitman $\mathbf{\Pi}_{\alpha,\theta}$ process. If the special value $\alpha = 0$ is chosen, the resulting urn model reduces to that of the Hoppe urn (Hoppe 1984), and the predictive probabilities are those of the Ewens family $\mathbf{\Pi}_{\theta} =: \mathbf{\Pi}_{0,\theta}$.[10]

Remarks In the Johnson-Carnap setting (prior known categories), symmetry among categories is purchased at the price of assuming that the total number of possible species $t < \infty$ (because $P[X_1 = x] = \dfrac{1}{t} > 0$, t must be finite). This does not happen in the case of the new continuum; it is possible for the number of species to be unbounded because no distinction is made between species at the time of the first observation. (Of course, *once* the first individual is observed, its species is then identified to the extent that subsequent individuals are classified as either LIKE or UNLIKE.)

In the Johnson-Carnap continuum, the confirmation of universal generalizations is not possible: if it were, then the predictive probability $f(0, n)$ for a known but unobserved category would depend on whether the observed number of species was greater than one. (That is, if $t = 1$ and $n_i = n$ for some cell i). But in the new continuum, *there are no* "unobserved" categories; thus, the confirmation of universal generalizations becomes possible.

Given a sample of size n, the probability that an old species occurs on the next trial is $1 - \dfrac{t\alpha + \theta}{n + \theta} = \dfrac{n - t\alpha}{n + \theta}$; thus the conditional probability, given that an old species occurs, that it is of type i is $\boxed{\dfrac{n_i - \alpha}{n - t\alpha}}$. It is interesting to note that this is of the same form as the Carnap continuum, but uses a negative value for the parameter α. Such predictive probabilities can arise in the classical setting in the case of a finite sequence (see, e.g., Zabell 1982) and also arise in Kuipers's work (see Kuipers 1978, chap. 6). Note also that in the special case $\alpha = 0$ the predictive probabilities reduce to Carnap's *straight rule*.

In the Pitman family $\Pi_{\alpha,\theta}$ the number of observed species $t \to \infty$ as the sample size $n \to \infty$ (see my final remark in the last section of this essay). To avoid this, it is necessary to relax Assumption 1, that all cylinder sets have positive probability. To simplify the discussion, such questions are not examined in this paper.[11]

There are (at least) two very different ways in which both the classical Johnson-Carnap continuum and the new continuum proposed here may be viewed. One of these, initially that of Carnap (his later views were more complex), is that the probabilities in question are *objective*: that is, the three assumptions enunciated at the beginning of this section accurately capture the concept of ignorance regarding possible outcomes, and the continuum, therefore, gives the logical probabilities, credibilities (to use Russell's terminology), or rational degrees of belief regarding those possible outcomes. The other, polar extreme is to view the probabilities in question as *subjective* or personal; this is the view of the present author. In this case the three assumptions may be regarded as possible descriptors of our current epistemic state. If the three *qualitative* assumptions do in fact accurately describe our actual degrees of belief, then the force of the result is that our *quantitative* personal probabilities are uniquely determined up to three parameters.[12]

The characterization given in this section completes a parallelism in

TABLE 12.2 Two Theories of Inductive Inference

	Multinomial Sampling	Sampling of Species
Types	t types $\{S_1, S_2, \ldots, S_t\}$	initially unknown
Sample	random sequence $X = (X_1, X_2, \ldots, X_n)$	random partition Π of $\{1, 2, \ldots, n\}$
Sufficient statistics	sample frequencies $n = (n_1, n_2, \ldots, n_t)$	allelic partition $a = (a_1, a_2, \ldots, a_n)$
Exchangeability	$n(x_1) = n(x_2) \Rightarrow P[X = x_1] = P[X = x_2]$	$a(\pi_1) = a(\pi_2) \Rightarrow P[\Pi = \pi_1] = P[\Pi = \pi_2]$
Representation theorem	De Finetti representation theorem	Kingman representation theorem
Atomic constituents	i.i.d. sequences	paintbox processes
Canonical processes	Dirichlet priors	Pitman family $\Pi_{\alpha,\theta}$
Urn model	Polya urn	Hoppe urn ($\alpha = 0$)
Sampling formula	Bose-Einstein statistics ($\alpha = 1$)	Ewens sampling formula ($\alpha = 0$)
Predictive probabilities	$\dfrac{n_1 + \alpha}{n + t\alpha}$ ($\alpha > 0$)	$\dfrac{n_i - \alpha}{n + \theta}, \dfrac{t\alpha + \theta}{n + \theta}$ ($0 \leq \alpha < 1, \theta > -\alpha; t > 1$)
Characterization of	$f(n_1, n)$; W. E. Johnson (1932)	RAM characterization; Pitman (1992b). $f(n_1, n), g(t, n)$; see Section 4.

the theories of inductive inference for multinomial sampling and the sampling of species problem discussed in Zabell 1992 (see table 12.2).

3. Prior Philosophical Literature

The continuum of inductive methods discussed here has the advantage that it simultaneously meets two of the most important objections to the original Johnson-Carnap continuum: that the categories are not empirical in origin, and that universal generalizations cannot be confirmed. Both of these objections are in fact quite old, and in this section some of the past attempts and criticisms that have been made concerning these two points are discussed.

Richard Price and His Appendix

It is interesting to note that perhaps the first discussion concerning the origin of the categories used in the probabilistic analysis of inductive inference goes back to Bayes's original essay (1764) or, more precisely, to an appendix to that essay penned by Bayes's friend, intellectual executor, and fellow dissenting Presbyterian clergyman, the Reverend Dr. Richard Price (1723–1791).

Bayes had considered "an event concerning the probability of which we absolutely know nothing antecedently to any trials made concerning it" (1764, 143). At the heart of Bayes's analysis is his famous (or infamous) postulate that in such cases all values of the probability p of such an event are equilikely. Given this assumption, it is easy to see that the chance that p falls between the limits a and b,[13] given the further information that the event has occurred n times in unfailing succession, is

$$P[a < p < b] = (n + 1) \int_a^b p^n \, dp = (n + 1)\left[\frac{b^{n+1} - a^{n+1}}{n + 1}\right]$$

$$= b^{n+1} - a^{n+1}.$$

Gillies (1987, 332) terms this *Price's rule of succession*, to distinguish it from the usual *Laplace rule of succession*: if there have been k successes in n trials, then the probability of a success on another trial is $\frac{k + 1}{n + 2}$. In particular the chance that the probability p lies between $\frac{1}{2}$ and 1 is $1 - \left(\frac{1}{2}\right)^{n + 1}$; in other words, the odds in favor of this are

$2^{n+1} - 1$ to 1. Price's rule of succession is an immediate consequence of Bayes's results; his intellectual executor Price carried the analysis further in an appendix to Bayes's essay. Curiously, Price's appendix is often neglected in discussions of Bayes's essay, despite its great interest for students of inductive inference; two notable (and excellent) exceptions are Gillies 1987 and Earman 1992, chap. 1.

Price's analysis consists of several stages. In these the roles of the first and second observations play key roles. Price begins by considering the case of an uninterrupted string of successes, "a given number of experiments which are unopposed by contrary experiments." First, Price argues, prior to observation, all possible outcomes must have infinitesimal probability:

Suppose a solid or die of whose number of sides and constitution we know nothing; and that we are to judge of these from experiments made in throwing it.

In this case, it should be observed, that it would be in the highest degree improbable that the solid should, in the first trial, turn any one side which could be assigned beforehand; because it would be known that some side it must turn, and that there was an infinity of other sides, or sides otherwise marked, which it was equally likely that it should turn.

A little further on, Price adds:

I have made these observations chiefly because they are all strictly applicable to the events and appearances of nature. Antecedently to all experience, it would be improbable as infinite to one, that any particular event, beforehand imagined, should follow the application of any one natural object to another; because there would be an equal chance for any one of an infinity of other events. (Bayes 1764, append.)

There are already several interesting issues that arise in these passages. First, we have the resort to the "urn of nature": the argument is initially framed in terms of an objective chance mechanism (here the many-sided die), followed by the assertion that the analysis is "strictly applicable to the events and appearances of nature."[14] Note also Price's repeated emphasis that the events in question must be ones that can come to mind *prior* to observation: the side of the die is one which "could be assigned beforehand"; the event in nature is one which could be "beforehand imagined."

Such introspection reveals that there is an infinite spectrum of equipossible outcomes: the die might have an "infinity of other sides," each of which is "equally likely"; there is "an equal chance for any one of an infinity of other events." But how can an infinite number of such events be equally likely? Price avoids saying that such events have zero probability: they are instead "in the highest degree improbable" or "improbable as infinite to one." Perhaps he thought that if an event has zero probability, then it is impossible; perhaps he even thought of the probabilities at issue as being instead infinitesimals.[15]

Price argues that the first observation of a species or type has a special significance:

The first throw [of the solid or die] only shews that *it has* the side then thrown, without giving any reason to think that it has it any one number of times than any other. It will appear, therefore, that *after* the first throw and not before, we should be in the circumstances required by the conditions of the present problem, and that the whole effect of this throw would be to bring us into these circumstances. That is: the turning the side first thrown in any subsequent single trial would be an event about the probability or improbability of which we could form no judgment, and of which we should know no more than it lay somewhere between nothing and certainty. With the second trial our calculations must begin; and if in that trial the supposed solid turns again the same side, there will arise the probability of three to one that it has more of that sort of sides than of *all* others. (Ibid.)

Thus — according to Price — the first observation of an event results in a belief change of a very non-"Bayesian" type indeed! The observation of the first event transforms the status of its probability in future trials from the known — but infinitesimal — to the unknown but finite; the probability lies "somewhere between nothing and certainty."

This point of Price links up with a recurring issue in later debates on probability and induction. In the subjective or personalist system of Ramsey and de Finetti, the status of initial probabilities is straightforward: such probabilities summarize our present knowledge prior to the receipt of further information.[16] The subjective account does not tell us what these initial probabilities should be, nor does it provide a mechanism for arriving at them. It is just a theory of consistency, plain and simple. But if one hopes instead for more, for a theory of probability as a unique system of rational degrees of belief, then it is natural to demand the basis for the initial probabilities that are used. In a purely Bayesian framework, however, these must come from other, earlier initial probabilities; these must in turn come from other, still earlier initial probabilities, *und so weiter*; it is "turtles all the way down." The usual move to avoid such an infinite regress is to assume that at some stage one reaches a state of "total ignorance" and then pass to a uniform prior on a set of alternatives or parameters by appealing to the so-called principle of insufficient reason. But this is absurd: our ability to even *describe* an event in our language (or understand the meaning of a term used to denote an event) already implies knowledge — considerable knowledge — about that event.

The argument that Bayes employs instead to justify his choice of the uniform prior is much more subtle. His concern is "an event concerning the probability of which we absolutely know nothing antecedently to any trials made concerning it." That is, our ignorance pertains not to a knowledge of the circumstances of the event itself, but to its *probability*. It is sometimes thought that at this point Bayes then immediately passes to a uniform prior on the probability p. But, in fact (in an often overlooked scholium), he argues that in cases of absolute ignorance, if S_n denotes the number of times the event has thus far occurred in n trials, then

$$P[S_n = k] = \frac{1}{n+1} \text{ for } 0 \leq k \leq n.$$

This is in effect a precise quantitative translation of the informal qualitative formulation that we "absolutely know nothing antecedently to any trials made concerning" the probability of the event. It then follows as a direct mathematical consequence of this assumption (for all $n \geq 1$) that the prior distribution for the unknown probability p must be the uniform distribution (although Bayes did not himself so argue; see Zabell 1988).

Price takes Bayes's argument one step further. He asks: Just when are we in such a state of ignorance concerning the probability of an event? How can we pass from our sense impressions, a knowledge of prior events experienced (this happened and this did not), to an ignorance of their probabilities? Price argues that Bayes's formula is only applicable to types that have already been observed to occur at least once; and that the correct value of n to use in the formula is then is *one less* than the total number of times that that type has thus far been observed to occur. (To check this, note that Price asserts the odds to be 3 to 1 after the same side turns up a *second* time; this corresponds to taking $n = 1$ in Price's rule of succession.)

From the vantage point of the new continuum, this corresponds to using the value $\alpha = 1$ in the continuum; that is, the *entire* weight of the first observation is given over to the prediction of further new species to be observed, and none is given to the prediction that the particular type observed will recur. There is, however, an obvious problem here. If $\alpha = 1$, then the probability of observing a second member of a species, given that one has already been observed, is

$$\frac{n_i - \alpha}{n + \theta} = \frac{1 - 1}{1 + \theta} = 0.$$

(As a result, the value $\alpha = 1$ partitions the set $\{1, 2, \ldots, n\}$ into the n singleton sets $\{1\}, \{2\}, \ldots, \{n\}$.) This difficulty does not arise in the new continuum proposed here, due to the constraint $\alpha < 1$; Price in effect circumvents such difficulties by forbidding the computation of probabilities at the first stage. Thus for Price it is only with "the second trial our calculations must begin"; and it is the observation of a *second* member of the species that permits calculation (for now $n_i - \alpha = 2 - 1 = 1$, and the predictive probabilities do not vanish).

Of course such an apparently ad hoc procedure requires justification. Price argues:

The first experiment supposed to be ever made on any natural object would only inform us of one event that may follow a particular change in the circumstances of those objects; but it would not suggest to us any ideas of uniformity in nature, or give us the least reason to apprehend that it was, in that instance or in any other, regular rather than irregular in its operations. (Ibid.)

This statement also has a natural interpretation within the context of exchangeable random partitions. The support of a paintbox process has both a discrete and a continuous component: The discrete component corresponds to the different species that occur (and recur) with positive probability; the continuous component corresponds to the *hapax legomena*, the species that occur once and then disappear, never to be seen again. In effect Price is saying that the first observation of a species tells us only that it lies in the support of the underlying process, but not whether it lies in its discrete component (and hence is "regular in its operations") or in its continuous component (and hence is "irregular").

Price illustrates the process of inductive inference in the case of natural phenomena by a curious hypothetical:

Let us imagine to ourselves the case of a person just brought forth into this world, and left to collect from his observation of the order and course of events what powers and causes take place in it. The Sun would, probably, be the first object that would engage his attention; but after losing sight of it the first night he would be entirely ignorant whether he should ever see it again. He would therefore be in the condition of a person making a first experiment about an event entirely unknown to him. But let him see a second appearance or one *return* of the Sun, and an expectation would be raised in him of a second return. . . . But no finite number of returns would be sufficient to produce absolute or physical certainty. (Ibid.)

This is, in fact, a direct attack on Hume. To see the close relation, consider the corresponding passage from Hume's *Enquiry Concerning Human Understanding*:

Suppose a person, though endowed with the strongest faculties of reason and reflection, to be brought on a sudden into this world; he would, indeed, immediately observe a continual succession of objects, and one event following another; but he would not be able to discover anything further. He would not, at first, by any reasoning, be able to reach the idea of cause and effect. (Hume 1748, 42; see also 27)

The image that Hume conjures up of a philosophical Adam first experiencing the sights and sounds of nature soon became a commonplace of the Enlightenment: It later appears in one form or another in Buffon's *Histoire naturelle de l'homme* of 1749, Diderot's *Lettre sur les sourds et muets* of 1751, Condillac's *Traité des sensations* of 1754, and Bonnet's *Essai de psychologie* of 1754 and *Essai analytique sur les facultés de l'âme* of 1760.[17] (Readers of Mary Shelley's *Frankenstein* [1818] will also recognize here the origin of the opening lines of part 2, chap. 3 of that book, when Frankenstein's creation first awakens to see the sun.)

Despite these later discussions (concerned primarily with issues in associationist psychology), it is clear that Price has Hume in mind: His discussion is a point-by-point attack on Hume's skeptical philosophical stance.[18] Hume denies that experience can give (immediate) knowledge of cause and effect; Price believes that the calculus of probabilities provides a tool that enables us to see how a person can "collect from his observation of the order and course of events" in the world, the "causes take place in it." Nor is this the only question at issue. In his *Treatise on Human Understanding*, Hume had written,

In common discourse we readily affirm, that many arguments from causation exceed probability, and may be receiv'd as a superior kind of evidence. One wou'd appear ridiculous, who wou'd say, that 'tis only probable the sun will rise to-morrow, or that all men must dye; tho' 'tis plain we have no further assurance of these facts, than what experience affords us. (Hume 1739, 124)[19]

Using Hume's own example of the rising of the sun, Price argues that in the case of uniform experience,

instead of proving that events will *always* happen agreeably to [uniform experience], there will be always reason against this conclusion. In other words, where the course of nature has been the most constant, we can have only reason to reckon upon a recurrence of events proportioned to the degree of this constancy; but we can have no reason for thinking that there are no causes in nature which will *ever* interfere with the operations of the causes from which this constancy is derived, or no circumstances of the world in which it will fail. (Bayes 1764, append.)

Thus Price argues that one can never achieve certitude regarding a *single* outcome on the basis of the finite experience at our disposal. (This should be distinguished from the assertion that it is unreasonable to confirm a universal generalization on the basis of a finite segment of experience.)

Augustus De Morgan

The one other classical student of the calculus of probabilities who appears to have considered the question of the origin of the categories used in inductive inference is the English mathematician Augustus De Morgan (1806–1871). (De Morgan was the leading enthusiast of Laplacean probability in England during the first half of the nineteenth century.) But while Price had, in effect, advocated the use of the parameter $\alpha = 1$, De Morgan introduced a different approach, corresponding to a choice of $\theta = 1$!

In a lengthy encyclopedia article that is in large part an exposition of Laplace's *Théorie analytique des probabilités* of 1812, De Morgan concludes his discussion of Laplace's rule of succession by noting:

There remains, however, an important case not yet considered; suppose that having obtained t sorts in n drawings, and t sorts only, we do not take it for granted that these are all the possible cases, but allow ourselves to imagine there may be sorts not yet come out. (1845, 414)[20]

In his little book *An Essay on Probabilities*, De Morgan gives a simple illustration of how he believes one can deal with this problem:

When it is known beforehand that either A or B *must* happen, and out of $m + n$ times A has happened n times, and B n times, then . . . it is $m + 1$ to $n + 1$ that A will happen the next time. But suppose we have no reason, except what we gather from the observed event, to know that A or B must happen; that is, suppose C or D, or E, &c. might have happened: then the next event might be A or B, or a new species, of which it can be found that the respective probabilities are proportional to $m + 1, n + 1$, and 1; so that though the odds remain $m + 1$ to $n + 1$ for A rather than B, yet it is now $m + 1$ to $n + 2$ for A against either B or the other event. (De Morgan 1838, 66)

De Morgan's prescription can be understood in terms of a Hoppe urn model in which initially there are three balls, one labeled "A," one "B," and one "black," the *mutator*. Balls are then selected from the urn as follows: balls labeled "A" or "B" (or any other letter or symbol) are replaced together with another of the same label; if the mutator is selected, then it is replaced by a ball labeled by a new letter or symbol not yet encountered. The resulting exchangeable random partition corresponds to a conditional Pitman process (after the observation of one "A" and one "B"), having parameters $\alpha = 0$ and $\theta = 1$.[21]

Recent Literature

It is an important historical footnote that Carnap thought that the sampling of species problem could be dealt with by introducing a predicate relation R: IS THE SAME SPECIES AS. But correctly recognizing the considerable increase in complexity this would introduce into the problem, Carnap did not pursue this idea further. (The information in this paragraph is due to Richard Jeffrey.)

There have been few attempts in the recent philosophical literature to deal with such problems since Carnap; this is not entirely surprising if one accepts the basic thesis of this paper, that the machinery of exchangeable random partitions is crucial in coming to grips with them. Hintikka 1966 and Hintikka and Niiniluoto 1980 consider cases where the predictive probabilities depend, not just on the sample size n and number of instantiations k, but also the number of species observed. Their results assume, however, that the total spectrum of possible species is both known and finite; for further information, see Kuipers 1978, which contains a careful and detailed analysis of these systems.

(It would be interesting to derive results parallel to theirs for exchangeable random partitions.)

Kuipers himself dealt with the problem in an early paper (1973). Kuipers's proposal interweaves two continua. One is binomial: on each trial, a new species does or does not occur with probability $\dfrac{t + \lambda}{n + 2\lambda}$; the other is multinomial: conditional on a new species not occurring, if k instances of a species have already been observed, then the probability that the species recurs is $\dfrac{k + \mu}{n + t\mu}$. (Note that the machinery of random partitions is implicit in Kuipers's proposal: a random partition, rather than a random sequence, is generated because the character of the new species to appear is not stated; it is just a new species.) Unfortunately, the random partitions so generated are not exchangeable: the probability of seeing an old species and then a new species does not equal the probability of seeing a new species and then an old one.

Kuipers (1978, chap. 6, sec. 10) also considered the possibility of extending Hintikka's results to the case of an infinite number of alternatives, and discusses what he describes as a "reformulation of an H-system." This turns out to be nothing other than the "delabeling" process described earlier.[22] Because delabeling is equivalent to passing to the underlying partition generated by a sequence, Kuipers's approach is equivalent to viewing matters from the perspective of random partitions. It is of particular interest to note that Kuipers proposes as the predictive probability of a new species, given n observations to date *and supposing that an infinite number of species are ultimately observed*, $\dfrac{\lambda_\infty}{n + \lambda_\infty}, 0 \le \lambda_\infty < \infty$. These conditional probabilities correspond in the present continuum to the special case $\alpha = 0$ (and $\theta = \lambda_\infty$), that is, the *Ewens subfamily*.

Confirmation of Universal Generalizations

The failure of Carnap's original continuum to confirm universal generalizations is too well known to require any but the briefest discussion here. (It is perhaps worth noting, however, that the issue itself is actually quite old [see Zabell 1989, 308–09 for a number of references prior to Popper and going back to the nineteenth century].) Barker (1957, 84–90) summarizes a number of the early objections. Popper's primary assault (1959, 363–77) was effectively rebutted by Howson 1973; other later criticisms include Essler 1975. There are a number of good discussions that can serve as entries into this literature; these include Kuipers (1978, 96–99) and Earman (1992, 87–95).

4. Derivation of the Continuum

Let us begin by using Johnson's original argument (Johnson 1932; Zabell 1982; Costantini and Galavotti 1987) and see where it leads, *assuming (as we do throughout this section) that the three postulates of the second section, "The New Continuum," hold.* The first step in the argument is to prove that for each $n \geq 1$, $f(n_i, n)$ is linear in n_i. This turns out to be nearly true here too. Recall that $\delta_n(k) = 0$ for $1 \leq k \leq n$, and $\delta_n(n) = 1$.

Lemma 1: For each $n \geq 1$, there exist constants a_n, b_n, and c_n such that

$$f(k, n) = a_n + b_n k + c_n \delta_n(k)$$

for $1 \leq k \leq n$.

Proof: If $n = 1$, 2, or 3, it is immediate that the desired equation holds for a suitable choice of coefficients a_n, b_n, and c_n (since the number of constraints is at most three). Thus, we may assume $n \geq 4$. Let $f(n_i, n) = \mathrm{f}(n_i)$ and $g(t, n) = g(t)$. Choose $c_n =: f(n) - f(n - 1)$. It suffices to prove

$$f(n_i + 1) - f(n_i) = f(n_i) - f(n_i - 1), \quad 1 < n_i < n - 1.$$

Suppose that at stage n there are t species, and that the frequency count is $(n_1, n_2, \ldots, n_t)$. Consider the ith species, and suppose that $1 < n_i < n - 1$. Because $n_i < n$, there exists at least one other species j; suppose that $n_j > 1$ (this is possible because $n_i > 1$). Then one has:

$$f(n_i) + f(n_j) + \sum_{k \neq i;j} f(n_k) + g(t) = 1.$$

Because $n_i > 1$, one can remove an individual from species i without extinguishing the species, and use it to create a new species; one then has for the resulting partition:

$$f(n_i - 1) + f(n_j) + \sum_{k \neq i;j} f(n_k) + f(1) + g(t + 1) = 1.$$

Equating the two and subtracting then gives:

$$f(n_i) - f(n_i - 1) = g(t + 1) - g(t) + f(1).$$

Likewise, by taking one element from j and creating a new species, we get

$$f(n_j) - f(n_j - 1) = g(t + 1) - g(t) - f(1).$$

Finally, take one from j, and put it into i; then

$$f(n_i + 1) + f(n_j - 1) + \sum_{k \neq i;j} f(n_k) + g(t) = 1,$$

and

$$f(n_i + 1) - f(n_i) = f(n_j) - f(n_j - 1),$$

hence

$$f(n_i + 1) - f(n_i) = f(n_i) - f(n_i - 1).$$

This concludes the proof.

Next, let us consider the effect of the sample size n. Suppose $b_n = 0$; then $f(n_i, n)$ is independent of n_i (except in the case of a universal generalization). In the Johnson-Carnap setting, one separates out such possibilities by showing that if b_n vanishes at a single stage n, then it vanishes at *all* stages. In the present setting, however, this never happens.

Lemma 2: For all $n \geq 3$, $b_n \neq 0$.

Proof: Consider a given sequence of observations up to stage n. Because the random partition is exchangeable, the observation of an old species once more at stage $n + 1$ and then a new species at stage $n + 2$ generates a partition having the same probability as the partition that arises from observing the new species at stage $n + 1$ and the old at stage $n + 2$. Suppose that $b = 0$. If $n_i < n$, then $t \geq 2$ (at stage n) and

$$(1 - ta_n)(a_{n+1} + b_{n+1}n_i) = a_n(1 - ta_{n+1} - b_{n+1}(n + 1)),$$

hence

$$(1 - ta_n)b_{n+1}n_i = a_n(1 - b_{n+1}(n + 1)) - a_{n+1};$$

thus $(1 - ta_n)b_{n+1}n_i$ is constant (as a function of n_i) for n_i in the range $1 \leq n_i < n$. It thus follows that if $n \geq 3$ (so that both $n_i = 1$ and $n_i = 2$ are possible), then either $1 - ta_n = 0$ or $b_{n+1} = 0$. If $1 - ta_n = 0$, then $a_n = t^{-1}$. But if $n \geq 3$, then both $t = 2$ and $t = 3$ are possible. Thus $a_n = t^{-1}$ is impossible (since a_n is a constant), hence $b_{n+1} = 0$. Thus: if $b_n = 0$ and $n \geq 3$, then $b_{n+1} = 0$. (Note that $b_n = 0 \to b_{n+1} = 0$ need not hold for $n = 1$ or 2, because b_n is not uniquely determined in these two cases.)

It immediately follows that if $b_n = 0$ for $n = n_0 \geq 3$, then $b_n = 0$ for all $n \geq n_0$. But it is easy to see that this cannot happen. For if $b_n = 0$, then $f(n_i, n) = a_n$ and $g(t, n) = 1 - t_n a_n$ for $t \geq 2$. Then, arguing as before, we see that

$$(1 - t_n a_n)a_{n+1} = a_n(1 - t_n a_{n+1}) \to a_{n+1} = a_n.$$

Thus if b_n vanishes from some point on, then a_n is constant from this point on. Let a denote the resulting common value of a_n, $n \geq n_0$. Since $a > 0$, it follows that $na > 1$ for n large. But this is impossible, because $1 - na = g(n, n) > 0$. Thus $b_n \neq 0$ for all $n \geq 3$.

Because b_n does not vanish, one can normalize a_n and c_n relative to it; thus let

$$a_n =: -\frac{a_n}{b_n}, \quad \gamma n =: \frac{c_n}{b_n}, \text{ and } \theta_n =: \frac{g(t_n, n)}{b_n} + t_n\frac{a_n}{b_n}.$$

Since

$$ta_n + b_n n + g(t, n) = \sum_{i=1}^{t}(a_n + b_n n_i) + g(t, n) = 1 \text{ for } t \geq 2,$$

$$b_n^{-1} = n + t\frac{a_n}{b_n} + \frac{g(t, n)}{b_n} = n + \theta_n,$$

hence

$$f(n_i, n) = a_n + b_n n_i + c_n \delta_n(n_i) = \frac{n_i + \dfrac{a_n}{b_n} + \dfrac{c_n}{b_n} \delta_n(n_i)}{b_n^{-1}}$$

$$= \frac{n_i - \alpha_n + \gamma_n \delta_n(n_i)}{n + \theta_n},$$

$$g(t, n) = 1 - t a_n - b_n n - c_n \delta_1(t) = \frac{t \alpha_n + \theta_n - \gamma_n \delta_1(t)}{n + \theta_n}.$$

Such normalization is also possible even in the special cases $n = 1$ and $n = 2$ not covered by Lemma 2: if $n = 1$, the probability of observing at the second trial the same species as on the first is $a_1 + b_1 + c_1$; if $n = 2$ the probabilities of observing a species, given it has been observed once or twice are $a_2 + b_2$ and $a_2 + 2b_2 + c_1$; and it is clear that in both cases we are free to choose b_1 and b_2 so that neither vanishes.

Lemma 3: For $n \geq 3$, α_n does not depend on n and θ_n does not depend on either n or t.

Proof: Step 1: $\alpha_n = \alpha_{n+1}$. If $n_i < n$, then there exist at least two categories: label these i and j. Consider two possibilities: you observe (1) a member of species i at time $n + 1$, j at time $n + 2$; (2) a member of species j at time $n + 1$, and i at time $n + 2$. Because the random partition is exchangeable, the conditional probabilities of these two possibilities are the same, hence

$$\left(\frac{n_i - \alpha_n}{n + \theta_n} \right) \left(\frac{n_j - \alpha_{n+1}}{n + 1 + \theta_{n+1}} \right) = \left(\frac{n_j - \alpha_n}{n + \theta_n} \right) \left(\frac{n_i - \alpha_{n+1}}{n + 1 + \theta_{n+1}} \right),$$

hence $n_i(\alpha_n - \alpha_{n+1}) = n_j(\alpha_n - \alpha_{n+1})$. Because $n \geq 3$, we can choose $n_i = 1$, $n_j = 2$; and it thus follows that $\alpha_n = \alpha_{n+1}$.

Step 2: $\theta_n = \theta_{n+1}$. Next, consider the two possibilities; you observe (1) a member of a new species at time $n + 1$, and a member of the old species i at time $n + 2$; (2) a member of the old species i at time $n + 1$ and a member of a new species at time $n + 2$. Equating the conditional probabilities of these two events gives us:

$$\left(\frac{t \alpha_n + \theta_n}{n + \theta_n} \right) \left(\frac{n_i - \alpha_{n+1}}{n + 1 + \theta_{n+1}} \right) = \left(\frac{n_i - \alpha_n}{n + \theta_n} \right) \left(\frac{t \alpha_{n+1} + \theta_{n+1}}{n + 1 + \theta_{n+1}} \right).$$

Because we already know that $\alpha_n = \alpha_{n+1}$, it follows that $\theta_n = \theta_{n+1}$.

The two predictive probabilities $f(n_i, n)$ and $g(t, n)$ are therefore seen to be of the form

$$f(n_i, n) = \frac{n_i + \alpha + \gamma_n \delta_n(n_i)}{n + \theta}; \quad g(t, n) = \frac{t\alpha + \theta - \gamma_n \delta_1(t)}{n + \theta}.$$

for some α and θ, and all $n \geq 3$. It is not difficult to see, however, that these two formulas continue to hold in the special cases $n = 1$ and 2 (using the same values for α and θ), provided appropriate choices are made for γ_1 and γ_2. This is trivial if $n = 1$; one just chooses an appropriate value for γ_1 given the already determined values of α and θ. For the case $n = 2$, note that one can choose $\alpha_2 = \alpha$ because there are three degrees of freedom (α_2, θ_2, and γ_2), but only two constraints involving predictive probabilities (the values of $f(1, 2)$ and $f(2, 2)$). Thus it remains to show that $\theta_2 = \theta$; but this follows using the same argument in Step 2 of Lemma 3 (using $n = t = 2$, together with the observation that $1 - \alpha$ does not vanish because $f(1, 2)$ does not, by assumption, vanish).

Note, however, that the formulas derived state only that the desired conditional probabilities must have the given form for *some* θ and α (and sequence γ_n); they do not assert that for each possible pair $<\alpha, \theta>$ an exchangeable random partition exists that satisfies our assumptions for all such values. Indeed, it is not hard to see that certain constraints are essential:

Lemma 4: The parameters α and θ satisfy $0 \leq \alpha < 1$ and $\theta > -\alpha$.

Proof: For any fixed value of θ, $n + \theta > 0$ for all n sufficiently large. Because all possible finite sequences have positive probability, $0 < g(t, n) < 1$, hence

$$0 < \frac{t\alpha + \theta}{n + \theta} < 1 \Rightarrow -\frac{\theta}{t} < \alpha < \frac{n}{t}.$$

Letting $n \to \infty$ and taking $t = n$, it follows that $0 \leq \alpha < 1$. Because $0 < f(1, 1) < 1$,

$$0 < \frac{1 - \alpha}{1 + \theta} < 1;$$

but $1 - \alpha > 0$, hence $1 + \theta > 0$, hence $\theta > -\alpha$.

On the other hand, exchangeable random partitions do exist for all possible $<\alpha, \theta>$ pairs in the ranges given in Lemma 4 (see Pitman 1992). It thus remains to identify the constant γ_n. In order to do this, however, a technical result is required.

Lemma 5: (Basic recurrence relation):

$$\boxed{\gamma_{n+1} = \frac{\gamma_n(n + \theta)}{(n - \alpha + \gamma_n)}.}$$

Proof: By the same argument as in Lemma 3, observe that

$$\left(\frac{n - \alpha + \gamma_n}{n + \theta}\right)\left(\frac{\theta + \alpha - \gamma_{n+1}}{n + 1 + \theta}\right) = \left(\frac{\theta + \alpha - \gamma_n}{n + \theta}\right)\left(\frac{n + 1 - \alpha}{n + 1 + \theta}\right);$$

canceling denominators and some simplification then gives the result.

Let $\gamma =: \gamma_1$, and for $n \geq 1$, let $\Pi_n =: \left(\frac{1 - \alpha}{1 + \theta}\right)\left(\frac{2 - \alpha}{2 + \theta}\right) \cdots \left(\frac{n - \alpha}{n + \theta}\right)$

and $d_n =: \gamma + (\alpha + \theta + \gamma)\Pi_{n-1}$ (by convention, $\Pi_0 = 1$).

Lemma 6: For all $n \geq 1$, $\gamma_n = \frac{\gamma(\alpha + \theta)}{d_n}$.

Proof: It suffices to prove that (1) $d_n \neq 0$ and (2) $\gamma_n d_n = \gamma(\alpha + \theta)$ for all $n \geq 1$. The proof is by induction.

Note first that $d_1 = \gamma + (\alpha + \theta - \gamma) = \alpha + \theta > 0$, and $\gamma_1 d_1 = \gamma(\alpha + \theta)$; thus the two assertions hold for $n = 1$. Next, suppose that $d_n \neq 0$ and $\gamma_n d_n = \gamma(\alpha + \theta)$ for a given value of $n \geq 1$. Then

$$d_{n+1} =: \gamma + (\alpha + \theta - \gamma)\Pi_n$$

$$= \gamma + (\alpha + \theta - \gamma)\left(\frac{n - \alpha}{n + \theta}\right)\Pi_{n-1}$$

$$= \frac{(n - \alpha)(\gamma + (\alpha + \theta - \gamma)\Pi_{n-1}) + \gamma(\alpha + \theta)}{n + \theta}$$

$$= \frac{(n - \alpha)d_n + \gamma_n d_n}{n + \theta}$$

$$= \frac{(n - \alpha + \gamma_n)d_n}{n + \theta}.$$

Thus $d_{n+1} = \dfrac{(n - \alpha + \gamma_n)d_n}{n + \theta}.$

But then $d_{n+1} \neq 0$ immediately follows from the inductive hypothesis (because $n - \alpha + \gamma_n > 0$ and $n + \theta > 0$ for all $n \geq 1$); and

$$\gamma_{n+1}d_{n+1} = \frac{\gamma_n(n + \theta)}{(n - \alpha + \gamma_n)}d_{n+1} = \gamma_n d_n = \gamma(\alpha + \theta)$$

(by the fundamental recursion formula, the preceding formula, and the inductive hypothesis).

Lemma 7: The parameter γ satisfies the inequalities $0 \leq \gamma < \alpha + \theta$.

Proof: Suppose $\gamma < 0$; since $n + \theta > 0$ and $n - \alpha + \gamma_n > 0$ for all n, it follows from the fundamental recursion formula that $\gamma_n < 0$ for all n. But $\lim_{n \to \infty} \Pi_n = 0$, hence

$$\gamma_n = \frac{\gamma(\alpha + \theta)}{\gamma + (\alpha + \theta - \gamma)\Pi_{n-1}} \to \alpha + \theta > 0,$$

which is impossible. Thus $\gamma \geq 0$. The inequality $\gamma < \alpha + \alpha$ follows from the inequalities $1 + \theta > 0$ and

$$\frac{1 - \alpha + \gamma}{1 + \theta} = f(1, 1) < 1.$$

The basic result stated in the second section, "The New Continuum," now follows immediately from Lemmas 1–7. The following theorem restates this in terms of a mixture of two partitions, and summarizes the primary technical contribution of this paper regarding the characterization of exchangeable random partitions.

Theorem 1: Let $\Pi = \Pi_1, \Pi_2, \ldots, \Pi_n, \ldots$ be an infinite consistent sequence of exchangeable random partitions. If the sequence satisfies the three Assumptions (1), (2), and (3) for all $n \geq 1$, then there exist three parameters ϵ, α, and θ ($0 \leq \epsilon < 1$, $0 \leq \alpha < 1$, and $\theta > -\alpha$), such that

$$f(n_i, n) = (1 - \epsilon_n)\left(\frac{n_i - \alpha}{n + \theta}\right) + \epsilon_n \delta_n(n_i)$$

and

$$g(t, n) = (1 - \epsilon_n)\left(\frac{t\alpha + \theta}{n + \theta}\right) - \epsilon_n \delta_1(t),$$

where

$$\epsilon_n =: \frac{\epsilon}{\epsilon + (1 - \epsilon) \prod_{j=1}^{n-1} \dfrac{j - \alpha}{j + \theta}}$$

is the posterior probability of the partition Π_∞ (all observations are of the same species), given that the first n observations are of the same species, an initial probability of ϵ in favor of Π_∞ and $1 - \epsilon$ in favor of the Pitman alternative $\Pi_{\alpha,\theta}$.

Proof: Note that if $z > x$, then

$$\frac{x + y}{z} = (1 - r)x + r \leftrightarrow r = \frac{y}{z - x}.$$

Thus letting $x = n - \alpha$, $y = \gamma_n$, $z = n + \theta$, and $r = \dfrac{\gamma_n}{\alpha + \theta} =: \epsilon_n$ gives

$$\frac{n - \alpha + \gamma_n}{n + \theta} = (1 - \epsilon_n)\left(\frac{n - \alpha}{n + \theta}\right) = \epsilon_n.$$

Let $\epsilon =: \dfrac{\gamma}{\alpha + \theta}$; it then follows from Lemma 6 that ϵ_n has the stated form. Because $0 \leq \gamma < \alpha + \theta$ (Lemma 7), it follows that $0 \leq \epsilon < 1$. That ϵ_n is the stated posterior probability is an immediate consequence of Bayes' theorem.

Remarks. Suppose further that $g(t, n) = g(n)$ for all $n \geq 1$; that is, that the probability of seeing a new species depends only on the sample size n, and *not* the number of species observed. Then necessarily $\epsilon = \alpha = 0$, and one has a characterization of the Ewens subfamily $\mathbf{\Pi}_{0,\theta}$.[23] To be precise:

Corollary: Let $\mathbf{\Pi}$ be an infinite exchangeable random partition, and let $\mathbf{\Pi}_n$ denote the exchangeable random partition induced by $\mathbf{\Pi}$ on the set $\{1, 2, \ldots, n\}$. Suppose that for each $n \geq 1$, (1) $P[\mathbf{\Pi}_n = \boldsymbol{\pi}_n] > 0$ for all partitions $\boldsymbol{\pi}_n$ of $\{1, 2, \ldots, n\}$; (2) the conditional probability of observing an old species i at time $n + 1$, given $\mathbf{\Pi}_n$, the past history up to time n, depends only on n and n_i, the number of times that the species i has occurred in the past (but not on i or n_j for $j \neq i$); and (3) the conditional probability of observing a new species at time $n + 1$, given $\mathbf{\Pi}_n$, depends only on n. Then the random partition $\mathbf{\Pi}$ is a member of the Ewens family for some value $\theta > 0$; that is, $\mathbf{\Pi} = \mathbf{\Pi}_{0,\theta}$ for some $\theta > 0$.

2. Consider the Pitman process $\mathbf{\Pi}_{\alpha,\theta}$. If E_n is the event that a novel species is observed on the nth trial, and t_n is the number of distinct species observed as of that trial, then

$$\sum_{n=1}^{\infty} P[E_{n+1}|t_1, t_2, \ldots, t_n] = \sum_{n=1}^{\infty} \frac{t_n \alpha + \theta}{n + \theta} \geq \sum_{n=1}^{\infty} \frac{\theta}{n + \theta} = \infty;$$

it then follows from the extended Borel-Cantelli lemma (see, for example, Breiman 1968, 96, Corollary 5.29) that

$$P[E_n \text{ occurs infinitely often}] = P\left[\omega: \sum_{n=1}^{\infty} P[E_{n+1}|t_1, t_2, \ldots, t_n] = \infty\right] = 1.$$

Thus, the total number of species observed in an infinite number of observations of the Pitman $\mathbf{\Pi}_{\alpha,\theta}$ is almost surely infinite. (In fact, Pitman [1992c] shows that the number of species t_n grows almost surely as the power n^α: the random limit $Z =: \lim_{n \to \infty} t_n/n^\alpha$ exists almost surely and has a distribution that depends on θ.)

NOTES

I thank Jim Pitman for drawing my attention to the $\mathbf{\Pi}_{\alpha,\theta}$ family of random partitions; his conjecture that some form of the Johnson theorem should apply to it led to the present essay. Theo Kuipers was also very generous in providing information regarding his 1973 paper. Persi Diaconis, Warren Ewens, Theo Kuipers, and Jim Pitman made helpful comments on a first draft. It is a particular pleasure to acknowledge the hospitality of the Instituto di Statistica of the Università degli Studi di Genova, and that of its director, Domenico Costantini, during a visit to Genova in April 1992, when my initial research began.

1. The confirmation of a universal generalization means that if only one species is observed, then an increased positive probability is assigned to the possibility that only one species exists.

2. Both the concept of an exchangeable random partition, and the closely allied concept of *partition structure*, are due to the English mathematician J. F. C. Kingman; see Kingman 1980, Aldous 1985, and Zabell 1992 for further information and references.

3. An early example of the use of the partition vector in the sampling of species problem can be found in letters of R. A. Fisher to his Cambridge colleague Sir Harold Jeffreys (see Bennett 1990, 151, 156–57, 160).

During the second world war the English mathematician and logician Alan Mathison Turing (1912–1954) recognized the importance of such "frequencies of frequencies" and used them to break key German military codes at Betchley Park (see Good 1965, chap. 7, and 1979). Turing's statistical interests in such problems are less surprising than they might at first seem: his 1935 undergraduate King's College fellowship dissertation proved a version of the Lindeberg central limit theorem, and this experience led him in later years to be on the alert for the potential statistical aspects of a problem (see Zabell 1995).

4. Strictly speaking, the term *partition structure* refers to the consistent sequence of random partition vectors generated by a consistent sequence of exchangeable random partitions.

5. How could it? In the Johnson-Carnap continuum, the fact that $f(n_i, n)$ does not depend on i is an *assumption*, but here it is a *consequence* of the framework: the ith species is not known to exist prior to sampling!

6. Thus with probability ϵ, all animals are of the same species as the first animal, and with probability $1 - \epsilon$, the predictive probabilities are $\dfrac{n_i - \alpha}{n + \theta}$. Theorem 1 in the final section of this essay, "Derivation of the Continuum," states the continuum in the alternative format of a mixture.

7. The two cases $\alpha = 1$ and $\gamma = \alpha + \theta$ are excluded because of Assumption (1): the case $\alpha = 1$ corresponds to the random partition where each species occurs only once; the case $\gamma = \alpha + \theta$ to the random partition where only one species occurs.

8. As noted at the end of the first section of this essay, Pitman has also investigated the more general family discussed here: Corollary 3 in Pitman (1992b) characterizes the distribution of the "size-biased permutation" of the atoms in the Kingman representation of exactly such partition structures. The ϵ in Theorem 1 of this essay's last section corresponds to Pitman's $P(P_1 = 1)$.

9. The role and interpretation of the α and θ parameters become much more complex at the level of the corresponding partition structures: for $\alpha > 0$ fixed, the laws of $\Pi_{\alpha,\theta}$ are mutually absolutely continuous as θ varies, but for θ fixed, the $\Pi_{\alpha,\theta}$ are mutually singular as α varies (see Pitman 1992c).

10. Such urns are the "delabeled" versions of urn models for Dirichlet processes that first appear in Blackwell and MacQueen 1973.

11. Thus, the classical Johnson-Carnap continuum for multinomial sampling and the continuum of inductive methods for the sampling of species discussed here represent two extremes, neither entirely credible. In one case (Johnson-Carnap), it is assumed that because of (but it is, in fact, in spite of) our supposed ignorance all possible categories are a priori known and equiprobable. In the other case (the sampling of species, considered here), it is assumed that one is ignorant of *all* categories in advance but, in spite of this, knows that an infinite number of them must occur over time. A more realistic continuum for the sampling of species would eliminate assumption (1). I hope to return to this question in joint work with Jim Pitman.

12. This in essence was Johnson's viewpoint: In his (posthumous) 1932 paper in *Mind*, he wrote:

The postulate adopted in a controversial kind of theorem cannot be generalized to cover all sorts of working problems; so it is the logician's business, having once formulated a specific postulate, to indicate very carefully the factual and epistemic conditions under which it has practical value. (1932, 418–19)

13. In Bayes's terminology, probabilities pertain to events, chances to probabilities of events.

14. The urn of nature, in this sense, goes back to James Bernoulli's *Ars conjectandi*; the cogency of the analogy was a primary target for the later critics of inverse probabilities (see, e.g., Zabell 1989, 302–03).

15. Price states in a footnote, "There can, I suppose, be no reason for observing that on this subject unity is always made to stand for certainty, and ½ for an even chance." But if unity stands for certainty, then presumably zero stands for impossibility. For a modern attempt to interpret a related species of inverse probabilities in terms of infinitesimals, see Sobel 1987.

16. For a discussion of Ramsey's system, as set forth in his 1926 essay, "Truth and Probability," see Zabell 1991.

17. For Buffon's philosophical Adam and Condillac's criticism of him, see Fellows and Milliken (1972, 125–31); for discussion of Condillac's alternative, the "statue-man," and the cited work of Diderot and Bonnet, see Knight (1968, chap. 4). These discussions do not cite Hume as a precursor; and it seems unlikely, given

the proximity in dates, that Buffon, the earliest of them, drew the image directly from Hume a mere year after the appearance of the *Enquiry.* Perhaps there is a common intellectual ancestor at work here, but I have not as yet been able to find one. All of these French narratives reflect interests in associationist psychology; for the more general link between associationism and probability, see Daston (1988, chap. 4). These fanciful narratives reflected a more general Enlightenment fascination with persons reared in the wild or initially deprived of certain sensory abilities (see Gay 1969, 174–76).

18. Gillies 1987 presents an able and convincing argument of the case (see also Daston 1988, 264–67, 326–30). There is some evidence that Bayes himself intended his essay as an answer to Hume (see Zabell 1989, 290–93).

19. Hume's interpolation of "proofs" as third species of reasoning, intermediate between that of "knowledge" and "probabilities" and consisting of nondemonstrative arguments that are "entirely free of doubt and uncertainty," is similar to Cardinal Newman's concept of "assent" in his *Grammar of Assent* (1870).

20. I have modified De Morgan's notation to conform with mine. Although much of De Morgan's article is closely based on Laplace's book, I have not been able to find either a paper or book of Laplace in which this point is made. It appears to be original with De Morgan.

21. Strictly speaking, the urn model is not specified by De Morgan. Indeed, his discussion elsewhere suggests that he thought the appropriate denominator to use in the predictive probabilities to be $n + t + 1$, where n is the (nonrandom) sample size and t the (random) number of species observed, rather than t the number of specied known to exist a priori.

22. See Kuipers (1978, chap. 6, sec. 10). In Kuipers's system the predictive probabilities for old species can depend on the number of species observed. The infinite system is derived by first delabeling a finite system, and then passing to the limit as the number of species increases to infinity. It would be of considerable interest to have a direct axiomatic derivation of this system in the infinite case.

23. For other characterizations of the Ewens subfamily, see Kingman (1980, 38) and Donnelly (1986, 279–81).

REFERENCES

Aldous, D. J. 1985. "Exchangeability and related topics." In P. L. Hennequin, ed., *École d'Été de Probabilités de Saint-Flour 1983, Lecture Notes in Mathematics* 1117:1–198.

Barker, S. F. 1957. *Induction and Hypothesis.* Ithaca, N.Y.: Cornell University Press.

Bayes, Rev. T. 1764. "An essay towards solving a problem in the doctrine of chances." *Philosophical Transactions of the Royal Society of London* 53:370–418. Reprinted in E. S. Pearson and M. G. Kendall, eds., *Studies in the History of Statistics and Probability*, vol. 1. London: Charles Griffin, pp. 134–53. Page references are to this edition.

Bennett, J. H., ed. 1990. *Statistical Inference and Analysis: Selected Correspondence of R. A. Fisher.* Oxford: Clarendon Press.

Blackwell, D., and J. B. MacQueen. 1973. "Ferguson distributions via Polya urn schemes." *Annals of Statistics* 1:353–55.

Breiman, L. 1968. *Probability.* New York: Addison-Wesley.

Carnap, R. 1952. *The Continuum of Inductive Methods.* Chicago: University of Chicago Press.

Costantini, D., and M. C. Galavotti. 1987. "Johnson e l'interpretazione degli enunciati probabilistici." In R. Simili, ed., *L'Epistemologia di Cambridge 1850–1950,* Società Editrice il Mulino, Bologna, pp. 245–62.

Daston, L. 1988. *Classical Probability in the Enlightenment.* Princeton: Princeton University Press.

De Morgan, A. 1838. *An Essay on Probabilities, and Their Application to Life Contingencies and Insurance Offices.* London: Longman, Orme, Brown, Green, & Longmans.

———. 1845. "Theory of probabilities." In *Encyclopedia Metropolitana,* vol. 2: *Pure Mathematics.* London: Longman et al.

Donnelly, P. 1986. "Partition structures, Polya urns, the Ewens sampling formula, and the ages of alleles." *Theoretical Population Biology* 30:271–88.

Earman, J. 1992. *Bayes or Bust? A Critical Examination of Bayesian Confirmation Theory.* Cambridge, Mass.: MIT Press.

Essler, W. K. 1975. "Hintikka vs. Carnap." In J. Hintikka, ed., *Rudolph Carnap, Logical Empiricist,* Dordrecht, Holland: D. Reidel Publishing Co.

Fellows, O. E., and S. F. Milliken. 1972. *Buffon.* New York: Twayne Publishers.

Gay, P. 1969. *The Enlightenment: An Interpretation,* vol. 2: *The Science of Freedom.* New York: W. W. Norton.

Gillies, D. 1987. "Was Bayes a Bayesian?" *Historia Mathematica* 14:325–46.

Good, I. J. 1965. *The Estimation of Probabilities: An Essay on Modern Bayesian Methods.* Cambridge: MIT Press.

———. 1979. "A. M. Turing's statistical work in World War II." *Biometrika* 66, 393–96.

Hintikka, J. 1966. "A two-dimensional continuum of inductive methods." In J. Hintikka and P. Suppes, eds., *Aspects of Inductive Logic.* Amsterdam: North-Holland, pp. 113–32.

Hintikka, J., and I. Niiniluoto. 1980. "An axiomatic foundation for the logic of inductive generalization." In R. C. Jeffrey, ed., *Studies in Inductive Logic and Probability,* vol. 2. Berkeley, Calif.: University of California Press, pp. 157–81.

Hoppe, F. 1984. "Polya-like urns and the Ewens sampling formula." *Journal of Mathematical Biology* 20:91–94.

Howson, C. 1973. "Must the logical probability of laws be zero?" *British Journal for Philosophy of Science* 24:153–63.

Howson, C., and P. Urbach. 1989. *Scientific Reasoning: The Bayesian Approach.* La Salle, Ill.: Open Court Press.

Hume, D. [1739] 1978. *A Treatise of Human Nature.* In the L. A. Selbe-Bigge text, revised by P. H. Nidditch. Oxford: Clarendon Press.

———. [1748] 1975. *An Enquiry Concerning Human Understanding.* In the L. A. Selbe-Bigge text, revised by P. H. Nidditch. Oxford: Clarendon Press.

Johnson, W. E. 1932. "Probability: the deductive and inductive problems." *Mind* 41:409–23.

Kingman, J. F. C. 1980. *The Mathematics of Genetic Diversity*. Philadelphia: SIAM.

Knight, I. F. 1968. *The Geometric Spirit*. New Haven and London: Yale University Press.

Kuipers, T. A. F. [1973] 1975. "A generalization of Carnap's inductive logic." *Synthese* 25:334–36. Reprinted in J. Hintikka, ed., *Rudolph Carnap, Logical Empiricist*. Dordrecht: D. Reidel Publishing Co.

Kuipers, T. A. F. 1978. *Studies in Inductive Probability and Rational Expectation*. Dordrecht: D. Reidel Publishing Company.

Newman, Cardinal J. H. 1870. *An Essay in Aid of a Grammar of Assent*, fifth ed. 1885. London: Longman & Green's.

Pitman, J. 1992a. "Partially exchangeable random partitions." Technical Report 343, Department of Statistics, University of California, Berkeley. Revised version to appear in *Probability Theory and Related Fields* 102(1995):145–58.

———. 1992b. "Random discrete distributions invariant under size-biased permutation." Technical Report 344, Department of Statistics, University of California, Berkeley. To appear in *Advances in Applied Probability* 28(1996): 525–39.

———. 1992c. "The two-parameter generalization of Ewens' random partition structure." Technical Report 345, Department of Statistics, University of California, Berkeley.

———. 1992d. "Random partitions derived from excursions of Brownian motion and Bessel processes." Technical Report 346, Department of Statistics, University of California, Berkeley.

Popper, K. [1959] 1968. *The Logic of Scientific Discovery*. New York: Basic Books. 2nd English ed., New York: Harper & Row.

Shelley, M. [1818] 1992. *Frankenstein*. 2nd ed. 1831. Reprinted, London: Penguin Books.

Sobel, J. H. 1987. "On the evidence of testimony for miracles: A Bayesian interpretation of David Hume's analysis." *Philosophical Quarterly* 37:166–86.

Wrinch, D., and H. Jeffreys. 1919. "On certain aspects of the theory of probability." *Philosophical Magazine* 38:715–31.

Zabell, S. L. 1982. "W. E. Johnson's 'sufficientness' postulate." *Annals of Statistics* 10:1091–99.

———. 1988. "Symmetry and its discontents." In *Causation, Chance, and Credence*, vol. 1, W. L. Harper and B. Skyrms, eds., Dordrecht: Kluwer, pp. 155–90.

———. 1989. "The rule of succession." *Erkenntnis* 31:283–321.

———. 1991. "Ramsey, truth, and probability." *Theoria* 57:211–38.

———. 1992. "Predicting the unpredictable." *Synthese* 90:205–32.

———. 1995. "Alan Turing and the central limit theorem." *American Mathematical Monthly* 102.

13

Science Without Induction

Frederick Suppe
Department of Philosophy, University of Maryland

Among my most treasured professional memories is the semester Wilfrid Sellars and I were colleagues. The two of us met an afternoon a week to discuss basic philosophical issues. Fueling that highly productive series of discussions[1] was the shared belief that philosophy of science was that branch of epistemology and metaphysics concerned with our most sophisticated forms of knowledge. Thus, any separation of philosophy of science from epistemology and metaphysics was artificial. That vision seems largely to have disappeared. A recent citation analysis of the epistemology and philosophy of science literatures (Kreuzman 1990) indicates only the most minimal concern with each other's literatures — reflecting, I believe, how insular the two areas have become. Epistemologists are more concerned with the arcane implications of possible papier mâché barns in Wisconsin for basic perceptual knowledge than they are in understanding the far more challenging epistemic achievements of contemporary physical, biological, and social science, which ought to serve as paradigm cases of a posteriori knowledge. And philosophers of science attempt to bracket or finesse the epistemological issues raised by, or presuppositional to, their analyses of science, though not always as blatantly as Larry Laudan (1977) in *Progress and Its Problems*.[2]

I believe such insularity has cost philosophy dearly, and for the past twenty years I have been attempting to work out a fine-detailed analysis of scientific knowledge embedded into a comprehensive epistemology of the a posteriori and associated metaphysics. Despite the fact that the two fields have been working in near isolation from each other, recent developments in both fields lay the basis for an effective rapprochement. To that end I want to "triangulate" from these separate developments in the attempt to address the question, "What sort of epistemology is adequate to accommodate contemporary science?" To anticipate: My answer will be one that construes scientific knowledge as *noninductive* — hence the title of my essay.

What Sort of an Epistemology?

My concerns here are to understand scientific knowledge, not to engage pervasive skeptics in dialogue, and so I ignore their doubts and concerns. Van Fraassen observes,

Disagreement with the skeptic is in the view of life. To disarm the skeptic's final move, one must say that, on one's own view of life, genuine epistemic *engagement* has its own value, and is to be preferred to a life of utilitarian calculation and prudence. . . . There is no escape from skepticism into theory; skepticism has no theoretical limits. . . . At some point, reactions to skepticism . . . equally become matters of decision, attitudes, and self will. . . . There is no need to counsel us to live dangerously; we do. The question is how to live with that danger — seeking to remain in the safest position possible, the skeptic's solution, is no more *theoretically* justifiable than any other. (van Fraassen 1988, 152)

My starting point is that the sorts of claims, including theoretical ones, that contemporary science yields on the basis of experiment and observation are paradigmatic both of knowledge and some means whereby it can be acquired. Initially at least I will follow much of the epistemological literature in assuming that "S knows that ϕ" can be analyzed as justified true belief, with possibly an additional condition added. I understand the notion of "justified" sufficiently broadly as to not beg questions against externalist epistemologies.

Considerations from Recent Philosophy of Science

Experimental science contrives circumstances to enable observers to interact perceptually with phenomena that usually are hidden from perception. Thus, we observe electrons by contriving a cloud chamber apparatus filled with supersaturated vapor, illuminated by a bright light, and placed within an electromagnetic field. Charged particles ionize the droplets of the vapor which glow under intense illumination, leaving iridescent trails; interaction with the electromagnetic field causes different kinds of particles to assume characteristic trajectories, thereby enabling us to identify the particle types. The vapor trail is here a *decisive indicator* that, say, an electron has been emitted. If we are to obtain knowledge from observations under such contrived experimental setups, *in some manner or another the various empirical regularities involved in the production of that decisive indicator must bear the brunt of evidential burden, of the justification required by a justified true belief analysis.*

Whether we look at the team efforts of "big science" or the individual efforts of "little science," the production of knowledge is not a solitary enterprise: Knowledge is produced, published, and thereby transmitted to other scientists, who built upon the contributions of others. The design of the cloud chamber rested on a huge body of prior scientific work, including findings on supersaturated vapors, ionization, the interaction of light with ionized supersaturated molecules, and the effects of electromagnetic fields on charged particles (see Hooker 1975, append.). Moreover, the lines of transmission of such knowledge from original researcher to consumer typically are quite long and convoluted, involving original articles, abstracting services, review articles, textbooks, and many other potential sources of epistemic contamination. Many such pieces of transmitted knowledge underlie even relatively simple experimental designs. Thus, a crucial test of an epistemology will be its ability to accommodate with realistic assumptions combinations of such extended chains of knowledge transference. Particularly telling will be matters of evidence dilution over such chains. For example, epistemologies which impose a modest probability threshold on epistemic justification will be particularly vulnerable, since, whatever the threshold, only a few transference steps will drop the evidence below the threshold for knowledge.[3]

In 1976, weak neutral currents were discovered using the Gargamelle bubble chamber at CERN and Fermilab. A large number of other studies were done, culminating in the August 1978 SLAC experiments of Prescott, et al. 1978. *Collectively* they were able to establish the correctness of the Weinberg-Salam electroweak theory. The total body of experimental evidence required to establish that theory is summarized in table 13.1. Such aggregation of evidence to establish a theory involves similar problems of knowledge transference and evidence dilution to those just mentioned. And if one looks, as Galison has, at the complexity of individual big science experiments such as the Gargamelle ones (see Galison 1987, esp. ch. 4), one finds essentially the same problems. Probability threshold accounts of epistemic evidence are vulnerable in both kinds of case. Thus, *an adequate epistemology must allow for long chains of knowledge transmission and complex aggregation of knowledge claims in manners that make realistic assumptions about the transmission process and do not succumb to excessive diminution of evidence.*

Sir Peter Medawar's 1963 charge that the scientific paper is a fraud because it purports to describe the actual processes whereby knowledge is gained, but in fact provides a different post hoc account, stimulated sociologists of science to investigate the private and the public discourse of science; they found two quite different forms of discourse, and also found that the write-ups of papers indeed do not provide accurate descriptions of the sorts of reasoning that went into the production and evaluation of the results presented in the resulting scientific papers. (See, e.g., Gilbert and Mulkay 1984.) Whatever else a scientific paper is, its arguments are *not* an account of the belief formation processes of its scientist authors.

Hacking tells us that experiments involve the production of phenomena in response to experimenter interference, the detection of resulting responses, and data analysis producing results which are presented under some ("theoretical") interpretation.[4] Yet Galison, who is well aware of the other ingredients, virtually identifies experiments with the construction of *interpretative arguments*: "Experiments are about the assembly of persuasive arguments" (Galison 1987, 277; see also Galison 1985, 356–59). His examples make it amply clear that the thrust of these arguments is the blocking of competing interpretations of one's results. Coming from other, quite different perspectives, Pickering 1989 and Shapere 1982 similarly stress the centrality of interpretive arguments in experiment and observation. Shapere analyzes such arguments in terms of the *removal of specific doubts* against the author's favored interpretation.

Recently I did a fairly comprehensive examination of articles reporting experimental studies in a variety of disciplines, ranging from particle physics to psychology and sociology. There was remarkable consistency in the functional and argumentative structure of the papers: They present the *reduced data* or *results of* the experiment. They make a case

TABLE 13.1 Weak Neutral Current Experiments 1976 through August 1978

Process	Comparison with Theory	Experiment	Total Sample	Events Observed	Background	Cross section $10^{-42} Ev$ cm²	Experimental Results	$\sin^2\theta$	W-S prediction at $\sin^2\theta = 0.23$
1. Purely leptonic $\nu_\mu + e^- \to \nu_\mu + e^-$	Very clean (no hadrons involved)	Gargamelle CERN PS		-1		-3			
		Aachen-Padova CERN PS spark chamber	32		21	1.1 ± 0.6			
		Gargamelle CERN SPS	41,000	9	0.4 ± 0.4	$3.7 \begin{smallmatrix}+2.0\\-1.3\end{smallmatrix}$ to $4.2 \begin{smallmatrix}+2.2\\-1.7\end{smallmatrix}$			
		Columbia-NNL. Fermilab 15 ft. chamber	160,000	11	0.7 ± 0.7	1.8 ± 0.8	(1.7 ± 0.5) $\times 10^{-42} E_\nu$ cm²*	$0.21 \begin{smallmatrix}+0.09\\-0.06\end{smallmatrix}$	1.5
$\bar\nu_\mu + e^- \to \bar\nu_\mu + e^-$		Gargamelle CERN PS		3	0.4 ± 0.1	$1.0 \begin{smallmatrix}+2.1\\-0.9\end{smallmatrix}$			
		Aachen-Padova CERN PS spark chamber		17	7.4 ± 1.0	2.2 ± 1.0			
		BEBC Wideband neon. CERN PS	7,500	-1	0.4 ± 0.2	-3.5			
		Fermi-Mich-IHEP-ITEP Fermilab 15 ft. neon	6,300	0		-2.9			
		Gargamelle CERN SPS	4,000	0		-3.3	(1.8 ± 0.9) $\times 10^{-42} E_\nu$ cm²*	$0.3 \begin{smallmatrix}+0.10\\-0.30\end{smallmatrix}$	1.3
$\bar\nu_e + e^- \to \bar\nu_e + e^-$		Savannah River Fission					(5.7 ± 1.2) $\times 10^{-42} E_\nu$ cm²	0.29 ± 0.05	5.0
2. Elastic scattering $\nu_\mu + p \to \nu_\mu + \pi$	Relatively straightforward. Some uncertainty due to proton form factors (M_A)	Harvard-Penn-BNL. BNL counter detector		255	88	$\nu_\mu + p \to \nu_\mu + \pi$ 0.11 ± 0.02			
		Columbia-Ill-Rock. BNL spark chamber		71	30	0.20 ± 0.06			
		Aachen-Padova CERN PS S.C.		155	110	0.10 ± 0.03			
		Gargamelle CERN PS		100	28	0.12 ± 0.06	(0.11 ± 1.2) $\times \sigma(\nu_\mu + n \to \mu^- + p)$*	0.26 ± 0.06	0.12
$\bar\nu_\mu + p \to \bar\nu_\mu + \pi$		Harvard-Penn-BNL. BNL counter detector		69	28	0.19 ± 0.05	(0.19 ± 0.08) $\times \sigma(\bar\nu_\mu + p \to \bar\nu_\mu + n)$*	≤ 0.5	0.11

	Model notes	Experiment	Ratio Measured		

3. Single Pion production — Model dependent due to hadronic vertex.

Experiment	Ratio Measured	Value	Prediction
CIR			
Aachen-Padova	$((\nu + X + \pi^0))/(2(\mu^- + X + \pi^0))$	0.21 ± 0.07	$0.24\pm$
Gargamelle complex nuclei	$((\bar{\nu} + X + \pi^0))/(2(\mu^+ + X + \pi^0))$	0.46 ± 0.10	$0.30\pm$
Argonne 12 ft. B.C. deuterium	$((\nu + n + \pi^+))/(2(\mu^- + \pi + \pi^+))$	0.13 ± 0.06	$0.07\pm$
	$((\nu + n + \pi^0))/(2(\mu^- + p + \pi^+))$	0.40 ± 0.22	$0.17\pm$
	$((\nu + \nu + \pi^-))/(2(\mu^- + p + \pi^+))$	0.12 ± 0.04	$0.17\pm$
Gargamelle CERN PS Propane	$((\nu p\pi^0) + (\nu n\pi^0))/(2(\mu^- + p\pi^0))$	0.45 ± 0.08	$0.42\pm$
	$((\bar{\nu} p\pi^0) + (\bar{\nu} n\pi^0))/(2(\mu^+ + n\pi^0))$	0.57 ± 0.11	$0.60\pm$
	$((\nu + p + \pi^0))/((\mu^- + p + \pi^0))$	0.56 ± 0.10	0.42 ± 0.13
	$((\nu + n + \pi^0))/((\mu^- + p + \pi^0))$	0.34 ± 0.09	0.42 ± 0.13
	$((\nu + p + \pi^0))/((\mu^- + p + \pi^0))$	0.45 ± 0.13	0.28 ± 0.08
	$((\nu + n + \pi^0))/((\mu^- + p + \pi^0))$	0.34 ± 0.07	0.28 ± 0.08

4. Inclusive $\nu_\mu + N \rightarrow \nu_\mu + \dots$ [**] — Quark parton model dependent

Experiment	Corrected Ratios	Value	Prediction
Gargamelle CERN PS	0.26 ± 0.04		
HPWF Fermilab	0.30 ± 0.04		
CITF Fermilab	0.27 ± 0.02		
CDHS CERN SPS	0.295 ± 0.01		
BEBC Narrowband CERN SPS neon	0.31 ± 0.04	$(0.29 \pm 0.01) \times \sigma(\nu_\mu + N \rightarrow \mu^- + \dots)$ 0.24 ± 0.02	0.30[***]

$\bar{\nu}_\mu + N \rightarrow \bar{\nu}_\mu + \dots$

Experiment	Corrected Ratios	Value	Prediction
Gargamelle CERN PS	0.39 ± 0.06		
HPWF Fermilab	0.33 ± 0.09		
CDHS CERN SPS	0.40 ± 0.08		
BEBC Narrowband CERN SPS neon	0.34 ± 0.03		
	0.37 ± 0.08	$(0.35 \pm 0.025) \times \sigma\bar{\nu}_\mu + N \rightarrow \bar{\nu}_\mu^+ + \dots$ 0.3 ± 0.1	0.38[***]

5. Atomic physics $e^- + Bi \rightarrow e^- + Bi$ — Large uncertainties due to atomic physics calculations

Experiment	Value	Prediction
Seattle	$(-0.5 \pm 1.7) \times 10^{-8}$	-10 to -18×10^{-8}
Oxford	$(-5 \pm 1.6) \times 10^{-8}$	-13 to -23×10^{-8}
Novosibirsk	$(-19 \pm 5) \times 10^{-8}$	-13 to -23×10^{-8}

6. Electron scattering $\vec{e} + d \rightarrow e^- + \dots$ — Quark-parton model dependent

Experiment	Value	Prediction
SLAC (Prescott et al 1978)	$(-9.5 \pm 1.6) \times 10^{-5}(GeV/c)^2$ 0.20 ± 0.03	$\sim 9.5 \times 10^{-5}$[****]

NOTES:

* Average of experiments reported.

** Two experiments for which corrected ratios were not obtained have been omitted.

*** Improved predictions are to be found in Abbott and Barnett 1979.

**** Exact predictions are not given in the published report [Prescott et al 1978], the data being presented in their Figure 4 for a number of values of $\sin^2\theta$. The data given above are for the point of best agreement with Weinberg-Salam.

(The above table is based for the most part on Baltay 1978. See also that work for references to the specific experiments.)

for the *relevance* of the experiment and its results to the concerns of the target scientific community. They provide at least some of the detail about the experimental setup or *method* — the design and circumstances of the experimental observations — needed to *replicate* or evaluate the study. They provide *an interpretation of the reduced data* (results) which yields the specific *experimental claims*. For the most part, the write-up is descriptive, not argumentative. Beyond the experimental results, one finds relatively little positive additional argument in favor of the proffered interpretation. Virtually all the marshaling of evidence and associated argumentation is aimed toward anticipating and erasing specific doubts that might appropriately be raised against the proffered interpretation. Often these take the form of rebutting competing interpretations. They also identify and acknowledge other specific doubts that appropriately might be raised against the study's claims and might affect the epistemic status accorded those claims within the discipline. To the extent that there are no unrebutted objections or competing interpretations, the typical form of the interpretative argument is "ϕ because not-A_1 and . . . and not-A_n." (See Suppe 1997; 1998b, chap. 2, sec. 4).

Such arguments are not what one would expect on hypothetico-deductive method and are hard to make sense of on Bayesian models, unless one assumes the prior probabilities are very close to correct on the first iteration. More fundamentally, if R is the uninterpreted experimental result, then $P(\phi, R \;\&\; \sim A_1 \;\&\; \ldots \;\&\; \sim A_n)$ typically will be inadequately low to qualify as evidence for knowledge under most extant epistemologies. Yet, typically, they are adequate for rational belief formation by scientists in the discipline.

One might be tempted to construe the argument form here as an "inference to the best explanation" (IBE). Doing so would require construing 'explanation' very broadly, since what is at stake is the descriptive *interpretation* of experimental results. But the proposal has plausibility if we grant such usage. For then the argument form in both cases is to argue that since E_1 is the best among the lot $E_1, \ldots, E_n$, it most likely is true. Van Fraassen has argued — persuasively to my mind — that by themselves such arguments have little probative value, since the selection $E_1, \ldots, E_n$ "may well be the best of a bad lot."[5] Some further evidence that the correct interpretation is likely to be among the $E_1, \ldots, E_n$ is needed.[6] And it is precisely such additional evidence that we do *not* find in scientific papers. The IBE model thus fails. Indeed, the similarity of these argument forms to IBE patterns, coupled with the failure of the IBE model, lends further credence to my observation above that such arguments do not provide adequate evidence and, hence, are not sufficient justification for knowledge, under most extant epistemologies.

Both epistemologists and philosophers of science have tended to make ideal rationality assumptions that are manifestly incompatible with actual human cognitive capabilities. Christopher Cherniak has done an especially compelling job of showing the disastrous consequences of such practices and delimiting the sorts of *minimal rationality* conditions that should constrain adequate epistemologies (see Cherniak 1986, esp. ch. 5). In particular, humans are incapable of real-world reasoning in regular conformity with philosophers' usual canons of argument. Rather, humans are heuristic imbeciles who have the ability to identify some apparently relevant counterpossibilities or specific doubts, and in argumentation one is to be held "responsible for some, but not all, counterpossibilities whose seriousness is implied by his current belief set" (ibid., p. 113). The sorts of interpretive arguments we find scientists, in fact, do make in their papers are exactly what we might expect from human "heuristic imbeciles." A further minimal rationality constraint results from the observation that human belief sets cannot realistically be assumed to be self-consistent.

Thus, an adequate epistemology for science must accommodate the fact *that the arguments adduced in scientific papers are aimed more at the elimination of specific doubts against proffered interpretations than they are at adducing positive evidence in favor of that interpretation of the reported results, and that the belief sets underlying such arguments need not be consistent.* This feature of science proves troublesome for most epistemologies and accounts of scientific knowledge. It is most readily accommodated by divorcing belief-formation processes from the evidential basis for knowledge, denying the probative force of such arguments in scientific articles, and relegating them instead to belief-formation processes.[7]

In his *Particles and Waves* (1991), Peter Achinstein has teamed his formidable ability to do close philosophical analysis with extremely careful and detailed historical analysis of episodes in 19th century experimental physics. He makes compelling cases for the inadequacy of both the hypothetico-deductive and Bayesian analyses to handle his cases. Thus, he reinforces a number of the observations I make elsewhere in this paper. Of particular importance is his finding that the experimental claims made on the basis of particular experiments often are generalizations that go beyond the particular experiment. Indeed, in J. J. Thomson's 1897 cathode ray experiments, the experimental results are put forth in both their singular and general forms, and the very same evidence is said to establish the one as the other (Achinstein 1991, 308). Moreover, this is the series of experiments which physics subsequently credited with establishing the existence of electrons — i.e., with having established the generalized experimental claim.

Such a view goes contrary to deeply held intuitions about how science is really inductive. But to Pickering's orthodox claim that "it is a commonplace of particle physics that a single event cannot prove the existence of a new phenomenon" (Pickering 1984, 93), Galison rightly rejoins that "Actually, many processes in particle physics have been accepted by experimentalists after only one or two events have been found. Some . . . examples . . . are the muon, the omega-minus, the cascade zero, and the first 'V' particles" (Galison 1987, 260). Thus, to establish the existence of weak neutral currents, the single golden leptonic event found in Gargamelle was sufficient. But, then why conduct the other, more difficult and problematic hadronic experiments? Why take on all the problems associated with measuring background level? Because the leptonic experiments were dead ends: They were incapable of establishing the nonexistence of weak neutral currents. And they could not be built upon to determine the experimental $\sin^2\theta$ results needed to confirm the Weinberg-Salam electroweak theory. But for the mere establishment of the weak neutral currents, the golden leptonic events were superior. Thus, the preferred experimental practice was to do both the leptonic and the hadronic experiments.[8] An adequate epistemology must be able to accommodate the fact that *major scientific discoveries, including generalizations that go beyond the experimental instances, often can be established and known directly on the basis of single experiments or observations.* This is one sense in which science is noninductive.

Bayesian analyses are particularly ill-equipped to do so except when the prior probabilities are virtually on target — in which case it is the nonprobabilistic plausibility assessments used to assign the prior probabilities, and not the Bayesian probabilities, that bear the epistemic burden. Some Bayesians acknowledge the necessity of Bayesian prior probabilities ultimately resting on nonprobabilistic probability assessments.[9] Such an admission is not fatal if one is *only* maintaining that the

Bayesian account is a good model for selected facets of scientific practice. But if one maintains that science is intrinsically Bayesian in the sense that all evidence for a posteriori general knowledge ultimately is probabilistic in accordance with the Bayesian account, then the admission is fatal. For, as I have argued in *The Semantic Conception of Theories and Scientific Realism* (1989, chap. 13), the underlying plausibility assessments bear the brunt of epistemic evidential burden. This entails the inability of the Bayesian account to be a comprehensive general account of scientific evidence; hence, that scientific knowledge fundamentally cannot be Bayesian. Difficulties accounting for the assignment of atomic probabilities and nonzero probabilities for generalizations on other approaches to probabilistic inductive logic led to essentially the same conclusion: At bottom, *probabilistic inductive logic is incapable of providing the evidential basis for scientific knowledge.* This strongly suggests that a nonprobabilistic evidential condition for knowledge is to be preferred.

Since induction today is almost always construed probabilistically, we have found a second sense in which an adequate epistemology for science requires a *noninductive* view of scientific evidence. Of course, if by 'induction' one merely means "nondeductive" then science *is* inductive — as is virtually all real-world human reasoning; Cherniak's heuristic imbeciles typically are noninductive in the probabilistic sense, but in practice almost always are nondeductive.

Considerations from Recent Epistemology

I argued above that, in some manner or another, the various empirical regularities involved in the production of decisive indicators in experiments must bear the brunt of evidential burden for knowledge in such cases. Recently a number of *externalist* epistemologies have been advanced which make the empirical regularities governing perceptual processes and associated cognitive operations bear the evidential burden for a posteriori knowledge. The perceptual and cognitive regularities are of a piece; both are causally remote from our sensory experiences. Indeed, in the case of experiments such as our cloud chamber example, the two are combined into a single, complex process whereby sensory states become decisive indicators that, say, an electron has been emitted (see Suppe 1989, chap. 9; Dreske 1969). The track record of the many attempts to develop *internalist* epistemologies since World War II[10] strongly indicates that they are too impoverished to exploit evidentially the sorts of empirical regularities that underlie contemporary scientific experimentation. So I will confine my attention to externalisms.

Earlier, Bertrand Russell attempted to exploit such regularities in accounting for knowledge in the external world (1927, chaps. xii, xiii). But his efforts collapsed into an infinite regress by virtue of demanding that these regularities be known scientifically prior to their epistemic exploitation. To avoid such collapse into regress, externalists must deny that such regularities be known before they can bear their epistemic burden. This is accomplished by denying the KK Thesis

'S knows that ϕ' entails 'S knows that S knows that ϕ'

or at least the subentailment that

'S knows that ϕ' entails 'S knows that S's evidence for ϕ is adequate.'

I have argued elsewhere (Suppe 1989, chap. 10) that the KK thesis is at the heart of skeptical challenges to knowledge. Further, it manifestly is incompatible with the routine epistemic achievements of unsophisticated young knowers. Moreover, whenever one attempts to exploit empirical regularities evidentially as Russell did, the KK thesis results in a vicious regress that precludes a posteriori knowledge. Finally, the only attempts to defend the KK thesis have turned on unrealistic rational assumptions, such as deductive omniscience or the referential transparency of knowledge and belief operators.[11] Overall, I think the arguments against the KK thesis are overwhelming.[12] The way is clear to defend an externalism.

Two main sorts of externalisms can be distinguished: most externalisms are *reliabilisms*, which focus on the mechanisms of belief formation and attempt to analyze epistemic justification in terms of empirical and cognitive regularities governing belief formation that properly discriminate between justified and unjustified belief. Nozick's (1981) "methods" approach and the later Dretske's (1981) information-theoretic analyses are among the better such attempts. Such approaches require an especially tight connection between evidence for knowledge and belief formation processes. The recent epistemology literature is rich with criticisms of and counterexamples to reliabilisms, both in general and against specific attempts.[13] Collectively the weight of criticism seems formidable. To my mind, the move of tying belief formation and evidence so tightly together harkens back to internalisms, sacrifices much of the potential of externalisms to internalist demands, and underlies much of the difficulties to which they succumb.

Earlier we saw the desirability of separating epistemic evidence from belief formation processes if we are to work within minimal rationality constraints and avoid unrealistic assumptions about the consistency of belief sets. Nonreliabilisms are better able than reliabilisms to effect such a separation. Yet relatively few nonreliabilisms have been developed. Fred Dretske made an early attempt in the "conclusive reasons" account embedded into his *Seeing and Knowing* (Dretske 1969; see also Dretske 1971). It generated a fair amount of controversy, and eventually he abandoned it in response to counterexamples. In chapter 12 of Suppe 1989 I show how most of Dretske's counterexample trouble was due to overly stringent independence conditions imposed on the analysis and could be avoided with refined conditions.

One troublesome counterexample, Martin's 1975 racetrack case, remained. I urged that Dretske should have bit the bullet on it and gave theoretical reasons for the appropriateness of doing so. Dretske's conclusive reasons account involved an unanalyzed causal possibility operator. Recently, when I was working out the metaphysics and truth conditions for a causal possibility operator I need for my own epistemology, I was surprised to discover, as an unexpected corollary to that analysis, that it blocked Martin's racetrack case on my modified Dretskian analysis.[14] The racetrack counterexample thus proves to be an artifact of Dretske's recourse to an unanalyzed primitive causality operator rather than a result of fundamental difficulties with the epistemology. The modified Dretskian analysis I presented in *The Semantic Conception of Theories and Scientific Realism* (1989) thus proves to be a very close approximation to a nonreliabilist externalist epistemology adequate for science.[15]

Opting for a nonreliabilist externalism permits (but does not require) loosening the connection between belief formation processes and the evidential basis for knowledge. There are a number of reasons for doing so: first, minimal rationality considerations (see Cherniak 1986) suggest that the evidence one, in fact, can adduce or be held responsible for in belief formation is quite limited — so limited as to grossly underdetermine the truth of the beliefs. Second, private and public discourse considerations (e.g., Gilbert and Mulkay 1984) indicate that scientific belief formation and the sorts of arguments put forth in scientific journal articles are different. Insofar as those arguments are evidential, this suggests that the scientist's belief formation processes and the evidence they give often are different. However, we saw that these arguments typically are extremely weak and greatly underdetermine the claimed findings of a paper. So, too, are the considerations raised in the private discourse during the experiments. But then the evidential basis must be something other than the public or private discourse about the experiments.

Third, from the perspective of a nonreliabilist externalism, the empirical regularities underlying the experimental design, augmented by the empirical and cognitive regularities governing the observers' interactions with the experiment, are the most plausible alternative source of evidence. Thus, the evidential basis for the knowledge might be that the scientists are in cognitive states that are decisive indicators of the truth of the experimental claims. If this is the evidential basis for scientific knowledge, it is no surprise that most of it does not figure consciously in the belief formation process. Fourth, on such a view, what is the point of the descriptions and interpretative arguments given in published write-ups of the experiment? I suggest it is to get the reader to accept or believe the experimental claims made *and* to put the reader in a cognitive state that could be a decisive indicator of the truth of the experimental claims found in the paper. That is, experimental write-ups are crafted to display knowledge already obtained and to facilitate transfer of that knowledge to others in the discipline.[16] We see, then, that *there are, indeed, a number of reasons for driving a rather large wedge between evidential and belief formation processes.* (I will give additional reasons after I have put forth my own nonreliabilist externalist epistemology.)

The notion that knowledge is justified truth belief was challenged in 1963 by the Gettier paradoxes. These paradoxes apply to justified true belief epistemologies that further assume that

(a) one can meet the justification requirement with respect to a false belief; and
(b) if one is justified in believing that ϕ and ϕ entails φ, then one is justified in believing that φ.

A huge body of literature followed Gettier's 1963 original little paper. Almost all were attempts to block the paradoxes by imposing restrictions on the justification transference condition (b). These attempts ranged from additional fourth (or more) conditions to the development of relevance logics (since it was believed that the counterexamples allowed transference of justification from φ to irrelevant yet entailed φ). These attempts were met with depressing rounds of new paradoxes or counterexamples, followed by new proposals for restricting (b), ad nauseam.

What is astonishing is that virtually no attention was paid to (a). Only with Dretske's *Seeing and Knowing* (1969) do we find anybody rejecting (a) as a means of avoiding the Gettier paradoxes. This is especially surprising since denying (a) by imposing some sort of contextual necessity evidential condition automatically blocks the Gettier paradoxes and all known successors. Although Dretske's work garnered a lot of attention by epistemologists, few followed his lead in avoiding the Gettier paradoxes in this manner. I suspect this is because if one is to avoid a pervasive skepticism, doing so requires denying the KK thesis. Only in 1962 had epistemologists begun to examine the KK thesis, with most discussions defending it.[17] The thesis is congenial to internalists — and internalisms then were dominating epistemology. Externalist considerations favor the rejection of the thesis, but externalisms were not then in vogue. So epistemologists were slow in coming to understand how pernicious the KK thesis is.

But even if one does reject the KK thesis, one has to opt for a truth-evaluating certainty evidential condition for knowledge if denial of (a) is to block the Gettier paradoxes and their ilk. Philosophical wisdom of the past sixty years or so has been so wedded to the idea that justification admits of degrees, that such a move has seemed highly implausible — as it is on all but a nonreliabilist externalism that exploits denial of the KK thesis to drive a deep wedge between evidence and belief-formation processes. Thus, it is no surprise how slow epistemologists have been to appreciate the plausibility of denying (a) as a means of avoiding the Gettier and related paradoxes. Nearly three decades of failure by approaches focusing on (b) to solve these paradoxes ought to make one pause. *The Gettier and related paradoxes must be avoided by an adequate justified true belief analysis and the fact is that the only known successful way of doing so is to deny (a) by imposing a truth-evaluating certainty evidential requirement on knowledge.*

A Nonreliabilist Externalism Epistemology of the A Posteriori

So far I have been raising considerations from the recent philosophy of science and epistemology literatures that suggest we need an epistemology that is a nonreliabilist externalism;

denies the KK thesis;

imposes a truth-evaluating certainty evidential condition and thereby avoids Gettier-like paradoxes;

allows that evidence be in the form of empirical regularities in virtue of which cognitive states are decisive indicators of the truth of beliefs;

allows arguments to function in gaining knowledge in ways that do not require either consistent belief sets/premises or that arguments confer even moderately high likelihoods on their conclusions, but can make sense of the typical interpretative arguments found in scientific papers;

allows major scientific discoveries, including generalizations, to be known on the basis of single observations or experiments.

Such an epistemology will yield a noninductive view of science, in which probabilistic inductive logic plays no fundamental epistemological role, and in which generalizations can be known directly on the basis of single experiments or observations.

The preliminary version I put forward in chapter 12 of *The Semantic Conception of Theories and Scientific Realism* (Suppe 1989) is a close approximation to such an epistemology. In volume I of my new book, *Facts, Theories, and Scientific Observation* (Suppe 1998a), I develop a refined version of it and demonstrate its superiority over other extant epistemologies. Then in volume II I show, in fine detail, how it applies to, and makes sense of, scientific knowledge and practice.

My epistemology is a nonreliabilist externalist justified true belief account in which empirical regularities bear the evidential or justificatory burden. In order to specify how they do, it is necessary to make recourse to a causal possibility operator. A possible-worlds semantic interpretation of that operator has been given in terms of a class U of *actual-world uniformities*, which are generalizations over nonindexical partial world state descriptions specified in a highly nonrecursive language with uncountably many terms for naming entities and designating attributes. There is no presumption that actual world uniformities can be given fully explicit finite, or even countable, representations. The analysis is compatible with a wide range of ontological commitments concerning possible worlds — including those van Fraassen, Plantinga, and David Lewis would be willing to countenance. Intuitively the causally possible worlds will be all logically possible worlds one is willing to countenance that are consistent with U.

The analysis embodies a deep insight of Richard Montague (1974) that modal operators are a kind of universal generalization that collapses quantifiers and their variables of quantification together. It provides a noncircular incarnation of the intuitive notion that causally possible worlds are those logically possible worlds consistent with the laws of nature operant in our world. At the same time, it takes seriously van Fraassen's idea that there is nothing more to laws of nature than regularities in nature (van Fraassen 1989). In particular, the range of uniformities countenanced by U goes beyond what most would count as laws. For example, it allows "noncausal" temporal patterns that do not conform to our intuitions about laws.

Note: Numbers in braces refer to *Facts, Theories, and Scientific Observation* (Suppe 1998a,b). In the case below, the condition is identical to that displayed as number (35) in chapter IX of volume 1. Other references in braces are analogous. All subsequent references are to this work unless indicated otherwise.

In order to handle knowledge of causal generalizations, I invoke a relativized version of the causal possibility operator:

$\diamondsuit^\phi(\varphi \ \& \ \chi)$ is satisfied $=_{df} \diamondsuit(U' \ \& \ \varphi \ \& \ \chi)$ is satisfied for every maximal $U' \subseteq U$ such that $\diamondsuit(U' \ \& \sim\phi)$.   {1-IX-35}

While not a general solution to the problem of interpreting iterated modalities it suffices for the cases I unavoidably encounter in doing my epistemology (see Suppe 1998a,b).

My epistemology is restricted to *a posteriori propositional knowledge*, being developed first for *seeing that*, then generalized. I summarize here some of the key developmental steps, but do not provide supporting arguments other than to point out that the analyses conform to the desiderata presented above. Full defense is the task of vol. I of *Facts, Theories, and Scientific Observation* (Suppe 1998a).

Given the close connection between seeing that and observing that, I first provided the following analysis:

S propositionally sees that θ if and only if
   (i) S simply sees one or more particulars $E_1, \ldots, E_m$ $(m \geq 1)$; hence S undergoes a visual process V as a result of looking at $E_1, \ldots, E_m$;
   (ii) S, knowing how to use φ and knowing how to use θ with the same propositional intent, as a result of undergoing V entertains the proposition φ with that propositional intent as being factually true or false;
   (iii) φ is factually true;
   (iv) there exists a conjunction C of partial world state descriptions and

optional knowledge K that S possesses such that C & $\sim \diamond^\phi(C$ & V & K & $\sim\phi)$ & $\diamond^\phi(C$ & K & $\sim\phi)$ & $\diamond V$;
(v) as a result of undergoing V, S believes that ϕ.[18]   {1-X-1}

The underlying idea here is that when one looks at something one goes into a sensory state or (a visual process comprised of a series of such states), and this is evidentially crucial to seeing that. But knowledge of the phenomenological contents of those sensory states does not provide the evidential basis for visual knowledge. Rather, such states help trigger the cognitive moves of entertaining and believing a proposition ϕ. If, at the same time, the visual process composed of those states is a decisive indicator of the truth of the proposition ϕ — in the sense that one couldn't be undergoing that process under the circumstances unless ϕ were true — then one sees that, hence knows that, ϕ. It is these empirical regularities (and not knowledge of phenomenological contents of sensory states together with their use in the rational evaluation of other beliefs) that carry the evidential burden in seeing that, and thus knowing that, P.

The analysis of seeing that ϕ is radical in its denial of the KK thesis: one need not have any knowledge of, beliefs about, or awareness of either those evidential regularities or the cognitive processes involved in belief formation in order to see that, hence know that, ϕ. Further, the belief formation process — the cognitive processes involved in satisfying clauses (ii) and (v) — generally play little or no role in providing the evidential basis for knowledge.[19] This feature of the analysis is crucial in accommodating the fact that the arguments employed in the interpretation of observational and experimental data often are invalid, may not be truth-preserving, and frequently employ inconsistent assumptions. It also is a key feature that enables the analysis to satisfy the further minimal rationality constraint that the cognitive and rational capabilities presupposed in an adequate epistemology must be compatible with known human cognitive capabilities, characteristics, and limitations.[20]

Evidential condition (iv) requires that visual process V be a decisive indicator of ϕ in the sense of conferring truth-evaluating certainty on ϕ. Differently put, satisfying (iv) entails the satisfaction of condition (iii). This feature denies one of the key assumptions underlying Gettier-type paradoxes. Thus, the analysis avoids such paradoxes without having to impose a further condition on justified true belief analyses (ibid., vol. 1, chap. X, sec. 8).

Analysis {1-X-1} next is generalized to include various other modes of perceptual knowledge as well as experiential knowledge via introspection (chap. VII, sec. 6) and other means not involving usual sensory receptors. To do so a class of *experiential processes*, construed as series of states of persons, is postulated (ibid., chap. X, sec. 1–2).

S experientially knows that θ if and only if

(i_e) S undergoes a visual process V;

(ii_e) S, knowing how to use ϕ and knowing how to use θ with the same propositional intent, as a result of undergoing V, entertains the proposition ϕ with that propositional intent as being factually true or false.

(iii_e) ϕ is factually true;

(iv_e) there exists a conjunction C of partial world state descriptions and optional knowledge K that S possesses such that $C \,\&\, {\sim}\,{\diamond}\phi(C \,\&\, V \,\&\, K \,\&\, {\sim}\phi) \,\&\, {\diamond}\phi(C \,\&\, K \,\&\, {\sim}\phi) \,\&\, {\diamond}V \,\&\, {\diamond}(V \,\&\, {\sim}\phi)$;

(v_e) as a result of undergoing V, S believes that ϕ.   {1-X-1$_e$}

Next, the analysis is extended to knowledge that involves memory (ibid., chap. X, sec. 3). *Memory processes* construed as sequences of states of persons are postulated.

S propositionally knows via memory that ϕ if and only if

(i_m) S undergoes a memory process M over the time interval $[t, t']$ such that *either*

S knows that θ at t by virtue of undergoing some experiential process V and stores the fact that θ in M

or

S undergoes experiential process V that is stored as memory process M;

(ii_m) S, knowing how to use ϕ and knowing how to use θ with the same propositional intent, as a result of undergoing V or M, entertains the proposition ϕ at some $t'' \le t'$ with that propositional intent as being factually true or false.

(iii_m) ϕ if factually true over $[t, t']$.

(iv_m) There exists a conjunction C of partial world state descriptions and optional knowledge K that S possesses such that $C \,\&\, {\sim}\,{\diamond}\phi(C \,\&\, V \,\&\, M \,\&\, K \,\&\, {\sim}\phi) \,\&\, {\diamond}\phi(C \,\&\, K \,\&\, {\sim}\phi) \,\&\, {\diamond}(V \,\&\, M)$.

(v_m) As a result of undergoing V or M, S believes that ϕ at t'.   {1-X-6}

The analysis is presented for tenseless ϕ only. If tense changes are required in ϕ over $[t, t']$, complications of it will be required to accurately reflect those changes. In clause (iv_m) one, but not both, of the V and M can be deleted. The third conjunction in (iv_m) precludes knowledge that ϕ from being among the prior knowledge K to which the contextual necessity evidential requirement is relativized. The analysis allows that both V and ϕ could be stored (separately) in memory to provide knowledge at some subsequent t'.

Epistemic processes construed as series of states of persons are postulated to account for being in a state of prior knowledge. Inferential knowledge is subsumed under a more general class of epistemic achievements (ibid., chap. X, sec. 5):

S propositionally knows that θ *on the basis of knowing that* $\varphi_1, \ldots, \varphi_n$ *if and only if*

(i_k) S undergoes epistemic processes $K_1, \ldots, K_n$;

(ii_k) S, knowing how to use ϕ and knowing how to use θ with the same propositional intent, as a result of undergoing $K_1, \ldots, K_n$ entertains the proposition ϕ with that propositional intent as being factually true or false;

(iii_k) 'ϕ' is factually true;

(iv_k) there exists a conjunction C of partial world state descriptions such that $C\ \&\ \sim\!\diamondsuit^\phi(C\ \&\ K_1\ \&\ \ldots\ \&\ K_n\ \&\ \sim\!\phi)\ \&\ \diamondsuit^\phi(C\ \&\ \sim\!\phi)\ \&\ \diamondsuit(K_1\ \&\ \ldots\ \&\ K_n)$;

(v_k) as a result of undergoing $K_1, \ldots, K_n$, S believes that ϕ.   {1-X-25}

Note that inference is relegated to the cognitive processes involved in satisfaction of clauses (i), (ii), or (v). They generally play no evidential role in (iv). This indicates how thoroughly my epistemology denies the KK thesis.

Now that the detailed analyses have been laid out, I can say more about my decision to handle inference this way, rather than in ways that would make arguments evidential. To do so in a way that paralleled other cases of the analysis I would have had to postulate a set of inferential processes that functioned in analogue to memory processes, where undergoing the inferential processes was a decisive indicator that ϕ. The only ways I could make that work required assuming consistent belief sets and/or unrealistic logical capabilities for S. In short, allowing inference to be evidential ran afoul of minimal rationality constraints, whereas {1-X-25} does not.

Regardless of one's stance on the epistemic relevance or irrelevance of probabilistic inductive logic, *any epistemology adequate for science must allow for probabilistic descriptive and statistical knowledge.* Laying the basis for empirical probabilistic knowledge requires a physical interpretation of empirical probabilities. Over the actual world history, various relative frequencies of events obtain. In Suppe 1998a, chapter IX, section 9, I adapted Van Fraassen's modal frequency theory of probability to the general ontology and metaphysics underlying my account of actual world uniformities and the causal modalities.[21] Intuitively this consists of embedding the relative frequencies actually exemplified over the actual world history into richer probability distribution spaces having the compactness and other mathematical properties required to yield probabilities. The ontological commitments associated with this interpretation of empirical probabilities can be relatively minimal.

In order to enable probabilistic knowledge under my epistemology it is necessary to expand the C satisfying evidential clause (iv) to include such probability distribution spaces, as well as partial world state descriptions. The result is that the clauses (iv) of these various epistemological analyses are generalized along the following lines:

There exists a conjunction C of partial world state descriptions and probability distribution spaces[22] such that C & $\sim\!\diamond^\phi(C$ & R & $\sim\!\phi)$ & $\diamond^\phi(C$ & $\sim\!\phi)$ & $\diamond R$. $\quad\quad\quad\quad\quad\quad\{1\text{-IX-}40'\}$

which, unpacked in accordance with $\{1\text{-IX-}35\}$ is equivalent to

There exists a conjunction C of partial world state descriptions and probability spaces such that for every maximal $U' \subseteq U$ satisfying $\diamond(U'$ & C & $\sim\!P)$, C & $\sim\!\diamond(U'$ & C & R & $\sim\!\phi)$ & $\diamond(U$ & $R)$. $\quad\quad\{1\text{-IX-}39'\}$

So generalized analyses $\{1\text{-X-}1_e\}$, $\{1\text{-X-}6\}$, and $\{1\text{-X-}25\}$ can be combined into a general analysis of a posteriori knowledge by postulating a class R of cognitive processes that includes experiential processes, memory processes, epistemic processes, and possibly other sorts of cognitive processes or states (ibid., chap. X, sec. 6).

My general analysis of a posteriori knowledge becomes:

S propositionally knows that θ if and only if
  (i) S undergoes a cognitive process R, or S has prior knowledge that K;
  (ii) S, knowing how to use ϕ and knowing how to use θ with the same propositional intent, as a result of undergoing R or having prior knowledge that K entertains the proposition ϕ as being factually true or false;
  (iii) 'ϕ' is factually true;
  (iv) there exists a conjunction C of partial world state descriptions and probability spaces such that C & $\sim\!\diamond^\phi(C$ & R & K & $\sim\!\phi)$ & $\diamond^\phi(C$ & $\sim\!\phi)$ & $\diamond R$ & $\diamond(R$ & $\sim\!\phi)$;
  (v) as a result of undergoing R or K, S believes that ϕ. $\quad\{1\text{-X-}29\}$

Either R or K but not both may be deleted from the analysis.

The analysis is formal. Many epistemologies attempt to provide formal definitions and conditions. But it is relatively infrequent for such epistemologies to yield nontrivial theorems or corollaries that are philo-sophically significant. In the next section we will see that {1-X-29} yields extremely powerful theorems that enable a robust noninductive ac-count of scientific practice and knowledge.

How good is the analysis? In Volume I (Suppe 1998a) I show the analysis coheres with our best scientific understanding of perception and cognition, that it avoids the main difficulties that have plagued other epistemologies, and that when it is run through the gauntlet of counterexamples that have scuttled other competing epistemologies it performs exactly as it should — hence, that it is superior to the competi-tion. It also should be clear that analysis {1-X-29} satisfies, in the ab-stract at least, most of the conditions we uncovered above. It is a nonreliabilist externalism that exploits empirical regularities eviden-tially, denies the KK thesis, avoids Gettier paradoxes in virtue of a truth-evaluating certainty evidential condition, and allows (by virtue of {1-X-25}) arguments to yield knowledge without presupposing consis-tent belief sets or violating minimal rationality constraints. It remains to be shown that it allows general knowledge to be obtained noninduc-tively, and allows for efficient transmission of knowledge through pub-lications, etc.

Noninductive Scientific Knowledge

I now show that the noninductive approach allowed by my epistemol-ogy corresponds sufficiently with actual scientific practice that it can handle real scientific knowledge.

General Knowledge

In Suppe 1998a and 1998b, I argue that humans regularly are able to experientially know that ϕ where ϕ is a singular causal proposition. (See vol. 1, chap. VII, sec. 5a; vol. 2, chap. IV, sec. 1.) Assuming that observation is a species of experiential knowing that, it follows that one is able to observe that ϕ where ϕ is a singular causal proposition.[23]

From the classical inductivist view, generalizations are obtained in-ferentially from a variety of singular observations. Such a view is not particularly faithful to actual scientific practice. As Torretti notes, ob-servations are implicitly general. "As soon as one begins to observe . . . one does in effect subsume . . . that . . . which one is observing . . . under a general concept. . . . The observer conceives of them as physical systems of a certain kind, which interact according to certain laws with objects of the class under investigation" (1986, 6). Thus, there is at least an implicit generality to observational claims.

Observations result in *observation reports* that make claims about interpreted data and observational results. The higher the level of the interpretation, the more likely that the observation report will be explicitly general. Sophisticated observations rarely issue in the "swan #1 is white," "swan #2 is white," etc., specificity assumed in induction-by-simple-enumeration accounts. Rather than claiming

(1) Using experimental design d, $m_1, \ldots, m_n$ were subjected to c and effects $e_1, \ldots, e_n$ were recorded,

instead we often find issued reports such as

(2) Whenever m are subjected to c under observational design d, effects of sort e occur.

When I examine J. J. Thomson's cathode ray experiments, observational claims of form (2) will be encountered.

In deterministic or classical causal contexts the point can be put as follows: Under circumstances where conventional inductive wisdom would expect scientists to issue singular observational claims, in sophisticated branches of science we instead find general observational claims being made. Instead of dismissing such practices out of hand as careless, excessive, or confused, suppose they are legitimate and warranted under ordinary observational circumstances. If so, it becomes plausible to suppose that the very same circumstances that enable singular observational knowledge (often) just as well permit general observational knowledge. I want to take this idea very seriously. Exploring it will be the task of this and the next section.

The following are particularly important corollaries to {1-X-29}:

Let ϕ and θ be propositions and suppose that S satisfies clauses (i)–(iii) and (v) of {1-X-29} with respect to them, K, and cognitive process R. Then S knows that θ if there is a proposition φ such that
   (α) φ is factually true.
   (γ) There is a C satisfying clause (iv) of {1-X-29} with respect to R, K, and φ; $\Diamond^{\phi}(C \,\&\, \sim\phi)$; and $\Diamond(R \,\&\, \sim\phi)$.
   (δ) φ causally entails ϕ.   {1-X-31}

This shows that individuals can know some, but not all, implications of what they know or could observe.

This result underlies the next two results which employ a *nonparadoxical causal implication* operator due to Burks 1977:

$$\varphi\beta \textbf{ npc } \chi\beta = {}_{df}[c](\varphi\beta \supset \chi\beta) \;\&\; \diamond(\varphi\beta \;\&\; \chi\beta) \;\&\; \diamond(\varphi\beta \;\&\;$$
$$\sim\chi\beta) \;\&\; \diamond(\sim\varphi\beta \;\&\; \chi\beta) \;\&\; \diamond(\sim\varphi\beta \;\&\; \sim\chi\beta) \;\&\;$$
$$\diamond\varphi\beta \;\&\; \sim[c]\chi\beta$$

This operator proves to be a sensitive vehicle for modeling scientific causal conditionals.[24] The following is a quite powerful result about knowledge of singular **npc** conditionals and their generalizations.

Let ϕ be a singular proposition of the form $\varphi\beta$ **npc** $\chi\beta$. Suppose that S knows that ϕ and that S satisfies clauses (i)–(iii) and (v) of {1-X-29} with respect to "knowing that ϕ" cognitive process R, prior knowledge K, and $(\alpha)(\varphi\alpha$ **npc** $\chi\alpha)$. Then S knows that $(\alpha)(\varphi\alpha$ **npc** $\chi\alpha)$. {1-X-32}

The next result is an even stronger version that does not require that S *ever* know the singular instance $\Psi\beta$ **npc** $\chi\beta$ prior to knowing the corresponding generalization.

Let ϕ and θ be propositions where ϕ is a closed singular proposition of the form $\psi\beta$ **npc** $\chi\beta$. Suppose there are R, K and C such that
    (i) S satisfies clauses (i)–(iii), (v) of {1-X-29} with respect to R, K, C, θ and $(\alpha)(\psi\alpha$ **npc** $\chi\alpha)$ (where '$\psi\alpha$ **npc** $\chi\alpha$' is just like '$\psi\beta$ **npc** $\chi\beta$' except that the variable α occurs free in the former wherever B occurs free in the latter);
    (ii) S satisfies clause (iv) of {1-X-29} with respect to R, C, K, ϕ, and θ.
Then S knows that $(\alpha)(\psi\alpha$ **npc** $\chi\alpha)$. {2-IV-5}

This result is pivotal in enabling us to have robust general knowledge that is noninductive in both senses articulated above: it does not depend on probabilistic induction, and it does not require moving from knowledge of the singular to knowledge of the general by an additional ampliative inference.

These two results, which initially will strike many as counterintuitive, hold only for causally necessary conditionals. They do not hold for nonmodal universal conditionals. Even so, they may seem surprising. Their proofs turn crucially on the way in which modal operators are like multiply-general quantifiers on my physical interpretation of the causal modalities. (See also Montague 1974.) These two results together with the fact that we *do* perceptually know singular causal propositions that appropriately are modeled by $\varphi\beta$ **npc** $\chi\beta$ conditionals entails that we can have knowledge of causal generalizations *noninduc-tively* in both senses indicated above.

Although {2-IV-5} allows one to know generalizations on the basis of single observational episodes, it does not require that all observational knowledge be based on single episodes. For analysis {1-X-29} allows one's ability to perceptually know in a given situation to be enhanced by prior knowledge. Thus, we may come to the observational situation arrived with prior experimental and observational knowledge $K_1, \ldots, K_m$. It may be that the setup is such that, absent that prior knowledge, our experiential process R would not be a decisive indicator that O, but given that we have prior knowledge $K_1, \ldots, K_m$ it is a decisive indicator.[25] Thus, it may be that the observational knowledge that O must be built up out of a complex of observations. However, the analysis does require that there be a specific episode in which one does come to experientially know that O.

Whether an observational result depends on a single or on multiple observational episodes, theorem {2-IV-5} indicates that it is epistemologically legitimate for scientists to present their observational results in the form of generalizations such as (2) rather than restricting themselves to the singular (1) analogues. For when those singular claims are **npc** conditionals, an evidential basis adequate for sustaining (2) is also adequate to sustain (1). Thus, the practice of generalizing one's observations is consistent with {2-IV-5}.

Knowledge Transmission

One final corollary concerns sufficient conditions for the transference of knowledge via testimony, journal articles, and the like.[26]

Suppose that S knows that ϕ under circumstances C in virtue of cognitive process R_S. Suppose further that as a result of process R_S, in context C' compatible with C S transmits the information that ϕ so as to produce the cognitive process R'_{S*} in S^*. If S^* believes that ϕ as a result of R'_{S*} in conformity with the remaining clauses of {1-X-29}, then S^* knows that ϕ on the basis of S's testimony, write-up, or whatever.[27]   {2-I-7}

The empirical assumptions here about the transmission process are quite weak. For any person who reads a scientific report, undergoes a cognitive process, and as a result believes the report, will obtain the knowledge it contains, provided that the original knower's decisive indicator cognitive state was a causal factor in writing the paper and transmitting it to the reader so as to produce the latter's cognitive process. Robust testimonial knowledge without dilution of evidence is allowed by {1-X-29} in this manner.

I now have shown that the analysis {1-X-29} has the remaining features we found desirable in an epistemology for science. However, it does so in a manner that makes problematic the function of interpretative arguments found in scientific papers. I now turn to this issue.

Credentialing General Claims

In virtue of the wedge driven between the evidence or justification for knowledge and belief formation processes, it is a corollary of {1-X-29} that obtaining knowledge and the rational defense of knowledge claims are two distinct enterprises.[28] A major theme of mine in *Facts, Theories, and Scientific Observation* is that, all too often, failure to keep distinct these two enterprises results in level confusions that repeatedly have sabotaged otherwise sophisticated epistemological efforts.[29] The KK thesis is a particularly pernicious vehicle for such level confusions. Having insisted that, I also want to claim that both dimensions must be taken into account if one is to properly understand scientific knowledge.

Science is a communal knowledge-seeking activity done within disciplines possessing shared conceptions, interests, and assumptions. It is a cooperative enterprise that (unlike philosophy) depends essentially on being able to reliably build upon the work of others — especially those within the discipline. Corollary {2-I-7} above to {1-X-29} shows how the production of a cognitive process in the recipient via the process of transmitting known information routinely can produce knowledge, provided that the recipient believes the transmitted information as a result of being in that process. For *knowledge* to qualify as *scientific knowledge*, it must be admitted by the scientific community (discipline) into its body of putative knowledge (its domain). Science typically demands that knowledge be *credentialed* before admitting it into the domain. The credentialing process is reflected in the standardized ways that observational, experimental, and theoretical results are written up for journal publication. The credentializing always is done against the background of a discipline's shared domain, background beliefs, presuppositions, and evidential standards or canons of reasoning. There is a strong social dimension to this credentialing process.

Such credentialing amounts to the demand that before knowledge is admitted into a discipline's domain of putative knowledge, it must be augmented by a defense of the associated knowledge claim that meets with the standards and practices of the discipline. In short, scientific knowledge involves the production of a posteriori knowledge *and* the rational defense of knowledge claims. It consists in beliefs that meet the evidential standards of analysis {1-X-29} and *also* meet the justificatory standards for claims to know imposed by the socially constituted discipline. Thus, to understand scientific knowledge it is absolutely crucial that we can distinguish and keep separate these two key constituents. Level confusions threaten any real understanding of scientific knowledge, and invite such hopelessly inadequate accounts as the view that knowledge *überhaupt* is *just* congeries of group opinion (Kuhn 1970) or that epistemology should become the sociology of knowledge (Fuller 1988, 1989; the Strong Programme — Bloor 1976 and Ashmore 1989). The relativisms endemic to such views *are* appropriate to the credentialing process. But they do not accrue to knowledge itself, which, in accordance with {1-X-29}, *is* highly objective.

The processes of credentialing and transmitting scientific knowledge involve belief formation, knowing that one knows, defending or justifying claims to know, social influences on knowledge, and bootstrapping. Note that these include those aspects of the epistemological/doxological landscape I have charged epistemologists tend to conflate with knowing that φ. Thus I acknowledge their importance in understanding human cognitive access to the external world; but at the same time I am concerned to understand what that influence is — how these other separable enterprises connect with the enterprise of knowing that φ.

My discussion here of these issues ancillary to understanding knowing that is confined, for the most part, to scientific contexts. Work on rationality and the growth of scientific knowledge makes it clear that different branches of science employ different approaches to the credentialing aspect of scientific knowledge and that within disciplines there is considerable evolution and change of such canons of rationality for defense of knowledge claims (see Suppe 1998b, chap. VI). Furthermore, science is not the only branch of knowledge that imposes credentialing standards. Law and theology are two other obvious examples. And despite occasional attempts to tie one to the other (Toulmin 1972; Peacocke 1984), it is an open question whether there are any profound, or even very interesting, common credentialing standards that cut across cooperative disciplines within or without science.

What distinguishes ordinary perception from scientific observation is these credentialing dimensions. In *Facts, Theories, and Scientific Observation*, I offer the following analysis:

> Relative to research group or discipline G, S *scientifically observes that* ϕ *only if*
>
> a. S satisfies $\{1\text{-}X\text{-}29\}$ with respect to ϕ and a cognitive process R that contains an experiential process E such that $\{1\text{-}X\text{-}29\}$ would not be satisfied were E deleted from R.
>
> b. ϕ satisfies the relevance conditions associated with G, its domain, and its background.
>
> c. S's claim that ϕ was produced in accordance with the replicability standards imposed by G.
>
> d. S's claim that ϕ is transmitted in an observation report which provides arguments addressing the actual or potential specific doubts about ϕ legitimized by G, and those arguments meet the adequacy standards for such arguments operant in G. $\{2\text{-}III\text{-}3\}$

Clauses (b) through (d) concern credentialing of scientific claims. Clause (d) parallels the interpretative argument forms we uncovered previously. Note that $\{2\text{-}III\text{-}3\}$ only specifies necessary, not sufficient, conditions; for some scientific disciplines impose further conditions in the credentialing process. The development and defense of $\{2\text{-}III\text{-}3\}$ is long and involved.[30] Here I will assume the analysis without argument. My use of it is primarily illustrative of how obtaining knowledge and the credentialing of knowledge claims come together in scientific observation.

My thesis is that the discursive and argumentative content of a scientific paper largely concerns the credentialing practices.[31] More specifically, the interpretative arguments found in a paper have as their focus the credentialing of scientific claims. My focus thus will be on clause (d) of $\{2\text{-}III\text{-}3\}$.

Clause (d) requires that there be arguments that adequately undercut actual or potential legitimate doubts surrounding the generalized claim. One question is whether any singular episode of observing that is capable of undercutting the specific doubts that can be raised about the corresponding generalization. To do so, one would need to show that the form (1) results were not representative of the general situation, and, hence, that the corresponding form (2) generalization cannot be true.

Sometimes this is possible. The observed connection between $m_1, \ldots, m_n$ and $e_1, \ldots, e_n$ under conditions d and c may be an accidental correlation. In such cases the generalization would be inappropriate. However, if the singular observation report is that subjecting $m_1, \ldots, m_n$ to c under circumstances d *caused* $e_1, \ldots, e_n$ to happen, then it is quite a different story. For, in that case, the form (1) observation report embodies the claim

$$(3) \; d \; \& \; m_1 \; \& \; \ldots m_n \; \& \; c \; \textbf{npc} \; e_1, \ldots, e_n$$

And the import of {1-IX-29} is that if (3) is factually true, then so too must be its generalization.[32] Hence, any doubts sufficient to undercut the generalization also should be sufficient to undercut the form (3) singular observation claim as well. Both are equally secure, equally problematic. The issue thus is simply a matter of fact: Does science with regularity produce observations resulting in generalized observation reports that are free from specific doubts? Does it do so in a way that exemplifies my noninductive view of science, wherein the arguments found in a scientific paper are directed mostly at credentialing claims rather than on providing the evidence whereby those claims do constitute knowledge? My answer to both questions is, "Yes." It is time I consider some specific facts. I will try to render my analysis plausible by looking in detail at a specific case.

Case Study: J. J. Thomson's Cathode Ray Experiments

Thomson's experiments were undertaken in the area of cathode ray studies, which was a subarea of electrical physics studies concerning the discharge of electricity through gases. At the time of Thomson's experiments, the following were key elements in the cathode ray domain: When the gas pressure in a gas-filled tube is lowered below .01 mm of mercury, greenish phosphorescent glows are observed near the cathode (negative electrode), which are caused by straight-line rays emanating from the cathode. The rays are perpendicular to the cathode, do not depend on the specific metal used to construct the cathode, and they cause silver salt to change color. Cathode rays can be deflected by magnetic fields. Hertz in 1883 had undertaken experiments in which cathode rays failed to be deflected by electrical fields. In 1891 he and his student Phillip Lenard discovered that cathode rays could penetrate thin layers of metal foils which were too thick for molecules or atoms to pass through. Exploring the idea that cathode rays might be composed of negatively charged particles, in 1890 Arthur Schuster had experimentally determined upper and lower bounds on mass-to-charge ratios that would be characteristic of such particles. The ratio of the hydrogen atom was known and fell within these limits.

All of the above had been determined experimentally. Motivating many of the experiments was a theoretical debate over the nature of cathode rays. Based on his own cathode ray experiments, William Crookes postulated in 1879 that cathode rays were comprised of gas molecules and showed how his charged molecules theory could explain much of the behavior of cathode rays. Against Crookes's charged particle theory there was the German ether wave theory of Hertz, Wiedemann, and Goldstein. The Hertz and Lenard experiments mentioned above had been undertaken to discredit Crookes's charged particle theory. Schuster's observations were offered in defense of a variation on Crookes's theory. The standard electromagnetic theory was Maxwell's, but its correctness was becoming problematic due, among other things, to the 1887 Michelson-Morley results. Beginning in 1890, H. A. Lorenz began developing his electron theory that postulated the existence of a negatively charged particle, which he presented in papers published in 1892 and 1895.

It was against this experimental and theoretical background that J. J. Thomson began his cathode ray experiments, beginning with an erroneously low determination of the velocity of cathode rays. In 1895, Jean Perrin performed experiments showing that the collector became negatively charged when cathode rays entered it, which he interpreted as showing that cathode rays carry a negative charge. Against this interpretation, specific doubts can be raised in terms of the possibility that the rays might initially be accompanied by, but not constituted of, negatively charged particles. In 1897, Thomson performed a variant replication of Perrin's experiment, where the variant experimental setup (using the tube in figure 13.1) was designed to undercut this specific doubt against Perrin's observational interpretation. On the charged particle theory, both magnetic and electric fields should result in deflection of cathode rays. Such deflections were known to be produced by magnetic fields, but in Hertz's experiments no such deflection occurred as a result of electrical fields. Against Hertz's experiments, Thomson raised the specific doubt that the gas in the cathode ray tube might act like a conductor that screens off the electric force from the

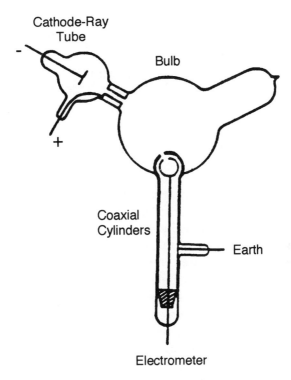

Cathode-Ray
Tube

Bulb

Coaxial
Cylinders

Earth

Electrometer

Figure 13.1 J. J. Thomson's cathode-ray apparatus used in his first series of 1897 experiments. This apparatus was a modification of that used by Jean Perrin. (Original drawing by Thomson, 1897, with text modified.)

charged particle. If so, a higher rate of exhaustion of gas from the tube should reduce the effect and allow one to observe the predicted deflections in the presence of electrical fields. Thomson achieved the higher levels of evacuation needed using the tube in figure 13.2, and the deflections were observed. These two experiments convinced him that cathode rays are negatively charged particles.

Next Thomson asked, *What sorts of particles?* To answer this, he used two sorts of tubes, one of which is a variation on the tube in figure 13.2 in which two plates are omitted, to calculate the ratio of the mass (m) of these particles to their charge (e) on the basis of observationally determined measures of cross-sectional electric charge of the beam (Q) subjected to a uniform magnetic field of strength H deflecting the beam with curvature r, so as to strike a solid body, thus converting the ray's total kinetic energy W into body heat. He undercut vari-

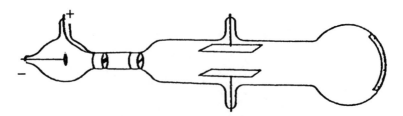

Figure 13.2 J. J. Thomson's cathode-ray apparatus used in his second series of 1897 experiments. A similar apparatus with the two parallel plates omitted was one of two types of tube used in his third 1897 series to determine m/e. (Original drawing by Thomson, 1897.)

ous potential specific doubts by experimentally showing that m/e is independent of the nature of the gas in the tube and of the cathode metal. In all of his measurements under variant gases and cathode compositions, m/e has magnitudes of order 10^{-7}. He noted that this is very small compared to the smallest m/e value then known (10^{-4}) — that of the hydrogen atom. Only three hypotheses accounting for the 10^{-7} figure are plausible under his experimental setup: the largeness of e or the smallness of m or a combination of both. But Lenard's results on cathode ray penetration of metal foil layers were incompatible with the largeness of e, and so he concluded that the charged particles must be small compared with ordinary molecules.

In May 1897, Thomson published a preliminary report of this series of experiments. A fuller report was given in October in Thomson 1897. In the latter paper, he resorted to a floating magnet model due to Mayer, to undercut specific doubts as to whether a chemical atom could be stable if composed of mutually repellent negatively charged particles.

But his 1897 arguments did not undercut all the specific doubts legitimized in the cathode ray research community, and few physicists then were willing to accept the existence of negatively charged particles on the basis of his arguments and experiments. (See Heilbron 1970, 367.) For example, FitzGerald 1897 commented on Thomson's 1897 report, finding the hypothesis that cathode rays are negatively charged particles plausible enough to be worth pursuing, but claimed it was not established; referring to "an *embarras de richesse* in the way of [other] possible explanations," including the possibility that cathode rays are composed of "free electrons," which are not a constituent part of the atom and not produced by disassociation of the atom (see FitzGerald 1897, 104). Specific doubts remained.

Thomson went on to perform experiments with negatively charged particles produced by ultraviolet light falling on an electrified metal plate and with particles produced by heating carbon filaments in a hydrogen atmosphere. In both cases, he determined m/e to be of 10^{-7} order of magnitude. In the ultraviolet light experiments, he determined a separate value of the charge e for the particles, which was of the same magnitude as the charge of the hydrogen atom in electrolysis. These results (Thomson 1899) thus provided variant replications of his 1897 determinations of m/e and laid to rest any doubts as to whether those values were due to the particles being smaller than atoms.[33] In particular, the separate determination of the charge e in the ultraviolet case rendered implausible nonparticle interpretations such as FitzGerald played with. This was in part due to the argumentative force of "triangulation" arguments provided by variant replications that I discussed in Suppe 1993.

With his 1899 results, the experiments and observational arguments Thomson had advanced in support of his charged particle interpretation collectively were quite convincing at undercutting legitimized specific doubts. Indeed, they were so convincing that between 1897 and 1900 persons of all theoretical persuasions, including the ether wave partisans, came to accept Thomson's theory (see Galison 1987, 31). Considering that his variant replications of the determination of m/e were only published in December 1899, the acceptance of his experimental results was virtually instantaneous. Clearly, all specific doubts legitimized by the cathode ray disciplinary area had been laid to rest.

Thomson consistently refers to these particles as "corpuscles." Although the term "electron" had been introduced by Stoney in 1891 to refer to a natural unit of electricity, FitzGerald, Rutherford, and others applied the term "electron" to Thomson's particles and the name stuck (see Hacking 1984, 261). Thus, Thomson is credited with the discovery of the electron.

Not all careful experiments are so successful as Thomson's. For example, Einstein's 1905 photoelectric paper put forth his quantum theory, but because of experimental difficulties it was not put to the test until R. A. Millikan's painstaking photoelectric measurements in 1915–1916. But in the interim, alternate theoretical accounts of the photoelectric effect had been produced. Thus, "when the long-awaited experimental confirmation came, it was not accepted as such! The fact that Lorentz, Thomson, Sommerfield, and Richardson had been in varying degrees successful in developing alternate interpretations of the photoelectric effect goes a long way in accounting for this remarkable situation. There were of course other factors, perhaps the most significant being von Laue's very striking 1912 discovery of the crystal defraction of x-rays" (Stuewer 1970, 260). In my terms, Millikan's careful experiments could not then be augmented with observational arguments against the specific doubts raised by these other developments. It wasn't until roughly 1924, after Compton's X-ray scattering experiments, that these specific doubts about the quantum hypothesis or the quantum interpretation of Millikan's photoelectric experiments were laid to rest (ibid., 261).

My interest in the Thomson cathode ray case is with what light it sheds on the ability of observations to establish generalized observation claims of form (2) rather than mere singular claims about what was observed in one's experimental or observational runs (form-[1] reports). The foregoing historical sketch is not sufficient. It is necessary to look more closely at the specifics of Thomson's experiments and arguments. I begin with Thomson's variant replication of Perrin's experiments. The specific doubts concerning those experiments were that what carried the negative charges might not be cathode rays, but rather something else temporarily associated with it (like the flash of a bullet exiting a muzzle). Thomson's variant apparatus (figure 13.1), which has a bulb containing two coaxial cylinders with slits, is connected to the cathode ray tube. The cathode rays travel through a slit in a metal plug in the neck of the cathode ray tube, and when the rays are deflected by a magnet, they enter into the inner cylinder of the bulb. An electroscope is connected to that cylinder. The path of the ray is indicated by a phosphorescent trace. When the ray enters the bulb and cylinder, a strongly negative charge is registered on the electroscope. When it does not enter the bulb and cylinder, the charge recorded by the electroscope is small and irregular. As the negative charge thus accompanies the cathode ray over its entire path, the observed results are incompatible with the alternate account that provided specific doubts against the charged particle interpretation of Perrin's experiments. He argues that his results show that the "negative electrification is indissolubly connected with the cathode ray" (Thomson 1897, 295). That is, the experimental setup has been contrived so that *only* the cathode ray could have *caused* the electroscope readings, not some other accompanying whatever.

The description already given of Thomson's electric field deflection variant replication of Hertz's unsuccessful ones is adequate for the purposes at hand. For it is clear that his experimental setup was able to show that the electric fields were the cause of the deflections in the cathode rays. So far as Thomson is concerned, his improved replications of Perrin's and Hertz's experiments are sufficient to undercut specific doubts raised against them, and so he concludes in addition that electrons are negatively charged particles. As we saw, however, other specific doubts were raised and only his 1899 experiments succeeded in undercutting those.

His m/e ratio experiments proceed on the assumption he has shown that cathode rays are composed of negatively charged particles. His concern is to determine their size. Unlike the previous two series of experiments he does not proceed to produce clear causal effects. Rather, he makes measurements of other quantities and then uses those to make a theoretical calculation of m/e. Achinstein provides a particularly illuminating analysis of Thomson's procedure here (1991, essay 11, sec. 2). His analysis employs the following analytic notions: In schematic form, *experimental result claims* are propositions asserting that

when "substance" s is made to interact in such and such a way with an experimental apparatus, which is also described, changes indicating so and so are produced in some specified detector or measuring device(s) associated with the apparatus. (Ibid., 302)[34]

The claims need not be true or based on actual experiments. Note that interpretations of the results enter such claims at the point "changes indicating so and so." One and the same experiment admits of different levels of interpretation (see Suppe 1998b, chap. 2, sec. 8), and different levels of interpretation admit of different experimental result claims. A *theoretical consequence* is any proposition that "follows from the theory together possibly with other auxiliary assumptions the scientist is making when he proposes the theory" (Achinstein 1991, 305).

Thomson's 1897 experimental determination of m/e depends on the following derivation of a theoretical consequence of his negatively charged particle theory. Thomson begins with the assumption that

$$(4)\ Ne = Q$$

where N is the number of particles in a beam of cathode rays crossing any section of the beam in unit time, e is the charge on an individual particle, assumed to be the same for all particles, and Q is the quantity of electricity carried by the N particles. Next he characterizes the total kinetic energy W in terms of the kinetic energy of individual particles.

$$(5)\ \frac{1}{2}Nmv^2 = W$$

Next he assumes total conversion of the kinetic energy of cathode rays into heat when they strike a solid body. He also assumes that if a cathode particle is subjected to a magnetic force, then the magnetic force is equal to the centripetal force for circular motion:

$$(6)\ \frac{mv}{e} = Hr = I,$$

where m is the particle mass, v its velocity, e its charge, H the strength of the magnetic field, and r the radius of curvature of the particle. (I is just an abbreviation for Hr.) From (4), (5), (6) he easily derives

$$(7)\ m/e = I^2Q/2W,$$

which Achinstein observes "is a theoretical consequence. However, Thomson's aim is not to test (7) experimentally, but simply to determine values for I, Q, and W experimentally. This, together with (7), will then yield a value for m/e, the ratio of mass to charge of the cathode ray particles" (Achinstein 1991, 306).[35]

Achinstein goes on to note, "Thomson clearly believes that m/e should be a constant" across gasses, cathode material, etc. (ibid.). From that assumption and (7) he "at least implicitly" draws the conclusion

$$(8)\ I^2Q/2W = \text{constant},$$

another theoretical consequence.

The point of the experiments is to determine the value of that constant. He uses a variety of different cathode ray tubes and measures temperature rise in solid target bodies, strength of magnetic field, and radius of curvature of the rays. He presents the results of these measurements in tables that have columns for measured value of W/Q and I and for the value of m/e calculated via (7). Rows of the table indicate the gasses filling the tubes for the various runs (Thomson 1897, 306). The values for the m/e so obtained range from $.32 \times 10^{-7}$ to $.57 \times 10^{-7}$. This results in two forms of experimental result claims Thomson makes, which Achinstein analyzes as:

(9) When cathode rays from a cathode ray tube of such and such a kind were deflected by a magnetic field produced in a way that is specified, then measurements of the quantities W/Q and I taken using devices of the sort Thomson describes yielded results given in the . . . table [described above].

and

(10) When cathode rays from a cathode ray tube are deflected by a magnetic field produced in such and such a way, then measurements of the quantities W/Q and I made using devices of the sort Thomson describes indicate that m/e has a value whose order of magnitude is 10^{-7}. (Achinstein 1991, 308)

Note that (9) and (10) respectively are a singular claim of form (1) and a general claim of form (2) above. Achinstein notes that Thomson uses both forms of experimental result claim. In both cases the conditionals ('then') appropriately are analyzed as being **npc** ones.

I noted above FitzGerald's and others' doubts about Thomson's claims that cathode rays are composed of charged particles smaller than atoms. These doubts are against the charged particle claim, and do not differentially effect (9) and (10). The point of contention was not the general vs. the singular version. This is exactly as it should be. For the specific doubts raised against the charged particle interpretation of the first two experimental results do not call into question the claim that cathode rays carry a negative charge or that the cathode rays caused the recorded negative charges. It is the unproblematic causal character of the experimental results (suitably interpreted as **npc** conditionals) — and not the disputed particle claims — that makes (9) and (10) equally legitimate. For, via {1-IX-29}, whenever the singular causal claim is true, so too must be its universal generalization.

To be sure, there are specific doubts about both (9) and (10) stemming from the fact that the m/e values given in both depend upon Thomson's derivation of (7). But (9) and (10) depend equally on that derivation, and so doubts about the derivation do not favor one over the other. More importantly, (9) and (10) can be easily unloaded so as to not depend on any particle assumptions for their interpretation. As

Achinstein notes, "these experimental results could just as well have been recorded using *increase in temperature* produced by the cathode rays . . . and using $I^2Q/2W$ instead of m/e" and so "one might well formulate and accept Thomson's experimental result claims without necessarily being committed to the theory that cathode rays are negatively charged particles whose ratio of mass is a constant" (1991, 308).

The results of so unloading (9) and (10) — call them (9') and (10') — are parallel singular and generalized experimental result claims that are exempt from all specific doubts that might be legitimized by the 1897 cathode ray research community. Recall that what initially prompted me to look at the Thomson case was the need for a historical case that would illustrate and give credence to my assertion that observation reports that make generalized, as opposed to singular, observation claims can meet the specific doubt requirement (d) of my analysis {2-III-3} of observing that. Unloaded claims (9') and (10') are precisely the sort of example I need. Further, it is clear from the details of the Thomson case that, with respect to (9') and (10') the relevance and replicability conditions of that analysis also are met. Further, we know that Thomson's observations were, in fact, decisive indicators of the causal claims made on the basis of the first two experimental series (see above), and thus that unloaded claims such as (9') and (10') can satisfy the epistemological requirement (a) of {2-III-3}.[36] Thus, with respect to (9') and (10') we have a clear, realistic example how the full analysis {2-III-3} of observing that can be satisfied with respect to generalized observation claims having form (2). Further, I maintain that the Thomson case is archetypal of how specific doubts in fact can be eliminated with respect to form (2) observation claims.

We saw that the arguments Thomson puts forward in his papers are on a par with respect to the singular (form [1]) and general (form [2]) versions of the observational claim. It follows that the point of such arguments cannot be the ampliative inference of the general form the singular. In short, actual scientific practice here coheres better with my analysis than it does an inductive one.

Further Considerations

A full defense of my noninductive approach to scientific knowledge will involve showing its adequacy for theoretical knowledge and for probabilistic knowledge. I do not have time to explore these matters in detail. But let me sketch how I handle them in Suppe (1998b).

For theoretical knowledge I assume the quasi-realistic version of the semantic conception I present in detail in *The Semantic Conception of Theories and Scientific Realism* (1989). On that view to propound a theory $\mathfrak{I}$ is to claim that $\mathfrak{I}$ describes how systems within its scope would behave were they isolated from outside influence. Let L be the law of $\mathfrak{I}$, let I be the intended scope of $\mathfrak{I}$, let N be an isolation ("non-interference") condition. Then $\mathfrak{I}$ will be empirically true if and only if

$$(x)(Ix \ \& \ Nx \ \mathbf{npc} \ Lx) \qquad \{2\text{-III-}33\}$$

is true. Thus all that is required to know that $\mathfrak{I}$ is empirically true is to know that {2-III-33}. Thus, theoretical knowledge is a special case of the general knowledge afforded by {2-IV-5}.

For the purposes of science (as opposed to inductive logic) only empirical probabilities are required. I construe such probabilities in terms of a slightly more realistic version of van Fraassen's modal frequency theory (1980, chap. 6) and allow such probability spaces to be included in the C of {1-X-29}. Knowledge of samples is straightforward on {1-X-29}. I argue that the applications of analytically derived statistics in science are derived from samples and features of probability distribution spaces. Exploitation of corollary {1-X-31}, together with the fact that probability distribution spaces can be included among the C to which clause (iv) of {1-X-29} is relativized, enables me to know any legitimate applications of analytically derived statistics applied to sample results.

This does not exhaust the use of statistics in science. Increasingly science makes recourse to nonanalytic statistics that are "derived" from data analysis. Examples are the exploratory statistics that Tukey has championed and the use of Monte Carlos and other simulations to establish the robustness of statistics. An adequate epistemology for science has to accommodate recourse to robust statistics.[37] However, very little attention has been paid to such matters by philosophers. Key to dealing with them is an adequate epistemology of simulation modeling—a problem to which philosophers only recently have given much attention.[38] In vol. 2, chap. VIII of *Facts, Theories, and Scientific Observation* (Suppe 1977b) I show that {1-X-29} has the resources to make epistemological sense of simulation modeling and then develop an epistemology of robust statistics on the basis of that simulation modeling analysis.

Conclusions

I began with the observation that philosophy of science and epistemology have been insular with respect to each other's literature, to the detriment of both fields. I have urged that, when brought together, recent developments in both fields strongly suggest that a noninductive account of scientific knowledge is plausible. I have sketched, but not defended, some key features of such an epistemology and showed how it leads to a noninductive view of scientific knowledge. Through detailed examination of J. J. Thomson's discovery of the electron, I have argued that science, in fact, exemplifies the sort of noninductive view I have been proposing. I have no illusions that I can do more than endow this view with some plausibility in a short essay. Adequate defense of the position requires book-length treatment—a very big book it turns out—my forthcoming two-volume *Facts, Theories, and Scientific Observation*.

More important, I have tried to make a case for the advantages of philosophy of science confronting head-on the epistemological issues raised by science. I have tried to convince the reader that epistemology and philosophy of science ignore each other's concerns (possibly even each other's literatures) only at great peril. Perhaps I have even convinced some readers to return to that vision that so animated many of the productive discussions we had with our late colleague Wilfrid Sellars.

NOTES

This essay was presented on October 18, 1991, in the 32nd Annual Lecture Series of the University of Pittsburgh Center for the Philosophy of Science. Drafts were subsequently presented at the University of Maryland and Columbia University. I am grateful for helpful comments made in discussion at these presentations and also for comments by Bas van Fraassen. This essay, together with Suppe 1993, provides a compact presentation of the view of scientific observation, experimentation, and associated scientific knowledge developed in chaps. 1–4 of Suppe 1998b.

1. He was working on revising the early draft of Sellars 1975 and I was working

on my theory of facts and factual truth first presented in Suppe 1973, refined and expanded in later writings (see esp. chap. 8, 9 of my forthcoming 1998a).

2. See also the symposium on the book at the 1978 PSA meetings in Asquith and Hacking 1979.

3. I am assuming that the net effect of such transference chains is multiplicative of the probabilities at each step.

4. The analysis of the anatomy of an experiment in Hacking 1988 is finer-grained than what I have reported here.

5. See van Fraassen 1989, 143; his chap. 6 is a sustained attack on the notion that inference to the best explanation has probative value — either in the bare-bones form I am considering or in various extant sophistications of the notion.

6. Achinstein 1991, chap. 3, analyzes the role of eliminative arguments between competing explanations in the establishment of hypotheses and argues that they must be augmented by independent warrant.

7. In such a view, concerns over the rhetoric of science take on more importance. The rather excessive claims of Latour 1987 thus become more germane to the epistemology of science.

8. Doing so had the further advantage of allowing a cross-check between the two. The history of the episode clouds the epistemology because of the confusions resulting from unstable, sometimes conflicting, claims coming out of the Fermilab spark-chamber project.

9. See, e.g., Salmon 1967, chap. vii, sec. 3, esp. 118, 125–29.

10. See, e.g., Chisholm 1982; BonJour 1985; Lehrer 1974.

11. Examples are Hintikka 1962 and Hilpinen 1970.

12. For the whole battery, see Suppe 1977, 717–27; 1989, chaps. 10–12; Suppe 1998a, *passim.*

13. See, e.g., Plantinga 1993, chap. 9; Pollock 1986, chap. 4.

14. The analysis of the causal possibility operator is in chap. 9 of Suppe 1998a. Its use in resolving the Martin 1975 racetrack case is detailed in chap. 11, sec. 1.

15. Remaining modifications concern obtaining general and probabilistic knowledge.

16. We will see below that it also provides a basis for *credentialing* that knowledge into a scientific domain so that others in the discipline may draw upon it with confidence. See also Suppe 1993.

17. The literature is stimulated by Hintikka 1962.

18. The basic analysis of seeing that for propositions of the form '*D* is *P*' is given in Suppe 1998a, chap. VI, sec. 1–5. Those developments draw heavily from earlier discussions of *simple seeing* and *seeing that* in chap. IV (esp. sec. 5–6) and chap. V (esp. sec. 1–6). In chap. VII, the analysis is extended to encompass relations (sec. 1), states of affairs and processes (sec. 2–3), complex propositions (sec. 4) generalizations and counterfactuals (sec. 5). The version of the analysis just presented in the text is an amalgamation of those various extensions and the factual truth analysis, and is drawn from the very beginning of chap. X.

19. Various aspects of the need for radical denial of the KK thesis are discussed in (Suppe 1998a) chap. I, sec. 1, 2, 5; chap. VI, sec. 7; chap. VIII, sec. 7; chap. X, sec. 10; and chap. XI, *passim.*

20. Minimal rationality considerations figure especially prominently in Suppe 1998a, chap. I, sec. 5; chap. X, esp. sec. 3 and 5; chap. XI, *passim*. They figure prominently in Suppe 1998b, chap. II (esp. sec. 3) and chap. III (esp. sec. 1).

21. See van Fraassen 1979, sec. 4.3; and 1980, chap. 6, sec. 4.4. My adaptations are done in Suppe 1998a, chap. IX, sec. 9.

22. Some restrictions need to be imposed on the spaces in accordance with {1-IX-67} here and in {1-IX-39′} and {1-X-29}.

23. I argue this in Suppe 1998b, chap. II.

24. For a comparison of its adequacy compared to other analyses of causal, subjunctive, or counterfactual conditionals, see Uchii 1990.

25. While this possibility is packed into {1-X-29}, it does not show up that overtly since the prior knowledge is collapsed into the cognitive processes S undergoes. The ability to exploit prior knowledge is explicit in analysis {1-X-1} of seeing that, of which {1-X-29} is a generalization.

26. See Suppe 1997a, chap. X, sec. 7.

27. Although not presented as a numbered result in Suppe 1998a, {2-I-7} is a straightforward summary of considerations raised in chap. X, sec. 7.

28. For fuller treatment of credentialing issues, see Suppe 1993.

29. See especially vol. 1, chap. XII. The notion of level confusions, due to Alston 1980, concerns levels resulting from the iteration of pistic operators such as 'knows that', 'believes that', and 'is justified in believing that'; the confusions involve assuming that what is true of a proposition, belief, or state of affairs ipso facto is true of correlated propositions, beliefs, or states of affairs on another level.

30. See Suppe 1998b, chaps. II–III for the defense of the analysis.

31. For elaboration and defense of this claim, see Suppe 1993, 1997.

32. The point here is a generalization of Suppe 1998a, chap. VII, sec. 5c that the ability to see that $(\alpha)(\phi\alpha \supset \psi\alpha)$ typically depends on our ability to see that $(\alpha)(\phi\alpha$ **npc** $\psi\alpha)$ under the same circumstances.

33. The foregoing account is heavily indebted to the account in Achinstein 1990, essays 10 and 11. I follow his account not only because it is well done, but also because I am going to discuss his analysis of the case below. The crucial differences between Achinstein's and my perspectives on the case will emerge below. Besides the original sources, other useful discussions of the Thomson case are Anderson 1964; Heilbron 1970; Falconer 1987.

34. Notational changes have been made here and subsequently.

35. See Thomson's derivation (1897, 302–03).

36. Whether they do depends solely on whether Thomson's measured results are accurate. Neither his specific results nor procedures for measuring them are germane to the point being made here, so it can be assumed for the sake of argument that they are correct.

37. Included are both nonanalytic and analytic statistics applied in violation of their assumptions; in both cases their legitimacy depends on establishing their robustness for the intended applications.

38. Arthur Burks (1975) did so early on. More recently, see the symposium on simulation modeling at the 1990 PSA meetings in Fine, Forbes, and Wessels 1991 and Peter Galison's talk, "Artificial Reality: Computer Simulation Between Ex-

periment and Theory," given in the 1991–92 University of Pittsburgh Philosophy of Science Lecture Series.

REFERENCES

Abbott, L. F., and R. M. Barnett. 1979. "Neutrino-current Quark Coupling Determination Using Neutrino-neutron and Neutrino-proton Deep-inelastic Cross Sections." *Physical Review D* 19/11 (June 1):3230–35.

Achinstein, P. 1991. *Particles and Waves*. New York: Oxford University Press.

Alston, W. 1980. "Level Confusions in Epistemology." 5: 135–50.

Anderson, D. L. 1964. *The Discovery of the Electron*. Princeton: Van Nostrand.

Ashmore, M. 1989. *The Reflexivity Thesis: Wrighting Sociology of Scientific Knowledge*. Chicago: University of Chicago Press.

Asquith, P., and I. Hacking. 1979. *PSA 1978, Vol. II*. East Lansing: Philosophy of Science Association.

Baltay, C. 1978. "Neutrino Interactions II: Neutral Current Interactions and Charm Production." In S. Homma, M. Kawasuchi, and H. Miyazawa, eds., *Proceedings of the 19th International Conference on High Energy Physics*, Tokyo, August 23–30, 1978. Tokyo, Physical Society of Japan, 882–903.

Bloor, D. 1976. *Knowledge and Social Imagery*. London: Routledge & Kegan Paul.

BonJour, L. 1985. *The Structure of Empirical Knowledge*. Cambridge: Harvard University Press.

Burks, A. 1975. "Models of Deterministic Systems." *Mathematical Systems Theory* 8:295–308.

———. 1977. *Cause, Chance, Reason*. Chicago: University of Chicago Press.

Cherniak, C. 1986. *Minimal Rationality*. Cambridge: MIT Press.

Chisholm, R. 1982. *Foundations of Knowing*. Minneapolis: University of Minnesota Press.

Dretske, F. 1969. *Seeing and Knowing*. Chicago: University of Chicago Press.

———. 1971. "Conclusive Reasons." *Australasian Journal of Philosophy* 49: 41–66.

———. 1981. *Knowledge and the Flow of Information*. Cambridge: MIT Press.

Falconer, I. 1987. "Corpuscles, Electrons, and Cathode Rays: J. J. Thomson and the 'Discovery of the Electron.'" *British Journal for the History of Science* 20: 241–76.

Fine, A., M. Forbes, and L. Wessels. 1991. *PSA 1990: Proceedings of the 1990 Biennial Meeting of The Philosophy of Science Association, Volume 1*. East Lansing, Mich.: Philosophy of Science Association.

FitzGerald, G. F. 1897. "Dissociation of Atoms." *The Electrician* 39: 104.

Fuller, S. 1988. *Social Epistemology*. Bloomington: Indiana University Press.

———. 1989. *Philosophy of Science and Its Discontents*. Boulder, Colo.: Westview Press.

Galison, P. 1987. *How Experiments End*. Chicago: University of Chicago Press.

———. 1985. "Bubble Chambers and the Experimental Workplace." In P. Achinstein and O. Hannaway, eds., *Observation, Experiment, and Hypothesis in Modern Physical Science*. Cambridge: MIT Press, 309–74.

Gettier, E. 1963. "Is Justified True Belief Knowledge?" *Analysis* 23: 144–46.

Gilbert, G. N. and M. Mulkay. 1984. *Opening Pandora's Box: A Sociological Analysis of Scientist's Discourse*. Cambridge: Cambridge University Press.

Hacking, I. 1984. "Experimentation and Scientific Realism." In J. Leplin, ed., *Scientific Realism*. Berkeley: University of California Press, 154–72.

———. 1988. "On the Stability of the Laboratory Sciences." *Journal of Philosophy* 85: 507–14.

Heilbron, J. 1970. "J. J. Thomson." In C. Gillispie, ed., *Dictionary of Scientific Biography*, vol. 13. New York: Scribners, 362–72.

Hilpinen, R. 1970. "Knowing That One Knows and the Classic Definition of Knowledge." *Synthese* 21: 109–32.

Hintikka, J. 1962. *Knowledge and Belief*. Ithaca, N.Y.: Cornell University Press.

Hooker, C. 1975. "On Global Theories." *Philosophy of Science* 42: 152–79.

Kreuzman, H. 1990. *Theories of Rationality in Epistemology and Philosophy of Science*. Ph.D. diss., University of Notre Dame.

Kuhn, T. 1970. "Reflections on My Critics." In I. Lakatos and A. Musgrave, eds., *Criticism and the Growth of Knowledge*. Cambridge: Cambridge University Press.

Latour, B. 1987. *Science in Action: How to Follow Scientists and Engineers through Society*. Cambridge: Harvard University Press.

Laudan, L. 1977. *Progress and Its Problems*. Berkeley: University of California Press.

Lehrer, K. 1974. *Knowledge*. Oxford: Oxford University Press.

Martin, R. 1975. "Empirically Conclusive Reasons and Skepticism." *Philosophical Studies* 28: 215–17.

Medawar, P. 1963. "Is the Scientific Paper a Fraud?" *The Listener & BBC Television Review* 70: 377–78.

Montague, R. 1974. "Logical Necessity, Physical Necessity, Ethics and Quantifiers." In R. Thomason, ed., *Formal Philosophy: Selected Papers of Richard Montague*. New Haven: Yale University Press, 71–83.

Nozick, R. 1981. *Philosophical Explanations*. New York: Oxford University Press.

Peacocke, A. 1984. *Intimations of Reality: Critical Realism in Science and Religion*. Notre Dame: University of Notre Dame Press.

Pickering, A. 1984. "Against Putting the Phenomena First: The Discovery of Weak Neutral Current." *Studies in the History and Philosophy of Science* 15/2: 85–117.

———. 1989. "Living in the Material World." In Gooding, Pinch, and Schaffer, eds., *The Uses of Argument*. Cambridge: Cambridge University Press, 275–98.

Plantinga, A. 1993. *Warrant: The Current Debate*. New York: Oxford University Press.

Pollock, J. 1986. *Contemporary Theories of Knowledge*. Totowa, N.J.: Rowman & Littlefield.

Prescott, C. Y. et al. 1978. "Parity Non-Conservation in Inelastic Electron Scattering." *Physics Letters* 77B/3: 347–52.

Russell, B. 1927. *The Outline of Philosophy*. London: George Allen & Unwin.

Salmon, W. 1967. *Foundations of Scientific Inference*. Pittsburgh: University of Pittsburgh Press.

Sellars, W. 1975. "The Structure of Knowledge." In H. Castañeda, ed., *Action, Knowledge, and Reality*. Indianapolis: Bobbs-Merrill, 295–398.

Shapere, D. 1982. "The Concept of Observation in Science and Philosophy." *Philosophy of Science* 49: 485–525.

Stuewer, R. 1970. "Non-Einsteinian Interpretations of the Photoelectric Effect." In R. Stuewer, ed., *Minnesota Studies in the Philosophy of Science*, vol. V, *Historical and Philosophical Perspectives of Science*. Minneapolis: University of Minnesota Press, 246–63.

Suppe, F. 1973. "Facts and Empirical Truth." *Canadian Journal of Philosophy* 3:197–212.

———. 1977. *The Structure of Scientific Theories*. 2nd ed. Urbana: University of Illinois Press.

———. 1993. "Credentialing Scientific Claims." *Perspectives on Science* 1:153–203.

———. 1997. "The Structure of a Scientific Paper." *Philosophy of Science*, forthcoming.

———. 1998a. *Facts, Theories, and Scientific Observation*, Vol. I: *A Posteriori Knowledge and Truth*. (forthcoming).

———. 1998b. *Facts, Theories, and Scientific Observation*, Vol. II: *Scientific Knowledge*. (forthcoming).

Thomson, J. J. 1897. "Cathode Rays." *Philosophical Magazine*, ser. 5, 44(October): 293–316.

———. 1899. "On the Masses of the Ions in Gasses at Low Pressures." *Philosophical Magazine* 48, 547–67.

Torretti, R. 1986. "Observation." *British Journal for the Philosophy of Science* 37: 1–23.

Toulmin, S. 1972. *Human Understanding*. Vol. I. Princeton: Princeton University Press.

Uchii, S. 1990. "Burks' Logic of Conditionals." In M. Salmon, ed., *The Philosophy of Logical Mechanism: Essays in Honor of Arthur W. Burks with his Responses*. Dordrecht: Kluwer, 191–207.

Van Fraassen, B. 1980. *The Scientific Image*. New York: Oxford University Press.

———. 1988. "The Peculiar Effects of Love and Desire." In B. P. McLaughlin and A. O. Rorty, eds., *Perspectives on Self-Deception*. Los Angeles: University of California Press, 123–56.

———. 1989. *Laws and Symmetry*. New York: Oxford University Press.

14

That Just Don't Sound Right

A Plea for Real Examples

David L. Hull

Department of Philosophy, Northwestern University

> If thought experiments in science are ultimately supposed to increase our understanding of the world by contributing to the process of changing the framework of concepts in terms of which we think about the world, then thought experiments in philosophy of science should be directed towards changing and clarifying the concepts that pertain to science itself.
> — Forge, "Thought Experiments in the Philosophy of Physical Science"

Although Kuhn published a paper in 1964 arguing for an important function for thought experiments in science, his fellow philosophers did not find this topic all that interesting until recently. Of the papers, books and anthologies that have appeared in the last half dozen years, most deal with thought experiments in science, primarily physics (Gooding 1990, 1993; Brown 1991, 1993; Nersessian 1993; Hacking 1993). However, a few also deal with thought experiments in philosophy, especially the philosophy of science (Wilkes 1988; Hull 1989; Horowitz and Massey 1991; Sorenson 1992). To my way of thinking, philosophers have done an excellent job investigating the role and justification of thought experiments in *science*. We have been much less successful in setting out the relevant issues with respect to *philosophy*.

Hypothetical reasoning has been part of Western thought since the pre-Socratics (Rescher 1991). More narrowly, Western intellectuals frequently make use of made-up examples. Some of these fictitious examples could occur but have not. Others are so described that they could not possibly occur. Some examples of hypothetical reasoning also count as thought "experiments" in a narrow sense. The authors cited above tend to concentrate on thought experiments, narrowly construed. In this essay examine the role of fictitious examples in both science and philosophy, specifically the philosophy of science, without limiting myself to thought experiments.

Recently a few philosophers, while acknowledging the positive role that thought experiments have played in certain well-articulated areas of science, logic and mathematics, question the legitimacy of appeals to thought experiments and science fiction examples in philosophy. According to Wilkes:

> Personal Identity has been the stamping-ground for bizarre, entertaining, confusing, and inconclusive thought experiments. To my mind, these alluring fictions have led discussion off on the wrong tracks; moreover, since they rely heavily on imagination and intuition, they lead to no solid or agreed conclusions, since intuitions vary and imaginations fail. What is more, I do not think that we need them, since there are so many actual puzzle-cases which defy the *imagination*, but which we none the less have to accept as facts. (1988, vii)

Fictitious examples in philosophy have led to wild speculations that are all but impossible to evaluate, but such examples can also have the opposite effect. In both philosophy and science, thought experiments can inhibit innovation, relying as they often do on well-entrenched intuitions. From the perspective of old-think, new ideas just don't sound right. Another problem with thought experiments, especially of the cuter variety, is that they are incoherent. Too often, in the face of the ubiquitous question "What would you say if . . . ?" I am forced to respond that I don't have the slightest idea (Fodor 1964). Many people seem to be able to picture bizarre hypothetical situations with amazing ease. Others, and I am among them, are unable to replicate this enthusiasm because we simply can't conceive of the situations described. Either we are psychologically deficient — sort of philosophical dyslexics — or else we have very different standards of conceivability. Just flitting before the mind's eye is not good enough. As Goldman complains, "Some of the most recent discussion of the analysis of knowledge has taken on a glass-beads-game quality — counterexamples dazzling in their originality, designed to capture intuitions few of us recognize as such, generating additional criteria for knowledge so complex in their application as to be of no use in deciding when and what we know" (1988, 19).

If, as Horowitz and Massey (1991) argue, thought experiments in the broadest sense have supplanted meaning analysis in analytic philosophy, then one would think that those philosophers who appeal to thought experiments would have set out in great detail standards for their use in philosophy. With the exception of Sorenson 1992, they have not. As Massey asks in connection with thought experiments such as Putnam's Twin-Earth example:

How do *I* know whether you have been successful? How for that matter do *you* know whether you have been successful? Well, if you feel satisfied with your conception, if it seems to you that you have managed to conceive such a state of affairs, you simply rest your case. And if it seems to me that you have succeeded, perhaps because of profiting from your tutelage I seem now able to conceive it myself, I capitulate. Either I abandon my thesis or, like Hume, I dismiss your counterexample as too esoteric to do serious damage. There is, of course, a third alternative, one surprisingly rarely taken, namely, to dismiss your conception as somehow bogus or counterfeit. (1991, 292)

Critics of the use of science fiction examples in philosophy find that such appeals are too "free and easy." In a sort of Protestant rebuff, they reason that anything *that* easy cannot be of much value, and I agree. The sort of science fiction examples to which analytic philosophers seem peculiarly prone have three glaring deficiencies. First, they lack sufficient detail. We are rarely told enough to understand the state of affairs being described (Dancy 1985). Second, they lack a theoretical context to enable us to fill out the description on our own (Wilkes 1988, 7, 21, 44–45; Massey 1991, 294; Horowitz 1991, 306; Irvine 1991, 159; Bunzl forthcoming). In real examples, the background knowledge is already known and widely shared. If some item in the background knowledge is brought into question, it is there to be investigated. In made-up examples, no such background knowledge exists. Real examples pose as fundamental problems as do made-up examples. They also have the advantage of presenting ways by which these problems can be solved. Third, in using such examples we trade on a notion of conceivability that remains all but unexplicated.

One way around this lack of detail is to avoid making one's examples too outlandish, too different from actual states of affairs (Wilkes 1988, 46; Forge 1991, 216; Gale 1991, 23). As Gooding remarks, legitimate thought experiments "posit a world — neither so familiar as to foreclose change nor so strange as to provide no footholds or handles on reality" (1993, 283). Thought experiments "persuade when there is enough strangeness to disturb and enough familiarity to be accessible." Thus, in our hypothetical examples, we can trade off what we already know about related real examples. In the limiting case, our hypothetical examples are not in the least science fiction. For example, the Ship of Theseus is an ordinary ship. The only strange thing about it is that it has to be rebuilt while it stays afloat at sea.

In scientific thought experiments, we are asked to countenance the modification of a single feature of the situation while everything else remains the same. Such examples conform to the "laws we know and trust." Put differently, "the 'possible world' is *our* world, the world described by our sciences, except for one distinguishing difference" (Wilkes 1988, 8).

One problem with this interpretation of hypothetical examples is that certain areas of science are tightly organized. Modifications ramify. As John Herschel remarked in his early work on the study of natural philosophy:

The liberty of speculation which we possess in the domains of theory is not like the wild licence of the slave broke loose from his fetters, but rather like that of the freeman who has learned the lessons of self-restraint in the school of just subordination. (1841, 190)

Present-day physicists agree with Herschel about constraints on the liberty of speculation. As Feynman, Leighton, and Sands remark:

The whole question of imagination in science is often misunderstood by people in other disciplines. They try to test our imagination in the following way. They say, "Here is a picture of some people in a situation. What do you imagine will happen next?" When we say, "I can't imagine," they may think we have a weak imagination. They overlook the fact that whatever we are *allowed* to imagine in science must be *consistent with everything else we know*: that the electric fields and the waves we talk about are not just some happy thoughts which we are free to make as we wish, but ideas which must be consistent with all the laws of physics we know. We can't allow ourselves to seriously imagine things which are obviously in contradiction to the known laws of nature. And so our kind of imagination is quite a difficult game. One has to have the imagination to think of something that has never been seen before, never been heard of before. At the same time the thoughts are restricted in a strait jacket, so to speak, limited by the conditions that come from our knowledge of the way nature really is. The problem of creating something which is new, but which is consistent with everything which has been seen before, is one of extreme difficulty. (1964, 2:20.10–11)

In the face of all these objections to using hypothetical examples in philosophy, philosophers might well be led to specify their examples more fully, supplying more detail and indicating the context in which these examples are to function. If no such context exists, philosophers need to construct one. Although these requirements might sound excessive to philosophers, authors of fiction have been fulfilling them for centuries. If Jane Austen can do it, so can Hilary Putnam.

Another alternative is to use real examples. In his "defense" of thought experiments in philosophy, Sorenson complains that too often philosophers use thought experiments as a mental crutch:

Instead of taking the extra time to find a robust actual case, they fall back on anemic hypotheticals. Thought experiment is an easy way of getting short-term results. Philosophers already know how to do it, it's cheap, it's quick, it's accepted. (1992, 256)

What is so wrong with philosophers using actual cases? One answer is that real examples can be difficult to explain, and in the midst of all the empirical detail, the philosophical points at issue can get lost. Goodman uses precisely this justification for resorting to grue and bleen ([1954] 1983, XIX). He thinks that commonplace and trivial illustrations "attract the least interest to themselves" and hence are "least likely to divert attention from the problem or principle being explained." Perhaps so, but his own example of grue emeralds is a major exception to this generalization. It has diverted massive attention to itself. For forty years, philosophers have discussed what a world would be like in which the predicate "grue" would be in some sense "natural."[1] In addition, if the calls for greater detail in fictitious examples are heeded, then hypothetical examples are likely to become as complex as their real counterparts. To make matters worse, no one has suggested how we are to go about adding details to fictitious examples.

One reason philosophers give for resorting to made-up examples is that the real world is too limiting. We need to go beyond the actual world to test and possibly expand our concepts. We need to conjecture possible worlds. I agree that if knowledge is to increase, we have to go beyond what we already know, but I have one suggestion for my fellow philosophers — *proceed more slowly from the real world to possible worlds. Fully exploit the world as we know it before conjuring up exotic possible worlds.* This is the strategy that Wilkes recommends for the huge literature on personal identity. According to Wilkes, "basing our arguments on actual cases allows us to check our imagination against the facts, and our intuitions get strengthened and rendered more trustworthy" (1988, 48). Once armed with these strengthened and trustworthy intuitions, we can then risk the construction of hypothetical situations if need be.

In biology, the area of science with which I am most familiar, biologists rarely go beyond the real world. They already have such long lists of bizarre actual counterexamples that they rarely feel the need to make up additional hypothetical problems. (Although Lennox [1991] has shown that Darwin resorted to thought experiments to argue for the possibility of species evolving, he also inundated his readers with real examples.) For instance, species in sexual organisms are determined by reproductive gaps. If two populations of organisms breed freely with each other when they come in contact and produce fertile offspring, then they belong to the same species. However, there are two species of fish that fulfill this criterion but are nevertheless considered separate species. When they meet, they breed freely with each other and produce offspring that are both fertile and very vigorous, so vigorous that they rapidly out-compete the organisms of the original species. When the last male belonging to the parental species dies, all reproduction ceases because all the hybrids are female. Even though these two species mate freely and produce fertile offspring, they are nevertheless reproductively isolated. It takes just a few generations before this reproductive isolation manifests itself. The supply of similar examples is inexhaustible.

Starting with real examples has the added advantage that one need not have a general theory of conceivability in order to work with them. A major reason philosophers resort to conceivability is to get at possibility. According to Massey's thesis of facile conception, casually alleged conceivability is invoked to establish possibility, and casually alleged inconceivability is invoked to demonstrate impossibility (1991, 291). Too often, the decisions that philosophers make rest heavily on intuitions about what sounds right. For example, logical empiricists have long attempted to distinguish between genuine laws of nature and accidentally true universal statements. Although many philosophers think that this attempt is misplaced, there does seem to be an intuitive difference, for example, between the claim that all species evolve through natural selection and that all terrestrial organisms use levo rather than dextro amino acids. The former is fundamental to the evolutionary process; the latter is most likely the result of an historical accident.

One suggestion for explicating this felt difference is by means of counterfactual conditionals. So the story goes, laws can support counterfactual conditionals while accidentally true universal generalizations cannot. If A were the case, then B would be the case. A is not a species, but if it were, it would evolve through natural selection. Laws also support subjunctive conditionals. If A should come to pass, then so would B. If a species becomes genetically homogeneous, then it will go extinct. The trouble is that these examples get us nowhere. *The intuitions that lead us to accept counterfactual and subjunctive conditionals are precisely the intuitions that lead us to distinguish between genuine laws of nature and accidentally true universals in the first place.*

One virtue of real examples is that they allow us to go directly to possibility without detouring through conceivability. Anything that is actual had better be possible. Whether or not everything that is actual is also conceivable depends on one's general theory of conceivability — a theory that we currently do not possess even in a rudimentary form. As Horowitz and Massey argue:

Philosophers should muster enough intellectual integrity to eschew conceivability arguments that fail to measure up to the standards that have been developed piecemeal in science. Second, philosophers ought not to rest content with piecemeal developments but should instead turn their talents to the construction of a general theory of conceivability. (1991, 23)

Such a general theory of conceivability is likely to be a long time in coming. In the interim, philosophers can get around all the problems generated by a largely tacit notion of conceivability by exploiting real examples to their fullest potential.

Although many real examples are as weird and counterintuitive as the sorts of examples that philosophers make up, they have the virtue of always occurring in a particular context against a wealth of background knowledge — knowledge that can be made explicit when the occasion demands. Although sometimes background knowledge is found wanting when we appeal to it in real examples, at least partially explicit standards exist that stipulate how we should go about supplementing it. As numerous authors have observed, in philosophy such standards are notable for their absence. Typically the examples preferred by philosophers "take us too far from the actual world, and from the 'other things' that hold roughly 'equal' here. . . . This means that we are left with no clue as to what has been varied in thought and what left (supposedly) untouched" (Wilkes 1988, 45).

What if another planet existed that was identical to earth in every respect except that the clear, tasteless, colorless liquid that covers so much of the earth's surface did not have the chemical constitution of water — H_2O? To avoid problems that knowledge of the molecular structure of these two substances introduces, Putnam (1981) assumes that the inhabitants of both planets are just ignorant enough to be oblivious to the chemical and molecular differences between these two substances and the possible ramifications of these differences. The former inhabitants of these two worlds to one side, how about *Putnam's readers now*? Perhaps earlier generations on these two planets were unaware of difference between the two substances that they call "water," but some of us at least know what the chemical structure of water is as well as some general features of the relation between chemical structure and both macro and micro properties. I can conceive of people who cannot tell all sorts of substances apart. I need not conjecture such people on different planets. We have plenty here on earth. But I cannot conceive of two different chemicals having *all* the same characteristics, both macro and micro. Even the rare earths and inert gases exhibit some differences in their salient features. Water also plays very particular and precise roles in the lives of all terrestrial organisms. Life on earth is a very complex, highly interdependent system. As such it is very sensitive to the slightest changes, especially in something as important to metabolism as water.

"But couldn't you envisage a system in which all the ramifications of 'water' having a different chemical makeup are fully worked out?" Putnam claims that it is "easy" to modify his example so as to avoid these objections (1981, 23). I disagree. Perhaps hundreds of scientists working for several decades might work out the details of such a system, but I do not know what it would be like for them to conjure up such a system in the blink of an eye, let alone my doing so all by myself in the space of my reading about Twin Earth (Fodor 1964). "But didn't something flit across your mind?" Something, but not much. Not enough to justify a philosophical analysis of reference.

How about brains in a vat? Mightn't I be a brain in a vat with my nerves connected to a computer presided over by an evil neuroscientist who stimulates my grey matter to maximize false beliefs? Couldn't we all be brains in vats, including the neuroscientist himself? This time Putnam (1981) introduces his example to show that, contrary to a lot of people's intuitions, no one can really think that he or she is a brain in a vat. Such a belief is "self-refuting" and "incoherent" because the preconditions of reference are absent, that is, *if* we accept Putnam's causal theory of reference. Sorenson disagrees. "How would things seem under such conditions? Just as they do seem! We know all too well what it is like to be a brain in a vat" (1992, 7). Hence, we can have legitimate doubt about any of our beliefs (see also Collier 1990). Once again, I must be missing some widespread ability because, quite independent of any causal theory of reference, I do not know what it would be like to be a brain in a vat.

Norton (1993) claims that all thought experiments can be recast as arguments, either deductive or inductive. Within this context, he goes on to claim that the picturesque details in descriptions of thought experiments give them an experimental flavor but are strictly irrelevant to the argument. Nersessian (1993, 296) disagrees. These "details usually serve to reinforce crucial aspects of the experiment" (see also Collier 1990). I am of two minds on this issue. If the examples that philosophers use are intended only as expository devices of no intrinsic importance, then I have no objection to them. They might help some readers see the point; others not. Since nothing much rides on them, a brief characterization is good enough. But too often the examples take over, becoming the chief topic of discussion (see note 1). They are what persuade the reader; not the arguments.

When the examples become the issue, then I for one would like to see more detail, not less. In such situations, I would like to see us take our examples seriously. However, I would much prefer that we begin with real examples, turning to hypothetical examples only when we run out of real ones. Even so, I agree with Norton (1993) that the underlying argument, not the "picturesque" part of the example, is what should carry conviction. In spite of the significant role that Maxwell's demon plays in this literature, Hacking (1993, 302) is forced to conclude that the "demon does not, for me, prove even the possibility of anything."

Because fictitious examples play a central role in analytic philosophy, my objections to them and my preference for real examples are likely to make analytic philosophers a bit defensive. How are we to do analytic philosophy without science fiction examples, the more exotic the better? If we deny ourselves appeals to such things as Twin Earth, brains in a vat, violinists hooked up to us while we sleep, etc., all is lost. We will be struck dumb. But are made-up examples really central to analytic philosophy? In thirty years of publishing, I have never attempted to clarify a concept or support a position by reference to fictitious example, except when I have been forced to respond reluctantly to the examples made up by others to criticize my views (Hull 1981, 1991). Either I am not engaged in analytic philosophy or else use of such examples is not essential to this way of doing philosophy.

In this essay I illustrate the virtues of real examples by reference to (what else?) two real examples: one in science (the species concept), the other in philosophy (natural kinds). Although both the terms and the conceptions have varied considerably through the ages, biological species have been viewed as paradigm natural kinds for over two thousand years, even after Darwin convinced us that species evolve. For over twenty years, Michael Ghiselin (1974) and I (Hull 1976) have argued that insufficient attention has been paid to the implications of Darwin's modification of the species concept. If species are the things that evolve, then they cannot be spatiotemporally unrestricted kinds but must be construed as spatiotemporally restricted particulars (or individuals).[2] Like individual organisms, particular species are spacetime worms. In the interim, a substantial number of biologists and philosophers have been persuaded that there is something to this bizarre idea and have incorporated it in their own research.

In the following two sections, I discuss the species concept and natural kinds in the context of evolutionary biology and then turn to some exotic examples that have been introduced to "clarify" these notions. After contemplating these examples, the reader can decide whether they really help or simply introduce a host of extraneous considerations that do nothing but cloud the issues. The issue is *not* whether I am right about species being in some sense "individuals" or even whether the distinction between individuals and kinds can bear up under philosophical analysis but the function of real versus fictitious examples in helping to decide these questions.

Biological Species

From Aristotle to the present, people in all cultures have conceived of biological species as kinds possessing core characteristics—a set of characteristics that all and only members of a particular species possess, a set of characteristics that make a species what it is — an occasional abnormal monster notwithstanding (Atran 1990). The trouble is that, as the biogeographic distributions of species were traced, more and more individuals had to be dismissed as monsters. Species frequently vary throughout their geographic range. Any attempt to subdivide such continua by means of sets of characteristics results in as many abnormal as normal individuals. In addition, some species exist in which males and females differ from each other so dramatically that any attempt to group them according to character covariation results in their belonging to separate species. The same can be said for various castes in social insects and clonal forms. The various organisms that make up a Portuguese man-of-war are phenotypically very different, yet they all belong to the same species.

In response to such character covariation, taxonomists have resorted to what Wittgenstein came to term "family resemblances." Species do not form discrete groups but only "clusters." All that is necessary for organisms to belong to a particular species is for them to possess enough variably weighted characteristics. The general notion of cluster analysis has seemed intuitively satisfying to many scientists and philosophers alike, but actual application is quite another matter. Traits have to be identified and weighted. One clustering algorithm must then be selected amongst the welter of competing computer programs. However, once the amount of labor necessary to use cluster analysis is expended, it works only for *contemporaneous time-slices of those species that exhibit a unimodal distribution* — a single bell curve around a single mean. But many species exhibit multimodal distributions. Which characteristics are "typical" varies from geographic location to geographic location. Averaging this variation to form a single cluster obliterates an important feature of biological species.

Species have a temporal dimension as well. If species are traced through time, they can be seen to split, undergo anagenetic change, go extinct and even merge. As a result, if species are held to have a temporal dimension, then temporal sequences of clusters collapse into a single distribution. The result is one huge smear. In biological evolution something more basic than character distributions is at work. Although natural selection is not the only mechanism that influences the evolution of species, it is primary. Selection is not a matter of picking balls from a jar and tossing them back in. Selection requires the differential build-up of changes through time. Such sequential differential change might occur by a variety of mechanisms, but the only mechanism that has evolved here on Earth requires descent. In a single generation, organisms of certain types produce more offspring than do others. They in turn produce even more offspring relative to their competitors in the next generation, and so on. If species are to evolve by natural selection, then they must be treated as space-time worms. Organisms that belong to a particular species belong to it, not *because* they share the same cluster of characteristics but *because* they are part of the same genealogical network. Because they belong to the same genealogical network, they will tend to possess similar characteristics, but they need not and sometimes do not.

The preceding view of species has proven to be counterintuitive. Some critics have objected that there has to be at least a common core of traits or underlying set of genes that characterize the members of a species or they could not mate with each other (Caplan 1980, 1981; Kitts and Kitts 1979). As plausible as this conviction may seem, it is false. In order for two organisms to mate, they must be sufficiently similar to each other, especially in certain key aspects, but it does not follow from this fact that a single common core of characteristics or genes exist that permit mating. Instead, there are numerous alternative sets of characteristics and/or genes that will do the trick. Any two organisms selected from the same species will have exactly the same alleles at almost every one of its loci, but in different pairs of organisms, *these need not be the same alleles or the same loci*. In addition, interbreeding is not an all-or-nothing affair. Organisms that belong to the same species are *more likely* to mate with each other than with members of other species, but not all members of the same species are interfertile, and sometimes organisms belonging to different species succeed in producing fertile hybrid offspring. It all depends on how much crossbreeding there is. In fact, interspecific hybridization is a common way of producing new species among plants.

Another common objection to the preceding view of species is that this is not how "we" conceive of species. I find this objection misplaced on a host of counts. First, who is the "we" being referred to? Terms are dynamite. Because modern genetics was termed "Mendelian genetics," we tend to think that Mendel had something to do with it. But just as meteors have nothing to do with meteorology, Mendel might not have had that much to do with the science that bears his name. In philosophy there is a movement termed "ordinary language philosophy." Because of its name, one might expect that it has something to do with ordinary language. "What do we mean when we say . . . ?" The "we" is, one would think, ordinary people. The trouble is, with the possible exception of Austin, ordinary language philosophers have never been inclined to do the sort of empirical research necessary to discover how ordinary people use the terms under investigation. Instead they plumb their own usage, assuming that it is "typical." The justification seems to be that if there were no common usage widely spread throughout a language community, people belonging to that community could not communicate with each other. Since they can, there must be some core meaning to every term, if we just work hard enough to uncover it.

The preceding argument should sound familiar. It is the same one used to insist on a core set or cluster of genes and/or traits for biological species. If all organisms belonging to the same species do not share at least a common core set of genes, they could not mate with each other. Similarly, if all members of the same language community do not share at least a common core set of meanings, they could not communicate with each other. Just as this argument is fallacious with respect to mating, it is fallacious with respect to communicating. People speaking the "same" language are broken down into partially overlapping language communities, each speaking its own dialect. Any two English-speaking people chosen at random will agree in what they mean by a vast number of English terms, but it does not follow that there is a common core of terms about which all English-speaking people agree.

But don't ordinary language philosophers, by introspection, discover such core meanings? I don't think so. I think they "construct" such core meanings much more than they "discover" them. As S. Thomason (1991) notes, linguists are never satisfied with hypothetical examples. They insist on real examples as well, because linguistic introspection is so unreliable. For example, according to Chomsky, the English sentence *We received plans to kill me* should be grammatical, while the sentence *We received plans to kill each other* should be ungrammatical (Thomason 1991, 255). When tested, Chomsky's students agreed with Chomsky's linguistic intuitions, but when ordinary people were presented with these two sentences, they came to just the opposite conclusion. In the *Meno* did Socrates really get the slave boy to *recollect* geometry, or did he teach the boy geometry through a series of leading questions? From my own experience in graduate school, professors force their students to adopt their linguistic intuitions under the guise of discovering common conceptions (Sorenson 1992, 272, terms this process "coaching").

More importantly, what difference does ordinary language and/or conceptions make? If an example shows that a particular scientific theory is internally inconsistent, well and good, but too often the incompatibility is between a technical area of investigation and common sense. As Brown argues, Schrödinger's cat "did not show that the Copenhagen interpretation is logically inconsistent" but rather that it is "in flagrant violation of well-intrenched common sense" (1991, 123). However, many of the major advances in science have come at the expense of common sense. Space does not seem to be "curved," but it is. According to ordinary usage, tomatoes are not fruit, as botanists insist, but vegetables. A division of linguistic labor is clearly necessary. Even though we all speak some ordinary language or other, it does not follow that ordinary languages are somehow prior to the technical languages developed for special purposes — and vice versa. Botanists have a reason for grouping tomatoes with fruit. Even so, I do not want tomatoes mixed into my fruit salad. Philosophers have as much right to develop technical languages as do scientists, literary critics, and what have you. I only wish that this school of philosophers had not chosen the title "ordinary language philosophy," when what they are doing is constructing languages that are anything but ordinary.[3]

In response to the claim that species are spatiotemporal particulars, Bradie (1991) set out a series of science fiction fantasies designed to challenge this conception. His first scenario is designed to challenge contemporary biological understanding. What if spontaneous generation were much more common than we think it is? Organisms sprout up all over the place from inanimate substances. Would evolutionary biologists still insist on a genealogical criterion for species? This scenario strikes me as similar to J. J. Thompson's people-seeds. Suppose "people-seeds drift about in the air like pollen, and if you open your windows, one may drift in and take root in your carpets or upholstery" (1971). Would you be responsible for taking care of such babies or could you mow them down with the vacuum cleaner? In both cases, I think that the answer is the same. The conception of species as individuals depends on the world being basically the way that present-day evolutionary biologists think it is. If natural phenomena were very different from what they are, if natural selection played no role in biological evolution, then all bets are off. Evolutionary biologists have fashioned their species concept to reflect the world as it is, not as it might be. Similarly, our morals depend to a large extent on the world being the way it is. In a very different world, we would have very different morals.

Bradie's other scenarios trade on the difference between our ordinary conception of what it is to be "human," assuming that there is such a thing, and someone's belonging to the biological species Homo sapiens. Scenario two concerns a mysterious stranger who comes to a small town in the Old West. If this stranger looked and acted pretty much as other people do, then he would be accepted as an ordinary human being. The inhabitants of this small town would have no reason to suspect otherwise. But what if he were a visitor from another planet? What if an extraterrestrial landed here on Earth and mated successfully with human beings, and these hybrids were fertile and mated with other human beings? As strange as it might seem, the answer depends on how common such occurrences turn out to be. Forget about small towns in the Old West and visitors from outer space. Biologists are already aware of cases in which an occasional member of one species mates successfully with members of another species. If such matings become sufficiently common, then the two species merge into one. If they remain rare and sporadic, the species remain distinct.

But what of the particular individuals involved? Wouldn't I be willing to term this stranger "human" if he looked like, sounded like, tested out as, etc., ordinary human beings even if he didn't have the appropriate pedigree (Bradie 1991, 249)? From a variety of ordinary perspectives, possibly so. I know what people are talking about when they say that certain people are "inhuman" and organisms that belong to other species are "human" or "almost human," but from the perspective of evolutionary biology, other considerations are more fundamental. Apparently this stranger was part of one genealogical network. Now he has become part of another — regardless of what he might look like. "But which network does he belong to?" He used to belong to one network; now he belongs to another. As disconcerting as this answer may seem, nothing more can be said on the subject.

Prior to Darwin, biologists found both the splitting and merger of species unpalatable. Darwin forced us to acknowledge that species split, but resistance to merger persists to the present. (Although the splitting and merging of people pose serious problems for the notion of personal identity, evolutionary biologists must consider it because species as lineages both split and merge.) One reason biologists, especially those interested in reconstructing the past, have tried to avoid acknowledging the existence of hybrid species is that they make reconstructing the past by the comparative method impossible. The comparative method is designed to uncover sequential splitting by means of character covariation. Any merger destroys the data necessary to reconstruct the past. Unfortunately, hybrid species do exist today, and we have no reason to expect that they were any rarer in the past (Masterson 1994). As Thomason notes, linguists have resisted the idea of genuinely hybrid languages for the same reason — the problems they pose for the comparative method (1991, 251). The arguments that linguists present to prove that hybrid languages are impossible seemed conclusive, but Thomason was able to find a genuinely hybrid language. As unlikely as it may seem, she found a language whose vocabulary came largely from one language and its grammar largely from another.

In his third scenario, Bradie asks what I would say if we gradually replaced the parts of a human being with plastic parts that somehow worked as well as the original parts (1991, 249). The net result looks and acts in every respect like the original. Isn't it still human? Two questions are involved: is it the *same individual* and does it continue to belong to the *same kind*? The first question turns on the old problem of the ship of Theseus. With respect to ordinary organisms, there is no problem. We remain numerically the same individuals even though we exchange our constituent parts several times over during the course of our existence — just so long as the process is sufficiently gradual and we remain sufficiently cohesive during such change (Wiggins 1967). Organisms can also change their kinds. Certain organisms start off as saprophytic. Later they become parasitic. In this process they remain numerically the same individual. However, "saprophytic" and "parasitic" refer to *ecological kinds*, not *genealogical networks*. Changing one's kind is a different sort of process from moving from one network to another (see Hull 1987).

But what if we made a baby from scratch using plastic parts? It would lack the appropriate genealogical connections, but might it not be considered human? If it matured and mated successfully with other unproblematic human beings, then, yes, it would have become part of the human genealogical nexus. However, there are many problems with these examples for which Bradie gives no hint of an answer. DNA is not a plastic. If the adult in the preceding example eventually comes to be made entirely of plastic as is the baby right from the start, then the genetic material of these individuals cannot be DNA. How fetuses are going to develop when half of their chromosomes are composed of DNA and the other half some sort of plastic, I do not know. But such "details" to one side, any entity that joins in a genealogical nexus is part of that nexus. If species are identified with such chunks of the genealogical nexus, then plastic entities belong in the same species with their organic brethren. "But that don't sound right." That cannot be helped.[4]

Fictitious examples can always be made to fit one's general thesis, while real examples always run the risk of counting against it. For example, Wilkes (1988, 15) takes for granted that "human being" and "Homo sapiens" denote one and the same natural kind. She argues at considerable length that "person" does not denote a natural kind of any sort (see also Churchland 1982). Even so, she thinks there is "very substantial intersection" between the class of persons and these other two coincident classes (Wilkes 1988, 22). This coincidence in turn allows us to "bring to bear the (relative) clarity of a natural-kind-term — *homo sapiens* — on to the puzzling intricacy of the everyday notion 'person' " (Ibid., 28; see also Churchland 1982). If species were natural kinds, Wilkes's strategy would look promising, but if species are historical entities (space-time worms) and not natural kinds, then her line of reasoning does not begin to get off the ground. Because thought experiments are devised to support a particular position, they are unlikely to challenge it — although sometimes authors are unaware of the implications of their own brain children. Calling for philosophers to make greater use of real examples is a double-edged sword. Real examples are as likely to count against a view as for it.

If thought experiments and science fiction examples are supposed to expand and clarify our conceptual systems, then the reader will have to decide how well these examples served. I for one find general explications of the issues, minus the fictitious examples, much more explanatory.[5] Such examples trade on intuitions, but I am suspicious of intuitions, my own as well as others. Can't I conceive of species as natural kinds in a significant sense of "natural kind"? Certainly, but what I cannot conceive of is species as natural kinds evolving by natural selection. I cannot conceive of species being both spatiotemporally *restricted* and spatiotemporally *unrestricted*. People who think that they can conceive of this state of affairs do not know enough about the evolutionary process. They think they are conceiving of natural kinds evolving by natural selection, but they are mistaken. All they need to do to discover this mistake for themselves is to learn a little bit about biological evolution.

To the extent that intuitions have any warrant, this warrant resides in the conceptual systems which give rise to them. Given Putnam's theory of reference, the claim that any of us is a brain in a vat is incoherent. "The sense in which I cannot be a brain in a vat is the sense in which it is impossible for me to make a true statement that I am" (Collier 1990, 415). But someone might be able to construct a system in which the idea would be unproblematic. Given Chomsky's analysis, certain sentences will seem grammatical, others not. Given the context of the comparative method, hybrids seem impossible. But frequently what is involved is the *replacement* of one conceptual system with another or a major *revision* of an existent system. In such cases, intuitions drawn from one conceptual system are not dependable in evaluating another.[6]

Natural Kinds

The preceding discussion concerned scientific concepts — species and Homo sapiens. Physicists get to decide what counts as pions, black holes, and charm. Biologists get to decide what counts as cistrons, hybrid sinks, and biological altruism. But "natural kind" is a philosophical concept, and here philosophers are the relevant experts. Philosophers through the centuries have developed the distinction between individuals and classes, primary and secondary substances, etc., and they have done so in the midst of a half dozen or so paradigm examples. Among the stock examples of individuals are particular organisms such as Sea Biscuit and Gargantua. The three most commonly cited examples of kinds are geometric figures, biological species and, a poor third, physical elements.

Scientists decide what is to count as a planet or a biological species in the context of the theories that they develop, and these theories inevitably depart from common sense or ordinary experience. Hence, some examples that sound perfectly cogent in the context of ordinary experience must be dismissed as mistaken regardless of what ordinary people might think. But, conversely, because scientists possess very general, highly articulated theories, they can afford to appeal to fictitious examples with little fear that they will do too much harm, and they might even do some good. The same observation holds for philosophers who, in the good old days, produced full-fledged, well worked out philosophical systems. Within the context of these systems, concepts can be explicated by trotting out a string of examples, including hypothetical examples.

But these days are past, at least among analytic philosophers. Global philosophical systems are out of fashion; we limit ourselves to bits and pieces. In the absence of general systems, however, whether scientific or philosophical, how are we to evaluate hypothetical examples? How can we decide what the implications of a particular example actually are? Ordinary language will not do because it is too various, heterogeneous and untechnical. We philosophers have our own technical vocabularies, and we are not about to bow to ordinary usage if it gets in the way of our explications. The distinction between primary and secondary substances, individuals and classes, etc., has been fairly prevalent in philosophy throughout its history. Of course, no two philosophers meant the same thing by "kind" or "natural kind" and their cognates in other languages. Is there an essence of "natural kind" as philosophers have used this phrase through the years? Quite obviously not. In the absence of such universal usage, I am forced to set out what I take to be a fairly prevalent usage to see the effect that hypothetical examples have on this conception.

Natural kinds has been commonly understood to have three defining characteristics. They are eternal, immutable and discrete. In what sense are geometric figures, biological species, and the physical elements eternal, immutable and discrete? By "eternal" some philosophers meant that every natural kind is always exemplified. Horses always were and always will be. However, many philosophers have had a weaker notion in mind. Natural kinds are built into the structure of the universe. At times they may be exemplified; at times not. During the first few minutes after the big bang, none of the heavier elements existed. If sexual reproduction involving reduction division is necessary for biological species, then for most of life on Earth, there were no biological species. Meiosis is a relatively recent innovation. Natural kinds can be "eternal" in this sense even though they periodically are not exemplified. Gold can come and go.

Some philosophers have held a very strong notion of "immutability." All individuals belong at any one level to one and only one natural kind, and no individual can change its kind. Water snakes never become vipers, nor do frogs ever become toads. Other philosophers are more liberal. A sample of lead might be transmuted into a sample of gold. Generations of alchemists tried. As it turns out, such transmutations are possible. The main drawback is that producing gold in this way is much more expensive than mining it. Similarly, a wire circle can be reshaped into a square. But in both of these cases, individuals are changing their kinds (although philosophers disagree about whether or not they remain numerically the same individuals). The kinds themselves remain unchanged. Lead qua lead is not evolving into gold qua gold, nor are we squaring the circle.

Finally, philosophers have frequently claimed that natural kinds in conceptual space are discrete; that is, each natural kind can be defined by a set of necessary and sufficient conditions. The entities that exemplify these kinds, however, need not do so perfectly. Some deviation is only to be expected. No triangle is perfectly triangular, no sample of gold is totally pure, and no horse perfectly exemplifies the essence of horse. The assumption is, however, that borderline cases are relatively rare. If mapped in conceptual space, most entities cluster around their norm, and each kind has only one norm. In a very real sense, any *variation* is necessarily *deviation*. A horse lacking a tail is a deviant horse, women are deviant males, blacks are deviant whites, and homosexuals are deviant heterosexuals. In this context, neutral references to variation are inappropriate.

The assumption in the preceding discussion is that geometric figures, biological species, and the physical elements are equally natural kinds, but if we look at these three examples with care, we can see that they form a heterogeneous hodgepodge. If the distinction between an uninterpreted formal system and a physical application of that system is taken seriously, then Euclidean geometry is very different from applied Euclidean geometry. Euclidean parallel lines never cross, but light rays always cross no matter which way they are directed. Euclidean triangles and triangles made of light rays are very different entities. They may be natural kinds, but they are natural kinds in two different systems. Physical elements do seem to be natural kinds, albeit derivative of the more fundamental natural kinds at the subatomic level. Given the subatomic particles, the physical elements are built into the basic structure of the universe.

But biological species as the things that evolve exhibit *none* of the characteristics of natural kinds. They are not eternal. They come and go, and once gone, can never return. A creature that looks like a dodo, exhibits the same array of genotypes as the dodo, etc., may come into existence. But, so we are told, the dodo is extinct, and "extinct" does not mean temporarily not exemplified. Conservationists are right to worry about endangered species; like herpes (and unlike true love), extinction really is forever. The dodo can no more re-evolve than Hitler can come into existence again. Nor are biological species in any sense immutable. One way of going extinct is to evolve into two or more descendant species. Finally, they are not discrete. To the extent that evolution is gradual, species when mapped onto conceptual space have numerous intermediates. In physical space, most species are reasonably discrete. Their ranges have fairly sharp limits, albeit limits that can change with the seasons. The relevant "gradualness" concerns one species changing into another species, either through anagenetic or cladogenetic modification. Even if speciation is usually punctuational, intermediates still exist because the mechanism set out by Eldredge and Gould (1972) is populational. Some species, however, do come into existence saltationally. For example, in cases of allopolyploidy, a new species can come into existence in the space of a single generation.

At this juncture, several alternatives are possible. Some authors, such as Dupré (1993), retain traditional conceptions of natural kinds and use the failure of biological species to exemplify this conception as one more argument against the existence of natural kinds. Another alternative is to loosen the notion of natural kind to include species. The trouble is that the degree of loosening necessary to include species results in organisms and just about everything else becoming kinds. The alternative I prefer is to leave the notion of natural kind as it is and admit that we got one of our examples wrong. On this alternative, biological species no longer pose any problem for the notion of a natural kind because they are not natural kinds.

One reason an adequate definition of "natural kind" seems so difficult to come by is that we have tried to formulate this definition on the basis of a motley collection of examples. Perhaps ordinary people think of geometric figures, physical elements and biological species as the same kind of thing, but ordinary conceptions are not good enough for either science or philosophy. At times they are not good enough for everyday life either. Just as philosophers have tried to distinguish between genuine laws of nature and accidentally true universal generalizations, we have also tried to explicate the parallel distinction between natural kinds and artificial collections. In both cases, there is a felt difference, but can we set out any criteria to justify these feelings?

One way to mark the latter distinction in science is to argue that natural kinds are the kinds that function in natural regularities. (I am not *endorsing* these distinctions but pointing out how they might possibly be salvaged.) A kind term is "natural" if it occurs in the statement of a natural law. On this analysis, the statements of uninterpreted calculi do not count as laws of nature. Hence, geometric figures do not form natural kinds. Some other justification is needed for them. Because biological species are not kinds at all, they cannot function as natural kinds. Statements such as "All swans are white," even if true, would not count as natural laws. Lawful regularities in the living world occur at higher levels of analysis. Although the physical elements are not especially fundamental kinds, they are nevertheless natural kinds on this analysis because the names of natural kinds function in natural laws.

Distinguishing between natural kinds and accidental collections via natural laws points up, quite obviously, the need for an adequate analysis of natural laws. (Once again I am not endorsing the legitimacy of either natural kinds or laws of nature but describing how they might be related if one accepts them.) None of the usual criteria that have been suggested works very well. The best is that natural laws are the generalizations that function in scientific theories. Any generalization that remains isolated, as true as it may be, is likely to be an accidentally true generalization. Even if this criterion turns out to be adequate, the next move is, of course, to demand an analysis of scientific theory. One way is to explicate the more inclusive notions in terms of their constituents. Scientific theories are made up of scientific laws, and laws are those generalizations that refer to natural kinds, never accidental collections. As always, the two ultimate alternatives are infinite regresses and circular reasoning. Outside these two alternatives resides pragmatic justifications — how well these analyses accommodate examples, and if this essay has any message, it is that these examples have to be real. They have to be part of some well-articulated system, whether scientific or philosophical.

How about fictitious examples? Don't they work just as well? In a celebrated paper, N. Goodman (1954) presented a contrary-to-fact example to pose a new riddle of induction. Such color predicates as green and blue seem like perfectly natural ways to divide up the world. Goodman suggests other predicates that seem decidedly unnatural — grue and bleen. "Green" applies to everything that is green — past, present and future, whether examined or not, and the same for "blue." "Grue," however, applies to things examined before a particular time, say, 2000 A.D., just in case they are green but to things not examined before 2000 A.D. just in case they are blue. Thus, anything examined before the year 2000 A.D. is both green and grue. What justification do we have for preferring talk of green emeralds to grue emeralds? That "emerald" is currently *defined* as green beryl is unlikely to give anyone pause.

Dozens, possibly hundreds, of papers have been written about grue and bleen without making much headway. My response will no doubt seem totally wrong-headed.

Genuine change does occur in nature. For example, in the early millennia of life on earth, respiration was entirely anaerobic. Somewhere along the way, aerobic forms of life gradually replaced anaerobic forms. In addition, life on Earth has been punctuated by mass extinctions in which upwards of 95 percent of all species went extinct. Given our current understanding, such occurrences are not anomalies. They are the sorts of thing that can happen, even though scientists are currently arguing about the mechanisms that actually produced them. However, according to current physical theory, predicates do not and cannot act the way that grue is supposed to act. There is nothing special about the year 2000 A.D. If things change their color wholesale on New Year's Eve of the second millennium (or, more accurately, if we change how we conceive of colors), then we have seriously misunderstood nature and have to reevaluate our beliefs from the ground up. Since our intuitions are informed by what we believe, in this new world we can no longer depend on our intuitions, but this is precisely what Goodman asks us to do.

"But can't you conceive of an alternative theory in which the year 2000 A.D. is as natural as absolute zero?" Once again, I have to say that I cannot. I can conceive of several generations of physicists producing such a theory but not me off the top of my head — and that is what appeals to such examples require.[7]

Conclusion

In this essay I have urged the advantages of real examples over fictitious examples. Both real and fictitious examples can be presented briefly, both can challenge our conceptions, but if the occasion demands, we can expand upon real examples because they appear in a larger context that supports such investigations. Made-up examples that occur outside of any context, including a well-formulated fictional world, can be modified any way one pleases. The considerations that I have sketched in this essay have forced philosophers and scientists alike to reevaluate some fundamental beliefs of Western thought from the ancient Greeks to the present. These considerations concerned the nature of species if species are supposed to be the things that evolve. Perhaps fictitious examples could also have motivated these changes, but I doubt it.

In this essay I have presented both a negative thesis and a positive thesis. The negative thesis is that made-up examples, especially of the sillier sort, have done massive damage to philosophy, in particular the philosophy of science. The chief deficiency of hypothetical examples in philosophy is that no one has bothered to set out a list of criteria by which they can be evaluated. What makes for a good hypothetical example? Why hasn't the Ship of Theseus done as much damage as grue and bleen?

The positive thesis is that philosophers should use real examples whenever they can because they have all of the virtues of made-up examples and none of the vices. While we are at it, we might set out a list of criteria for the evaluation of real examples as well. Another reason for my urging real examples on philosophers is that real examples, unlike fictitious examples, can be used as evidence — but that is another story.

NOTES

I owe a note of appreciation to several friends and colleagues who encouraged me in the writing of this essay: Martin Bunzl, John Collier, Arthur Fine, Tom Ryckman, and Ken Waters. However, I periodically got the feeling that they were encouraging me to be the first penguin off the ice floe.

1. As Norton sees it, the problem is that case studies tend to come in two polar types. "Either they are contrived 'toy' models, whose logical relations are clear but

whose connection to real science is dubious. Or they are instances or real science of such complexity that one must be disheartened by the task of mastering the scientific technicalities let alone disentangling its logical structure" (1993, 412). One characteristic which real experiments possess and thought experiments lack that I have not discussed in this paper is historicity. Hacking (1993) argues that real experiments "have a life of their own" in a way that made-up examples do not (see also MacIntyre 1981 and Sorenson 1992). However, some made-up examples have been discussed so frequently, later commentators modifying them for their own purposes, that they have developed a life of sorts.

2. Of course, with a sufficiently broad notion of set (or class, or kind) everything counts as a set from bare particulars to Richard Nixon. My response to this maneuver is to distinguish between two sorts of sets — those that are spatiotemporally restricted and those that are not (see Hull 1978, 340). Species are instances of the first sort of "set." According to current physical theory, the universe is spatially finite and probably temporally finite as well. If so, then all classes are "spatiotemporally" restricted in this sense (Hull 1981, 148).

3. In response to Sober's (1984) analysis of selection in evolutionary biology, Waters objects that Sober "bases his own acceptance of the harsh Pareto-style requirement on intuitions about phenomena that seem to have little relevance to evolution by means of spatially dependent selection. Rather than wage a battle of philosophical intuitions about phones connected to Rube Goldberg devices and baseball fans jumping up in stadiums, I am basing my case against applying the Pareto-style requirement *to selection theories* on the claim that we must abandon this principle to maintain scientific realism. If this means we must compromise some intuitions about population-level causation, then so be it. Philosophers of science have not held intuitions sacred in trying to preserve realism in the face of modern physics. I see no reason why intuitions about nonevolutionary phenomena should be considered gospel when it comes to preserving realism about the force of natural selection" (1991, 570).

4. One reason philosophers may be so reticent to treat Homo sapiens as a historical entity is that it threatens a central notion of certain ethical systems — human nature. If human beings have no nature, no essence, then all is lost. But just because biological species, including the human species, have no essence, it does not follow that persons, rational agents, or bearers of immortal souls have no essence. The problem is the identification of these kinds with something that is not a kind — Homo sapiens (see also Hull 1988).

5. I am afraid that the acclaim accorded J. J. Thompson's 1971 article stems more from her arresting examples than from her arguments. I find myself in large agreement with her conclusions in spite of her examples, not because of them. Nor does Thompson expect her readers to take her examples realistically, as if they really were to wake up connected to a famous violinist. The typical American reaction would be to get a lawyer and sue the Society of Music Lovers. A few million dollars would be ample compensation for a few months of inconvenience. But Thompson would no doubt dismiss this response, as realistic as it might be, as inappropriate.

6. In two of the examples that I discuss, the conceptual tensions one feels can be

reduced to contradictions. Species cannot be both spatiotemporally restricted and spatiotemporally unrestricted. One cannot simultaneously accept a particular set of criteria for reference and accept a term that lacks these criteria as actually referring. Norton (1993) thinks that all reliable thought experiments in science involve nothing but inference. Bunzl goes even further, arguing that all reliable thought experiments turn on straightforwardly deductive considerations.

7. Akeroyd (1991) has presented a realistic example of a grue-like predicate. According to Akeroyd, "grue" is defined as green in the past but blue in the future. He suggests "regulatic" phenomena as more practical examples: those that "exhibit regularities before time t and erratic behaviour thereafter" (Akeroyd 1991, 535). He then proceeds to give a real example of such regulatic phenomena from economics. Prior to 1969, the relationship between the percent rate of unemployment and the percentage change of money wage rates exhibited regular behavior; thereafter, it did not. The economic system had evolved into quite a different system. In certain economic systems stagflation is impossible. In others it is perfectly possible.

REFERENCES

Akeroyd, F. M. 1991. "A Practical Example of Grue." *British Journal for the Philosophy of Science* 42: 535–37.

Atran, S. 1990. *Cognitive Foundations of Natural History: Towards an Anthropology of Science*. Cambridge: Cambridge University Press.

Bradie, M. 1991. "The Evolution of Scientific Lineages." In A. Fine, M. Forbes, and L. Wessels, eds., *PSA 1990*, vol. 2. Ann Arbor, Mich.: Philosophy of Science Association, 245–54.

Brown, J. R. 1991. *The Laboratory of the Mind: Thought Experiments in the Natural Sciences*. New York and London: Routledge.

———. 1993. "Why Empiricism Won't Work." In D. L. Hull, M. Forbes, and K. Okruhlik, eds., *PSA 1992*, vol. 2. Ann Arbor, Mich.: Philosophy of Science Association, 271–79.

Bunzl, M. Forthcoming. "The Logic of Thought Experiments." *Synthese*.

Caplan, A. 1980. "Have Species Become Déclassé?" In P. D. Asquith and R. N. Giere, eds., *PSA 1980*, vol. 1. Ann Arbor, Mich.: Philosophy of Science Association, 71–82.

———. 1981. "Back to Class; A Note on the Ontology of Species." *Philosophy of Science* 48: 130–40.

Churchland, P. 1982. "Is the Thinker a Natural Kind?" *Dialogue* 21: 223–38.

Collier, J. 1990. "Could I Conceive Being a Brain in a Vat?" *Australasian Journal of Philosophy* 68: 413–19.

Dancy, J. 1985. "The Role of Imaginary Cases in Ethics." *Pacific Philosophical Quarterly* 66: 141–53.

Dupré, J. 1993. *The Disunity of Things: Metaphysical Foundations of the Disunity of Science*. Cambridge: Harvard University Press.

Eldredge, N., and S. J. Gould. 1972. "Punctuated Equilibria: An Alternative to Phyletic Gradualism." In T. J. M. Schopf, ed., *Models in Paleontology*. San Francisco: Freeman, Cooper & Co., 82–115.

Feynman, R. P., R. B. Leighton, and M. Sands. 1964. *The Feynman Lectures on Physics*, vol. 2. Reading, Mass.: Addison-Wesley.

Fodor, J. 1964. "On Knowing What We Would Say." *The Philosophical Review* 73: 198–212.

Forge, J. 1991. "Thought Experiments in the Philosophy of Physical Science." In T. Horowitz and G. J. Massey, eds., *Thought Experiments in Science and Philosophy*. Savage, Md.: Rowan & Littlefield, Publishers, Inc., 209–22.

Gale, R. M. 1991. "On Some Pernicious Thought-Experiments." In T. Horowitz and G. J. Massey, eds., *Thought Experiments in Science and Philosophy*. Savage, Md.: Rowan & Littlefield, Publishers, Inc., pp. 297–303.

Ghiselin, M. T. 1974. "A Radical Solution to the Species Problem." *Systematic Zoology* 23: 536–44.

Goldman, A. 1988. *Empirical Knowledge*. Berkeley: University of California Press.

Gooding, D. 1990. *Experiment and the Making of Meaning*. Dordrecht: Kluwer Academic Press.

———. 1993. "What Is *Experimental* about Thought Experiments?" In D. Hull, M. Forbes, K. Okruhlik, eds., *PSA 1992*, vol. 2. East Lansing, Mich.: Philosophy of Science Association, 280–90.

Goodman, N. [1954] 1983. *Fact, Fiction, and Forecast*. Cambridge: Harvard University Press.

Hacking, I. 1993. "Do Thought Experiments Have a Life of Their Own? Comments on James Brown, Nancy Nersessian, and David Gooding." In D. Hull, M. Forbes, K. Okruhlik, eds., *PSA 1992*, vol. 2. East Lansing, Mich.: Philosophy of Science Association, 302–08.

Herschel, J. F. W. [1841] 1987. *A Preliminary Discourse on the Study of Natural Philosophy*. Chicago: University of Chicago Press.

Horowitz, T. 1991. "Newcomb's Problem as a Thought Experiment." In T. Horowitz and G. J. Massey, eds., *Thought Experiments in Science and Philosophy*. Savage, Md.: Rowan & Littlefield, 305–16.

Horowitz, T., and G. J. Massey. 1991. "Introduction." In T. Horowitz and G. J. Massey, eds., *Thought Experiments in Science and Philosophy*. Savage, Md.: Rowan & Littlefield, 1–26.

Hull, D. L. 1976. "Are Species Really Individuals?" *Systematic Zoology* 25: 174–91.

———. 1978. "A Matter of Individuality." *Philosophy of Science* 45: 335–60.

———. 1981. "Kitts and Kitts and Caplan on Species." *Philosophy of Science* 48: 141–52.

———. 1987. "Genealogical Actors in Ecological Plays." *Biology & Philosophy* 1: 44–60.

———. 1988. *Science as a Process: An Evolutionary Account of the Social and Conceptual Development of Science*. Chicago: University of Chicago Press.

———. 1989. "A Function for Actual Examples in Philosophy of Science." In M. Ruse, ed., *What Philosophy of Biology Is: Essays for David Hull*. Dordrecht, Holland: D. Reidel, pp. 313–24.

———. 1991. "Conceptual Evolution: A Response." In A. Fine, M. Forbes, and L. Wessels, eds., *PSA 1990*, vol. 2. East Lansing, Mich.: Philosophy of Science Association, 255–64.

Irvine, A. 1991. "On the Nature of Thought Experiments in Scientific Reasoning." In T. Horowitz and G. J. Massey, eds., *Thought Experiments in Science and Philosophy*. Savage, Md.: Rowan & Littlefield, 149–65.

Kitts, D. B., and D. J. Kitts. 1979. "Biological Species as Natural Kinds." *Philosophy of Science* 46: 613–22.

Kuhn, T. 1964. "A Function for Thought Experiments." Reprinted in *Essential Tension*. Chicago: University of Chicago Press, 1977, 240–65.

Lennox, J. G. 1991. "Darwinian Thought Experiments: A Function for Just-So Stories." In T. Horowitz and G. J. Massey, eds., *Thought Experiments in Science and Philosophy*. Savage, Md.: Rowan & Littlefield, 223–45.

MacIntyre, A. 1981. *After Virtue*. Notre Dame: University of Notre Dame Press.

Massey, G. J. 1991. "Backdoor Analyticity." In T. Horowitz and G. J. Massey, eds., *Thought Experiments in Science and Philosophy*. Savage, Md.: Rowan & Littlefield, 285–96.

Masterson, J. 1994. "Stomata Size in Fossil Plants: Evidence of Polyploidy in Majority of Angiosperms." *Science* 264: 421–24.

Nersessian, N. J. 1993. "In the Theoretician's Laboratory: Thought Experimenting as Mental Modeling." In D. Hull, M. Forbes, K. Okruhlik, eds., *PSA 1992*, vol. 2. East Lansing, Mich.: Philosophy of Science Association, 291–301.

Norton, J. D. 1993. "A Paradox in Newtonian Gravitation Theory." In D. Hull, M. Forbes, K. Okruhlik, eds., *PSA 1992*, vol. 2. East Lansing, Mich.: Philosophy of Science Association, 414–22.

Putnam, H. 1981. *Reason, Truth and History*. Cambridge: Cambridge University Press.

Rescher, N. 1991. "Thought Experiments in Presocratic Philosophy." In T. Horowitz and G. J. Massey, eds., *Thought Experiments in Science and Philosophy*. Savage, Md.: Rowan & Littlefield, 31–41.

Sober, E. 1984. *The Nature of Selection*. Cambridge: MIT Press.

Sorenson, R. 1992. *Thought Experiments*. Oxford: Oxford University Press.

Thomason, S. G. 1991. "Thought Experiments in Linguistics." In T. Horowitz and G. J. Massey, eds., *Thought Experiments in Science and Philosophy*. Savage, Md.: Rowan & Littlefield, 247–57.

Thompson, J. J. 1971. "A Defense of Abortion." *Philosophy & Public Affairs* 1: 47–66.

Waters, C. K. 1991. "Tempered Realism about the Force of Selection." *Philosophy of Science* 58: 553–73.

Wiggins, D. 1967. *Identity and Spatio-Temporal Continuity*. Oxford: Blackwell.

Wilkes, K. 1988. *Real People: Personal Identity without Thought Experiments*. Oxford: Oxford University Press. Routledge.

15

A Logical Framework for the Notion of *Natural Property*

J. Michael Dunn

Departments of Philosophy and Computer Science, Indiana University

Animals are divided into (a) those that belong to the Emperor, (b) embalmed ones, (c) those that are trained, (d) suckling pigs, (e) mermaids, (f) fabulous ones, (g) stray dogs, (h) those that are included in this classification, (i) those that tremble as if they were mad, (j) innumerable ones, (k) those drawn with a very fine camel's hair brush, (l) others, (m) those that have just broken a flower vase, (n) those that resemble flies from a distance.   — From Borges 1964

Introduction

1.1. Dividing Nature by the Joints

The problem of sorting out "real" or "natural" properties is as old as philosophy itself. Thus in Plato's *Statesman* (287c), the Stranger, in talking to young Socrates about dividing up the arts so as to separate the kingly art, says: "Let us divide them according to their natural divisions as we would carve a sacrificial victim. For we must in every case divide into the minimum number of divisions that the structure permits" (Greek translations come from Hamilton and Cairns 1963).

And Socrates, in *Phaedrus* (265d, e), talks of two procedures which we might now call synthesis and analysis. "The first is that in which we bring a dispersed plurality under a single form, seeing it all together — the purpose being to define so-and-so, and thus to make plain whatever may be chosen as the topic for exposition." He says that the second procedure is "the reverse of the other, whereby we are enabled to divide into forms, following the objective articulation; we are not to attempt to hack off parts like a clumsy butcher."

Let us first consider the problem of analysis (Plato's "division"). Given a natural property, can its instances be divided up into two or more natural subdivisions? In *The Statesman* (262d) Plato gives as a contrary example the division of humans into "Greeks and barbarians," saying that "barbarian" means nothing other than "non-Greek," and that it is a mistake to think that non-Greeks have some common property that characterizes them. This citation of Plato must be taken as motivation only in a global sense, for, as will be made clear, "relevant properties" are closed under negation (assuming the dubious premise that *being a Greek* is a relevant property — that Greeks can be distinguished from non-Greeks by intrinsic properties; cf. Fact 3 of sec. 1.3 below). This distinguishes relevant properties from "natural kinds" (cf. sec. 5 below).

The example of the division of animals in the *Celestial Emporium* provides a humorously exaggerated example of a problematic analysis, an example more pertinent to the theory of relevant predication.

There is also the converse problem of synthesis (composition). Given two or more natural properties, how can they be combined logically so as to form further natural properties? This essay will focus more on the problem of synthesis than on the problem of division (though it will indirectly cast light on the latter as well, saying something about what formulas determine properties, and hence can be used in analysis). Classical logic is not a very good tool for either analysis or synthesis, for it gives no distinction between formulas that determine properties and those that do not.

Indeed, even if a lambda operator (λx) is added to first-order classical logic so as to be able to form names for properties, the standard rules of formation allow it to be prefixed to any formula whatsoever. Thus λx can be prefixed to a formula P that does not even have a free occurrence of the variable x so as to form the vacuous "property term" $\lambda x P$. And trying to require that x occurs free in P falls flat because of the logical equivalence of P and $P \wedge (P \vee Fx)$. Thus classical logic may be viewed as providing a *maximalist* account of the synthesis of properties.

Some contemporary philosophers have given radically *minimalist* accounts of synthesis. For example, R. Grossmann (1973) denies that there are any complex properties at all, and in particular says there are no conjunctive or disjunctive properties and no negative properties (though he admits complex facts). D. M. Armstrong (1978) also denies the existence of disjunctive properties and negative properties, although he does admit conjunctive properties.

We shall be presenting a theory that hopefully sails between the Scylla and Charybdis of maximalist and minimalist theories of properties, steered by the compass of an independently motivated natural logic: the relevance logic system **R** of Anderson and Belnap. This system originated in the late 1960s, with no thought of the problems of predication (see Anderson and Belnap 1975, and Anderson, Belnap, and Dunn 1992 for "encyclopedic" accounts of relevance logic). As will be stated more precisely later, while not every formula determines a "relevant property," still relevant properties are closed under conjunction, disjunction, and negation.

1.2. Relevant Properties

Dunn 1987 presented a formal theory that gives us one way to sort out the natural properties from the "hokey" properties (as Fodor 1987 wonderfully refers to them). But I should be clear as to what issues this theory of predication addresses, since what I have proposed by no means does everything that anyone might have wanted to do under this heading.

In Dunn 1990a, a metaphysical interpretation was placed upon the formal theory: relevant properties (and relations) "make a difference" to their subjects. Relevant nonrelational properties are intrinsic to their subject, and relevant relations affect the intrinsic properties of their subjects. In every case there is a "modification" of their subjects, which can *often* be tested by thought experiments involving change. I say "often" because as was made clear in Dunn (1990a, 1990b), relevant predication can be applied to eternal (e.g., abstract) objects. See end of sec. 2.1 below.

Considering the properties of animals from the *Celestial Emporium of Benevolent Knowledge*, we notice that some do not necessarily modify their subject at all. Thus *belonging to the Emperor* (a) is a relational property that need not modify an animal (unless being branded, for example, is made part of the very concept). And *being a stray dog* (g) similarly seems to be a relational property involving location in respect to where the owner expects the animal to be and to involve no intrinsic property (other than being a dog). *Being fabulous* (f) is a matter of being included in fables and has to do with the human practice of storytelling, not the animal itself. *Being included in this classification* (h) seems similarly to relate to the writer of the encyclopedia. And *being drawn with a very fine camel's hair brush* (k) seems to concern artists and their choice of tools. *Having just broken a flower vase* (m) would seem not to be a modification of an animal except on the most minute theory of causation by contact. *Being a suckling pig* (d) is ambiguous. If it means a pig who is not yet weaned, it seems to be a relational property (not just to the mother but to time) that involves no necessary modification of the animal itself (putting aside possible small nutritional differences). On the other hand, if it means "infant" pig, then this seems to be a modification of an animal.

Being embalmed (b) certainly modifies the animal (treating a dead animal as still an animal), and *being trained* (c) does modify the animal on any reasonable theory of potentialities as involving structures. *Being a mermaid* (e) presents philosophical difficulties because of the fictional status of mermaids. But if being a mermaid is treated as just a biological classification that happens not to be filled (as with *Tyrannosaurus Rex*), then it certainly would involve a modification of a thing if it were to become a mermaid (or cease to be one, as in the Hans Christian Andersen tale). *Trembling as if being mad* (i) certainly seems to be a modification of an animal, albeit described in a strange way. *Resembling a fly from a distance* (n) is a bit difficult, but it seems that this is a relational property in several senses, involving reference to human perceivers and also to flies, and is not a property of an animal in itself.

As for the property of *being innumerable* (j), it is hard to understand this as a property of an *individual* animal. If it is understood as the individual's being of a *species* that is "innumerable," this is not an intrinsic property of the individual. *Being other* (l) presumably means "none of the above." We shall see below that the disjunction of relevant properties is again a relevant property, and that the negation of a relevant property is also a relevant property. So *if* each of the properties in the above definition were really a relevant property, then so would be not having any of those properties. But as we have seen, very few of those properties are in fact relevant properties, so there is no reason to think that (l) defines a relevant property.

1.3. Relevant Predication

We here quickly sketch the formal theory adumbrated in Dunn 1987. For simplicity, we shall concentrate here on monadic predications.

Consider the pair of sentences

$$\text{Socrates is such that he is wise,} \qquad (15.1)$$

$$\text{Reagan is such that Socrates is wise,} \qquad (15.2)$$

They may be put into even more stilted English and then into symbols as:

$$\text{If anyone is Socrates, then he is wise,} \qquad (15.3)$$

$$\forall x(x = s \rightarrow Wx); \qquad (15.4)$$

$$\text{If anyone is Reagan, then Socrates is wise,} \qquad (15.5)$$

$$\forall x(x = r \rightarrow Ws). \qquad (15.6)$$

(15.1) can be put more colloquially as

$$\text{Socrates is wise.} \qquad (15.7)$$

But (15.2) resists colloquialism, unless one has become overly sophisticated by studying a bit of philosophy and/or a bit of logic and recognizes vacuous properties, or more formally, vacuous instances of λ-*conversion*:

$$(\lambda x P)r \leftrightarrow P. \qquad (15.8)$$

Then one sees it as a logician's trick way of saying again the slightly more colloquial (15.5).

I urge that untutored intuition is the better guide here, and that one should see (15.2) as involving an irrelevant predication, in line with (15.5) being an irrelevant implication. There is no connection, whether of implication or predication, between a thing's being Reagan and Socrates being wise. *Being such that Socrates is wise* is a hokey property.

This can be understood rather simply by remembering a well-known strategy for dealing with singular terms and predication within the context of the syllogism. Aristotle himself did not allow singular propositions, but by the Middle Ages it was standard to treat a statement such as (15.1) as "All *Socrates* are wise," where this last elevates "Socrates" to something like a general term. (15.3) can be understood as a cleaner way to the same end. Thus just as "All Greeks are Mortal" can be understood as a universal statement involving a hidden implication:

$\forall x(Gx \rightarrow Mx)$, the same is true of "Socrates is wise." Imagine an arrow drawn from the property of *being a Greek* to the property of *being mortal* for the general proposition, and in the singular case from the individuating property of *being Socrates* to the property of *being wise.*

All this suggests the following definition of *relevant predication*:

$$(\rho x \varphi x)a =_{\text{def}} \forall x(x = a \rightarrow \varphi x) \text{ (Relevant Predication).} \qquad (15.9)$$

The problem then with (15.2) is then that it is of the following form, where P does not contain occurrences of either x or a (and hence there is no connection between the antecedent and the consequent, as is required for relevant implication):

$$(\rho x P)a =_{\text{def}} \forall x(x = a \rightarrow P) \text{ ("Irrelevant Predication").} \qquad (15.10)$$

Let us turn to another, related topic: the indiscernibility of identicals. It is commonly expressed as follows:

$$x = y \rightarrow (\varphi y \rightarrow \varphi x) \text{ (Indiscernibility).} \qquad (15.11)$$

But in any logic that has permutation (as does the paradigm relevance logic **R**), this can be rephrased as:

$$\varphi y \rightarrow (x = y \rightarrow \varphi x) \text{ (φx determines a relevant property).} \qquad (15.12)$$

As is more thoroughly motivated in Dunn 1987, we shall take this as our definition of a formula φx *being of a kind that determines relevant properties.* But we do remark that saying that the formula φx satisfies Indiscernibility is a way of saying that it determines properties of objects themselves in a description-independent way. ("A rose by any other name would smell as sweet.")

Note that

$$Ws \rightarrow (x = r \rightarrow Ws) \qquad (15.13)$$

is just an instance of the following fallacy of relevance, meaning Ws is not a formula of the kind that determines relevant properties:

$$P \rightarrow (Q \rightarrow P) \text{ (Positive Paradox).} \qquad (15.14)$$

As is implicit I think in my earlier papers, and as Kremer 1994 emphasizes, definition (15.12), and not definition (15.9), is the fundamental notion (see also Kremer 1989). Definition (15.12) is stated in a semantical tone of voice, although I allowed myself also a metaphysical tone of voice from time to time, waffling between formulas "really" determining properties and formulas determining "real" properties (see Dunn 1987, 355; 1990a, 190–91, 200–02). Kremer, who tends to speak in the more forceful metaphysical tone of voice, characterizes (15.12) as a test for "a real property *tout court*."

The definition (15.12) is motivated by the following facts.

Fact 1. If φx is a formula of the kind that determines relevant properties with respect to x, then if φa then $(\rho x \varphi x)a$.

Fact 2. If $(\rho x \varphi x)a$ then φa.

Fact 3. If φx and ψx are formulas of the kind to determine relevant properties with respect to x, then so are $\sim \varphi x$, $\varphi x \wedge \psi x$, $\varphi x \vee \psi x$, and $\varphi x \rightarrow \psi x$. Also when P has no free occurrences of x, $P \rightarrow \psi x$ and $\psi x \rightarrow P$ determine relevant properties, but not necessarily $\varphi x \wedge P$, and $\varphi x \vee P$.

Fact 3′. If for every y, φxy is a formula of the kind to determine relevant properties with respect to x, then so are $\forall y \varphi xy$ and $\exists y \varphi xy$.

Fact 4. If φx relevantly implies ψx, then $(\rho x \varphi x)a$ relevantly implies $(\rho x \psi x)a$.

Fact 5. If φx and ψy are formulas of the kind to determine relevant properties with respect to x and y, respectively (and φx has no free occurrences of y, and similarly ψy has no free occurrences of x) then $\varphi x \wedge \psi y$, $\varphi x \vee \psi y$ do not necessarily determine relevant relations, but $\varphi x \rightarrow \psi y$ does.

Proof. All but Fact 3′ were proven in Dunn 1987. We note that Fact 3′ is an "infinite" analog of the fact that relevant properties are closed under finite conjunctions and disjunctions (part of Fact 3). We provide a sketch of the proof for the universal quantifier (the proof for the existential follows easily using the usual definition and properties of negation). Since we have permutation, we can take determination of a relevant property as equivalent to Indiscernibility. We also implicitly use transitivity of $\rightarrow$.

1. $\forall y[x = x_2 \rightarrow (\varphi xy \rightarrow \varphi x_2 y)]$ Hypothesis of Indiscernibility for every y

2. $x = x_2 \rightarrow \forall y(\varphi xy \rightarrow \varphi x_2 y)$ 1, confinement (IQ1)[1]

3. $x = x_2 \rightarrow [\forall y(\varphi xy) \rightarrow \forall y(\varphi x_2 y)]$ 3, distribution of $\forall$ over $\rightarrow$ (IQ4).□

Remark. Facts 1 and 2 together allow for the following (equivalent) definition: "*φx is a formula of the kind that determines relevant properties*": $\forall y([\rho x \varphi x]y \leftrightarrow \varphi y)$.[2] This can be given the metaphysical reading that a relevant property (*tout court*, as Kremer puts it) is one such that having it and having it relevantly are the same.

1.4. Applications

In Dunn 1987 there was a "punch list," not meant to be inclusive, of possible applications of the notions of relevant property and relevant relation:

1. Intrinsic Properties
2. Cambridge Change (Geach)
3. Internal Relations
4. Essential Properties
5. Intentional Relations
6. Fictional Objects and Nonexistent Objects
7. Existence, Truth, Exemplification
8. Russell's Paradox
9. Goodman's "Grue-Bleen"
10. Barwise-Perry "Slingshot"
11. Goodman's Problems About Aboutness

In case anyone is counting, I will make a brief progress report on where things stand with respect to carrying out these applications. Intrinsic properties (1) were discussed in Dunn 1990a and related to real change in an object, and so implicitly (though the term was not used) to "Cambridge change" (2). Also in the same paper internal relations (3) were discussed as those that involved modification of their terms. Essential properties (4) were the focus of Dunn 1990b, where it was suggested that an essential property is not just a necessary property, but a necessary relevant property. Intentional relations, e.g., Jane thinking about Paris, were tangentially discussed in Dunn 1990a. Fictional and nonexistent objects (6) are the subject of a paper still on my computer, "Is Existence a (Relevant) Property," which of course also relates to the first item of (7). Exemplification and Russell's Paradox (part of 7, 8) were dealt with in Dunn 1990b, though there is still much more to be said.

Here we shall appropriately focus on applications that have a philosophy of science flavor. (Other authors have also applied relevance logic to problems in the philosophy of science. Sylvan and Nola 1991 consider paradoxes of confirmation, and Waters 1987 considers problems of hypothetico-deductivism. I also consider the former below, but my remedy is different from Sylvan's.) This means focusing more directly on the problem of Cambridge change (2), the question as to whether truth is a (relevant) property (7), Goodman's "Grue-Bleen" paradox (9), and adding a discussion of Hempel's Paradox of the Ravens. We shall not here discuss the Dretske-Tooley-Armstrong account of natural laws, since it involves the ideal of a relevant higher-order *relation* between properties, and lies outside the scope of this essay. I also want to rehearse an application in Dunn 1990b to Fodor's "Fridgeon" and the frame problem in artificial intelligence as an illustration of Cambridge change. We shall not directly address the Barwise-Perry "Slingshot" (10) and Goodman's notions of "aboutness" (11), though we shall say something about "aboutness" in discussing Hempel's Paradox.

Cambridge Change

2.1. *Cambridge Change and Intrinsic Properties*

P. T. Geach introduced the notion of a "Cambridge change":

The great Cambridge philosophical works published in the early years of this century, like Russell's *Principles of Mathematics* and McTaggart's *Nature of Existence*, explained changes as simply a matter of contradictory attributes' holding good of individuals at different times. Clearly any change logically implies a 'Cambridge' change, but the converse is surely not true; there is a sense of "change," hard to explicate, in which it is *false* to say that Socrates changes by coming to be shorter than Theaetetus when the boy grows up, or that the butter changes by rising in price, or that Herbert changes by 'becoming an object of envy to Edith'; in these cases, 'Cambridge' change of an object (Socrates, the butter, Herbert) makes no 'real' change in that object. (1969b, 321–22)

Socrates would change posthumously (even if he had no immortal soul) every time a fresh schoolboy came to admire him; and numbers would undergo change whenever, e.g., five ceased to be the number of somebody's children. (1969a, 72)

Geach (1969b, 71–72) describes "Cambridge change": "The thing called 'x' has changed if we have '$F(x)$ at time t true' and '$F(x)$ at time t^1' false, for some interpretation of 'F', 't', and 't^1'."

Geach laments in several places that he has no idea how to explicate real change as opposed to mere Cambridge change.

But I do not know of any criterion, let alone a sharp one, that will tell us when we have a *real* change in Socrates and not just a 'Cambridge' change. The search for such a criterion strikes me as an urgent task of philosophy. (Ibid., 99)

I do not know why Geach does not explicitly suggest this, but it seems true to his idea to explicate a real change in an individual as a special case of the more abstract notion of a Cambridge change, but where one of the attributes in question is an intrinsic property of the individual:

An individual x *really changes* iff at one time x has some intrinsic property F and at another time it does not.[3]

Real change is thus a matter of an object's coming to have or losing an intrinsic property. But what is an *intrinsic property*? One natural response is that an intrinsic property is a purely monadic (i.e., nonrelational) property such that an individual's coming to have it, or coming to lose it, means *ipso facto* that the object has undergone a real change. There is clearly a kind of circle here, with "real change" being defined in terms of "intrinsic property," which in turn has been defined in terms of "real change." But this in no way makes the concepts any less important or any less intuitive. Nonetheless, there are reasons for not always linking intrinsic properties to change.

In Dunn 1990a, a metaphysical interpretation of relevant predication was made in terms of intrinsic properties. While the interpretation of relevant predication in terms of intrinsic properties is not the only interpretation possible, it is a very natural one. So we can explicate *real change*, as opposed to mere Cambridge change, in an individual x as a $\rho y(\varphi y)$ holding of x at a time t but not at another time t^1.

In the case of a monadic relevant predication, relevant predication was simply interpreted as intrinsic predication. Polyadic relevant predication was interpreted to mean that an intrinsic property was determined in some or all of the argument places. Taking the dyadic case as paradigmatic, φxy can establish a relevant property either in a definite one of its argument places (thereby determining the "relevant relational properties" $\rho y(\varphi a y)$ or $\rho x(\varphi x b)$), or else φxy can establish relevant properties in both arguments (determining the "relevant relation" $\rho xy(\varphi xy)$).[4]

The notion of a relevant relational property was thought of on the one hand as a "real relation" and on the other as an "internal relation."

This seems somewhat paradoxical, since it seems to require that relations are thus intrinsic to each of their arguments; the ghost of Absolute Idealism hovers overhead. But the explanation was offered that a real relation affects each of its terms in the sense that it determines some intrinsic properties of each term, and that the "internality" or "intrinsicality" in question is thus distributed among the terms. This was uncannily anticipated by Bosanquet, who said:

The phrase 'internal relation' seems to me to be not quite satisfactory, as suggesting relations between parts within a given term. At least the view which to me appears reasonable would be better expressed by some such term as 'relevant relations', i.e., relations that are connected with the properties of their terms, so any alteration of relations involves an alteration of properties, and vice versa. (1988)

In Dunn 1990a, the two types of relevant relational properties were motivated with reference to Aquinas's view "that creatures are really related to God Himself; whereas in God there is no real relation to creatures, but a relation only in idea, inasmuch as creatures are related to him" (Aquinas [1270] 1948, Q. 13, Art. 7). It was pointed out that a sentence such as "x created y" could determine a relevant property in y but not in x, and this was related to the fact that "on the traditional view of God as unchanging, God could not be modified, even by the relation of his creating the world."

Aquinas was quite clear as to his explanation as to how this apparent lack of a converse to the relation of *creating* can happen. A long discussion, with examples, concerns relations that are real in one or both extremes, as opposed to merely "ideas" ("logical") in both or one of their extremes. This was used in Dunn 1992a in motivating the different types of relevant predication that can occur in the dyadic case $\varphi x y$.

I was surprised then to find that Geach 1969b contains an almost identical discussion of the "well-known thesis of Thomistic theology that the relations of creatures of God are 'real' but the relations of God to creatures are not 'real'," though with no credit to Aquinas for providing the outlines of a solution and not just the problem. Thus Geach says:

If we may after all regard a relational proposition as making predications about the related things A and B, then it will make sense (whether it is true or not) to suppose that when we take the proposition as a predication about A, there is some actuality in *A* answering to the predication, but that when we take the same (or a logically equivalent) proposition as a predication about *B* there is no actuality in *B* answering to the predication. (1969b, 321)

Geach gives as a concrete example the proposition "Edith envies Herbert," and explains it by reference to the notion of "real change," saying "if Edith comes to envy Herbert, it is natural to regard this as a change in Edith rather than a change in Herbert."

Although he says he has no idea how to "offer a criterion for selecting, from among propositions that report at least 'Cambridge' change, those that report 'real' change," Geach is nonetheless "certain that there is no 'real' change of numbers." I agree with this, although it causes some difficulty for the identification of an intrinsic property with a property such that if an object were to come to have it or lose it, there would be a real change in the object. This would suggest that there are no essential intrinsic properties of an object, or at least eternal objects, whereas Dunn 1990c argues that the only essential properties worth the name are not merely necessary but also intrinsic.

Dunn (1990a, 181–82 and n. 15) makes clear that there can be relevant, as well as irrelevant, predications of eternal objects. Dunn makes

one qualification to the idea that a relevant predication involves a change in the terms. This is a nice intuitive handle on the notion, but 'change' is sometimes more of a metaphor. What are we to say regarding terms that are necessarily existent and have all of their intrinsic properties necessarily (depending on your metaphysics, numbers, pure sets, properties, whatever)? Rather than "change" *per se*, in such cases we mean "modification" in its technical metaphysical sense, where attributes 'modify' their subjects. (1990b, 81)

Kremer (1994) discusses this problem, and gives an intuitive test for a formula φa not determining a real property of *a*, which involves the idea that there can be two theories that disagree on the formula φa (in the usual sense that one asserts it while the other denies it), but which do not *disagree about a*. As Kremer is aware this involves making sense of theories "disagreeing about" an object; there is a kind of circularity afoot in that *disagreement about a* would quite naturally be cashed out in terms of disagreement about *a*'s real properties. We shall show at the end of sec. 3.2 how all this can be explicated using the apparatus of relevant predication. But first we turn to a concrete example of mere Cambridge change.

2.2. "Fridgeon" and the Frame Problem

This application has already been treated in Dunn 1990c, but we do it again because of its particular simplicity and subsequent usefulness to motivate intuitions about relevant properties. Fodor 1987 raises a certain interpretation of the famous "Frame Problem" in artificial intelligence. Fodor says:

> It has to work out on any acceptable model that when I turn the refrigerator on, certain of my beliefs about the refrigerator — and about other things, of course — become candidates for getting updated. . . . On the other hand, it should fall out of a solution of the frame problem that a lot of my beliefs — indeed, MOST of my beliefs — do not become candidates for updating. . . . Consider a certain relational property that physical particles have from time to time: the property of BEING A FRIDGEON. I define 'x is a fridgeon at t' as follows: x is a fridgeon at t iff x is a particle at t and my fridge is on at t. It is of course a consequence of this definition that, when I turn my fridge on, I CHANGE THE STATE OF EVERY PHYSICAL PARTICLE IN THE UNIVERSE; namely, every physical particle becomes a fridgeon.

We can give the following definition:

$$a \text{ has the relevant property of being a fridgeon} =_{\text{def}} \rho x (Px \wedge Fj)(a). \tag{15.15}$$

It is clear from Fact 3 that *being a fridgeon* need not be a relevant property, since the second conjunct of $Px \wedge Fj$ is apparently not a relevant property of x. But rather than just appeal to "facts," at least this first time, let us trace through the definition and see why $\rho x (Px \wedge Fj)$ is not a relevant property.

The definition (15.15) boils down to:

$$\forall x (x = a \rightarrow Px \wedge Fj). \tag{15.16}$$

But from (15.16) it clearly follows (instantiation, conjunction elimination) that

$$x = a \rightarrow Fj. \tag{15.17}$$

And this is a *prima facie* irrelevant conditional.

I say *prima facie* because there is no reason to think that there is any relevance between the antecedent and the consequent. There is no sharing of content that is discernible at either the syntactic level (no sharing of terms or predicates) or the semantic level, and I know of no reasonable physical theory that would postulate any physical connection, say causality, between distant particles of the universe and Fodor's refrigerator. I do not say that there can be no such theory, but I do not know of any such theories that are serious. Presumably Fodor does not either, and this is why he picked this example.

Paradoxes of Confirmation

3.1. Grue-Bleen

Having warmed ourselves up on fridgeons, so to speak, let us consider the more difficult case of "grue-bleen" predicates.

Every philosopher of science knows that N. Goodman, in *Fact, Fiction, and Forecast* (1955) says:

Suppose that all emeralds examined before a certain time t are green. At time t, then, our observations support the hypothesis that all emeralds are green. . . . Now let me introduce another predicate less familiar than "green". It is the predicate "grue" and it applies to all things examined before t just in case they are green but to other things just in case they are blue. Then at time t we have, for each evidence statement asserting that a given emerald is green, a parallel evidence statement asserting that the emerald is grue.

Writing out the definition of "grue" more explicitly, we get:

x is *grue* $=_{def}$ (x is green and x is examined before time t) or (x is blue and x is not examined before time t).

It is easy to get to think that one is stuck with *grueness* as a relevant property on the assumption that greenness and blueness are relevant properties. This is because the relevant properties of a given individual x are closed under conjunction and disjunction (Fact 3). But this overlooks the plausible conclusion that *being examined before time t* is not a relevant property. Of course, ignoring quantum mechanical niceties, one reason is that "examines" is an intentional relation. In general (putting aside the use of acids, hammers, etc.), whether an ordinary macro object, such as an emerald, has or has not been examined has nothing to do with the object itself; it is a mere relational property. Examining an emerald by eye, at least on a commonsense theory, need in no way change the emerald.

But I do not want to make too much of this because there are other versions of the grue-bleen paradox that are not "epistemological" in character, although they do still involve reference to time (or perhaps to some other external factor such as place). Consider the following "ontological version" in which there is no mention of "examination." This often appears in the literature as a seemingly unconscious simplification of Goodman's original definition.

x is *grue* $=_{def}$ x is green before time t and x is blue at time t and thereafter.

Let us first discuss this with reference to present tense sentences, understanding "x is green" as "x is presently green" (Quine's "occasional" as opposed to his "eternal" sentences). Then we can have the definition:

x is *grue* $=_{def}$ x is green and the present time is earlier than t, or x is blue and the present time is t or later. In obvious symbols:

$$Grue(x) =_{def} [Gx \wedge (p < t)] \vee [Bx \wedge (p \geq t)]. \quad (15.18)$$

Put this way, it is clear that the first disjunct ("being green before time t") is not an intrinsic property of x. The individual x does not wear a watch, so to speak. More formally, we appeal to Fact 3. The only reason to think that the disjunction determines a relevant property of x is that each disjunct determines a relevant property of x, and the only reason to think that the first disjunct determines a relevant property of x is that each of its conjuncts determines a relevant property of x. But $p < t$ does not even contain an occurrence of x, and so in instantiating the appropriate instance of (15.12) we obtain the following example of Positive Paradox (15.14):

$$p < t \rightarrow (x = y \rightarrow p < t). \quad (15.19)$$

Thus we are able to cash out the irrelevance of the grue predicate in terms of an irrelevant implication.

Things are not so easy to explain if we define "grue" using tenseless predicates with an explicit time parameter: Gxt: "x is green at time t," Bxt: "x is blue at time t." To begin with, there are several different ways of defining "x is grue," equivalent in classical logic, but nonequivalent in relevance logic. The symbolization that would be the closest parallel to (15.18) is:

$$Grue(x) =_{def} \forall p\{[Gxp \wedge (p < t)] \vee [Bxp \wedge (p \geq t)]\}. \quad (15.20)$$

An alternative would be to symbolize "*x* is green before time *t*" as $\forall p < t(Gxp)$, "*x* is blue at time *t* and thereafter" as $\forall p \geq t(Bxp)$, and finally, as the conjunction of these:

$$Grue(x) =_{def} \forall p < t(Gxp) \wedge \forall p \geq t(Bxp). \quad (15.21)$$

Alternatives are proliferating, and we shall start with the definition (15.20). The point is that the lessons we learned with (15.18) can be directly applied. In this case the disjunction is universally quantified, but presumably it is on the basis of Fact 3' that one would think that the formula determines a relevant property, and so we are really considering the disjunction. Again, the only reason for the disjunction in (15.20) to determine a relevant property of *x* would be for each disjunct to, and so

$$Gxp \wedge (p < t)$$

would have to determine a relevant property of *x*. But again the only reason we could have for thinking it does is that each conjunct determines a relevant property of *x*, but clearly the second conjunct does not (since it contains not even an occurrence of *x*).

We next discuss the definition (15.21). This definition is actually a schematic one and boils down to three separate definitions, depending on how restricted quantification is understood. At top level, all three definitions have the same form: a conjunction of two restricted quantifications. The only reason for thinking that the conjunction determines a relevant property is given by Fact 3, meaning each conjunct would have to determine a relevant property. This means

$$\forall p < t(Gxp) \quad (15.22)$$

would have to determine a relevant property. Let us then examine three different ways to understand (15.22).

In classical logic the standard way of expanding a restricted quantification uses the material conditional, so we obtain

$$\forall p(p < t \supset Gxp). \quad (15.23)$$

But when the material conditional $\varphi \supset \psi$ is understood in the standard way in relevance logic as $\sim \varphi \vee \psi$, the definition (15.21) (with restricted quantification expanded using the material conditional) is in fact equivalent to the definition (15.20), and so we have no more reason to believe that the one determines a relevant property than the other. This equivalence is left as an exercise for the reader, but the reader who does not want to do his or her homework can instead see that in order for (15.23) to determine a relevant property of x, its "disjunct" $\sim(p < t)$ would have to determine a relevant property of x, which it clearly does not do.

Let us next consider the natural alternative where the expansion of restricted quantification uses relevant implication:

$$\forall p(p < t \rightarrow Gxp). \tag{15.24}$$

While this is a natural alternative on formal intuitions, it in fact is not a natural alternative, once one stops and thinks about the meaning. This requires that there be a strong connection between times and the color of objects — indeed, a kind of law requiring that times earlier than t be times at which x is green. This was a close call. If (15.24) were an acceptable way to expand restricted quantification, then (by expanding [15.21] in this manner), "grue" would turn out to be a relevant property of x. This is because we could apply the part of Fact 3 that deals with conditionals only one component of which contain a free occurrence of x.[5]

Cohen 1983 argued convincingly that there is in fact no good way to define restricted quantification within the confines of the standard syntax for relevance logic. As we have just seen, using relevant implication is too strong to express accidental generalizations. More subtly, using the material conditional is too weak. The problem is that the rule of detachment ($\varphi, \varphi \supset \psi \vdash \psi$) does not hold for the material conditional (see Anderson and Belnap 1975), and so one could not argue from the premises "All Greeks are human" and "Socrates is a Greek" to the conclusion "Socrates is wise."

So Cohen proposes a Goldilocks third alternative that is not too weak and not too strong, but just right. He proposes to add Belnap's connective (/) of conditional assertion to relevance logic, with restricted universal quantification understood as $\forall x(Fx/\ Gx)$. Thus (15.22) would be expanded to

$$\forall p(p < t \mid Gxp). \tag{15.25}$$

Although Cohen (1983) has addressed the question of how to add conditional assertion to first-order **R**, to the best of my knowledge no one has directly addressed how to add it to first-order **R** with identity. This is not the place to discuss such a formalism. But one would have to be careful about Indiscernibility. Thus it is most implausible that a conditional assertion of the form P/Gx (P contains no free occurrences of x) would determine a relevant property of x (even if Gx does). The reason is that a logic with conditional assertion must be interpreted as either a three-valued (or partial) logic. Thus if P is true, then the value of the conditional assertion is the same as that of Gx, but if P is false, then it is as if nothing had been said and so the value of the conditional assertion is "neuter" (or undefined). Now the truth value of P may have nothing to do with any property of the object x. Consider examples such as "I assert that John was at the party on the condition that I am not confusing him with Joe," (or, "that his twin was not there," or "that the party was held last Friday," etc.). Thus John may have been at the party, but whether the formula P/Gx is true of him still depends on P. Thus John will have the "property" P/Gx or not (in this last case as "neuter," not just false) depending on the truth value of P. In one possible world (where I am not confused it is true), and in another possible world it is not true but "neuter."

In the conditional assertion (15.25) at question, it seems that time p being before a certain time t has little if anything to do with an object x being green at time p. The object can be green at time p, but whether the conditional assertion is true depends as well on the relation of time p to time t.

Returning to the "big picture," I have been arguing that on any plausible formal definition that I can think of, the predicate "grue" does not determine a relevant property. Goodman anticipated the kind of argument I have made, though he states it in terms of "qualitative" properties instead of "relevant" or "intrinsic" properties as I have done.

"Surely it is clear," the argument runs, "that the first two [green, blue] are purely qualitative and the second two [grue, bleen] are not; for the meaning of each of the later two plainly involves reference to a specific temporal position." . . . The argument that the former but not the latter are purely qualitative seems to me quite unsound. True enough, if start with "blue" and "green," then "grue" and "bleen" will be explained in terms of "blue" and "green" and a temporal term. But equally truly, if we start with "grue" and "bleen," then "blue" and "green" will be explained in terms of "grue" and "bleen" and a temporal term. . . . Thus qualitativeness is an entirely relative matter and does not by itself establish any dichotomy of predicates. (1965, 79)

I cannot argue with the purely formal point that Goodman has made. And I must be careful not to overstate what we have achieved. Metaphysics moves back and forth incrementally between the "rock" of relativism and the "hard place" of absolutism (as epistemology moves between skepticism and dogmatism). As stated in Dunn 1987:

It is not the business of logic, but rather of metaphysics (or perhaps of whatever field whose subject matter is being formalized, e.g. physics) to determine what formulas "really" determine properties, just as it is not the business of logic to tell us what formulas are true, but only what formulas follow from each other. Roughly speaking, logic should tell us only that *if* certain formulas are postulated to "really" determine properties, *then it follows* that certain other formulas "really" determine properties.

But the point is that we do not want logic to tell us that every compound formula determines properties, and this can be taken to be the problem with "grue." The choice of those atomic predicates that determine relevant properties lies outside of the realm of logic. In this limited sense the question as to what predicates "really" determine properties is relative. But once that choice is made, and one has a relevant predicate G ("green") and a relevant predicate B ("blue"), for example, one does not have to stand idly by and watch in horror as the logical machinery goes on by itself to manufacture compound relevant predicates for "grue" and "bleen." The relativism as to the initial choice as to which predicates determine natural properties is just as harmless as the relativism as to which sentences are taken as initial truths (axioms) of a theory.

A final point. Kremer 1994 points out that there is no reason to identify the relevant predicates with Goodman's "projectable" predicates (roughly those that support induction). This is right on the metaphysical interpretation that I have given of relevant predication being of the sort that makes a difference to its subjects. At least I can think of no argument for identifying relevant predicates (under this interpretation) with those predicates that do not mess up induction. But it is plausible that there is at least a one-way connection: projectable predicates should be relevant. Of course, the particular metaphysical interpretation I have given is not forced by the logic, and perhaps one might try to reinterpret relevant predication as projectable predication. But there seems no reason to think that projectable predicates are closed under negation. Since projectability is one thing that might be required of "natural properties," this seems to me one of the instances to at the start of the paper, in which the theory of relevant predication is not able to do everything that everyone might want for a theory of natural properties.

3.2. The Paradox of the Ravens

Hempel created his famous paradox of confirmation by use of the familiar logical equivalence of "All ravens are black" with "All nonblack things are nonravens"; in symbols, the equivalence between the following pair of sentences:

$$\forall x (Fx \rightarrow Gx), \qquad (15.26)$$

$$\forall x (\sim Gx \rightarrow \sim Fx). \qquad (15.27)$$

He remarked that just as observing a black raven confirms the statement "All ravens are black," observing a nonblack nonraven so too confirms the statement "All nonravens are nonblack," *and hence* also the equivalent "All ravens are black."

It might be thought that relevant predication would provide a straightforward solution to Hempel's paradox, perhaps by allowing us to say that the predicates "nonblack" and "nonraven" do not determine (relevant) properties. However, on the assumption that "black" and "raven" determine properties, Fact 3 tells us that their negations also determine relevant properties. So this move is ruled out. Unfortunately, things get a bit twisted as I explore various ways that one might try to apply the theory and show how mostly they do not work. But in the end we do find a use for relevant predication in adding precision to Hempel's own solution to his own paradox.

Various moves are tempting. One is to deny the logical equivalence itself. This is difficult since the equivalence goes back to Aristotle, and is accepted by all modern logics that symbolize universal affirmative propositions, "All F's are G" as $\forall x (Fx \rightarrow Gx)$, and grant the principle of contraposition (in both directions) $\varphi \rightarrow \psi \dashv\vdash \sim \psi \rightarrow \sim \varphi$.

A second move is to grant the logical equivalence, but to somehow deny that the two propositions, although logically equivalent, are in fact identical (see Grossmann 1983).

A third move is to grant the seemingly strange conclusion that observing a red apple does in fact confirm the statement "All ravens are black," *but only a teeny, teeny bit.* The idea is that the number of nonblack things is very, very large, and so finding one of them that is a nonraven does add some very, very small confirmation to the statement "All nonblack things are nonravens," and hence to its equivalent "All ravens are black."

The second move has never appealed to me because even if the two propositions are distinct, the mere fact that they are logically equivalent would seem to indicate that whatever confirms one must also confirm the other, and, as Hempel clearly saw (he dubbed this the "Equivalence Condition"), this is all that is needed for the paradox to take off. Nonidentity seems not to be the issue, unless it can somehow be (independently) linked to different confirmation conditions.

The first move is ruled out by orthodox relevance logics, e.g., **R**, for their negation behaves classically, aside from not making equivalent all contradictions (or all tautologies).[6] But formally speaking, it is possible to have a weaker, say constructive negation; and in fact this has been advocated by Urquhart (1972) and others. Thus one would have

$$(\varphi \to \psi) \to (\sim \psi \to \sim \varphi) \text{ (constructive)}, \qquad (15.28)$$

but not its converse

$$(\sim \psi \to \sim \varphi) \to (\varphi \to \psi) \text{ (non-constructive)}. \qquad (15.29)$$

It might be thought that much the same end can be accomplished in the standard relevance logic **R** by noting that the following pair of propositions are not equivalent:

$$\forall x[(\rho y R y)x \to (\rho y B y)x] \qquad (15.30)$$

$$\forall x[(\rho y \sim B y)x \to (\rho y \sim R y)x]. \qquad (15.31)$$

But in fact, on the assumption that Ry and By are formulas of the kind that determine relevant properties, then so (Fact 3) are $\sim Ry$ and $\sim By$. But (by the Remark of sec. 1.3), $(\rho y R y)x$ is equivalent to Rx,

$(\rho yBy)x$ is equivalent to Bx, and so (15.30) may be simplified (replacement of equivalents) to

$$\forall x(Rx \rightarrow Bx). \qquad (15.32)$$

Similarly, (15.31) may be put in the equivalent form

$$\forall x(\sim Bx \rightarrow \sim Rx). \qquad (15.33)$$

But it has already been observed that in **R** the two statements (15.32) and (15.33) are logically equivalent because of contraposition. And so (15.30) and (15.31) are equivalent, despite the initial hope that they might not be.

So if one is somehow going to avoid the Raven Paradox by the choice of an underlying logic, it seems the system **R** as it stands is not the appropriate choice. One would have to drop contraposition in at least one of its two directions.

We will go no further in exploring what kinds of modifications might need to be made to the system **R** to "solve" the Raven Paradox. Indeed I believe that Hempel himself provided the solution to the very paradox he proposed, and I will show how this can be illuminated by the apparatus of relevant predication.

Hempel said that the paradox is based on a "misleading intuition," and goes on to explain:

One source of misunderstanding is the view . . . that a hypothesis of the simple form 'Every *P* is a *Q*' . . . asserts something about a certain limited class of objects only, namely, the class of all *P*'s. . . . Our interest in the hypothesis may be focused upon its applicability to that particular class of objects, but the hypothesis nevertheless asserts something about, and indeed imposes restrictions upon, *all* objects. (1945, 18)

It seems as if the statement "All ravens are black" is about ravens, but in fact is a statement about all objects in the universe, saying of each and every one of them that if it is a raven, then it is black. Once one recognizes this, one is no longer bothered by, a red apple having some (positive) relevance to the hypothesis that all ravens are black.

But what does it mean to say that a statement φ is *about* an object a? Now is not the occasion to go into all the subtleties, but Goodman 1984 made a distinction between a proposition being *about a* and *being true about a* (see also Goodman 1965; Ullian and Goodman 1977). It seems plausible to cash out both of these using the apparatus of relevant predication as follows. First, replace any occurrences of *a* in φ by a variable x that is new to φ. Call the result φx. To say that φ is *about a* is to say that φx is a formula of the kind that determines relevant properties with respect to x, and to say that φ is *true about a* is to say that $(\rho x \varphi x)a$. Incidentally, we can define φ is *false about a* as $(\rho x \sim \varphi x)a$.

These definitions do not have quite the formal properties that Goodman wanted, but I submit that they are plausible. They in effect say that *aboutness* boils down to making a difference to the object, either in principle or in actuality (this last is *true about* or *false about*, depending on the "polarity").

Before applying these observations regarding *aboutness* to the Paradox of the Ravens let us briefly note that we can now make formal sense of Kremer's notion (sec. 2.1 above) of two theories *disagreeing about a*. It means that one asserts that some sentence ψ is true about *a*, and the other asserts that the same sentence is false about *a*.[7]

Returning to the Raven Paradox, let us renew our assumption that *being a raven* and *being black* are both relevant properties. Then (Fact 3) the formula $Rx \rightarrow Bx$ determines a relevant property, and so if the law $\forall x(Rx \rightarrow Bx)$ is true, then it can be rephrased so as to be *true about* each object y, i.e., (using Facts 1, 2, 3') it is equivalent to $\forall y[\rho(Rx \rightarrow Bx)(y)]$. (Since relevant properties are closed under negation and disjunction (Fact 3), the same holds true if the law is formulated with the material implication instead of a relevant implication, with $Rx \supset Bx =_{\text{def}} \sim Rx \lor Bx$). We take this as a formal explication of Hempel's view that the hypothesis that all ravens are black indeed asserts something about all objects. As Hempel (1945, 18) says: "The impression of a paradoxical situation is not objectively founded; it is a psychological illusion."[8]

Truth and the Paradox of the Liar[9]

4.1. The Liar Paradox

Truth has been a big topic in recent years, both in logic and the philosophy of science, and in general philosophy. Two of the biggest issues concerning truth have been (1) whether truth is an objective property, and (2) how to avoid the Liar and related paradoxes. In this section we discuss the Liar and in the next the question as to whether truth is in any serious sense a property.

To say that the truth predicate T is of the kind that determines a relevant property is to say that for all sentences s the following holds:

$$T(s) \to (x = s \to T(x)).$$

We know from the Remark of sec. 1.3 that this is equivalent to the following:

$$T(s) \leftrightarrow [\rho x T(x)](s). \tag{15.34}$$

This may be described as saying that "S is true" and "S is *relevantly* true" are equivalent.

Fact 3 tells us that since $T(x)$ determines relevant properties, then so does $\sim T(x)$, so we also have (again by the Remark of sec. 1.3) the following equivalence:

$$\sim T(s) \leftrightarrow [\rho x \sim T(x)](s) \tag{15.35}$$

From the pair of equivalences above, it easily follows, *on the assumption that truth is a relevant property*, that there is no distinction between *not* being relevantly true, and being relevantly *not* true (contrapose [15.34] and transitize with [15.35]):

$$\sim[\rho x T(x)](s) \leftrightarrow [\rho x \sim T(x)](s).$$

$[\rho x T(x)](s)$ might be naturally expressed as "s has the property of truth" and $[\rho x \sim T(x)](s)$ as "s has the property of falsity," and so there would be a kind of bivalence.[10]

A truth theory that thinks truth is a relevant property would presumably have the Tarski "T-sentence":

$$[\rho x T(x)](s) \leftrightarrow \varphi, \tag{15.36}$$

making the usual assumption that s is a logically descriptive name of the sentence φ (s could be φ in quotes, the Gödel number of φ, or some other description from which φ can be decoded). It is plausible to think that $\leftrightarrow$ in (15.36) is a relevant bi-implication.

We are now in a position to ask how the Liar Paradox fares on this treatment of truth. Let us continue our supposition for the moment that truth is a relevant property of sentences, and also assume that we have a language with devices sufficiently flexible for referring to all of its own sentences (e.g., quotation or Gödel numbering). With this last we can assume that (l) refers to the sentence that says that l is not true:

$$(l): \sim T(l) \tag{15.37}$$

As an appropriate instance of (15.36) we obtain

$$[\rho x T(x)](l) \leftrightarrow \sim T(l). \tag{15.38}$$

Now *if* truth were a relevant property, we could apply the equivalence (15.34) to (l), obtaining

$$T(l) \leftrightarrow \sim T(l), \tag{15.39}$$

from which a contradiction could be derived in the usual way. But rather than blindly concluding a contradiction, and calling "paradox," we can take this as a *reductio ad absurdum* to the hypothetical that truth is a relevant property.[11]

4.2. Is Truth a Relevant Property?

We have just shown on the theory being adumbrated here that one can take it as a consequence of the so-called Liar Paradox that truth is not a relevant property. But of course it is not satisfying to avoid a paradox merely by giving up one of the assumptions that lead to it. One wants independent reasons for giving up that assumption (as opposed to others). In this section we develop such reasons for thinking that truth is not a relevant property. These reasons are partly informal and partly formal, and in the end they reinforce each other.

Let us start with the informal, or metaphysical reasons. The common view that truth is an objective property has recently been under much attack from various forms of relativism. This is fundamentally an attack on objectivity itself, and I wish to separate the question as to whether truth is a relevant (real) property from the general question of objectivity. Whether a given predicate (simple or complex) holds of an object might be a completely objective matter, whether or not the predicate determines a relevant property.[12] And conversely, even in a fundamentally subjectivist framework it might be useful to separate out some predicates from others in terms of their "naturalness."

A more "relevant" attack comes from the so-called Redundancy Thesis, credited to Ramsey and importantly elaborated by D. Grover (1992) as the "Prosentential theory of truth." This attack can be interpreted as saying that there is fundamentally no such thing as a property of truth at all. We shall be arguing to the same conclusion, but on different grounds.

Among those philosophers who admit truth as a property, there has been much argument as to just what sort of entities bear it. Let us assume for the moment that there are only two choices for truth bearers: sentences and propositions. These represent the division linguistic vs. nonlinguistic, and if someone has another candidate, it will presumably fall on one side or the other (or maybe some mixture of both) and what we say can most likely be adapted. For simplicity, our default assumption shall be that truth bearers are sentences, but we shall at crucial points consider the option of their being propositions.

A further complication is that truth can be either viewed as a property *simpliciter*, or as a relational property involving sentences and the world. For truth to be a relational property presumably there would have to be some relation, let us call it "true in," between sentences and possible worlds. Viewed this way, whether true is a property depends on whether "true in" is a relation, or at least whether it determines relational properties on its left-hand pole. (The same points can be made about the relation φ *is satisfied by sequence s* in the technical Tarskian framework, pumped full of metaphysical significance by Davidson and his followers.)

We are of course talking about relevant properties and relations here, and it was argued in Dunn 1990a that these are to be identified metaphysically with internal (intrinsic) properties and relations, i.e., those which "make a difference" to their terms. Intuitively, the truth value of a sentence makes no (intrinsic) difference to the sentence itself. This is clearest for what Quine has labeled "occasional" sentences, such as "It is raining." A change in the weather makes a big difference in the world, but no difference at all in the sentence itself. Or even if we take an eternal sentence such as "Snow is white," we can imagine other possible worlds where snow is puce instead. While these worlds might differ widely from ours with respect to the color of snow and related matters, they need not differ one iota in linguistic resources. One has to be careful here and acknowledge that the point is most easily made with sentences abstractly conceived as universal types. (And the same point can be made just as well if one prefers propositions as truth bearers, since these are commonly supposed to be Platonic and unchangeable anyway.)

However, even if one thinks that truth bearers are the actual tokens of sentences, that is, concrete utterances and inscriptions, it still seems that their intrinsic character does not depend on their truth value. It depends on such physical characteristics as pitch, shape, etc., and not on truth-value. Even if one has a view that truth bearers are some exotic combination of sentences (whether types or tokens) and propositions ("statements," i.e., sentences with meaning), it still seems that the intrinsic character of these will not change depending on truth value.

Let us now view things formally for a while, and again consider the T-sentence (15.36), which we display again for the convenience of the reader:

$$[\rho x T(x)](s) \leftrightarrow \varphi, \qquad (15.40)$$

It should strike the reader immediately that ordinarily at least this should be an irrelevant bi-implication, since although s occurs on the left hand side, it will not occur on the right hand side except for the unusual case when φ is self-referential. We shall see that this irrelevant predication, and so the T-sentence itself should be rejected.

Let us assume that φ is in fact a true sentence, say "Snow is white." Using (15.36) we obtain:

$$\forall x(x = s \rightarrow Tx). \qquad (15.41)$$

But since T is assumed to be a relevant property, we have

$$Tx \rightarrow (s = x \rightarrow Ts). \qquad (15.42)$$

From (15.41) and (15.42) we can obtain (by universal instantiation and transitivity of implication)

$$x = s \rightarrow (s = x \rightarrow Ts), \qquad (15.43)$$

and from (15.43) using self-distribution of implication, we get:

$$(x = s \to s = x) \to (x = s \to Ts). \qquad (15.44)$$

But it is surely a postulate of relevant identity theory that identity is (relevantly) symmetric:

$$x = s \to s = x, \qquad (15.45)$$

and so we get from (15.44) and (15.45):[13]

$$x = s \to Ts. \qquad (15.46)$$

From (15.46), (15.34), and (15.36), using transitivity we obtain

$$x = s \to \varphi, \qquad (15.47)$$

and then by universal generalization:

$$\forall x(x = s \to \varphi). \qquad (15.48)$$

This last is just $(\rho x \varphi)(s)$. So a certain sentence s has the *vacuous and relevant* property of *snow being white*. This is very strange, even though the sentence in question is "Snow is white." Let us consider the implication (15.47) which founds this relevant implication, in English:

$$x = \text{"Snow is white"} \to \text{Snow is white.}$$

An implication then holds between x being identical to a certain item (a sentence) and a sentence in which that item is nowhere mentioned. This idea is visually clearer if a Gödel number, say 1001, is used in place of the quoted "Snow is white," so as to get the implication:

$$x = 1001 \to \text{Snow is white.}$$

This cannot be a relevant implication, and this is the formal counterpart to the intuition that truth is not an intrinsic property of the sentence. The fact that snow is white in no way depends upon ("modifies") the sentence "Snow is white."

Now the reader can legitimately be worried about whether there might be some relation between s and φ in the exceptional, self-referential cases, particularly since the Liar sentence is one of these. I have some anxiety myself, but we have shown at least that there is no reason to believe in general that truth is a relevant property (at least not for the ordinary non-self-referential cases, and this must undermine the intuition in general). And this provides independent evidence for the result of the previous section, that in the most notorious self-referential case (the Liar sentence), we must conclude that truth is not a relevant property. Admittedly there is some uncovered middle ground. It might be that truth is not a property in the non-self-referential cases, nor in the self-referential cases that lead to "paradox," but there may still be some harmless cases of self-reference where truth is a relevant property. We leave this subtlety unexplored.

Natural Kinds

What is a *natural kind*? Various answers have been given; for example, Putnam 1970 talks of "structure" and Armstrong 1978 talks of "hidden real essence."

The theories of Kripke and Putnam in the early 1970s helped rehabilitate the notion of a natural kind. Putnam, after saying that some general names (e.g., "bachelor") are capable of explicit definition, says:

An important class, philosophically as well as linguistically, is the class of general names associated with *natural kinds* — that is, with classes of things that we regard as of explanatory importance; classes whose normal distinguishing characteristics are 'held together' or even explained by deep-lying mechanisms. *Gold, lemon, tiger, acid*, are examples of such nouns. (1970)

We will not concern ourselves here with the details of the Putnam-Kripke theory of natural kind terms (which emphasizes their ultimate undefinability, in analogy with proper names and a causal account of reference). But I think we have a good enough grasp of the notion to raise the question as to whether natural kinds and relevant properties should be identified with each other.

First notice that the union of two natural kinds is not itself necessarily a natural kind (consider human beings and lizards, an example of Hardegree's which we will discuss below). Yet the disjunction of two relevant properties is again a relevant property (Fact 3). Also, the negation of a relevant property is also a relevant property (again Fact 3). But it would seem, usually at least, that the negation of a natural kind would not itself be a natural kind. Plato's point from *The Statesman* (mentioned above) about there not necessarily being something in common to non-Greeks would seem to apply to natural kind terms. What structure is common to nonlemons, or to nondogs? Thus the closure properties of relevant properties are broader than those for natural kinds.

R. H. Thomason (1969) has proposed that the disjunction of two natural kinds would be the smallest natural kind that contains them both. Thus with the case of dogs and cows this would be mammals (or perhaps some subclasss of the mammals). Thomason postulates that *taxonomic systems* are lattices (partially ordered sets in which each pair of elements a, b has a least upper bound $a \lor b$ and a greater lower bound $a \land b$). Thomason also postulates an empty kind Λ (as well as a universal kind).[14] Thomason does not talk of negation, but on his account, while natural kinds are closed under a kind of disjunction (as well as conjunction), on structural grounds alone it is not possible to identify natural kinds with relevant properties. This is because Thomason gives examples of "real life" taxonomic systems, showing that they do not satisfy the distributive law. But relevant properties do form a distributive lattice; indeed, taking negation into account, they form a "De Morgan lattice" (see Anderson and Belnap 1975 for a definition).

Although not all relevant properties are natural kinds, it would seem reasonable that all natural kinds will be relevant properties. Or, if there are ontological scruples that somehow make one want to distinguish natural kinds from properties (say because they are taken to be classes), at least natural kinds will somehow be intimately associated with relevant properties. The intuition behind relevant properties is that they make a difference to their subjects. Certainly this is true of natural kinds, though on one view nothing can change from one natural kind to another — one substance would simply cease to exist while another one begins. But to me this is the biggest kind of difference one can imagine.

There are reasons for thinking that there are relevant properties that are not natural kinds besides the closure and structural reasons mentioned. One reason is linguistic. Natural kind terms are common nouns, whereas many relevant properties are given by other linguistic types, e.g., predicates. This at least shows that there are relevant *predicates* that are not natural kind terms. But there is more to it than this. Relevant properties are those that make a difference to their terms, and one can find predicates that make a difference to their terms that no one would regard as natural kind terms. This is particularly true of relational properties. And if you do not believe this, there is a hammer (the one now on my workbench) that I can hit you over the head with, so that (on your terms) you can join the "natural kind" of *objects dented by that hammer*. So the framework of relevant properties seems to be a more general framework than the framework of natural kinds.

It is possible, using notions developed by R. Wille in the early 1980s (and independently by G. Hardegree), to give a formal modeling of natural kinds. (See Davey and Priestley 1990 for a good exposition of Wille's work.) We here follow in the main the terminology and notation of Hardegree 1982. In this framework a *natural kind* (NK) *structure* is a structure (I,T,Δ), where I and T are nonempty sets (thought of, respectively, as a set of "individuals" and a set of "traits"), and Δ is a relation between I and T, i.e., $\Delta \subseteq I \times T$. Given $A \subseteq I, B \subseteq T$,

$$tr(A) =_{\text{def}} \{t \in T : a\Delta t, \text{ for all } a \in A\},$$

$$in(B) =_{\text{def}} \{i \in I : i\Delta b, \text{ for all } b \in B\}.$$

Obviously $tr(A)$ is the set of traits common to all individuals in A, and $in(B)$ is the set of individuals that share all the traits in B. It is plausible to define a *natural kind* as a set of individuals A such that $A = in(tr(A))$.[15] This means that if one starts with a set of individuals A, and then considers the set of traits $tr(A)$ that they share, and then the set of individuals that share those traits $in(tr(A))$, one gets back to the same set A where one started. As Hardegree illustrates:

The set H of humans is a natural class because there is (presumably) an associated set tr(H) of human-determining traits, that is traits that (collectively) demarcate humans from non-humans — every human has every trait in tr(H), and no non-human has every trait in tr(H). Similarly the set L [of lizards] is a natural class, because there is a corresponding set tr(L) of lizard-determining traits, traits which demarcate lizards from non-lizards. The situation is quite different in the case of H ∪ L. Consider the set of all traits shared by all members of H ∪ L, that is, traits shared by all humans and all lizards; call this set tr(H ∪ L). Next, consider all those individuals that possess every trait in tr(H ∪ L). Claim: in[tr(H ∪ L)] contains many individuals besides those in H ∪ L; for example, it probably includes all birds, all mammals, and in fact all vertebrates. Thus, although there is a natural class that includes both humans and lizards (probably vertebrates), the class consisting of humans and lizards, and nothing more, is not a natural class. (1982, 124)

Mathematically speaking (as Hardegree and the other authors note), the structure above is an example of defining a *Galois connection on a polarity*. Birkhoff 1940 originated this as an abstract construction, and he showed that the *Galois closed sets* (in Hardegree's notation, those of the form $A = in(tr(A))$) form a lattice. Hartonas and Dunn (1993) have a representation theorem for lattices showing that (up to abstraction) any lattice can be embedded into a lattice of Galois closed sets on a polarity, and hence can be interpreted as a lattice of "natural kinds." (Hardegree claims a similar representation for *complete* lattices, although to the best of my knowledge the proof has never appeared.)

One wonders whether there is some way of combining the framework of Wille and Hardegree with the framework of relevant properties to produce a "relevant" modeling of natural kinds, by finding some formal requirement on Δ that gets at the idea that traits "make a difference" to the individuals that instantiate them. (The relation Δ would not seem to be a relevant relation on the usual Platonic conception of properties — Hardegree's "traits" — since an individual can come to instantiate a property, or cease to instantiate it, without their being a change in the property. See Dunn 1990a.) Belnap, G. Helman, and others have had ideas about how to characterize *functions* that "depend on their arguments" (see Anderson, Belnap, and Dunn 1992, secs. 70–71), and perhaps one of these could be adapted to the present purpose (e.g., thinking of the relation Δ as a two-valued function, as would Frege). (There is thus the possibility of defining relevant implication in terms of relevant predication, rather than the other way around.)

Lakoff's "Jello" Theory of Properties

We take the opportunity to address a recent relativist attack by the linguist George Lakoff against the whole Platonic idea that there is such a thing as "dividing nature by its joints." One might briefly describe Lakoff's view by saying that Nature, rather than being like an animal with natural divisions, is rather more like jello, which we can carve up into whatever shapes we will (as a small child might do with a spoon).

Lakoff's attack on what might be called misplaced objectivity is not of course explicitly directed against relevant predication. I doubt he had even heard of it. But nonetheless I feel obligated to give some sort of response. My defense will be rather like the proverbial lawyer who argued that the alleged crime did not take place, but if it did the evidence against her client is fabricated, and he wasn't at the scene of the crime anyway. (I leave out the part that corresponds to his being an orphan.) I will argue that Lakoff overstates the case for the relativity of concepts, quarrel with his data, and finally point out that relevant predication need not be taken as an objective notion (in any absolute sense at least).

In Lakoff's *Fire, Women and Dangerous Things*, he utilizes the same quote from Borges that I used to indicate that there are unnatural classifications, but to the opposite purpose. He suggests that seemingly unnatural classifications from a Western point of view actually may make sense in the context of another culture. He extends the example from the Australian aboriginal language Dyirbal, from which a partial classification makes up the title of his book. The classification comes from R. M. Dixon.

I. *Bayai*: men, kangaroos, possums, bats, most snakes, most fishes, some birds, most insects, the moon, storms, rainbows, boomerangs, some spears, etc.

II. *Balan*: women, bandicoots, dogs, platypus, echidna, some snakes, some fishes, most birds, fireflies, scorpions, crickets, the hairy mary grub, anything connected with water or fire, sun and stars, shields, some spears, some trees, etc.

III. *Balam*: all edible fruit and the plants that bear them, tubers, ferns, honey, cigarettes, wine, cake.

IV. *Bala*: parts of the body, meat, bees, wind, yamsticks, some spears, most trees, grass, mud, stones, noises and language, etc.

Somehow the point is dulled for me when Lakoff reveals that "whenever a Dyirbal speaker uses a noun in a sentence, the noun must be preceded by a variant of one of four words" (the four words above). It is not as if the language contains common nouns such as the four above, but rather than these act as "classifiers" that are placed in front of common nouns. The words would seem to function much like the articles *der*, *die*, *das* in German, where a number of items are thrown more or less arbitrarily into either the masculine or feminine categories. The principal difference would seem to be the creation of what one might think of as two neuter categories.

Suppose one saw the following classification:

I. *der*: men, dogs, moon, shields, spears, trees, honey, wine, cake, wind, mud, stones, some noises, etc.

II. *die*: women, sun, bats, snakes, fish, birds, crickets, fruit, cigarettes, bees, language, etc.

III. *das*: kangaroos, ferns, limbs, meat, grass, some noises, etc.

Would one think from this that native Germans somehow have a markedly different way of categorizing the world from native English speakers? I do not think so. One realizes that the assignment of gender looks almost as arbitrary to a German as to those who are not German. The reason for the gender of nouns is frequently lost in the history of a language, perhaps reflecting an animistic and mythological world view of which the speakers are no longer conscious. I do not mean to say that such a world view has no "poetic" hold on current speakers, but surely the Germans are just as capable as English speakers of adopting a scientific world view essentially free of animism. And to the extent that Germans, or the Dyirbal, or any other people, cannot do this, they have simply another view of the world from ours and their classification may be just as objective, given their theories about the world. A classification of objects in terms of how much phlogiston they contain made perfect sense before the discovery of oxygen.

Finally, as was pointed out in Dunn 1987, "It may be only us (and not the universe) saying when a property (or relation) is 'natural,'" I left open the alternative that neither relevant predication, nor the relevant implication on which it is founded are "part of the ontological furniture of the universe, but rather [are] in some fundamental sense subjective and mind-dependent." Elaborating on this somewhat, from a perspective "internal" to a given conceptual framework, it may be possible, and even useful, to have an apparatus that allows us to isolate the "natural" properties. But from an "external" point of view, this naturalness may be a feature of our conceptual schemes and not of Nature. My own predilection is to believe that "these concepts do reside in the objective universe, and that it is the job of science not just to tell us what items there are in the universe, and what facts hold of them, but also to tell us what relevant implications there are among those facts, and what are the relevant properties that go into making up those facts" (ibid.). But then the very objectivity of science has also recently been under sustained attack from many quarters.

Lewis on Natural Properties

In closing, let me point out that Lewis has a good discussion of the notion of a "natural property". Lewis says: "The name is borrowed from the familiar term 'natural kind'; the contrast is meant to be with unnatural, gerrymandered, gruesome properties" (1986, 60, n. 44).

Lewis starts by distinguishing the "abundant properties" from the "sparse properties," saying that both exist and that the former are a subset of the latter. He identifies "sparse properties" with "natural properties," and says of the "abundant properties" that "there is one of them for any condition we could write down, even if we could write at infinite length and even if we could name all those things that must remain nameless because they fall outside our acquaintance" (ibid., 59–60).

The theory of relevant predication can be thought of as a theory of "sparse properties." I am perhaps too much of an "experimentalist" in formal ontology as to have a definite view about whether what Lewis calls "abundant properties" exist in addition to the "sparse" (relevant) properties. My theory of relevant predication and interpretive talk of "real properties" is not intended as a definite commitment to the view that the relevant properties are the only properties that exist. I hope the remarks of Dunn (1987, 371) make it clear that I share Kremer's disclaimer that the "use of the expression 'real property' is not meant to evoke the debate over the ontological standing of properties. . . . Even the ontologically parsimonious nominalists might accept the claim that some *predicates* are somehow special or fundamental" (Kremer 1993). It was just this evocation I wanted to avoid with the term of art "relevant predication." As Kremer says, my "project trades talk of 'real' properties in for talk of relevant *predication*." In general terms I am a "methodological platonist" (a term I learned from Nuel Belnap years ago) and am inclined to be liberal with ontological status. But in Dunn 1987 I have indicated that one might have a comprehension axiom only for relevant properties.

I am particularly sympathetic to Lewis's point that one natural division between properties is the *intrinsic* versus the *extrinsic* properties. Although I developed my theory of relevant predication before I read Lewis, much of my work on relevant predication can be taken as a formal explication of this distinction. While it is different from Lewis's in detail (and certainly uses apparatus which he is unlikely to endorse), it captures many of the same intuitions (see Dunn 1990a for a comparison with Lewis).[16] As my discussion of "natural kinds" shows, I am in agreement with Lewis when he says: "It cannot be said that all intrinsic properties are perfectly natural — a property can be unnatural by reason of disjunctiveness, as the property of being tripartite-or-liquid-or-cubical is, and still it is intrinsic if its disjuncts are. But it can plausibly be said that all perfectly natural properties are intrinsic" (1986, 61).

I do have one *potential* bone to pick with Lewis. I say "potential" because I am finally unclear about just what his view is. In one place he talks of physics having "its short list of 'fundamental physical properties,'" (charge, mass, spin, etc.), and goes on to say, "In other worlds where physics is different, there will be instances of different fundamental physical properties, alien to this world" (ibid., 60). He also says: "What physics has undertaken, whether or not ours is a world where the undertaking will succeed, is an inventory of the *sparse* properties of this-worldly things."[17]

However, in a footnote on the same page he says that the name "natural property" "has proved to have a drawback: it suggests to some people that it is supposed to be *nature* that distinguishes the natural properties from the rest; and therefore that the distinction is a contingent matter, so that a property might be natural at one world but not at another. I do not mean to suggest any such thing. A property is natural or unnatural *simpliciter*, not relative to one or another world."

I am not sure how to resolve this seeming tension in Lewis's views. But as I said in discussing Lakoff, I believe that "it is the job of science not just to tell us what items there are in the universe and what facts hold of them, but also to tell us . . . what are the relevant properties that go into making up those facts" (Dunn 1987). I thus endorse the view of Armstrong that "it is the task of total science, conceived of as total inquiry, to determine what universals there are" (1978, vol. I, *xiii*). But this personal predilection is not itself part of the theory of relevant predication, nor need it necessarily commit one to physicalism.

NOTES

1. The notations "IQ1" and "IQ4" refer to axioms from Anderson, Belnap, and Dunn 1992, 72.

2. There is the understanding here that φx and φy differ only in that where one has free occurrences of x the other has free occurrences of y, and vice versa.

3. Kremer 1994 investigates essentially the same proposal. Although he cashes things out in terms of "real properties," rather than "intrinsic properties," he too uses the apparatus of relevant predication with what would seem to be the same metaphysical interpretation of it.

4. There was also a more subtle alternative where φxy establishes an intrinsic property in one or the other of its argument places, but it is not determined which (determining a "property of the pair" $\rho(x,y)(\varphi xy)$).

5. Part of what makes Goodman's example of "grue" seem paradoxical is the arbitrary character of the choice of the time t. If there really were some reason to think that time t is special, and that there are laws connecting times previous to t with an object's being green and other times with the object's being blue, we might well think that each of these laws is grounded in a property of the object (something like a "rate of decay" grounded in some universal structural property of green objects), and hence so is their conjunction.

6. However, not every classical equivalence relevant to the Raven Paradox is supported by relevance logic. Thus Hempel uses for certain purposes the equivalence of the following to $\forall x(Rx \to Bx)$ (but there is no relevant equivalence):

$$\forall x[(Rx \wedge \sim Bx) \to (Rx \wedge \sim Rx)],$$

$$\forall x[(Rx \vee \sim Rx) \to (\sim Rx \vee Bx)].$$

7. For the record, Kremer 1994 suggests understanding "φx is a relevant context" as "φx makes a claim about x." Taking "relevant contexts" as those formulas that are of the kind that determine relevant properties, our definition of "aboutness" just reverses this proposal. Incidentally, my analysis of "aboutness" and "true about" predates Kremer's suggestion (as is evidenced by the fact that it was on the original "punch list" of Dunn 1987).

8. I do not want to appear to ignore that there are subtleties involving statistics and probability that must be discussed in defending Hempel's solution. The "semantical solution," showing how the statement "All ravens are black" is about all objects, does not by itself say everything that must be said. There is a large literature defending or attacking Hempel's solution on probabilistic grounds. See Korb 1994 for a recent example (which seems to show that Bayesian solutions have problems).

9. I wish to thank my colleague Anil Gupta for helpful comments on this section, which is not to suggest that he agrees with everything in it.

10. Indeed, given the **R**-theorem $T(s) \vee \sim T(s)$, it follows from the equivalences above that $[\rho x T(x)](s) \vee [\rho x \sim T(x)](s)$, i.e., each sentence s has either the property of truth or of falsity.

11. Note incidentally that by virtue of Fact 1, $[\rho x Tx](l) \to T(l)$. So from this and (15.38) we can derive by transitivity $\sim T(l) \to T(l)$, and so $T(l)$. I frankly do not know what to make of this, but we have just seen that the Liar Sentence (l) can be proven true. But it cannot be proven false, since this requires the converse of Fact 1, which holds only for relevant properties.

12. Let Fa be as absolute a statement as one can imagine and let P be the most absolute statement one can think of that is totally irrelevant to Fa. Then $Fx \wedge P$ would not determine a relevant property, though the question as to whether $Fa \wedge P$ holds is presumably objective.

13. The reader may believe that we have "gone around Robin Hood's barn" to get to the next step (15.46), thinking it follows immediately by universal instantiation from (15.41). But note that the instantiation gives $s = s \to Ts$ rather than the needed $x = s \to Ts$.

14. Thomason also imposes the following condition:

$$(D') \ a \wedge b = \Lambda \text{ or } a \wedge b = a \text{ or } a \wedge b = b.$$

(D') says that no two natural kinds overlap except when one is a subkind of the other. I tend to agree with Hardegree (1982, footnote 6) that "Although this principle is satisfied by taxonomic classification systems in biology, it is not true of natural kinds in general."

15. Or if one thinks of a natural kind as a set of traits (those characterizing the natural kind) it is plausible to define a *natural kind* as a set $B \subseteq T$ such that $B = tr(in(B))$. Both Hardegree and Wille actually take the pair of such sets (A,B) so as to avoid violating symmetry with an arbitrary choice.

16. There are of course sectarian differences. The way that Lewis (1986, 62) defines an *internal relation* corresponds to what I have called a *relevant property of a pair*, rather than to a *relevant relation*. Dunn 1990a discusses various "grades" of internality for the case of dyadic formulas.

17. It does not seem that Lewis believes compounds of sparse properties can themselves be sparse properties. His examples from physics do not suggest so. But he does allow such compounds as "somewhat natural in a derivative way, to the extent that they can be reached by not-too-complicated chains of definability from the perfectly natural properties" (1986, 61). Lewis says, "Probably it would be best to say that the distinction between natural properties and others admits of degree," and presumably that "degree" has something to do with the length of such chains of definability. Certainly this fits commonsense intuitions about what is more or less natural *to us*, but my distinction between relevant and nonrelevant properties is intended to be absolute (at least relative to a given conceptual framework).

REFERENCES

Anderson, A. R., and N. D. Belnap, Jr. 1975. *Entailment: The Logic of Relevance and Necessity*, vol. I. Princeton: Princeton University Press.

Anderson, A. R., N. D. Belnap Jr., and J. M. Dunn. 1992. *Entailment: The Logic of Relevance and Necessity*, vol. II. Princeton: Princeton University Press.

Aquinas, S. T. [ca. 1270] 1948. "Summa Theologica." In A. C. Pegis, ed., *Introduction to Saint Thomas Aquinas*. New York: Modern Library.

Armstrong, D. M. 1978. *Universals & Scientific Realism*, vol. I: *Nominalism and Realism*; vol. II: *A Theory of Universals*. Cambridge University Press.

———. 1983. *What Is a Law of Nature?* Cambridge: Cambridge University Press.

Birkhoff, G. 1940. *Lattice Theory, American Mathematical Society*, Providence.

Bonsaquet, B. 1888. *Logic, or the Morphology of Kings*, vol. 3. Oxford University Press, 2nd ed., 1911.

Borges, J. L. 1964. "The Analytical Language of John Wilkins." In *Other Inquisitions*, trans. R. L. C. Simms. University of Texas Press, 101–05.

Cohen, D. H. 1983. *The Logic of Conditional Assertions*, Ph.D. diss., Indiana

University. Relevant portions are summarized by Cohen as a contribution to Anderson, Belnap, and Dunn 1992, 472–87.

Davey, B. A., and H. A. Priestley. 1990. *Introduction to Lattices and Order.* Cambridge University Press.

Dunn, J. M. 1987. "Relevant Predication 1: The Formal Theory." *Journal of Philosophical Logic* 16: 347–81.

——. 1990a. "Relevant Predication 2: Intrinsic Properties and Internal Relations." *Philosophical Studies* 60: 177–206.

——. 1990b. "Relevant Predication 3: Essential Properties." In J. M. Dunn and A. Gupta, eds., *Truth or Consequences.* Kluwer, 77–95.

——. 1990c. "The Frame Problem and Relevant Predication." In E. Kyburg et al., eds., *Knowledge Representation and Defeasible Reasoning.* Kluwer, 89–95.

Fodor, J. 1987. "Modules, Frames, Fridgeons, Sleeping Dogs, and the Music of the Spheres." In Z. Pylyshyn, ed., *The Robot's Dilemma.* Ablex, 139–49.

Geach, P. 1969a. *God and the Soul.* London: Routledge.

——. 1969b. "God's Relation to the World." *Sophia* 8, no. 2. Reprinted in P. T. Geach, *Logic Matters.* Oxford: Basil Blackwell, 1972, 318–27.

Goodman, N. 1961. "About." *Mind* 70, no. 277: 1–24.

——. 1965. *Fact, Fiction, and Forecast.* 2nd ed. New York: Bobbs-Merrill.

——. 1973. *Ontological Reduction.* Bloomington: Indiana University Press.

——. 1983. *The Categorial Structure of the World.* Bloomington: Indiana University Press.

——. 1984. "About Truth About." In *Of Mind and Other Matters.* Cambridge: Harvard University Press, 90–107.

Grover, D. L. 1992. A Prosentential Theory of Truth." Princeton: Princeton University Press.

Hardegree, G. M. 1982. "An Approach to the Logic of Natural Kinds." *Pacific Philosophical Quarterly* 63: 122–32.

Hartonas, C., and J. M. Dunn. 1993. "Duality Theorems for Partial Orders, Semilattices, Galois Connections and Lattices." Indiana University Logic Group Preprint Series, IULG-93-26.

Hempel, C. 1945. "Studies in the Logic of Confirmation I-II." *Mind, LIV,* 1–26, 97–121; reprinted in *Aspects of Scientific Explanation.* New York: Free Press, 1965, 3–46.

Korb, K. B. 1994. "Infinitely Many Resolutions of Hempel's Paradox." In *Theoretical Aspects of Reasoning about Knowledge,* ed. R. Fagin. Proceedings of the Fifth Conference (TARK 1994). San Francisco: Morgan Kaufmann.

Kremer, P. 1989. "Relevant Prediction: Grammatical Considerations." *Journal of Philosophical Logic* 18: 349–82.

——. 1994. *Real Properties, Relevance Logic, and Identity.* Ph.D. diss., University of Pittsburgh.

Lakoff, G. 1987. *Women, Fire, and Dangerous Things: What Categories Reveal about the Mind.* University of Chicago Press.

Lewis, D. 1986. *On the Plurality of Worlds.* Oxford: Basil Blackwell.

Putnam, H. 1970. "Is Semantics Possible?" In H. E. Kiefer and M. K. Munitz, eds. *Language, Belief, and Metaphysics.* Albany: State University of New York

Press. Reprinted in *Philosophical Papers*, vol. 2. Cambridge University Press, 1975.

Sylvan, R., and R. Nola. 1991. "Confirmation without Paradoxes." In G. Schurz and G. J. W. Dorn, eds. *Advances in Scientific Philosophy.* Rodopi, 5–45.

Thomason, R. H. 1969. "Species, Determinates, and Natural Kinds." *Noûs* 3: 95–101.

Ullian, J., and N. Goodman. 1977. "Truth About Jones." *Journal of Philosophy* 74: 317–38.

Urquhart, A. 1972. "Semantics for Relevant Logics." *Journal of Symbolic Logic* 37: 159–69.

Waters, C. K. 1987. "Relevance Logic Brings Hope to Hypothetico-Deductivism." *Philosophy of Science* 54: 453–64.

16

Singular Causation and Laws of Nature

David M. Armstrong

Department of Philosophy, University of Sydney

Humean Causes and Humean Laws

This billiard ball collides with that billiard ball on a certain billiard table, setting the second ball in motion. This is singular causation. Hume used the case as a paradigm of singular causation, and a very good paradigm it is.

But what is it *on the side of the objects* that makes this a case of causation? As we know, analytical philosophers have long been enslaved by Hume's idea (or what is generally taken to be Hume's idea) that, setting aside the elements of contiguity and succession which may or may not be essential, there is nothing in the sequence which makes it causal except what happens elsewhere and/or elsewhen in the universe. The causality is not intrinsic to the individual sequence, but is relational. In particular, the sequence must be an instance of a regularity.

We all think that causes have something to do with laws of nature. Hume himself paid little attention to the notion of law. But when his successors did so, a Humean view of law was already at hand. On the side of the objects a law is nothing but a regularity in the behavior of things. The startling developments in fundamental physics that we have seen this century suggest that some or all of these regularities may be irreducibly statistical. But, after all, statistical regularities are regularities nonetheless.

Cause and law can then be brought together in a very simple manner. A case of singular causation, the billiard ball case for instance, automatically instantiates a law(s). What makes it a causal sequence and what makes it an instantiation of a law is the very same thing: its being an instance of a regularity spread throughout space and time.

Note, however, that this bringing together of cause and law does not force the Humean to maintain that all laws of nature are causal laws. He or she could well maintain that, among the cosmic regularities, only some deserve to be called *causal* regularities. The picking out of the subset of causal laws from the set of the laws, a picking out that could not of course appeal to the notion of causality, is no doubt tricky, and curiously little progress has been made with the problem. But, I suppose, the Humeans thought (if they thought about the matter at all) that solving this problem was not much more than a matter of tidying up.

All very simple. This is the way that I used to think about causes and laws when I was young and hopeful. Many, many, other analytical philosophers thought the same way, and many still do. However, it has become increasingly evident over the years that these accounts of cause and of law face profound difficulties. In the next section I discuss some of the difficulties that face the regularity theory of causation. These difficulties appear to support a *singularist* theory of singular causation. But a problem then arises concerning the connection between causes and laws. Can the link be a mere fortunate accident? I go on to argue in the third and fourth section that a way out of this dilemma appears if we abandon the implausible regularity theory of laws in favour of a conception of laws as connections between universals. Singular causes and laws can then be linked together in a satisfying and plausible manner.

Arguments for a Singularist Account of Singular Causation

We do associate causation with regularities. Fire burns, water suffocates, and bread nourishes. These are all causal regularities. But there are plenty of cases where we are tolerably sure that causation was present in the singular case, but we have no idea what regularity it is an instance of. Consider a nice case I once heard given by Douglas Gasking. A small piece of stuff is dropped into a beaker of liquid, and the beaker immediately explodes. One might easily recognize that the sequence is causal. Yet what regularity would one be bringing the sequence under?

In "Causal Relations" (1967), Donald Davidson tried to deal with

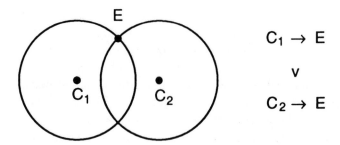

Figure 16.1

this difficulty for a regularity theory of causality by arguing that in such a case as that of Gasking's, though we know no regularity to subsume it under, we do, in recognizing it to be causal, commit ourselves to holding that there is some law (conceived by Davidson as a regularity) which the sequence falls under. But while this *saves* the regularity theory, it is not any sort of *argument* for it, as Elizabeth Anscombe pointed out very forcefully (1971, 81).

Here are two arguments which, while far from *proving* the Singularist case, do indicate its attractions. One is ontological, the other epistemological. The first of these arguments uses a certain type of case proposed by Michael Tooley (see Armstrong 1983, 133) and independently by John Foster (1979, 169–70). The cases may be no more than merely possible, but do bring out the force of a singularist position.

Suppose there are two particles of the same sort, C_1 and C_2, at a certain small distance from each other. There is an irreducibly probabilistic law that such particles have a certain chance, less than 1, of bringing a certain sort of effect into existence at one or another of the points at a fixed distance from the particle. We may suppose that this distance is some minimum quantum so that there is no intermediate chain of causes between cause and effect (see figure 16.1).

Suppose now that, on a particular occasion, the effect E happens to appear at one of the intersection points. There seems to be a definite question whether it was C_1 or C_2 that brought about E (nonexclusive "or" to allow for the possibility that E's occurrence was overdetermined), even if we have no particular resources for answering the question.

The following counterfactual queries appear to be in order. In the absence of C_1, would E still have occurred? What if C_2 had been absent? Or was it the case that E would have failed to occur only if *both* C_1 and C_2 had been absent—the overdetermination case? (Note that the counterfactuals involve no generality; they do not look beyond the particular situation.) A singularist account of singular causation can make sense of these queries without any difficulty. A regularity theory cannot.

Now for an epistemological argument that seems to support Singularism. The argument has to meet some deep prejudices, because it denies something that Hume is supposed to have proved; namely, that there is no impression (that is, perception or introspective datum) from which the idea (that is, concept) of causation is derived. Against this, I assert that there is an impression of causality. It is *not* a purely theoretical relation.

One important case here is our perception of pressure on our own body. When we are (tactually) aware of such pressure, we are aware of causal action on our body. Pressure on our body *is*, analytically, causal action. And there is no doubt that we often perceive (feel) such pressure, feeling it *as* pressure. Furthermore, it is phenomenologically plausible that this sort of perception is as direct, as untheoretical, as untheory-laden, as any other perception. It is naturally compared, not contrasted, with the perception of color, size, shape, and distance, heat and cold, tastes and smells. (Notice that this list is *not* confined to the secondary qualities.) (See Armstrong 1968, 97–99.)

The same conclusion was arrived at independently by Evan Fales (1990, chap. 1) who goes into the matter in much greater detail than I did. He offers evidence that when we have a sensation of pressure we are aware (not necessarily veridically) of a force vector: (1) the pressure is felt as having a location, an area of our body; (2) it is felt as having a magnitude; (3) it is felt as having a spatial direction; (4) where more than one force is applied to nearly the same part of the body, they can sometimes be distinguished; (5) the forces can be felt to add together in a way that depends upon their respective magnitudes and directions; and (6) the force is felt as repulsive rather than attractive.

Suppose, now, that these epistemological points are on the right track. They tell heavily against a regularity ontology of causality because, on that view, what makes a singular causal sequence *causal* is spread through space and time. For a regularity theorist a perception of causality has to be inferential, even if automatic and un-self-conscious inference.

The argument is not conclusive, of course. One can suggest that physical pressure on our body causes in us an idiosyncratic sensation with various features (what are these?) which, either in virtue of earlier experience of the world (but what sort of experience?) or else by triggering an innate disposition to infer, enables us to infer the physical nature of the particular pressure. The features of the sensation will be well correlated, but no more than well correlated, with features of the physical cause. But not a very plausible epicycle, I suggest.

Another place where it is plausible to think that we are directly, noninferentially, aware of causal action in the singular case is when we are introspectively aware of the successful operation of our will either in bringing about a bodily or a mental result. Fales also mentions this, although he does not go into details. But he claims that this view has been held by many including Reid, James, Whitehead, Stout, Keynes, Searle, and the psychologists Biran de Maine and Piaget (Fales 1990, 315, n. 12).

What is the nature of this relation of which we are immediately aware? In our direct experience of causation we no doubt fail to penetrate its full essence. But what we are given seems valuable, even if not entirely clear. We experience causation as a bringing about, a producing, a making something happen. Of that, I believe, we are as directly and securely aware as of anything that we are aware of.

These considerations, I think, constitute a strong case for adopting a singularist theory of causation, and for some years I did just this. The trouble is, though, that one then seems forced to say that it is a mere contingent fact that singular causes fall under laws, and that seems a very uncomfortable thing to say. Elizabeth Anscombe (1971) seems to be right to say, against Davidson, that there is no a priori or conceptual argument from causes to laws. But do not causes and laws have a more than de facto link?

The Regularity Theory of Laws of Nature

My present view is that we can accommodate the insights of a singularist theory of causation, and yet preserve a necessary link between causes and laws. First we must dispose of the regularity theory of *laws*. (So far we have been discussing the regularity theory of *causation*.) I will concentrate on three main difficulties (see Armstrong 1983 for a longer discussion).

First, there is the problem, originally raised by Kneale (1961), of unrealized physical possibilities. There seems to be no nomic objection to solid lumps of pure gold with volumes greater than a cubic mile. There is, perhaps, no biological objection why moas (an extinct New Zealand bird) should not have lived beyond fifty years, although by mischance none of them did. There can be no nomic objection to a race of snowy-plumed ravens (developed in the course of arctic living), although, perhaps, no race of ravens ever will be snowy-plumed. These cases show that there can be cosmic regularities that are not laws. The first case is particularly striking because it can be contrasted with the nomic impossibility of solid lumps of pure uranium 235 with volumes greater than a cubic mile.

Second, there is a problem about irreducible probabilistic laws. On a regularity theory, these have to be identified with statistical distributions. This at once raises the point that there are indefinitely many statistical distributions that are not laws. (A further turn of the screw of Kneale's argument.) But even if we have got hold of genuinely nomic statistical distributions, we cannot identify the law with the actual distribution because, given independence of chances, a probabilistic law does not *mandate* any distribution. It only assigns a certain probability to that distribution. A gap opens up between the law and the statistical regularity, which is bad news for a regularity theory.

Third, there is the well-known difficulty about counterfactuals. It seems clear that laws of nature "sustain" counterfactuals. Given the law, and the contrary-to-fact supposition that something instantiates the antecedent condition of the law, we are, in general at least, prepared to say that the thing would fall under the consequent. (Think how we sweat with relief when we fail to instantiate the antecedent of some baneful law of nature. We know the power of laws.) But given a regularity theory of law, we seem to have no warrant for bringing the new case under the regularity.

The trouble is that, given a true law statement, the truthmaker that a regularity theory of laws offers is the whole assemblage of the individual instantiations. (Together, perhaps with the state of affairs that these *are* the whole assemblage.) To imagine a new case, a new instance falling under the antecedent, as is done in a counterfactual supposition, is to imagine the assemblage enlarged. And then why should this new instance conform to the pattern set by the actual regularity? (I pass over John Mackie's ingenious, but ultimately I think unsuccessful, attempt to solve the problem for the regularity theory — a problem that he clearly perceived — by arguing that the warrant is inductive. See Mackie 1966, criticized in Armstrong 1983, chap. 4, sec. 4.)

Bringing Cause and Law Together Again

The first step is to move beyond a regularity theory of laws. The crisis in that theory has recently moved a number of philosophers to suggest that what a law of nature is, is a *connection of properties* (see Dretske 1977; Tooley 1977; Shoemaker 1980; Swoyer 1980, 1982; and Armstrong 1983). The suggestion is a very natural one. For instance, Boyle's Law, taking it to *be* a law for convenience, tells us of a certain connection between volume, temperature, and pressure properties of gas samples. This connection is lacking in mere uniformities, enabling the theory to escape Kneale's argument against the regularity theory. (There are the usual failures of full agreement between different philosophers. For instance, Dretske, Tooley and I think of the connection between properties as contingent, whereas Shoemaker and Swoyer think of them as necessary. I will not discuss that important issue further here.)

It is of the essence that these properties be taken to be universals. Only so will the law be a single thing — an atomic fact, as Tooley puts it — a thing that remains the same however many or few positive instances fall under the law. This explains why laws sustain counterfactuals. An imagined new instance of the antecedent condition is not an imagined enlarging of the law, as it is with the regularity theory. Hence there need be no worry that the law may fail to apply in the imagined new instance.

These connections between properties may be hypothesized to come in various *strengths*. With the greatest strength, strictly 1, we have a deterministic law. Concentrating upon causal laws, the instantiation of the complex of universals, a state of affairs that constitutes the antecedent condition of the law, brings about, produces, causes the consequent state of affairs. (Perhaps it only brings about that consequent in the absence of certain interfering conditions.) With less than highest strength we have a probabilistic law. In each instantiation of the antecedent conditions there is a certain *objective probability* (but no more than a probability) that these conditions will bring about, produce, cause the instantiation of the consequent. I think it is important to insist that the bringing about, producing, causing, is exactly the same in the deterministic and the probabilistic case. When they occur, deterministic causality and probabilistic causality are the same. All that differs is the prior probability that the antecedent conditions are going to bring the effect into existence. Strictly, there is no probabilistic causality. There is only a certain probability of causing. When causing actually occurs, when, as we may put it, the law "fires," then the effect occurs.

If we go along with all this, then the laws and their instantiations may seem to be well and truly separated. It is easy to form a picture of the laws, the connection of properties, as existing in a platonic realm apart from but "governing" their spatiotemporal instantiations. That would make the laws "distinct existences" from their instantiations, yet at the same time somehow connected to the instantiations by entailment or, in the case of probabilistic laws, by a logical probability. I agree with David Lewis, in particular, that this is hard to understand. Distinct existences ought not to be logically linked. In any case, I do not believe in any realm except the spacetime realm.

Here is a way to try to keep the law and its instantiations together. A law is a certain state of affairs: two or more universals connected in a certain way. But, the next step in the argument, may we not also say that the law is a universal? It will be a complex, in particular, a *structural* universal, having, perhaps, a structure of the following sort: something's being F bringing it about (producing, causing) that that same something (or a suitably related something) is a G. A less controversial structural universal would be: something's being F *succeeded by* that same something (or a suitably related something) being G.

There is, however, a major difference between the first case — the nomic case — and the case where there is mere succession. In the nomic case, the two universals, F and G, are directly connected. We have a second-order relation between the two first-order universals, and not one that supervenes upon the first-order states of affairs involved. If, on a particular occasion, a's being F brings about, produces, causes b to be G (or b which is suitably related to a to be G), then this occurs in virtue of the connection between F-ness and G-ness; that is, it is a manifestation of the $F \rightarrow G$ law, an instantiation of the $F \rightarrow G$ structural universal.

No doubt the content of a law is never given to us in direct experience. But if I am right that bringing about, producing, causing is given in direct experience, and given that we have experience of properties that are universals, then we can at least see, in good Lockean or empiricist style, how to *understand* what a nomic connection of properties is, at least in the causal case. This is my answer to the *Identification Problem* raised by van Fraassen in his important critique of my theory to be found in his *Laws and Symmetry* (1989). Van Fraassen believes that solving this problem (avoiding Scylla) makes it inevitable that I must fail to solve the *Inference Problem* (Charybdis). My answer to the latter problem comes at the end of this section.

Things have to be complicated a bit to allow for the possibility that the law is probabilistic only. We need something's being an *F* having a certain objective chance, *C*, less than 1, of bringing it about that that same something (or a suitably related something) is *G*. We then have two sorts of *F*s: those that produce the effect, *G*, and those that do not. It is only in the case of the *F*s that produce *G*s that we have a true instance falling under the probabilistic law. Only then do we have a *C*-strength connection between properties *F* and *G* that is actually instantiated.

What of *F*s that do not produce *G*s? Did they not have a potentiality to produce *G*s, even if an unrealized potentiality? We can say this with truth. But my belief is that the only truthmaker needed for such truths is the instantiation of the $F \to G$ connection in the cases where an *F* actually brings forth a *G*. We do not have to postulate irreducible powers or potentialities.

So far, then, I have argued that laws are structural universals instantiated by those cases in which possession of the antecedent properties actually brings about possession of the consequent properties. But now, in good Aristotelian fashion, I argue that the law, the structural universal, has no existence *except* in its instantiations. Where the law is instantiated, there the law is, and is *completely*. And it is nowhere else. Probabilistic laws continue to be troublesome. To instantiate them it becomes necessary to argue that every law, even a probabilistic law, must be fully instantiated on at least one occasion. Otherwise, given my earlier argument that laws are structural universals, there is no law. Certain difficulties in this position have been raised by Michael Tooley (see my Armstrong 1983, chap. 8) and, from another direction, by van Fraassen (1987). I try to answer Tooley in the chapter just referred to, and to van Fraassen in Armstrong 1988.

If this notion of a law as an (instantiated) structural universal can be made good, then an account can be given of singular causation that makes causes law-governed, yet does justice to singularist intuitions about causes. Singular causation is nothing but the instantiation of a law, that is, it is an instantiation of a certain sort of structural universal. Since, like any universal, it is complete in its instance, the law is complete in its instance. At the same time, because it is a universal, it is potentially general: It may be found in other instances, and so can function as a general law.

Note that I am *not* arguing that this identification of singular causation with instantiation of a law is a conceptual truth. Once one is seized of the point that one thing acts upon another in virtue of some, but only some, of its properties, and that the effect also is to be characterized in terms of properties, then it is natural to think in terms of a nomic connection of properties. To use a schematic oversimplification, if the state of affairs of a's being F brings it about that a is G, then it is natural to postulate some form of $F \rightarrow G$ law. But there seems to be no way in which this conclusion that singular causation is an instantiation of a law can be strictly proved. That is the force of Anscombe's challenge to Davidson.

A simple way out of the difficulty has been suggested to me by Adrian Heathcote (see Heathcote and Armstrong 1991). Why should not the identification of singular causation with instantiation of a law be made on empirical, a posteriori, grounds? It is an a posteriori necessity, to be compared with the necessity that Kripke has taught us to recognize in the identity of heat with motion of molecules and other identities. Cause and law are two in concept, but one in reality.

Now to try to answer van Fraassen's Inference Problem. The problem is simple enough. Laws, even probabilistic laws, have something to do with regularities in the world. I have given a somewhat elaborate account of laws as a certain sort of connection between universals. What unmysterious account can be given of the link between these connections and simple old universally quantified truths?

It must be conceded that those of us who hold the view that laws of nature are relations of universals have used a rather unfortunate symbolism. The connection was often written as $N(F,G)$, one property necessitating another. The "N" (necessitation) was not very felicitous, and in effect in this paper I am substituting "C" (cause) for "N", and so answering the Identification Problem. But it was really the "(F,G)" part of the formula that seriously underdescribed the situation.

Let us ask if there could be a law of the form $N(F,F)$? No, one may be led to say, because how could a universal have a nomic relation to itself? But we want such laws, do we not? After all, like may cause like. What seems to be required to solve this problem, and what in any case seems correct, is a "Fregean" conception of universals as *unsaturated* entities, entities that involve one or more "places" for particulars. A monadic universal, such as F, is more perspicuously represented as $\_F$, a dyadic universal as $\_R\_$, and so on. It is part of the essence of a particular universal that it has a certain definite number of places that cannot differ in different instantiations of that universal. If a's being F is a state of affairs, then the universal $\_F$ may be thought of as a *state of affairs type*. The word "type" is here a modifier of the phrase "state of affairs"; a state of affairs type is not a species of state of affairs.

Now we can understand what a causal law connecting like with like will itself be like, and with it any other causal law. We might have a law where the antecedent condition is that something's being F *and* having relation R to some further something CAUSES the further something to be F. This is a relation of states of affairs types, that is, it is a relation of universals. Mention is made of particulars, but no particular particular. Perhaps it could be written $\_1F$ & $\_1R\_2$ CAUSES $\_2F$, where the subscripts keep track of the particulars. Singular causation involving just this relationship will *be* instantiation of this particular law. The law is an atomic, if higher-order, state of affairs, as emphasized by Tooley, and if the relation CAUSES here is a deterministic and non-defeasible connection then the universally quantified state of affairs, the regularity, seems to be transparently entailed.

That is proposed as the solution to the Inference Problem.

What Marks Off Causal Laws?

So cause and law are brought together again, on better terms, I hope, than in the Humean theory. But I have left the same loose end that the Humeans left. What account am I to give of the distinction between causal and noncausal laws? The matter is fairly urgent for me, because I have in this essay used the notion of causality — bringing about, producing — in my analysis of what a law is.

What I wish to argue (or at any rate put forward as a hypothesis) is that all laws that are both *first-order* and are *fundamental* are causal. To say that a law is first-order is to say that it is not a law that itself governs laws (as may be the case with functional laws). To say that a law is fundamental is to say that it does not supervene on any other law or laws. Boyle's law, for instance, merely correlates quantities without saying anything about causes. But it is not a fundamental law. The correlations between pressure, volume, and temperature given by the law turn out to supervene, necessarily supervene, upon the causal interactions of the molecules of samples of gas.

Conservation laws may be thought to be noncausal laws, and presumably some at least of these laws are fundamental. Such laws, however, determine the way that certain features (properties) of systems evolve over time. As a result, where some quantity is conserved, past states of the system may be thought of as determining later states, and the determining may be taken as causing. It is an immanent and monotonous form of causing where the past spawns something like itself. See Russell's (1948) conception of a continuing thing as a *causal line.*

Are there any cases of noncausal laws that are first-order and fundamental? Michael Tooley (private communication) has suggested various possibilities. First, there is Newton's Third law. If we consider the *First* law, then it seems to be some sort of conservation law, and it is not implausible to think of it as causal also. The body is in a certain state of motion at a certain time, and *as a result of this*, it is in that same state of motion at a later time. But the principle that, when a force acts the action and reaction are equal and opposite, is a bit trickier. Can we think of the action as *producing* the reaction? This is unclear. But the Third law is a conservation law, though, and perhaps both action and reaction can be thought of as produced by the previous state of the system.

Second, there is the Einstein-Podolsky-Rosen paradox. Certain phenomena, such as electron spin, are nomically correlated in a certain way, although they are separated by a spacelike interval (which fact seems to rule out direct causal connection) and, furthermore, seem to lack a common cause. What we have here is, I suppose, some sort of conservation law, but I do not know whether the treatment already suggested for conservation laws can apply.

A third point raised by Tooley is that it may well be the case that every event, indeed every state of affairs, has causal antecedents. (This, of course, does not entail determinism. The laws governing the causal sequences could be one and all merely probabilistic.) If such universal causality is the case, then presumably it is a *law*, and not a mere Humean uniformity. Yet, though having causes as its subject, it would not itself be a causal law.

But might it, perhaps, be a supervenient, nonfundamental law, supervenient upon the laws which govern the entities that the world contains? Suppose, for instance, that everything that exists is governed in all its behavior by the elusive unified field equation, and that the equation can be rendered as a causal law. What need of a further law of causality? Do we need a *law* to tell us that everything there is is governed by this equation? A connection between the properties involved in the equation should be enough. A combinatorial theory of possibility, which I favor (Armstrong 1989), would not allow for even the possibility of *radically* different sorts of property, what David Lewis calls *alien* properties.

Tooley's fourth suggestion is to point to the possibility of laws which lay it down that certain conditions are *necessary* for the production of a certain sort of effect. Such a law would have, as it were, a direction of travel from the effect to the cause. But the effect cannot produce, that is to say, cause, the cause.

My doubt in this case is whether such laws can actually be found. I do not think that my argument is in danger from a merely possible case, since it begins from causation as we actually find it.

There is much further investigation to be done on this topic, an investigation that may produce surprises, but it is clear that it cannot be assumed (though it may turn out to be the case) that all first-order and fundamental laws are causal laws. How does this affect my argument? What I *think* ought to be argued is that we understand the lawlikeness that attaches to noncausal laws *by analogy*, but by analogy only, with causal laws. I have argued that we have direct, if unanalyzed, acquaintance with causality. In noncausal laws, antecedent and consequent conditions (properties) will have to be connected in some way that bears a likeness to causality, but is not causality.

NOTE

The views put forward in this essay are developed more fully in a book, *A World of States of Affairs* (Cambridge: Cambridge University Press, 1997), chaps. 14–16.

REFERENCES

Anscombe, G. E. M. 1971. *Causality and Determination*. Cambridge: Cambridge University Press. Reprinted in *Causation and Conditionals*, ed. E. Sosa. Oxford: Oxford University Press, 1975.

Armstrong, D. M. 1983. *What Is a Law of Nature?* Cambridge: Cambridge University Press.

——. 1988. "Reply to van Fraassen." *Australasian Journal of Philosophy* 66: 224–29.

——. 1989. *A Combinatorial Theory of Possibility.* New York: Cambridge University Press.

Davidson, D. [1967] 1975. "Causal Relations." In *Causation and Conditionals,* ed. E. Sosa. Oxford: Oxford University Press.

Dretske, F. I. 1977. "Laws of Nature." *Philosophy of Science* 44: 248–68.

Ducasse, C. J. [1926] 1975. "On the Nature and Observability of the Causal Relation." In *Causation and Conditionals,* ed. E. Sosa. Oxford: Oxford University Press.

Fales, Evan. 1990. *Causation and Universals.* London and New York: Routledge.

Foster, J. 1979. "In *Self*-Defence." In *Perception and Identity,* ed. G. F. MacDonald. London: Macmillan.

Heathcote, Adrian, and D. M. Armstrong. 1991. "Causes and Laws." *Noûs* 25: 63–73.

Hume, D. 1960. *Inquiry Concerning Human Understanding,* ed. L. A. Selby-Bigge. Oxford: Oxford University Press.

Kneale, W. C. 1961. "Universality and Necessity." *British Journal for the Philosophy of Science* 12: 89–102. Reprinted in *Philosophical Problems of Causation,* ed. T. L. Beauchamp. Belmont, Calif.: Dickenson, 1974.

Kripke, Saul. 1980. *Naming and Necessity.* Oxford: Blackwell.

Lewis, D. 1986. *Philosophical Papers,* Vol. II. New York: Oxford University Press.

Mackie, J. L. 1966. "Counterfactuals and Causal Laws." In *Analytical Philosophy: First Series,* ed. R. J. Butler. Oxford: Basil Blackwell.

——. 1974. *The Cement of the Universe.* Oxford: Oxford University Press.

Russell, B. 1948. *Human Knowledge, Its Scope and Limits.* London: Allen and Unwin.

Shoemaker, S. 1980. "Time and Properties." In *Time and Cause,* ed. P. van Inwagen. Dordrecht: Reidel. Reprinted in *Identity, Cause and Mind, Philosophical Essays by Sydney Shoemaker,* 109–35. Cambridge: Cambridge University Press, 1984.

Swoyer, C. 1982. "The Nature of Natural Laws." *Australasian Journal of Philosophy* 60: 203–23.

Tooley, M. 1977. "The Nature of Laws." *Canadian Journal of Philosophy* 7: 667–98.

——. 1987. *Causation: A Realist Approach.* Oxford: Clarendon Press.

Van Fraassen, B. 1987. "Armstrong on Laws and Probabilities." *Australasian Journal of Philosophy* 65: 243–60.

——. 1989. *Laws and Symmetry.* Oxford: Clarendon Press.

Action and Rationality

17

Action and Autonomy

Fred Dretske
Department of Philosophy, Stanford University

When Clyde, wanting to get to Chicago, turns north because he thinks that is the way to Chicago, it is surely obvious that what he believes — that *that* is the way to Chicago — explains, or helps explain, why he turns north. If we speak of what is believed as the meaning of the belief (treating referential differences as differences in meaning), then

(1) When beliefs figure in the explanation of behavior, it is their meaning (*what* is believed) that explains, or helps explain, why the behavior occurs.

Though not everything we do is explained by what we believe and desire (e.g., sneezing, snoring, blushing, hiccoughing), intentional behavior certainly is. Clyde turns that way because he thinks that is the way to Chicago. Had he thought otherwise, given what he wants, he would have behaved differently. His belief's semantic content or meaning — that north is *that* way — explains, or (together with collateral beliefs and desires) helps explain, why he turns in that direction. The same is true of the other propositional attitudes (assuming for these purposes that desire is a propositional attitude: wanting an apple is wanting to *have* an apple).

I will call (1) The Principle of *Explanatory Relevance*. It occupies an important place in our conception of ourselves as rational agents, as agents who (sometimes at least) do things for reasons — *because* we

515

want this and believe that. By combining the explanatory and rationalizing aspects of reasons, it constitutes what Bilgrami (1987) calls the integrity or unity of intentional states: when we act deliberately, the reasons we behave the way we do are our reasons for behaving that way. When behavior has this kind of explanation — an explanation in terms of the actor's beliefs and desires, intentions and plans — I refer to it as intentional behavior or, simply, action.

There is, however, another aspect of meaning, a perhaps less obvious part of our ordinary scheme of psychological attribution and explanation. Meaning, it seems, is not wholly a function of what is going on *in* the believer. Meaning is a relational property, constituted in part at least by objects and conditions in the believer's surroundings and the way the person is — or *was* — related to these objects and conditions. Just as the meanings of words are not to be found in the books in which the words are to be found, so the meanings of beliefs are not to be found in the place (presumably the head) where the beliefs themselves reside. We thus have a second principle, one I shall call the *Extrinsicness of Meaning*:

(2) A belief's meaning is an extrinsic property of the belief.

I shall not defend this claim here: Though the idea that beliefs are, in part at least, externally individuated has gained wide acceptance — largely through the arguments and examples of Kripke (1972), Putnam (1975), and Burge (1979) — it remains a controversial thesis (see Lepore and Loewer 1986). Nonetheless, I accept it without argument. My purpose here is to examine the implications of regarding explanatorily relevant meanings as extrinsic. What I hope to show is that the behavior such meanings help explain — intentional behavior — enjoys a special kind of causal autonomy, the kind of autonomy required of a free act.

Extrinsic Systems

As already indicated, an extrinsic property is a property a thing possesses, not because of the way it is, but because of the way things to which it stands in certain relations are (or were). My being an uncle depends, not (just) on me, but on whether my siblings have children. Whether the woman is a widow (or a divorcee) depends on whether she had a husband who died (or from whom she was divorced). The painting is a genuine Pissaro only if it was painted by Pissaro.[1]

According to this usage, then, the property of being money (i.e., legal tender) is an extrinsic property of certain pieces of paper. Whether the paper in my wallet is real money depends not on anything we could possibly observe concerning the current condition of the paper itself (even under the most powerful microscope), but on facts about where and when and under whose authorization this paper was printed. The money is genuine if it had a certain origin, if it was printed and distributed under the authority of the United States Government. Paper that lacks this (extrinsic) property, paper that does not stand in those relations, causal and otherwise, to the appropriate governmental agency — perfect counterfeit, say — is not money even if everyone treats it as such.

A property might be extrinsic to object O but not to a larger system S, of which O is a part. If the other objects to which O must stand in certain relations in order to have this property are themselves parts of S, then we can say that the property (or some closely related property), though extrinsic to O, is intrinsic to S. Movement is this sort of property. An object's movement, being relative, is extrinsic to the object we say is moving but intrinsic to the larger system that includes both the object and the frame of reference. The (extrinsic) movement of the planets (relative to the sun and the other planets) is an intrinsic property of the entire system — the solar system — of which the sun and planets are parts. If a light switch is wired to — and thus controls — a light, then *contains-a-light-switch* is an extrinsic property of the metal box in which the switch (but not the light) is installed, but an intrinsic property of the house containing both switch and light. All properties are intrinsic to the universe as a whole. When I speak in (2) of meanings being extrinsic, I mean they are extrinsic to the entire system (person, animal, plant or machine) whose behavior they help explain.

It will also prove useful to distinguish between extrinsic properties that require the existence of some other object at the time the property is possessed, and extrinsic properties (like being a widow or a divorcee) that require the existence of other objects (to which they stand in some special relation) at some earlier time. I will call a property temporally extrinsic (*t*-extrinsic) when an object's possession of this property requires the prior existence of some wholly distinct object to which the possessor stands (or stood) in some special relation. Being a widow is a temporally extrinsic property. Being a wife, though extrinsic, is not. Most theories of meaning (including my own: see Dretske 1981, 1988) make the meaning or content of a belief a temporally extrinsic (i.e., a developmental or historical) property of the internal structures that qualify as beliefs.

Some properties are hybrids — mixtures of extrinsic and intrinsic elements. Being a widow, for example, requires not just having had a husband — an extrinsic property — but being a woman — an intrinsic property. The possibility exists, therefore, that the component of meaning that makes the meaning of a belief extrinsic — that makes (2) true — is not the component that figures in the explanation of behavior — that makes (1) true. In this case, using extrinsic meanings to explain behavior would be like using the fact that Hilda is a widow to explain why she can't join the Men's Club, rather than the intrinsic fact that she is a woman. Should this possibility be realized in the case of beliefs and desires, then explanatory meanings could be extrinsic, but, *qua* extrinsic, irrelevant to the behavior being explained. In such a case the extrinsic components of meaning would be epiphenomenal.

To avoid this possibility, I define an *extrinsic system* as a system whose explanatory meanings are not only extrinsic, but essentially extrinsic — i.e., a system whose behavior (some of it anyway) is explained by the extrinsic components of the internal meanings. Extrinsic systems, then, are systems which not only have extrinsic meanings (2) that explain behavior (1), but whose meanings explain behavior precisely because they are extrinsic. Relative to the behavior being explained, the extrinsic aspect of these meanings is not epiphenomenal.

The statement that extrinsic systems enjoy a special kind of causal autonomy is worth noting because human beings (and presumably some animals) are extrinsic systems (Dretske 1988).

The Causal Assumption

If we ignore those parts of biology concerned with the history and development of biological systems and think of biology as concerned with the intrinsic activities and states of an organism and its parts, then the mental life of an extrinsic system does not supervene on the biological stuff of which it is composed. Biologically identical organisms (physical duplicates) can be psychologically different. Having a belief is like being a real dollar bill: two objects can differ in this respect while remaining intrinsically indistinguishable. If beliefs have their meanings extrinsically determined, if (2) is really true, then we can have counterfeit believers. And if (1) is true in the intended sense, then such perfectly forged believers, though behaving the same as genuine believers, can have much different reasons (or no reasons at all) for behaving that way.

For these and other reasons,[2] many philosophers regard our two principles, (1) and (2), as irreconcilable: extrinsic systems are impossible. Hence, they regard any theory (such as ordinary folk psychology) that allegedly embodies a commitment to both as fundamentally incoherent. Meanings can be extrinsic or they can help explain behavior, but they can't do both. Something has to give.

The perception of (1) and (2) as irreconcilable has driven philosophers in all directions. Some (for example, Kim 1979) accept (1) and take this as a reason to reject (2), for accepting what Stich (1978, 1983) has called the "Autonomy Principle": only intrinsic properties, those that supervene on the biology of a system, can be cited in explanations of its behavior. B. Loar (1988) concludes from the fact that belief ascriptions are extrinsic (and the assumption that explanatory contents must be intrinsic) that belief ascriptions do not faithfully reflect the explanatory content of a person's belief: in saying what a person believes (in order to explain what he does), we do not succeed in expressing a content (or meaning) that actually explains the behavior.

Others use (2) as a basis for rejecting (1). This is a strategy pursued by philosophers who are convinced that our familiar conception of one another as systems whose behavior is explained by reasons (by what we believe and desire) is a bankrupt "theory" doomed to eventual extinction as a maturing neuroscience, concentrating on the properties that are intrinsic to the system being studied (these may, of course, be extrinsic properties of the system's parts), becomes progressively better at telling us why we really behave the way we do. The most visible and vocal exponents of this extreme viewpoint are Paul Churchland 1981 and Patricia Churchland 1986. There are also elements of this view in Stich 1983, but Stich is typically more cautious in his conclusions and recommendations.

Another reaction to the perceived tension between (1) and (2) is to tinker with meaning — fashioning it, so to speak, nearer the heart's (or, in this case, the head's) desire. If different (extrinsic) meanings, call them *broad* meanings, can inhabit the same (intrinsic) heads, then for the purpose of doing cognitive science, for the purpose of explaining behavior, a different sort of meaning, call it *narrow* meaning, a meaning more closely tailored to the intrinsic properties of the head, must be developed. Such narrow meanings — supervening (as they will) on the intrinsic, the causally efficacious, properties of the biological substrate — will then become explanatorily (or at least predictively) relevant meanings. Such a strategy can be found most prominently in Fodor 1987, but also in Bilgrami 1987, who wants to "bifurcate" content (one content being extrinsic, the other — the intrinsic content — doing the explaining), and Brown 1986, which talks of "immediate objects" of belief — meanings that are determined by the intrinsic properties of the believer.

A fourth option for dealing with the tension between (1) and (2), a response closely related to the third, is to modify the intent of (1). Despite being extrinsic, meanings can be made relevant to a science of psychology, to the explanation of human behavior, if one takes the right and proper view about what qualifies as an explanation in the special sciences (everything but physics). Despite being extrinsic, meanings can be made explanatorily relevant to behavior if, by adverting to meaning, we can formulate useful generalizations about behavior, generalizations that would not otherwise be available to us (at least not in any practical sense) if we restricted ourselves to the bewildering array of neurobiological (i.e., intrinsic) properties that can realize these meanings. Whatever their ultimate ontological fate, meanings, on this strategy at least, are epistemologically indispensable for a science of psychology and, hence, for the primarily epistemological business of explanation. According to this story — a story that can be found, I think, in works as otherwise diverse as Fodor 1987 and Pylyshyn 1984, on the one hand, and D. Dennett 1987 on the other — meanings, though causally inert, are useful, perhaps even indispensable, devices for talking about behavior.

There are, finally, those whose desire to salvage (1) while acknowledging the truth of (2) forces them to interpret the kind of explanation meanings make possible as a unique, a *sui generis* form of explanation. Though meanings, being extrinsic, in no way enhance a belief's causal powers, they nonetheless rationalize behavior. If, then, we have a rational agent, an agent who (normally) does what he believes will promote his interests, then despite there being no causal laws connecting extrinsic (mental) properties to the intrinsic motor behavior, we get an explanation — call it a *rationalizing* explanation — for such behavior. One can find this view in Melden 1961, Dray 1957, and Gordon 1986.

The tension between (1) and (2), a tension that each of the above strategies is designed to relieve, is generated by a widely shared assumption about the requirements on a causal explanation. Roughly speaking, the assumption is that extrinsic properties are causally inert. This means that whether an object (or event) has such a property is irrelevant to explanations of why that object (or event) has the effects it has.[3] Other things remaining the same, it would have exactly the same effects if it lacked the extrinsic property. Uncles, dollar bills, and divorcees can cause lots of trouble, but the fact that they are (as opposed to being believed to be) uncles, dollar bills, and divorcees, does not explain their various effects on the world (though it might explain the effects that certain larger, composite, objects have on the world — see footnote 3). These effects are to be explained, instead, by the intrinsic properties of uncles, dollar bills, and divorcees — their physical strength, appearance, height, mass, movements, coloration, and markings. More generally, then, the assumption ($\mathbf{CA}$ = Causal Assumption) lying behind the perceived incompatibility of (1) and (2) is that

$\mathbf{CA}$: If C causes E, then the fact that C has extrinsic property M does not explain (or help explain), why E occurs.

Taken together with the Davidsonian idea that beliefs (desires, etc.) cause behavior, $\mathbf{CA}$ generates the tension between (1) and (2). For if the meaning of a belief is an extrinsic property of the belief, and if beliefs cause behavior, then the fact that the belief has this meaning cannot, according to $\mathbf{CA}$, explain why the behavior occurs — a violation of (1) (see Dretske 1988 for an ambiguity in $\mathbf{CA}$).

$\mathbf{CA}$ has an attractive plausibility. Whether or not the paper I hand the cashier is real money or not does not seem able to alter the effects of this transaction. Given a perfect forgery, I get the coat, the cashier gets the paper, and no one is ever the wiser. So, since counterfeit — good counterfeit (i.e., paper having the same intrinsic properties) — is as effective in obtaining coats and groceries as is real money, the property of being real money adds nothing to the paper's causal powers. Take away the t-extrinsic property (i.e., the paper's having the right causal history, the causal history that makes it real money), leaving everything else the same, and the same effects are produced. It seems the occurrence of these effects is not to be explained by the cause's possession of this extrinsic property. The same is true of being an uncle, a widow, and a divorcee.

There are, however, technical problems with **CA**. If *legally owning a coat* is an outcome of purchasing a coat with real (not counterfeit) money, then the property of being real money does make a causal difference: using real money gives one legal ownership of the coat while using counterfeit doesn't have this effect. So real money does make a difference. There are also counterexamples to **CA** that depend on the fact that *causing-E* is an extrinsic property of the cause (C), an extrinsic property that would seem to be relevant (explanatorily relevant?) to the effect (E).

I will not fuss about the exact expression of **CA**. I will asssume that, properly qualified, it is a valid principle. There is clearly a tension between (1), (2), and even if **CA** isn't quite the missing premise needed to formally exhibit this tension, even though **CA** needs fine tuning, the impression is irresistible that something like **CA** is true and the source of this tension. Additionally, the counterexamples to **CA** all seem to betray a common theme that most likely cannot be exploited in every case in which reasons function to explain behavior.[4] Though some behavior can be described in an "inflated" way, a way that logically requires appeal to extrinsic properties in its causal explanation (in the way that a description of the effect as "obtaining legal ownership of the coat" requires appeal to extrinsic properties in its causal explanation), it is surely implausible to suppose that all descriptions of behavior are of this sort. We can, after all, explain why a person moved his finger by citing extravagant beliefs and desires whose meanings are extrinsic if any meanings are extrinsic. Yet, "moving one's finger" is a description of behavior that does not require even an intentional (i.e., belief-desire) explanation (*a fortiori*, not an explanation in terms of beliefs and desires with extrinsic content). People can move their fingers with no intention or purpose at all — in their sleep or reflexively. Extrinsic meanings surely do not derive their explanatory relevance from this way of describing the behavior. There are loopholes in **CA**, yes, but not enough of them, and not of the right kind, to make sense out of the way so much of our ordinary behavior is explained by meanings that are — if (2) is true in the intended way — extrinsic. Furthermore, since I think that a correct understanding of extrinsic systems — and, hence, the compatibility of (1) and (2), in no way depends on the denial of **CA**, I am willing to grant this assumption without argument (and without fussing about the details) in order to explore the implications of accepting (1), (2), and **CA**. If we are extrinsic systems, and if we accept the Causal Assumption (in a form that still applies to the behavior that meanings are supposed to explain), what does this tell us about action, its explanation by reasons, and its autonomy?

The Autonomy of Action

Given the Causal Assumption, the only way (1) and (2) can be consistent, the only way an extrinsic system is possible, is if:

(3) The behavior of an extrinsic system is not caused by the reasons — the beliefs, desires, fears and purposes — that explain its occurrence.

When an extrinsic system does something, *B*, for reasons — because, say, it believed *M* and wants *W* — this belief and desire do not cause *B*. For if, contrary to (3), the belief that *M* was the cause, or part of the cause, of behavior *B*, then — according to (1) — the fact that the belief had this meaning would explain, or help explain, why *B* occurred. But given the extrinsicness of meaning — i.e., given (2) — this is contrary to **CA**: the meaning of a belief, being an extrinsic property of the belief, does not help explain the occurrence of whatever effects the belief has. So if both (1) and (2) are true, then, given **CA**, (3) must be true: the beliefs and desires do not cause the behavior they help explain.

It is important to understand exactly what behavior is included, and what behavior is not included, in (3). We are talking about behavior explained and rationalized by the actor's reasons — his or her beliefs and desires. Not only must the behavior occur because the actor believes *M* and wants *W*, but believing *M* and wanting *W* must be her reasons for, must give expression to her purpose in, doing *B*. By emphasizing that these beliefs and desires are her reasons for behaving that way, the intent is to exclude a variety of what Dennett calls "reactions" — behaviors that occur (in some sense) because of what is believed (or what is desired) but which are not purposeful, intentional or in any way rational (1973). Dennett points out, for example, that someone's desire for revenge may give them ulcers, their belief that a gun is loaded cause them to tremble, and their thought about a narrow escape make them shudder. People sometimes blush when they remember embarrassing incidents. In such cases, beliefs and desires figure in explanations of behavior (assuming one is willing to classify shuddering, trembling, getting ulcers, and blushing as behavior), but these beliefs and desires in no way rationalize the behavior they help explain. They do not exhibit the agent's reasons for behaving in that way. The thought about the embarrassing incident is the reason she blushed, but the fact that the incident was embarrassing (the content of her thought) is not her reason for blushing, nor her purpose in, blushing. Quite the contrary. Given her desires (to conceal her thoughts), it may, in fact, be a reason not to blush. Rational behavior is behavior in which the actor's reasons (beliefs and desires) not only provide the explanation (reason) why she did it, but also give her reasons for doing it, supply her purpose in doing it. It is this kind of behavior, what I call *action*, behavior that is both explained *and* rationalized by the agent's reasons, that (3) is meant to describe.

But how is this possible? How is it possible for a belief that M to be a part of the explanation of behavior without its being, at the same time, part of the cause of this behavior? How is it possible to understand the "because" of "He did B because he believed M" as the "because" of explanation and, yet, not understand the belief that M as the cause, or as some part of the cause, of B? Are we dodging an inconsistency by embracing a mystery?

Although my purpose here is only to explore the implications of having a scheme in which such explanations are supposed to occur, something should be said about the possibility of explaining behavior in the way mandated by the joint acceptance of (1), (2), and **CA**. Until one is convinced that extrinsic systems are actually possible, and there is some basis for supposing that (1) and (2) can be true of real objects in the world, the fact that (1) and (2) imply (3) will be of theoretical interest only. Understandably enough, a reader may be interested not merely in valid arguments, but in valid arguments whose conclusions might actually be true.

So I offer the following brief sketch, without argument. (For my argument, see Dretske 1988.)

Suppose that beliefs and desires are internal events (states, structures, or what have you) of a certain kind. These events, in virtue of their intrinsic properties, have effects on muscles, nerves, and glands — and, hence, on bodily movements and change. Suppose, furthermore, that we identify behavior not with the bodily movements and changes these internal events cause, but, instead, with the process in which these internal events cause them. If some internal event, X, causes Y, a finger movement, say, then the behavior is to be identified not with the movement that X causes, but with X's causing it. The fact that X causes Y is not itself caused by X. It is Y, not X's causing Y, that X causes. So if the behavior is X's causing Y, and, in the case of intentional behavior, the X is some belief, or some combination of belief and desire, then the behavior is not an effect of the reasons (the beliefs and desires) whose meanings explain it. It is Y, the bodily movement, that is an effect of the belief and, therefore, to be explained (according to **CA**) by facts about the intrinsic properties of the belief. But it may (who knows?) be facts about X's extrinsic properties — perhaps even facts about what X means — that explain why X causes Y, why the behavior occurs. By distinguishing the action or behavior, X's causing Y, from the bodily movements (the Y) that such actions (typically) involve, one thereby avoids the conclusion that "if certain movements are caused, then actions are caused, because the movements are the actions" (Honderich 1973, 199. See also I. Thalberg 1977; and R. Tuomela 1977, 1984). On this conception of behavior, although it may be true (and is true according to determinism) that both actions and movements are caused, it does not *follow* that actions are caused from the fact that the associated movements are caused.

In this picture, explanations of behavior are explanations of causal processes $(X \rightarrow Y)$ typically having (at least in the case of basic actions) bodily movements (Y) as their product. In explaining why he moved his finger (in contrast to explaining why his finger moved) we explain not why the finger movements (Y) occur, but why a process having finger movements as its product occurred — why X caused Y (finger movements) rather than something else.[5] The explanation of this fact may need to advert to the extrinsic properties, in fact the meaning, of X.

So much for the possibility of an extrinsic system, a system whose behavior, though not caused by internal states having extrinsic properties, is nonetheless causally explained by the fact that these states have these extrinsic properties. An immediate consequence of (3) is:

(4) Whatever causes beliefs (desires, etc.) does not thereby cause the behavior that occurs because of what is believed (desired, etc.).

Whatever causes Clyde to believe there's a beer in the fridge does not, *qua* cause of belief, cause him to go to the fridge even if he goes there because he thinks there's a beer there. This is so because his belief that there is a beer in the fridge does not itself cause him to go to the fridge. Hence, the causes of belief and desire are not (*qua* causes of belief and desire) among the causes of the behavior that the beliefs and desire explain — at least not when the behavior is intentional.[6]

Better than (3), (4) makes explicit an important respect in which intentional behavior enjoys an independence from external determination. It thereby reveals an aspect of autonomy. The freedom to *do B* is compatible with the complete determination of those factors — our beliefs and desires, our purposes and plans — that explain why we do it. Even if the will is not free (in the sense that our reasons for acting are completely determined by factors over which we have no control), it by no means follows (though it may in fact be true) that our actions are not free, that they are determined by factors over which we have no control. For what causes us to have the reasons that explain our actions are not thereby the causes of our acting that way. Intentional behavior may be causally determined, of course. Nothing I have said so far implies it isn't. (4) certainly doesn't imply that our actions are not causally determined. Nevertheless, (4) does tell us that even if freedom is (in some way) incompatible with causal determinism, our freedom of action is compatible with a causal determination of those factors — our reasons — that explain our acting that way. The fact that we are caused — perhaps even compelled — to believe and desire as we do does not imply that we are caused — certainly not compelled — to act as we do, even when we act that way precisely because we believe and desire what we do.

If folk psychology is the psychology of extrinsic systems, then we are now in a position to understand an otherwise puzzling fact about our ordinary explanations of behavior: to wit, the fact that we normally have no hesitation in admitting that voluntary behavior — Clyde's trip to the kitchen, for example — is explained by factors (his desire for a beer and belief that there is one in the kitchen) that are in no way voluntary. One doesn't choose to believe that there is a beer in the fridge. Typically, one is caused to believe this, by earlier seeing the beer itself (or something that looked like beer) or having someone tell you there was beer there. The causes of belief are not things over which one has control. Neither does one choose to be thirsty, to want a beer. One's thirst, and preferences in satisfying thirst, are (often enough anyway) neither rationally explicable nor voluntary. Still, if Clyde goes to the kitchen because he is thirsty and thinks there is a beer in the fridge, the behavior itself is perfectly rational and voluntary. If the reasons for which we act were the causes of our acting that way, then (given the transitivity of the causal relation) it would be hard to see how such actions could be voluntary. How can Clyde's trip to the kitchen be voluntary when he is caused to go there by factors (those which caused him to believe and desire as he does) over which he has no control?

If we assume, then, that folk psychology is an explanatory scheme that treats rational agents as extrinsic systems, as systems for which both (1) and (2) are true, then the result we have so far reached, far from constituting an embarrassment to folk psychology, far from constituting a paradox, neatly supports the idea that rational agents, when they are acting for reasons — when they are doing something they believe will get them what they want, and doing it, moreover, because they think it will get them what they want — are not *thereby* caused to behave that way. By removing behavior from the range of effects that beliefs and desires have, by taking it (as politicians are fond of saying) out of the (causal) loop, our ordinary mode of understanding why we do the things we do exempts such behavior from whatever causal determination the elements of that explanation — the beliefs and desires themselves — may be subject.

NOTES

My thanks to members of various philosophy departments — Syracuse, Georgetown, Princeton, Utah, Santa Barbara, and Colorado — for the help they gave me in fine tuning this essay. I want especially to thank the people at Wisconsin for mauling an early version. Berent Enc and Elliott Sober were particularly effective members of the wrecking crew. I am also indebted to John Fischer for helpful comments.

1. J. Kim (1982), adapting Chisholm, defines an *external* property as one that is "rooted" outside the object that has it. A property is rooted outside if and only if there must exist some object that is wholly distinct from the object possessing the property. A property can also be rooted outside *times* at which the object has the property (Kim 1982, 60). D. Lewis (1983) notes some technical flaws in this definition. My notion of an extrinsic property is closer to what Lewis calls a *positive* extrinsic property.

2. There are other problems not directly related to the explanation of behavior that extrinsic meanings are supposed to generate. If meanings are extrinsic, for example, it isn't clear how to explain how agents can introspectively know what they believe. Since the meanings depend on *external* conditions (conditions that are, presumably, inaccessible to *intro*spection), how can believers know *what* they believe? See D. Davidson (1987) and T. Burge (1988).

3. This is not to say that such properties are causally inert *period*. It only means they are inert relative to the causal efficacy of the objects for which they are extrinsic. One object weighing more than another (an extrinsic property of both) can make a big difference to the behavior of a beam balance on which both are placed. The fact that A weighs more than B explains why one arm of the balance goes down. In this case the property in question (A's weighing more than B), though it is extrinsic to both A and B, is intrinsic to a larger system (the combination, as it were, of A and B) which has this effect on the balance. It isn't A or B that has this effect on the balance, it is A and B, and what explains the effect on the balance is an intrinsic property of this larger, composite, system.

4. Although, if I understand him right, it is this kind of "loophole" in **CA** that T. Burge exploits in defending the idea that externally individuated (i.e., extrinsic) meanings can be explanatorily relevant in psychology and elsewhere (1986). If the behavior we are trying to explain with extrinsic meanings is, so to speak, *as extrinsic* as are the meanings we use to explain it, then it would be understandable why we should need (or at least be able to use) extrinsic meanings in the explanation of it. We would need them for the same reason we would have to mention the extrinsic properties of paper — the fact that it is money — to explain why its exchange gave us legal ownership of the coat.

5. In seeking a causal explanation of a causal process, there are several different things one can be looking for: (1) why X caused Y rather than something else; (2) why X, rather than something else, caused Y; (3) or why the X (which caused Y) occurred (and, therefore, by transitivity, why Y occurred). This is a difference in what I called a structuring cause (what caused X *to* cause Y) and a triggering cause (what caused the X *which* caused Y) of behavior (1988). We sometimes speak, incorrectly, of the triggering cause (what caused you to believe M) as the cause of behavior that is explained (in part) by the fact that you believe M.

6. It may be supposed that the causes of belief and desire cause the behavior, not (as it were) *through* the belief and desire that explains the behavior, but through some alternate causal route. From C causing something (a belief, say) that doesn't cause A, it doesn't follow that C doesn't cause A. Hence, (4) does not follow from (3).

This objection, correct as far as it goes, is irrelevant to (4). The claim in (4) is not that C does not cause the behavior, but that it does not *thereby* (as a cause of belief and desire) cause the behavior.

REFERENCES

Bilgrami, A. 1987. "An Externalist Account of Psychological Content." *Philosophical Topics* 15.1: 191–226.

Brown, C. 1986. "What Is a Belief State." In French, Uehling, and Wettstein, eds., *Midwest Studies in Philosophy*, vol. 10. Minneapolis: University of Minnesota Press, pp. 357–78.

Burge, T. 1979. "Individualism and the Mental." In French, Uehling, and Wettstein, eds., *Midwest Studies in Philosophy*, vol. 4. Minneapolis: University of Minnesota Press, pp. 73–121.

Burge, T. 1986. "Individualism and Psychology." *Philosophical Review* 95: 3–45.

———. 1988. "Individualism and Self Knowledge." *Journal of Philosophy* 85: 649–63.

Churchland, P. M. 1981. "Eliminative Materialism and the Propositional Attitudes." *The Journal of Philosophy* 78: 67–90.

Churchland, P. S. 1986. *Neurophilosophy*. Cambridge: MIT Press.

Davidson, D. 1987. "Knowing One's Own Mind." In *Proceedings and Addresses of the American Philosophical Association* 60:441–58.

Dennett, D. 1973. "Mechanism and Responsibility." In Honderich, ed., *Essays on Freedom of Action*. London: Routledge and Kegan Paul.

———. 1987. *The Intentional Stance*. Cambridge: MIT Press.

Dray, W. H. 1957. *Laws and Explanation in History*. Oxford: Clarendon Press.

Dretske, F. 1981. *Knowledge and the Flow of Information*. Cambridge: MIT Press.

———. 1988. *Explaining Behavior*. Cambridge: MIT Press.

Fodor, J. 1987. *Psychosemantics*. Cambridge: MIT Press.

Gordon, R. 1986. "Folk Psychology as Simulation." *Mind and Language* 1:158–71.

Honderich, T. 1973. "One Determinism." In Honderich, ed., *Essays on Freedom of Action*. London: Routledge and Kegan Paul.

Kim, J. 1979. "Causality, Identity and Supervenience in the Mind-Body Problem." In French, Uehling, and Wettstein, eds., *Midwest Studies in Philosophy*, vol. 4. Minneapolis: University of Minnesota Press, pp. 31–49.

———. 1982. "Psychophysical Supervenience." *Philosophical Studies* 41: 51–70.

Kripke, S. 1972. "Naming and Necessity." In Harman & Davidson, eds., *Semantics of Natural Language*. Dordrecht: Reidel, pp. 253–355.

LePore, E., and B. Loewer. 1986. "Solipsistic Semantics." In French, Uehling, and Wettstein, eds., *Midwest Studies in Philosophy*, vol. 10. Minneapolis: University of Minnesota Press, pp. 595–614.

Lewis, D. 1983. "Extrinsic Properties." *Philosophical Studies* 42: 197–200.

Loar, B. 1988. "Social Content and Psychological Content." In Grimm & Merrill, eds., *Contents of Thought*. Tucson: University of Arizona Press.

Melden, A. 1961. *Free Action*. London: Routledge.

Putnam, H. 1975. "The Meaning of Meaning." In Gunderson, ed., *Language, Mind and Knowledge: Minnesota Studies in the Philosophy of Science*, vol. 7. Minneapolis: University of Minnesota Press.

Pylyshyn, Z. 1984. *Computation and Cognition: Toward a Foundation for Cognitive Science*. Cambridge: MIT Press.

Stich, S. 1978. "Autonomous Psychology and the Belief-Desire Thesis." *Monist* 61: 573–91.

———. 1983. *From Folk Psychology to Cognitive Science*. Cambridge: MIT Press.

Thalberg, I. 1977. *Perception, Emotion and Action*. Oxford: Blackwell.

Tuomela, R. 1977. *Human Action and Its Explanation*. Dordrecht and Boston: Reidel.

———. 1984. *A Theory of Social Action*. Dordrecht and Boston: Reidel.

18

Explanations Involving Rationality

Peter Railton

Department of Philosophy, University of Michigan

I come to the topic of explanations involving rationality as a philosopher of science rather than a philosopher of mind. Nowadays, that is perilous, given that the philosophy of mind has become such a powerful export-oriented philosophical economy. Indeed, these exports have been a source of novelty in some otherwise fairly quiet corners of the philosophical globe — perhaps most notably in the philosophy of social science. Although this might be seen as the reawakening of old disputes, we should not underrate the extent to which these issues are now susceptible to more sophisticated and compelling formulations. I don't think a regiment of Diltheys armed with mighty tomes would ever have convinced a notable number of analytic philosophers that the social sciences are essentially interpretive, but Donald Davidson seems to have done so with a few short articles.

I will focus on the possibility in mentalistic psychology of finding serious nomological generalizations — general statements of probabilistic or nonprobabilistic regularities, which are confirmed by their instances (short of the exhaustion of cases), and which support counterfactuals, predictions, and explanations. This issue, which I think of as the question of the relevance of a *nomothetic ideal* of explanation and understanding to the psychology of intentional action, has a long and somewhat unfortunate history. It has sometimes been characterized the question of whether psychology can or should imitate natural science. But this is misleading.

First, for all their glory and prestige, the natural sciences would not provide an appropriate ideal for psychology unless their investigative and explanatory practices satisfied relevant methodological desiderata. What might these desiderata be in the case of nomothetic ideals of explanation and understanding? Broadly, it is too easy to give an explanation that possesses some intuitive credibility and relevance, especially when one invoke familiar schemata of cause-and-effect, type-and-trait, or goal-and-directed-conduct (in the case of both natural and social teleologies). Serious theorizing should allow us some empirically constrained way of questioning conventional, intuitively appealing understandings and of avoiding merely ad hoc explanations. This requires that we be able to discipline ourselves about what our theories do or do not actually say or imply about any given phenomenon. Moreover, it requires that we hold our predictions and explanations to standards of relevance, coherence, and consistency across cases.

Although there might be multiple ways of meeting these desiderata, one way that has been important in the emergence and flourishing of the natural sciences involves the formulation of general, lawlike hypotheses and the effort to apply them deductively to explananda. When we are able to find and deductively apply nomological generalizations, we see explanandum phenomena fitting into nonaccidental, projectible patterns. These patterns sometimes enable us to unify phenomena that might at first seem unpredictable, unrelated, or idiosyncratic, but also to distinguish phenomena that commonsensically strike us as similar but actually do not form a "natural kind." Even certain other ideals of explanation and understanding in the natural sciences, such as the ideal of finding causal mechanisms, draw upon the notion of a lawlike regularity. At least some of the distinctive glory and prestige of the natural sciences can perhaps be credited the prevalent demand for laws to underwrite explanatory claims. Asking whether a nomothetic ideal of explanation and understanding is relevant in psychology, is of a piece with asking how, if at all, these broad methodological desiderata can be satisfied in psychological inquiry, where one finds a special danger of too-ready acceptance of certain intuitively plausible explanations.

Second, there could be no one way that "psychology imitates natural science," as both areas of inquiry are highly diverse, and in some areas even overlap (for example, in neuropsychology and psychobiology). Systematic nomological structure is much more developed in some areas of the natural sciences (say, quantum mechanics) than others (say, geology). Within the domain of the sciences as a whole, it is hardly unusual to find practices that convey explanatory information about the causal origin, constitutive nature, or structural relations of various phenomena without explicitly invoking laws. Sometimes this is because, in the context, the laws are too well known to be worth stating. Sometimes it is because such laws aren't yet known.[1]

For example, we may learn a great deal of explanatory information about the causes and nature of earthquakes by being told:

(1) Earthquakes are violent movements of the outer layers of the earth that typically result when pressure along the lines of intersection of the large moving plates of the earth's surface, which are propelled primarily by convection currents in the earth's fluid outer core, becomes sufficiently great to overcome frictional resistance and slippage occurs.

Contrast (1) as an explanation of earthquakes (or rather of "typical earthquakes") with:

(2) Earthquakes are violent movements of the outer layers of the earth that typically result when sporadic turbulence in the earth's fluid outer core produces concentrations of stress on relatively weak portions of the stationary crust of the earth sufficiently great to fracture the solid mantle.

Neither (1) nor (2) explicitly states a law. Both invoke generalizations, though hedged with phrases like "typically" and "sufficiently great." Moreover, the generalizations are parochial. They concern the phenomenon of sudden movements of the surface of *earth* and suggest nothing about such phenomena on other planets or celestial bodies — unless, of course, those planets or other bodies are "sufficiently similar" to earth. Yet (1) and (2) are heavily laden with purportedly explanatory information, and there is a great difference in the explanations they afford. When geology came to accept something more like (1) than (2), this was a major reorientation of earth science.

Explanations like (1) and (2) exhibit the characteristic way in which more definite nomological regularities — known or unknown — are assumed in everyday scientific explanatory practice. Such explanations presuppose acquaintance with known principles of friction and resistance, the communication of force, cohesion, stress and strain, etc. But they also take the liberty of supposing that imperfectly understood phenomena such as fluid dynamics, inelastic deformation, and discontinuous "overcoming" of friction are subject to nomic principles. These presuppositions and suppositions are an important part of why (1) and (2) can function as explanations when heard by modern ears.

However, one must be careful to distinguish these sorts of presuppositions and suppositions from the stronger idea that the geological terminology of (1) and (2) is or will be neatly reducible to categories of fundamental physics. The explanatory informativeness of (1) or (2) does not depend upon that prospect. Indeed, purported explanations (1) and (2) have special explanatory interest precisely because they organize complex phenomena — impossibly complex from the standpoint of fundamental physics — at a level where it appears that some stable, confirmable, counterfactual-supporting generalizations are possible. Of course, (1) and (2) are sketchy, and geological science has refined them considerably by bringing to bear geological observation and articulating geological theory. Often such observations and theory development draw upon areas of science not part of geology proper — chemistry, mechanics, etc. But this infusion of geological generalization with greater scope, rigor, and precision by "mixing" geological categories with those other sciences, and subsequent better approximation of the nomothetic ideal, is not dependent upon a reduction as such.

Casual observation reveals wide variation within what we call "the natural sciences" in their current capacity for, or near-term expectation of, law-based explanation of the explicit, law-invoking kind rather than the inexplicit, law-gesturing kind. Although there is a long tradition in the most basic macroscopic and microscopic sciences of developing powerful nomological principles, these principles often can be applied rigorously only to relatively simple, isolated systems. Methods devised to deal with more complex, situated systems typically involve simulation or approximation, and require various simplifying assumptions to prevent computational difficulties from getting out of hand. Such applications — for example, quantum-mechanical models of materials with real engineering interest — are motivated by an assumption that the more complex systems are made up of parts of whose behavior we have a good nomic understanding. But it is well known that one generally cannot simply add such parts together in linear ways to achieve descriptive adequacy, to say nothing of explanatory adequacy.

Even if we assume that complex, situated systems are built from parts governed by the principles of fundamental physics, we should not assume that only those accounts of the behavior of such systems that are couched in the language of physics — or are neatly reducible to it — can provide significant explanatory information. The proper image of the relation of the "special sciences" such as biology, geology, or psychology to physics is not that of a house built upon a foundation, where the shape and scope of the house is fixed by the shape and scope of the foundation. If a single image of the special sciences is to be sought, one might think of tropical vines, which proliferate widely and often begin life in the upper canopy. Always searching for the wherewithal to propagate, they grope downward at one end toward the earth ("basic science") and at the other end upward toward the light ("experience" and "application"), constantly sending shoots and tendrils sideways to find support from other flora, including other vines ("collateral theories" and "auxiliary hypotheses"). The resulting convoluted tangle would not please the eye of the Enlightenment gardener who insists that a *proper* plant grow symmetrically upward from the ground, espaliered if necessary against the grid of a trellis to ensure this arboriform "ideal." But the tangle of the rainforest is considerably more productive biologically than the formal severity of an eighteenth-century garden.

Metaphors aside, our concern is with a general kind of explanation-seeking activity and with questions about where it might — or might not — be profitably pursued. The issue is not a matter of the *logical form* of explanations, or of the substantive unity of all sciences, or of reducibility as the archetype for all intertheoretic relations. The sorts of generalizations that geological or biological or psychological science develop, and the explanatorily informative stories they tell in part by appealing to these generalizations, might or might not be "emergent" in an interesting sense. Much of the current philosophical debate surrounding the arguments we will be considering below is indeed preoccupied with matters of the reducibility of psychology to natural science, where reduction is understood rather narrowly as a matter of matching the categories, entities, and events of one level of theorizing with another. But reduction in general and taxonomically matching reduction in particular are not our immediate concerns here, nor do they loom as the ultimate, underlying, "real" issue. Supervenience, which is granted on all sides of these debates, can be had without taxonomic matching; it supplies enough security against the special sciences undermining our fundamental conception of the world as constituted from natural ingredients.[2] Neat matching of categories, entities, or events would afford one important kind of theoretical explanation, but it is not the only kind. We eventually will face some questions about "mixed" laws involving both psychological and nonpsychological vocabularies, but not as candidates for "type-type" psychophysical reduction rather than as distinctive psychophysiological or psychobiological hypotheses that could be explanatory in their own right.

My particular concern is that philosophical fascination with (admittedly important) questions of reducibility not be allowed to obscure the potential explanatory contribution of "less basic" levels of nomothetic theorizing. I am interested in the possibility that, in cognitive social science as in many other special sciences, robust and informative generalizations can be found that are confirmable by their instances, support counterfactuals, and exhibit explanatorily relevant structures and patterns of causality. Even if we suppose that the history of evolution could be retold at the level of a blow-by-blow physiochemical account, such an account would lack crucial explanatory information. For it would not make explicit the ways in which the evolutionary process has lawful features that are independent of the particular physiology that realizes it. If, for example, so-called sexual selection takes place, then the distribution of certain molecules on the earth's surface will reflect patterns of mate-preferences of breeding organisms. So long as they play (the relevant part of) the role of genes, these molecules could be DNA or XYZ.

True methodological naturalism in philosophy is a matter of being responsive to the various kinds of knowledge — including modal or explanatory knowledge — empirical science can produce, not of an a priori commitment to reducibility. Naturalism is not physicalism. A naturalistic attitude toward "cognitive social science" therefore should be just as interested in the prospects for informative theorizing in which attributions of belief, desire, inference, intention, and rationality play a central explanatory role as it is in the prospects for reduction. Is there, in short, some reason to doubt whether in principle psychological theorizing could resemble much other theorizing in the special sciences, in providing distinctive patterns of explanation that can be refined nomically without thereby necessarily being revised out of existence or replaced wholesale?

A final preliminary: I will be focusing on so-called rational agent explanations, but not because I believe that methodological considerations — e.g., "methodological individualism" — require that all meaningful theory about social phenomena be reducible to a microtheory of individual rational choice. Many of the most interesting macrotheories and structural theories in the social sciences are part of larger stories that can, with some effort, be seen to *include* rational agent explanations at the individual level. This is so, I believe, even for some of the theories about which doubt on this score has traditionally been most prominent, including Durkheim's theory of the sacred and Marx's theory of ideology. The possibility of developing such a microtheory even in these seemingly recalcitrant cases perhaps encourages us to think that rational agent explanation will be ubiquitous in social science. But we should avoid confusing the desideratum of possessing a satisfactory microtheory with a requirement that the microtheory take a particular form. Especially, we should avoid confusing a desideratum with a condition of meaningfulness. The question remains whether nomothetic theoretical practice can be sustained in the social sciences.

I

The writings of Donald Davidson furnish a rich collection of considerations that militate against the possibility of a nomothetic psychology. So rich — as the explosion of interpretations of his writings now available in the literature attest — that I wonder whether any sane person would put himself or herself forward as discussing "Davidson's views." Sane or not, I only attempt to understand and assess some important arguments suggested by various texts of Davidson's. Those considered include "Actions, Reasons, and Causes" (referred to here as ARC), "Mental Events" (ME), "Psychology as Philosophy" (PP), and "Hempel on Explaining Action" (HEA), as reprinted in *Essays on Actions and Events* (1980).

I have detected in Davidson's work five apparently interrelated considerations in favor of skepticism about nomothetic psychology. Some are relatively a priori, while others are relatively a posteriori.

1. An unprejudiced survey of the current state of psychology will reveal no "laws of psychology" akin to the strict, well-formulated laws of basic physical science. This is true despite a long history of attempts to develop systematic psychological theory.

2. Davidson himself once tried to develop such a systematic, empirical theory with respect to decision making, but the attempt failed in a

way that convinced him the idea was hopeless. (Reasons of *this* kind are rarely found among philosophers, and call for special respect from those of us who have never tried to run an experimental program.)

3. Davidson holds that *genuine* or *serious* laws — the kind that, in his view, could support causal explanations — must characterize a closed system, and must characterize such a system strictly, without open-ended or otherwise unspecified "ceteris paribus" conditions. But the phenomena of belief-desire psychology do not form a closed system in this sense.

4. Belief-desire psychology is constrained by a set of *constitutive* principles that are distinctive, decisive, and effectively unrevisable. They separate psychology from other disciplines with different constitutive principles, such as physiology or physics. In particular, they give belief-desire psychology a fundamentally normative character.

5. The explanations offered by belief-desire psychology concern individuals essentially, and do not embody or point toward the generalized hypothetical knowledge of *kinds* characteristic of a genuinely nomothetic science.

Let us take up all five in order, noting the interconnections and mutual dependencies as they emerge.

II

1. *The current state of belief-desire psychology.* Certainly it is true that one does not find in contemporary belief-desire psychology a theoretical structure of nomological generalizations comparable to present-day physics. One instead finds great bundles of claims (often contested and always subject to multiple interpretations) concerning more or less robust and significant statistical correlations and, usually, associated theoretical "models" (see Ajzen and Fishbein 1980). Typically these models are idealizations, sometimes quite severe, of what are supposed to be (though again not without disagreement) the component processes of cognition and action. Behaviorism, with its self-consciously nomological formulation of psychological claims (e.g., "the law of effect"), has largely faded from view. Mentalistic cognitive psychology flourishes, which should be good news for someone interested in finding systematic theory involving the notions of belief and rationality. Some cognitive psychologists, however, explicitly disavow goals other than interpretive ones, and the large body of cognitive psychologists

who continue to search for nomic generalizations in theory development do not appear to be cranking out psychological equivalents of Ohm's law.

We certainly need an explanation for this. The Davidsonian position that nomothetic psychology is not possible would afford a principled explanation. However, we will be in a position to assess the advantages and disadvantages of this principled explanation only after we have looked further into what it involves and have compared it to alternative hypotheses. Let us pass to the four other considerations and return to this one afterwards.

For now, like Davidson we will assume the relatively anomic state of psychological theory as a datum to be explained.

III

2. Davidson's history as an experimental psychologist. In a 1986 essay on the state of rational choice theory, Jon Elster writes, "Once we have constructed a normative theory of rational choice, we may go on to employ it for explanatory purposes." Davidson might reply, "I've been there and back."

Clearly I am in no position to comment on Davidson's history as an experimental psychologist, since I have only his own brief accounts of it to go by (PP, 234–7; HEA, 268–72). But as these accounts are highly suggestive, let us discuss them.

Davidson begins by noting that *if* one could give "a clear behavioristic interpretation" to "S prefers A to B," decision theory would become "eminently testable," albeit "palpably false" (HEA, 270). However, Davidson believes there is no real danger of thus falsifying the theory. For our response to any negative general result concerning the applicability of rational decision theory to mankind would be to reject the "clear behavioristic interpretation" as inadequate and save the theory. It would not be vanity alone that would drive us to this. Rather, if we are intelligibly to attribute attitudes and beliefs to an agent at all, even as part of an account that shows the agent to be cognitively far from perfect, then we are committed to finding a large degree of rationality and consistency in that agent's behavior, belief, and desire (PP, 237).

There is, then, a holism about belief and desire attribution that precludes genuine behavioral testing. We are able to use behavior as a guide in the attribution of beliefs and desires precisely because they make sense of the agent's behavior. Were we to drop wholesale the assumption that the agent's behavior made sense relative to his or her beliefs, it would be unclear how such attribution could proceed. Less-than-wholesale departures from making sense are certainly not ruled out, but even to make progress in attributing to an agent an inappropriate belief that b, for example, we must assume that he does with b — in avowal, inference, deliberation, etc. — pretty much what any rational b-believer would do with it. Otherwise, how could his behavior count in favor of saying that the agent believes in b at all, rather than some other proposition, or none? How, for example, would we be able to see the compulsive's obsessive hand-washing as evidence of his belief that contamination is everywhere unless we view his behavior as rational *relative to* this belief — and to a host of others concerning the shape of his environment, the effects of his behavior, and so on.

Put more generally, if we attempted a "clear behaviorist interpretation" of possessing a belief or having a desire, we would perforce be led to modify our interpretation if it threatened the general hypothesis of rationality in agents. To give up the hypothesis of rationality would be to see human behavior in altogether different terms from those of belief and desire — as movement that, however predictable, is not intentional. In short, it would be changing the subject. As Davidson writes:

My point is that if we are intelligibly to attribute attitudes and beliefs, or usefully describe motions as behaviour, then we are committed to finding, in the pattern of behaviour, belief, and desire, a large degree of rationality and consistency. (PP, 237)

What are the implications of such holism for the prospects of a nomothetic psychology?

Initially, we might think that the seeming impossibility of devising independently credible empirical criteria of belief- and desire-attribution precludes any serious theory testing: rational decision theory is presupposed in the very formulation of the "test conditions." However, this would not distinguish the situation in psychology from that of science in general, where holism in theory testing is chronic. Biology does not really *test* the hypothesis that the world contains life-forms whose nature and behavior can be understood in terms of the organizing principles of self-sustaining, self-replicating organisms. To deny this would be to change the subject away from biology and back to, say, chemistry. Measurement theory in physics *presupposes*, rather than tests, basic principles such as transitivity and triangle inequality in length (as Davidson notes in ME, 220–21). What, then, is peculiar about the lack of independence of empirical criteria in psychology such that it would play hob with the possibility of a nomothetic theorizing?

Davidson offers several suggestions; following up on these will lead us through the third, fourth, and fifth considerations canvassed below.

IV

3. *The closed nature of a nomic system.* Belief-desire psychology, Davidson claims (ME, 224), does not have a closed system and so as its subject matter cannot be expected to yield laws specifiable without open-ended *ceteris paribus* conditions. Laws without such conditions can be *strict* without being deterministic, as probabilistic physics has shown. Strictness is important in laws, according to Davidson, since he believes it is necessary if causal claims are to be supported.

The lack of closure in psychology stands in contrast with, say, the character of physics, which is committed to comprehensiveness and generality, and which has yielded an impressive collection of strict laws. Indeed, in Davidson's view, it is strict physical law that makes good — if anything does — our causal claims on behalf of mental events.

Physics presents the aspect of a closed system only under certain idealizing and restrictive assumptions. Perhaps only at the level of the Grand Unified Theory do we find in physics claims to genuine comprehensiveness. The view that physical laws can be sharpened without limit, possess no open-ended *ceteris paribus* clauses, and strictly imply relations among particular events has been on the wane in philosophy of science for some time. Moreover, Davidson's view that causal relations must be backed by strict laws (in his sense) is demanding, and I am not sure how it would be defended.[3]

In any event, our interest in the possibility of nomothetic psychology is somewhat independent of these reservations. For Davidson challenges us to think about what kinds of generalizations psychology might reasonably aspire to attain, and to ask how much causal/explanatory weight they might bear. Have we any reason to think that psychology might arrive at a system of confirmable psychological hypotheses relating non–ad hoc kinds, and capable of supporting counterfactuals, causal claims, explanations, etc. — hypotheses not merely general in form, but plausibly seen as possessing nomic force?[4]

Let us ask instead why it might appear obvious to Davidson that psychology is not a closed system of the required sort. The most straightforward account would be that psychology does not even claim to sweep with its variables over all possible determinants of action. Davidson writes, "The mental does not . . . constitute a closed system. Too much happens to affect the mental that is not itself a systematic part of the mental" (ME, 224). Thus, as long as they enjoy the logical independence necessary for a genuinely causal relationship, my beliefs and desires, on the one hand, and my basic actions, on the other, can hardly be supposed to be connected in a way that no nonpsychological phenomena could disrupt. At any moment a cosmic ray might strike me, passing through my neurons and energizing a particular synapse, thereby precipitating a discharge that would not otherwise have occurred. This discharge might inhibit neural firings that would otherwise have instantiated in me the "normal" psychological connection between my current beliefs and desires and my intentions to act. I would act in a way that made little psychological sense, and yet the cosmic source of perturbation would never come within the predictive grasp of psychology. Thus the "normal" psychological connection will never be formulated as a strictly deterministic law (or law with perfectly precise single-case probabilities) within the vocabulary of psychology alone.

This seems unquestionably right, but it is also characteristic of every area of inquiry that does not purport to have the full generality of fundamental physics. If the kind of nomological understanding available in psychology is no more compromised than that afforded in biology, or chemistry, or geophysics, then little has been done to unsettle the nomothetic ideal for the social sciences.

Let us try, then, to locate a less generic way of understanding what might be problematic about "psychological systems" as a locus for nomological development. One possibility with which Davidson might be sympathetic turns on a feature that recently has been prominent in arguments concerning the status of psychology: multiple realizability. Certain psychological kinds are or appear to be realizable by different physical systems, perhaps as different as those studied by organic chemistry and condensed-matter physics. Various philosophers have argued that the possibility of multiple realization makes it unlikely that we can find strict nomological regularities either within psychology or connecting psychology to physiology.

But why should this be so? It seems likely that the most plausible account of mental states will characterize them *at least in part* functionally. To have a belief is, among other things, to have certain tendencies to avow, to notice, to infer, to act, to be surprised, etc. No physical state could be a suitable candidate to realize such a mental state unless it by and large "makes true" this functional characterization in the agent in question — an agent in that physical state must have the relevant tendencies. It therefore follows that, whatever variability there might be at the physical level due to multiple realization, possessors of given mental states must exhibit certain regularities in behavior. We can have confidence that such regularities will obtain among possessors of mental states not because we can establish strict laws relating psychological types to physical types, but rather because any individual to whom we could successfully attribute any given mental state would have to be disposed to behave in such a way as to satisfy the truisms about what it would be to realize such a state.[5] And though these behavioral dispositions secure truisms, there is nothing unempirical about them. For even if it were analytic that a possessor of a belief that p is normally disposed to assent to p, it would remain synthetic whether any given individual as such — i.e., you yesterday or me today — is actually so disposed.

A reply to this line of thought is readily available, however. The functional characterizations constitutive of belief and desire are famously loose tendency statements, not taut enough to provide the material for serious nomic theorizing. Exceptions are tolerable — a believer that p need not be disposed always (or with a certain definite probability) to assent to p, even in favorable settings. And since exceptions — or "unfavorableness" of setting — might be attributable to variations in the underlying physical state, it would be quite beyond the vocabulary of psychology to detail them and thereby sharpen and strengthen the theory. So despite assurances arising from the functional characterization of mental states, psychological theory would still have little prospect of approaching an empirical, nomic theory that exhibited the sort of strict, serious character that Davidson deems necessary to the formulation of genuine scientific laws.

What are we to make of this counterargument? One reason for thinking that this is not the kind of argument needed here is that the question before us doesn't really concern a possible "universal psychology"; rather, the issues appear to arise concerning a nomothetic psychology for *Homo sapiens*. Of course, there might be sufficient neuropsychological variation among human beings—consistent with the limits set by the nonfunctional as well as functional elements in the characterizations of the various mental states, to prevent any improvement of purely psychological "laws" beyond the loose tendency statements of folk psychology. In psychopharmacology, for example, one encounters situations in which a drug that is strikingly effective for some individuals suffering from a given disorder is seemingly ineffective or even counterproductive for others. Such situations generate real worries among psychobiologists about identifying disorders with physiological conditions. Given certain pharmacological evidence, should we distinguish two or more disorders when mentalistic psychology has found only one?

Were there no strategy for handling such difficulties, this possible variability might threaten the prospects for improving the nomic claims of psychology.[6] But there does appear to be a strategy for making the relevant nomic claims sharper, less exception-prone, and more plausibly applicable to individual cases. For example, differences in the effectiveness of certain medications can sometimes be given a unified physiological explanation: the onset of puberty alters hormone secretion, some individuals are chronically deficient in certain enzymes, and so on. In other cases, still finer-grained physiological distinctions are called for, or we might conclude that the mechanisms involved could embody elements of randomness.

Consider how things work in the many other areas of science where we also find phenomena at least partially functionally characterized. A characterization of what it is to be *gas*, for example, from *Van Nostrand Scientific Encyclopedia*, includes functional as well as substantive characteristics:

Gas. A state of matter, in which the molecules move freely and consequently the entire mass tends to expand indefinitely, occupying the total volume of any vessel into which it is introduced. Gases follow, within a considerable degree of fidelity, certain laws relating their conditions of pressure, volume, and temperature. Gases mix freely with each other, and they can be liquified. (1976, 1136)

The property of being a gas, thus characterized, is a role that can be realized by an array of physically different microsystems, all of which exhibit certain behaviors, including obedience "within a considerable degree of fidelity" to the ideal gas law

$$pv = nRT.$$

Any microsystem that failed even to approximate this law would fail to be a gas in the intended sense. Does this approximateness, due to "nonideal" physical features of the particular microsystems that play the role of being a gas, pose an obstacle to the development of serious laws for the macroscopic behavior of gases?

Chemists, who are interested in developing serious laws of this kind, use a class of equations called "characteristic equations" in order to achieve greater accuracy. The best-known characteristic equation is van der Waals equation,

$$(p + n^2a/v^2)(v - nb) = nRT$$

which contains two additional parameters, one a pressure correction term and the other a volume correction term. These parameters would be zero for a genuinely ideal gas, small for gases that are close approximations of an ideal gas (such as helium or hydrogen), and large for gases with larger molecules that approximate an ideal gas less well (such as nitrogen or ammonia). The Clausius equation further corrects the ideal gas law:

$$(p + n^2a/T[v + c]^2)(v - nb) = nRT.$$

It includes a further parameter and makes the pressure correction term introduced by the van der Waals equation temperature-dependent. As before, the values of the correction terms depend upon the particular gas in question. More elaborate characteristic equations carry the process of refinement yet further. Perhaps — if there are significant dimensions of physiological variation among those human beings sufficiently alike functionally to satisfy the truisms about belief and desire — a similar strategy might be pursued to refine psychological generalizations. Multiple physical realizability therefore need not in itself bar serious nomologicality.[7]

On the strength of what we have considered so far, nothing would distinguish psychological properties and roles from the many other (at least partially) functionally characterized properties or roles that exhibit multiple realizability scattered throughout the sciences: lever, heat sink, harmonic oscillator, gene, elastic body, and so on. How, for example, would we get a reasonably accurate theory of the behavior of solid-state computers? Presumably by placing side-by-side in our generalizations elements drawn from the mathematical vocabulary of "algorithm" and "addition," the computer-science vocabulary of "input" and "command," the electrical-engineering vocabulary of "register" and "or-circuit," and the physical vocabularies of "capacitor" and "resistor" on the one hand, and "silicon" and "temperature" on the other. Thus: "A silicon transistor circuit of this type will, given a digital input in the proper form, compute the arithmetic sum, at least as long as the line voltage remains within the range ca. 105–115 V at 50–60 Hz and the temperature of the circuit board remains within the range ca. 5–35 degrees Celsius."

Increasing the nomic force of the theory of gases or of computing machines by developing "characteristic equations" or elaborating "boundary conditions" leads to what Davidson has called *heteronomic* generalizations — generalizations drawing from the vocabulary of more than one domain. We find in Davidson's work the suggestion that heteronomologicality in itself is problematic from the standpoint of nomic development. That would count against the sorts of "mixed" strategies just mentioned for improving nomic force.

Davidson provides several examples in which it does seem problematic that a purported law invokes disparate kinds. In particular, he considers the idea that we might know a priori that certain predicates are "not suited to one another" — "grue" and "emerald," for example (ME, 218). Even granting this example, however, it would seem that other examples of "mixed" principles, e.g., in the prediction and explanation of the behavior of thermodynamic systems or electronic computers, do not exhibit such failures of fit. It would seem, then, that the difficulty for developing "serious" laws in any given case is not *the presence of disparate kinds*, but the *kind of disparateness present*. Have we been given a reason for thinking that we can know a priori that the predicates of psychology and those of physiology in particular are among those "not suited to one another"? Not yet, but the fourth consideration purports to provide just such a reason.

V

4. *Disparate constitutive principles*. Davidson has argued compellingly that belief-desire psychology, on the one hand, and the parts of physiol-

ogy or physics that would be imagined to figure in "characteristic equations" or "boundary conditions" for the mental, on the other, are governed by different constitutive principles. In this sense, then, we can see that they are "not suited to one another."

The notion of a *constitutive principle* could itself stand greater scrutiny and elucidation, but I will assume that we know roughly what is intended: an axiom, rule, definition, or even hypothesis which is so central to the content, character, or method of a domain that is applicability is a presupposition of any inquiry into that domain. To give it up would be, all else equal, to change the subject. This way of characterizing constitutive principles gives them an a priori flavor, but it should be kept in mind that they need not be "analytic" — at least if that means true-by-meaning. For some of these principles are substantive or prescriptive — synthetic a priori, one might say (compare ME, 216).

Let us return to psychology. First, three reminders.

1. Recall that we are interested in ways in which, in principle, psychological theory might strengthen its nomological character by developing generalizations framed in terms drawn from physiological and physical as well as psychological vocabularies. Such "mixed" nomological development might enhance our capacity to carry out controlled testing, increase our diagnostic and therapeutic capacities, and generally improve empirical adequacy and explanatory power. Davidson has famously argued against nomic psychophysical laws of reduction, and some of the arguments he had made on that score will enter our discussion below. But it should be re-emphasized that we are looking for *improved* nomic generalizations, not necessarily *reductive* ones, and so a number of issues emerge in a different light.

2. Recall that we are interested in a *seriously* nomothetic psychology, but not an *absurdly* serious one. To be seriously nomothetic it hardly seems necessary to aspire to perfectly strict laws, either deterministic or probabilistic. The process of improving the explanatory informativeness of psychological generalizations by discovering increasingly fine-grained and homogeneous reference classes for tendency statements, or by becoming more explicit about boundary conditions, need not be thought of as a process that can go on in any practicable sense without limit. It would seem sufficient if we could arrive at nontrivial and robust relations concerning relatively homogeneous classes of individuals and involving fairly stable probabilities significantly different from chance.

Perhaps Davidson would not dispute this in the end. For him the issue might simply be: Can psychological tendency statements, pure or "mixed," reasonably be expected to be perfectly strict? The short answer is that perfect strictness seems an unreasonable expectation. I would be delighted not to have a deep disagreement to explain, but things do not appear to be so simple.[8]

3. Recall finally that we are interested in the question of possible *in-principle* barriers to the development of psychological theory. The question of what might reasonably be attainable in psychological practice would call for a more empirical discussion than that undertaken here.

Let us return to the attempt to explain why Davidson views "mixed" generalizations involving both psychological and physical predicates as nomologically problematic. The rationale given is that psychological and physical predicates are "not suited to one another," owing to a difference in the constitutive principles of these domains. Now a difference in constitutive principle might indeed lead us to conclude that two classes of concepts are unfit to cohabit in serious nomic generalizations. But, as Davidson notes, a mere difference in constitutive principles will not suffice for this.

Chemists and physicists, for example, have their different "cultures" partly because chemistry has constitutive principles that differ from those of basic physics. To view a system chemically draws one's attention to a distinctive range of similarities and differences that cut up the world in unique ways. Consider some of the characteristic notions of chemistry: acid, solvent, exothermic, surface tension, catalyst, etc. Many of these notions are at least partially functional: they group phenomena together in terms of similarities in dispositions to behave despite fairly significant underlying differences in physical composition.

Acidity affords an example. The chemical property "acid" groups together a wide range of physically distinct substances — complex and simple molecules, organic and inorganic compounds — on the basis of certain similarities in chemical behavior. Chemical theory offers several physical "definitions" of acidity, meant to give a unifying account of what substances play the role of acids. It is the function of such physical definitions to explain, for example, why HCl and $CH_3COOC_6H_4COOH$ are both acids, why some acids are stronger than others, why acids and bases neutralize one another, and so on. Partly because our notion of an acid has functional as well as substantive aspects, chemists work with a variety of physical definitions depending upon the aspects of acidity made salient in a given context. At least four different physical definitions of acidity are in use — Arrhenius-Ostwald, Bronsted-Lowry, Lewis, Usanovich — each with a somewhat different extension and none entirely satisfactory.[9]

Not only do some of the principal terms of similarity and dissimilarity differ between chemistry and physics, but the same term may be interpreted in systematically different ways. A slight but telling example is given by Kuhn (1970, 50–51), in reporting a researcher's discovery that an eminent chemist and an eminent physicist each answered without hesitation, but in the opposite way, the question of whether a single atom of helium is a molecule. Characteristically, the chemist took a thermodynamic consideration as determinative of moleculehood, while the physicist took the emission spectrum as determinative.

Have we therefore reason to say that chemistry and physics are "not suited to one another," so that we cannot expect that the nomic status of chemical generalizations could be improved by judicious use of physical variables? No doubt this would strike us as a rather aprioristic or formalistic line of argument concerning a question that would be much better illuminated by study of the actual history of the two disciplines. Of course, there might not be neat physical categories answering to every important category in chemistry. For example, there does not appear to be a physical category that exactly matches chemistry's "acid." As a result, chemistry textbooks continue to introduce multiple physical definitions, sometimes speaking not of "acid" or "base" *tout simple*, but "Lewis acid" or "Bronsted-Lowry conjugate base," etc. And yet the predictive and explanatory power of the chemical theory of acids and bases has been enhanced, not eliminated, with the help of physics. Mere difference in constitutive principles, or cross-cutting categories, even irreducibility, do not seem enough to undermine the nomothetic ideal in chemistry.

Moreover, it is the *differences* in constitutive principle between physics and chemistry that afford a rationale for bothering to improve chemical theorizing rather than seeking to replace its scheme of similarities. For if the substances grouped together chemically as acids display a fairly wide and intercontextually robust set of interrelated similarities in behavior, then we can gain a distinctive kind of understanding of the world by invoking them even if "acid" does not have an unequivocal counterpart at the level of physics. Acid in the environment, of physically different kinds, present in naturally-occurring substances in the soil, in biological by-products or in "acid rain" can have similar effects, such as contributing to erosion or loss of biodiversity. Nonetheless, it might also be true that our theorizing about erosion or ecological effects could be improved by attending to physical differences among acids.

Mere difference in constitutive principle does not, then, seem to rule out possible refinement in pursuit of a nomothetic ideal by means of "mixed laws." As we discovered in the case of heteronomologicality itself, what seems to matter is the *specific character* of differences in constitutive principle involved, not their bare existence.

One long-influential idea about the specific character of the difference involved has emerged again in a new form, in Davidson's discussions. This idea has had many forms, which share the central claim that the constitutive principles of social science involve an irreducible *normativity*. This normative element differs from methodological norms we find throughout the sciences — use of controlled experiments, goal of predictive accuracy, honesty in reporting findings, and so on. Here is how Davidson describes it:

In our need to make [an agent] make sense, we will try for a theory that finds him consistent, a believer of truths, and a lover of the good (all by our own lights, it goes without saying). (PP, 222)

Such a "principle of charity" is thought to constitute partially the subject matter of social science and to have "no echo" in natural science. For despite the efforts of the rationalists, we no longer recognize any comparable, rationalizing norm of explanation in the natural sciences. First allegiance in natural science, it is said, is to descriptive adequacy, even when this has required changing or abandoning long-held beliefs about which sorts of natural phenomena are — or are not — rationally intelligible.

Perhaps natural science in the modern period has been dominated by concerns of prediction and control, but for the social sciences to reject the *interpretive stance* in favor of these concerns would constitute a radical shift in subject. It would mimic, in an exaggerated way, our response to certain profoundly psychologically disturbed individuals. Such individuals are sometimes more predictable than comprehensible: their behavior is highly routinized, with occasional bouts of violence, and their words are heard by us more as indicators of mood than as communicators of sense. In effect, we adopt toward such individuals an "objective" rather than subjective attitude. This seems inhuman because it replaces our ordinary norms of interpersonal conduct with a "coping strategy." We respond to their utterances and gestures in an effort to influence mood or behavior, not in order to reach shared ideas or information. In this denial of personhood, the ways we talk about how to act are not much different from the ways an expert in animal behavior might discuss advisable conduct around potentially dangerous animals: "Use a reassuring tone of voice. Try to follow familiar routines and to maintain steady eye contact. Back off gently if you hear an edge to his voice."

It would appear that for social scientists to abandon the interpretive stance in favor of an undiluted concern for prediction and control of the human organism would be to change the subject. However, something fishy about this line of thought is signaled by the introduction of the psychologically disturbed individual in place of a typical human subject. My impression is that social science has experimented pretty thoroughly with an approach — behaviorism — which transplanted the "animal model" into human psychology. The widespread abandonment of this approach — except for those areas of psychology which deal with exceptionally disturbed individuals or with various kinds of "nonrational" (e.g., sleep disorders) or "irrational" (e.g., phobic) conduct — has been driven by its notable failure to improve the predictive and manipulative adequacy of psychology. Explaining various reliably-reproducible behavioral phenomena (e.g., dissonance) required constituting the individual as a subjectivity, an agent with a point of view whose conduct can be understood only by including his own interpretation of his situation and behavior. Even behaviorism itself, as practiced on human subjects, deployed such basic notions as sameness of stimulus, locus of reinforcement, and assent based on imputing an interpretation to the agent (see Quine 1960, 1968). Therefore, we should be prepared to see an interest in prediction and control as allied to, rather than necessarily antagonistic with, the interpretive stance. After all, as cognitive psychology has recently emphasized, human beings themselves operate on the basis of interpretations, schemata, and assignments of meaning.

At the same time, it is not clear that the interpretive stance means we must see others as "believer[s] of truths, and lover[s] of the good." Interpretive charity, especially as applied across social and historical divides, often requires that we see others as possessing a great many false (but, in their context, often justified) beliefs, and significantly different (but, in their context, explicable) values. And interpretive charity, even as applied to ourselves, is tolerant — it has enough flexibility to bend the norms of rational decision without breaking. There remains in this notion of charity genuinely normative elements, for it involves seeing others as reasoning beings who exhibit certain kinds of coherence, who treat experience (in at least some contexts and some respects!) as evidence, and who infer from evidence to conclusion and from conclusion to action. We therefore see their attitudes and behaviors not perhaps as *rational* in the full-blooded sense, but as possessing *rationales*. Given what they have to work with, we can make a kind of sense of what they believe and desire, and why, even though we might disagree heartily with what they conclude or do.

This sort of rationalizing coherence is clearly not what we seek in rendering the behavior of elementary particles or planets scientifically intelligible — at least, not in mainstream contemporary physics. So the methods we employ when we interpret others as subjectivities differ in this fundamental respect from our methods when we predict or explain physical systems. But none of this precludes in principle more or less extensive collusion among these norms of inquiry. Not only does successful prediction and control of systems akin to human agents seem to call for interpretation, as noted above, but interpretation itself takes an interest in prediction and control. Many interpretations are possible for any given bit of behavior, and we appeal — in both "psychological science" and everyday practice — to success at prediction and control in helping to decide among them. Do individuals employ a "matching" strategy in choices because of the slight reward that comes from variability and from the satisfaction of predicting a rare event? Then increasing the rewards for correct answers and the costs for incorrect answers should induce a change toward a "purer" strategy (see Siegel, Siegel, and Andrews, 1964).

Moreover, it is important that we not think of the differences in constitutive principle between psychology and physics, for example, as a divide between a domain with constitutive norms and one not so constituted. Physical science has its own norms of intelligibility and explicability. Consider the historical example of the "principle of sufficient reason," and the role it had in setting a Laplacian ideal for physical theory. Consider, too, more recent notions of intelligibility and explicability — operationalism, symmetry requirements, and so on — and, indeed, consider the very norm of nomotheticity itself.

There are, then, constitutive principles of intelligibility on both sides of the distinction usually drawn between the science of the mental and the science of the physical. If physical kinds are not "made for" mental kinds, that is not because norms function constitutively on one side but not the other. The incompatibility must stem rather from a profound difference in constitutive norms, a difference that would preclude or inhibit the development of tight connections, even if occasional collusion is possible. Davidson suggests something like this:

> There are no strict psychophysical laws because of the disparate commitments of the mental and physical schemes. It is a feature of physical reality that physical change can be explained by laws that connect it with other changes and conditions physically described. It is a feature of the mental that the attribution of mental phenomena must be responsible to the background of reasons, beliefs, and intentions of the individual. There cannot be tight connections between the realms if each is to retain allegiance to its proper source of evidence.... We must conclude, I think, that nomological slack between the mental and the physical is essential as long as we conceive of man as a rational animal. (PP, 222–3)

But is a mere potential for conflict ground enough for dismissing the possibility of improving psychology's nomothetic prospects by developing "mixed" laws? After all, if one simply considers the various criteria of theory choice in the natural sciences — simplicity, generality, descriptive adequacy, informativeness, conservativism, etc. — it is clear that these have significant potential for pulling us in different directions. Yet scientists have not had to remove these possible conflicts, but only to evolve working equilibria among the criteria, in order to make nomothetic progress. Why might the different allegiances of the psychological vs. physical sciences be such that they could not co-participate fruitfully in an equilibrium-seeking process akin to this one?

To make vivid how "slack" (or is it "tension"?) between the allegiances of psychological vs. physical science might be thought to preclude the normal operation of such reflective equilibrium, let us briefly discuss a possible parallel.

Consider a species of natural theology, meant to be responsive both to scripture and to going empirical science, and therefore (we will suppose) possessing at least two potentially distinct norms of intelligibility — divine plan and nomic expectability. Natural theologians of this stripe are committed to seeing the world both as governed by empirical law and as rationally chosen. Let us develop a bit of made-to-order history, first concerning the course of "natural history," then concerning the demands of natural theology.

"Nature does not make leaps" was a regulative principle of intelligibility in the domain of natural history in the early nineteenth century, leading to a static view of geology and speciation. Discoveries of fossils of unfamiliar-looking organisms, and fossils of familiar-looking organisms distributed in unfamiliar places, along with other evidence of catastrophic geological and climatic change, broke the hold of this long-standing regulative principle. (Recall, this is made-to-order history!) Faced with a conflict among the principle "Nature does not make leaps," prevailing ways of treating certain kinds of evidence, and the central norm of predictability (or retrodictability), natural historians eventually moved away from — alternatively, one might say, very substantially reinterpreted — the "no leaps" principle.

Now consider the case of natural theology. Suppose that for natural theologians allegiance to the principle that "Nature does not make leaps" were tightly tied to deeply held and essentially non-negotiable religious doctrine. Discontinuities and novelties in the chain of being would be evidence of lack of divine foresight or power, and such faults would be unthinkable in God. Then we would have a regulative principle in natural theology that functioned differently from the regulative principle of natural history just discussed, and an example of a conflict arising from differential "allegiance to [their] proper source[s] of evidence" when the fossil reports reach London. The nature, source, and force of the theological regulative principle would simply be autonomous from the nature, source, and force of other regulative principles regarding the treatment of observation. We can see how a meliorist conception of sharpening and improving the generalizations of natural theology by the introduction of conditions couched in the language of empirical science would be out of place. To be sure, mere autonomy is not enough — the case must also be one in which the autonomous regulative principles *both* are in some sense non-negotiable (which may only mean that were they to be given up, we would view it as the subject changing) and both lay claim to the same explanatory domain. One can imagine two ways in which such a conflict might play itself out within natural theology.

According to the first script, theological doctrine might come to be unfalsifiable simply by excusing itself from the task of explaining natural phenomena — "Render unto physics what is physics'." No doubt this is the sort of reconciliation many religious scientists have made. Here autonomous *explanatory* contribution is foregone, or kicked upstairs like a superannuated executive to some "higher" level where it will do no harm. The introduction of physical "corrections" in the end really leads to *replacement* of the theological component as a functioning part of our going theory of the physical world, not to improvement of its nomic status within that theory.

But there is another possible script with a different ending. The theological core might instead operate in a more intransigent way, refusing to yield any explanatory turf. As undiluted natural theologians we would insist on making all adjustments in response to new evidence in favor of preserving the core at the expense of views about the relevance or import of observation at the periphery. Here, too, we would have a relatively clear case in which what happens is not an equilibrium process that improves the nomic status of natural-theological generalizations. Rather, all naturalistic ambition is lost as the theory ceases to possess any real adequacy to observation (perhaps this is the trajectory of creationism).

How would we make the comparison with the case of psychology? We could pit the purportedly autonomous normative conception of rendering action intelligible embodied in rational decision theory — "conceiv[ing] of man as a rational animal" — against the predictive/explanatory norms of intelligibility of physiology or physical science. Suppose that observations were made that seemed to challenge our rationalizing ways of understanding one another — not in the case of particular acts or an isolated individual or sect, but pervasively for that substantial fragment of humankind with which we are on otherwise "normal" terms, our actual and inevitable peers. It could then be claimed that *either* we would not allow ourselves to give up the core of rational decision theory — and so would always stand ready to adjust our interpretation of the evidence to protect it (this is the "intransigent" script) — *or* we would allow ourselves to modify the core of rational decision theory, but to that extent abandon the attempt to render human behavior intelligible *as rational action* (the "atheological" script). In the first case we would have no psychological *science*; in the second case we would have no *psyche*.

There seems to be no doubt in Davidson's mind as to which choice we would make. The second case would involve abandoning not only the core of rational decision theory, but our very conception of ourselves as persons and moral agents, subject to moral and rational evaluation.[10] And, in his view, we take the norms of rational decision theory to be regulative in our deliberation, and constitutive of our very notion of personhood. If we somehow were forced to see our scheme for rationalizing explanation as inadequate for descriptive/explanatory purposes, this would not really destroy our normative conception of what it makes sense for ourselves, or others, to do. Rather, it would lead us to drop any descriptive/explanatory pretensions made on behalf of decision theory, treating it instead as a normative scheme that at best allows us to put a handy label on causes of action (e.g., "Jason's reason for going to L.A."), but which affords no real insight into the nomological structure of things.

But why the forced choice? Why not say that, in the face of recalcitrant evidence, we would modify our conception of rationality more or less significantly so as to keep alive both normative and explanatory ambitions for our theory of rational action?

The analogy with natural theology is supposed to help here — given the divergent allegiances of two domains of discourse, there might be no *via media*. But let me hazard two observations. First, it is difficult to believe that there is so rich and definite an a priori constitution of our notions of personhood and moral agency as this argument assumes. Rather than speak only of the "slack" *between* the mental and the physical ways of thinking, I would also insist upon the slack *within* these two ways of thought. Just as physical thinking learned to live without what rationalists viewed as its central constitutive principle, the principle of sufficient cause, I suspect that mentalistic psychology could learn to live without any particular element of rational decision theory. The list of synthetic a priori necessities that have gone by the board in the course of the development of inquiry is impressive. To mention just a few: in physics, not only the principle of sufficient cause, but also teleological causation and the plenum; in geometry, the fifth postulate of Euclid; in biology, final causation and *non saltus*; and in some corners of logic and mathematics, the excluded middle. Is the independence axiom of rational choice theory, for example, any more central or indispensable than these? Have we seen no evolution in our thinking about what rational agency involves since Plato (with his bifurcation of reason and desire), or perhaps even Hobbes (with his surprising views about rational deliberation and risk-assessment), suggesting that contemporary rational choice theory represents but one form such thinking could take? One would want a much stronger argument than any I have seen before concluding that rational decision theory is in its present form entirely non-negotiable or free of internal tensions that might be resolved in various ways. (Cf. the "paradoxes" of rational choice theory.)

Rather than carry on further in this impressionistic vein, I will mention a second observation. This, too, concerns a way in which the argument — or the natural-theological analogy I have offered on its behalf — fails accurately to represent what *is* constitutive of the domain of imputations of rational agency. One could put the point in terms of autonomy. It is central to our *normative* as well as explanatory uses of rationality that there be a rather tight connection to our ideas about the prediction and explanation of behavior. These normative uses therefore must be responsive to our best thinking about the causes of human conduct.

How do we evaluate our attributions of beliefs and desires? What role do they play in the regulation of our conduct, and to what interests or concerns or constraints is this role responsive? We form our evaluations not in an autonomous "evaluative domain," but as part of our effort to formulate plans and take actions that meet our needs and satisfy our interests in the world. This involves expectations (predictions) and attributions of responsibility (explanations), and decisions about actions (e.g., praising or blaming) linked to both. All of this makes sense only if there are systematic, nonaccidental connections between this evaluative enterprise and the causes and consequences of actual behavior. It is hardly enough for this enterprise to be appropriate in our own eyes that we think our mentalistic categories might just happen to enable us to refer to causes of behavior; we must also think that the causal structure these categories impute reflect in some systematic way the causal structure of action. Talk of wood sprites and magic spells might, for example, have picked out a number of genuine causal events in forest ecology — the drying up of springs causes the death of certain flora and fauna even if it is not itself due to enchantment. But our practices of praise and blame, of childrearing and education, of holding people responsible, and so on, need more than this sort of sporadic, anomic referential success in order to make sense.

One bit of evidence for the strong linkage between mental attribution and the descriptive-explanatory ambitions of natural science is the widespread belief that the mental supervenes upon the physical. Indeed, whatever their beliefs about reduction, contemporary theorists about mind share some hefty assumptions not only about the strength of this relationship, but also about (roughly) *which parts* of nonmental reality the mental supervenes upon. By contrast, supervenience is not assumed in our discourse about gods and wood sprites.

A more compelling bit of evidence that there is a constitutive linkage between what are thought of as the normative realm of the intensional and the descriptive/explanatory realm of the causal is the principle that "*ought* implies *can*." Here we have an explicit, seemingly a priori tethering of normative ambition to causal possibility. An example might help.

Consider the recently awakened interest in virtue ethics — in moving away from emphasis upon moral duty and toward a central focus on developing traits of character that will lead to good actions across a wide range of circumstances. The virtue theorist typically believes that possession of such traits of character will also promote the quality of life of the agent. Thus, instead of harping on the duty to help others and the discipline needed to enforce this duty upon oneself in the face of one's less generous impulses, the virtue theorist encourages cultivation of a generous character, which would make it a "natural" and even rewarding part of one's life to be reliably responsive to the needs of others.

An attractive prospect. But consider now recent work in personality theory and cognitive social psychology which casts into doubt the existence of the sorts of stable, cross-situational behavioral dispositions that virtue theory assumes (see Ross and Nisbett, 1991). Except for a few dimensions, individuals (it is claimed) display much greater similarity in their responses to situational cues than the "character trait" view imagines. Virtually everyone can be led by certain rather trivial cues (such as being late) to behave ungenerously, and virtually everyone can be led by equally trivial cues to behave generously. Situational effects largely swamp personality variables in these experiments.

If this view of the actual and possible predictive/explanatory impotence of traits of character could be sustained empirically, it would constitute a powerful argument against virtue ethics. It has been the aim of virtue ethics to get at the springs of behavior; should the notion of stable virtues fail to illuminate the actual and potential causes of action, then virtue theory's *normative* standing would be undermined. " 'Ought' implies 'can' " is not merely a principle regulating the attribution of individual duties, but a generic constraint on practicability.

These challenges to the notion of virtue from psychological theory remain controversial empirically. However, it seems to many philosophers as if they *must* be wrong: it simply is obvious from personal experience that people differ in character in just the ways virtue ethics presupposes. But part of the theoretical force of contemporary cognitive social psychology is precisely in explaining why constructs like "character trait" have proven so persistent a feature of folk psychology — and, by implication, of the folk psychology upon which most moral philosophers rely — even when they do not stand up to experimental test. There is an elaborate story to tell about "attribution" and its errors, based on an extensive experimental literature (for a summary, see Nisbett and Ross, 1980).

Just as it is part of our first-order norms for imputation of moral responsibility that we take into account available causal/explanatory evidence concerning why an individual has acted as she did (e.g., was she under the influence of a drug?), so is it part of our higher-order norms for shaping our ideas of belief, desire, rationality, responsibility, and so on, that they be responsive to available causal/explanatory theories of behavior (e.g., which, if any, cross-situational constancies in behavior do individuals display?). This is manifestly the case in the application and evolution of civil and criminal law, where the prospect of punishment leads us to impose upon ourselves an obligation to be attentive to causal evidence. Whatever *autonomy* rationalizing explanation is said to have, this should not be confused with the *independence* of such explanation from sources of causal evidence based outside its own sphere. Indeed, the requirement of responsiveness to "outside" causal evidence is a regulative maxim that our practice of rationalizing gives unto itself, a maxim given for its own proper purposes and respectful of its own proper principles of assessing action as right or reasonable. The manner in which changes in our normative conceptions have "kept pace with" changes in our empirical, explanatory/predictive theories of behavior is, then, no accident or mere coincidence. It reflects our deepest commitments in the complementary tasks of explaining and criticizing actions. (It is, for example, natural that certain radical ways of rethinking the causal structure of action, such as depth-psychology, have led their proponents to some radical rethinking of the norms of childrearing, and of attributions of responsibility and guilt, as well.) Once we ask how a principle of charity ought plausibly to be formulated — what sorts of beliefs and desires it is (in some suitably broad sense) reasonable to attribute to others in light of their behavior and circumstances (and even physiologies) — we can see that this empirical tether is anchored in the heart of interpretivism.

Might not this very tether sink belief-desire psychology? After all, belief-desire psychology began life as a folk theory, and folk theories in other domains of science have often been heavily revised or even abandoned. Virtue theory in particular might be sunk by its tether to an untenable personality theory. However, virtue theory is but one family of views within an array of possible conceptions of human agency. It is much more difficult to convince oneself that psychological research might undermine this entire array, or that we even have a clear idea of what this might be like. It is, in any case, too early to tell what a better-established empirical psychology might suggest. And one thought can be registered even at this preliminary stage in the evolution of psychological theory: Isn't there some reason to think that human beings have been on more intimate and sustained terms with the sorts of experiences, practices, and interests that could shape the evolution of a workable basic scheme for the typical causes of human action than we are for the fundamental laws and elementary constituents of matter?[11] If so, then much of the basic belief-desire framework of folk theory about the psychological can be expected to find vindication, rather than refutation or abandonment, in a more developed psychology.[12]

VI

5. *Individual laws.* I have the uneasy sense that Davidson — or at any rate a Davidsonian — might agree with so much of what I have said, that I cannot have captured his argument concerning the possibility of a nomothetic psychology. Davidson does after all write in "Hempel on Explaining Action" that we *could* treat rational decision theory as empirical, though he also holds that we *will not* do this, out of a commitment to seeing ourselves a certain way. We will tinker with the connections linking evidence to theory rather than allow ourselves to give up the theory. But in that essay he also writes that he does not think of rational decision theory as analytic, or necessarily true — his point is "sceptical" and negative: We simply don't know what it would

be like to find this theory false (HEA, 272–3. Although he also speaks of decision theory as "constitutive of concepts assumed by further satisfactory theory"). One might want to disagree with even this more modest and negative position. Perhaps recent work in cognitive psychology has given us some idea, still limited, of what it might be like to find various elements of orthodox rational decision theory descriptively false in a fairly systematic way.[13]

I should not close without mentioning what appears to be another consideration Davidson raises concerning nomothetic psychology, especially since this consideration is self-consciously presented by him as indicating the real difference between his view of rational agent explanation and the deductive-nomological conception advocated by Hempel. He focuses on the question of what explanatory contribution is made by decision theory, over and above the explanation afforded by individual attributions of beliefs and desires:

I think I have an argument to show that the main empirical thrust of an explanation of an action in decision theory, or of a reason explanation, does not come from the axioms of decision theory, or from an "assumption of rationality", but rather from the attributions of desires, preferences, or beliefs. (HEA, 273)

The argument here seems different from arguments thus far canvassed, which had emphasized the constitutive role of rational decision theory in the attribution of desires, preferences, and beliefs. That emphasis makes it difficult to see how one might distill the explanatory contribution made by individual attribution from that made by general theory—they would be holistically bound together.[14] But instead of explicitly changing theoretical emphasis, Davidson takes us back to ground level, asking us to reflect on an example.

Consider this purported explanation of a piece of (then-President) Gerald Ford's political conduct:

(RA) Ford compromised on the energy bill because he wanted to curry favor with the voters.

For Davidson, this is a paradigm reason explanation. Hempel would have us understand it as a "sketch" of a properly deductive-nomological rational-agent account that would look something like this:

(RA′) L   Any rational agent in circumstances C1, . . . , Cn—which include wanting to curry favor with the voters, believing the energy bill to be controversial, etc.—will seek to compromise on the energy bill.

F1   Ford was a rational agent at *t*.
F2   Ford was in circumstances C1, . . . , Cn at t.

E   Ford sought to compromise on the energy bill at *t*. (Compare Hempel 1965.)

(RA') is meant to bring to the surface the nomological principle "tacitly asserted" by (RA), L, and to show how L is brought deductively to bear on the explanandum. But Davidson asks:

What conditional predictions follow from a reason explanation [like (RA)] . . . [?] If the assumption of rationality is the important empirical law [as in (RA')], then this explanation of Ford's action should tell us what (conditionally) to expect of all mankind. But if I am right, it mainly tells us what (conditionally) to expect of Gerald Ford. (HEA, 273–4)

Further:

The line I have been developing suggests that the laws implicit in reason explanations are simply the generalizations implied by attributions of dispositions. . . . [T]he relevant generalizations are just the ones that express dispositions like wanting to curry favor with the voters, or believing that compromising on the energy bill would make the voters like him. (Ibid., 265, 273–4)

This seems plausible, though at first glance it is puzzling how we could offer such dispositions as part of a rational agent explanation of Ford's action unless we see (RA) as presupposing that Ford at the time was — at least in this respect — also disposed to act upon his beliefs and desires in the way rational individuals characteristically do (see Hempel 1965). An emotionally overwrought or otherwise irrational politician who wanted to curry favor with voters just as much as Ford might nonetheless fail to act as a rational person with this desire would, and might miss the opportunity to compromise on the energy bill.

Even so, Davidson believes that Hempel's (RA') does not yield the proper understanding of the nomological assumption that does lie behind (RA):

If the assumption of rationality (or decision theory) had the empirical force that Hempel's account requires, the explanation of Ford's action would be like an explanation of why a particular small cube dissolved in warm coffee, an explanation that takes as its premise the facts that if something is soluble it dissolves in liquids of a certain sort, that warm coffee is such a liquid, that all sugar is soluble, and that this cube is sugar. This explanation implies something about all sugar and all coffee ("all mankind"); but there is a shorter and less informative explanation available. This cube was soluble, soluble things dissolve in coffee, this cube was in coffee. This explanation tells us nothing about all sugar-kind, though it tells us a lot about this cube. I think reason explanations are of the second kind. (HEA, 274)

So we have two different explanatory accounts not really in conflict with one another, but differing in informativeness. One conveys more explanatory information. It tells us of the existence of a kind, *sugar*, to which this cube belongs and which is by law soluble in certain circumstances — just as (RA′) tells us of the existence of a kind, *rational agents*, to which Ford belongs and who by law act thus-and-so when disposed to curry favor with the voters. The other is less informative. It tells us simply that *this lump* is soluble — just as (RA) tells us simply that *Ford* is disposed to curry favor with the voters. Even if we see (RA) as presupposing that Ford is not distracted, overwrought, or otherwise likely to ignore or work against his own goals, still, it might be said, we have not invoked his membership in a "rational kind" or even presupposed that there is such a kind or such laws as L governing it.

Let us grant for the sake of argument that in-the-street reason explanations "tacitly assert" no more than the less informative gloss of (RA). What would that tell us about the prospects for nomothetic psychology? Generally speaking, resorting to relatively less informative explanations does not preclude interest in, or success at, finding more informative ones. On the contrary: "I see, the cube was soluble. Was that because something special was done with it, or is it the kind of stuff that naturally dissolves in coffee?" Explanation is like salted peanuts, Quine once observed — *l'appetit vient en mangeant*. Davidson does not dispute this, and, like Hempel, bundles some further explanatory ambition into the very content of what is asserted in particular explanatory episodes. He even goes a step beyond Hempel in this respect, asserting that causal claims as well as laws are presupposed:

Solubility implies not only a generalization, but also the existence of a causal factor which accounts for the disposition: there is something about a soluble cube of sugar that causes it to dissolve under certain conditions. That this is so is not proved by the fact that it dissolves once. In the same way, no single action can prove that a disposition like a desire or belief exists; desires or beliefs, however short-lived, cannot be momentary, which is why we typically learn so much from knowing the beliefs and desires of an agent. This is the point, I suggest, where general knowledge of the nature of agents *is* important, general knowledge of how persistent various preferences and beliefs are apt to be, and what causes them to grow, alter and decay. Such knowledge is not used in giving reason explanations, but it is surely part of why reason explanations are so satisfying and informative. (Ibid.)

Davidson concludes his discussion by noting that his reflections actually "reinforce" Hempel's view that the "general logical character" of reason explanations is not different from that of explanations in physics. (Davidson rejects the older "autonomist" view that intensional explanation is unempirical or analytic.) The only dissent he registers from the Hempelian view is that "the laws that are implicit in reason explanations seem to me to concern only individuals" (ibid.). Presumably, this is meant to contrast claims like Hempel's L, "Any rational agent in the circumstances would . . . ," with claims like "Ford in the circumstances, if rational, would . . . ," where the latter is to be read as an "individual law" attributing a counterfactual-supporting disposition.[15]

Have we, then, located the nub of the dispute? Interestingly, Davidson remarks that he is not expressing skepticism about the existence of general "nomic wisdom concerning mankind and womankind," which in his view is part of the background that enables rational agent explanations like (RA) to be so informative (ibid.). His point is more restricted: Such general "nomic wisdom" does not "directly enter into reason explanations" even though it is part of how we fit individual dispositional explanations into a larger scheme of knowledge about "the nature of dispositions like desires and beliefs — particularly how long they are apt to last, and what causes them" (ibid., 274–75). The depth and significance of Davidson's disagreement with Hempel could be grasped only if one clearly understood what "direct entry" involves, and how it might contrast with Hempel's equally intuitive notion of "tacit assertion." But it seems unlikely that this disagreement could constitute the whole of the contrast Davidson wishes to draw between psychology and natural science. A dispute over how much of our "nomic wisdom" about rational agents is "tacitly asserted" in individual rational agent explanations and how much is part of a background necessary for understanding their full force, feels like a minor quarrel within the nomothetic camp, not a deep divide over the status of psychology.

Perhaps the real dispute is alluded to in the final sentence of Davidson's discussion of Hempel: "This [nomic] knowledge does not make our reason explanations any stronger, but it does make them more valuable by letting us fit them into a larger scheme." How could this larger scheme add value without strengthening explanation? The answer would appear to be that the broad "nomic wisdom" we possess about beliefs and desires does not, in Davidson's view, amount to a collection of serious laws that could underwrite or strengthen individual attributions or dispositions in the manner that serious laws in the natural sciences underwrite informal dispositional explanations. Perhaps we say of an individual cube that it is soluble. This is vague and only partly right—there are many conditions under which it would not dissolve. Serious science gives us a powerful scheme of categories ("the cube is made of sucrose") and laws ("sucrose is soluble in water under such-and-such conditions") that enables us to back up what we said about the individual cube and the coffee. But more general or higher-order psychology does not, in Davidson's view, afford such serious science—only somewhat grander and vaguer generalizations about tendencies, favorable circumstances, and so on, which cannot be much improved upon. These embody a practically valuable way of looking at one another and the world, but they do not strengthen more local generalizations about tendencies, favorable circumstances, etc.—they merely bring more and more such understanding into our purview. Not more strength, then, just more of the same.

This is not a view that receives further or independent defense from Davidson's contrast between his own and Hempel's account of rational agent explanation. It appears to be presupposed by this contrast, rather than validated by it. So I am inclined to think that we have not located an independent source of argument for this view in Davidson's remarks on Hempel.

VII

We now return in conclusion to consideration 1, the absence of serious-looking laws in belief-desire psychology. If, after inspection, we are not inclined to accept a Davidsonian explanation of this state of affairs, we must in fairness ask whether a better explanation is anywhere in the offing.

I suspect that a rather unexciting, unprincipled explanation is closer to the truth than the exciting suggestion found in Davidson's work that there might be a principled explanation. The unexciting explanation cites multiple factors: The domain of the social sciences is highly complex (how many *serious* laws are there in meteorology proper?), it is hedged about with often legitimate restrictions on experimentation, and it possesses (as rational expectation theory and deterrence theory have recently emphasized) a peculiar capacity for countersuggestiveness. Moreover, the social sciences are, inevitably it seems, characteristically more vulnerable to the effects of ideologies and interests present in the communities in which they are practiced.

Despite this explanation — which should be enough for the underdeveloped state of psychology — the claim that the nomothetic ideal is wholly unrealized in contemporary mentalistic psychology seems to me an exaggeration. Contemporary cognitive social psychology, for example, affords nontrivial, even surprising generalizations characterizing partial relations among variables under mentalistic descriptions ("belief," "desire," and so on). Indeed, this essay is partly motivated by a sense of the interest and informativeness of results in cognitive social psychology and neuropsychology. We seem to gain a distinctive, not-always-folksy, nomic kind of understanding from work on cognitive dissonance, pressures to conform, inferential strategies, memory, and perception. How much such generalizations can be or need be improved in the direction of strictness is an open question, as far as I can see. So is the question of how much such theories will put pressure on us to revise relatively aprioristic ways of thinking about belief, desire, preference, etc. Amos Tversky and Daniel Kahnemann's subjects, for example, are unquestionably holders of beliefs and desires, and our confidence about this does not depend upon anything so specific as confidence that these subjects' responses will someday be reinterpreted to fit the axioms of orthodox decision theory neatly.

Philosophers can, if they wish, operate with notions of belief, desire, and personhood that are "front-loaded" with very definite and strong a priori constitutive principles, but the effect of this strategy might be the opposite of the one intended. Far from generating interesting synthetic a priori truths about humankind, it might simply make it more difficult to establish that humankind are persons. One can specify a priori that personhood involves beliefs and desires, and that these, in turn, obey certain definite axioms. One can then wonder whether humankind fills the bill. Or, one can specify a priori that humankind are persons, so that anything constitutive of personhood must be something we have or do. But to make both specifications strictly and simultaneously is to engage — wittingly or unwittingly — in the dubious enterprise of trying to coin substantive empirical truth a priori.

And there is an alternative to heavy front-loading. It is to think of our notions of belief, desire, and personhood in broader terms, with more room for a posteriori science to tell us how the representational and affective functions of persons work and fit together, assuming, as seems overwhelmingly probable, that we do represent and do possess affect and conation, and that these together shape conduct in something like the ways we have always assumed.

I judge it to be unwise to attempt to settle the fate of psychology, or for that matter of psychobiology, or "the science of the intentional," by aprioristic argument. This has long been a philosopher's itch, but perhaps our best hope for not making it worse is not to scratch it.

Of the experiments he ran, Davidson writes:

There was . . . an entirely unexpected result. If the choices of an individual over all trials were combined, on the assumption that his 'real' preference was for the alternative of a pair he chose most often, then there were almost no inconsistencies at all. Apparently, from the start there were underlying and consistent values which were better and better realized in choice. I found it impossible to construct a formal theory that could explain this, and gave up my career as an experimental psychologist. (PP, 236)

It is difficult to believe that one so inventive and insightful as Davidson was so thoroughly discouraged. In any event, psychology's loss was philosophy's gain. Others, however, have gone on to attempt to improve our understanding of (and where necessary to devise experimental controls for) such phenomena as learning through participation in experiments.[16] One might indeed hope that the surprising bit of potentially nomic understanding Davidson uncovered will find its proper explanation in a developed psychological theory.

NOTES

Previous versions of this essay were presented at the University of Pittsburgh, Princeton University, Rutgers University, and the California Institute of Technology. I have benefited considerably from the discussion it received on these occasions. I would also like to thank David Lewis and Mark Johnston in particular for thoughtful comments.

1. On this view of laws and explanation, see White 1965. The view has also been advocated by Lewis 1986, "Causal Explanation." See also Railton 1981.

2. See Haugeland 1982. I am grateful here to Stephen Yablo.

3. See, however, note 8.

4. Russell, of course, pointed out long ago that if one is sufficiently relaxed

about what sort of generalization will count as a law, then there is no difficulty (except in some limiting cases) in finding a "nomic characterization" of universal form for any phenomenon. For the possibility of nomothetic psychology to constitute a nontrivial issue, we must have in mind some more restricted idea of law than this. Philosophy of science has not developed a ready and comprehensive definition of law, and so the lawlikeness of a generalization is typically discussed by looking at a constellation of interrelated features: the "natural" vs. ad hoc character of the kind-terms that figure in it; our willingness to project it to new cases or take incomplete data as confirmatory of it; our willingness to use it to support counterfactuals or explanations; its simplicity, generality, finite expressibility; whether it would figure in an ideally simple and comprehensive description of the world; and so on. Relative to any particular strategy for distinguishing (usually not sharply!) between laws and "accidental generalizations," there is a question about whether a system (of some particular degree of richness or scope) of generalizations of the kind it deems lawlike can be found in a given area of inquiry. It is this last sort of question we are asking here, so we need not take a stand on the question of which such strategy is to be preferred.

5. Compare Lewis's strategy for defining psychological terms using folk theory, in Lewis 1983, "An Argument for the Identity Theory."

6. How large or serious such worries are will depend upon empirical questions concerning human variability. The mere possibility of extensive variability need not be an obstacle to a nomothetic psychology for actual humans. Moreover, depending upon empirical details, actual extensive variation need not present much of a problem. In the treatment of disorders, however, we do find some genuine evidence of why variability might matter. In the remainder of the essay, I will proceed on the assumption that the worries are potentially large enough to warrant some response.

7. Whether it is worthwhile refining a generalization is a further question. The variability among gases under discussion here is still among types — hydrogen, oxygen, etc. — and among isolated systems considered as such. This, as it happens, leads to relatively simple and stable correction terms in the characteristic equations. Were we to consider forms of variability more idiosyncratic or inconstant, or nonisolated systems, the potential predictive or explanatory gain from relatively straightforward correction terms might be significantly less. Of course, one might well think it likely that the latter conditions more closely approximate the circumstances of most social science.

8. Despite his invocation of strictness, Davidson does in places seem to be questioning the possibility of something much weaker: *improving* the nomic seriousness of psychological theory in ways akin to those suggested herein. See, especially, his discussion of "individual laws" in the essay "Hempel on Explaining Action," to be discussed below. However, Brian McLaughlin has informed me that to his knowledge it is only in his commentary on Hempel that Davidson considers anything other than perfect strictness as his focus. It would be difficult, in my view, to read Davidson's discussion of the proper aspiration of psychological theory in his essay "Psychology as Philosophy" (to take one example) as concerned only with the issue of strictness. Moreover, if strictness were the central issue, then the

situation of psychology would in principle be no different from that of virtually any "molar" theory in the physical sciences—indeed, no different in principle (though still very different in practice) from that of classical thermodynamics. Admittedly, "Psychology as More Like Thermodynamics Than Quantum Mechanics" would be a less stirring slogan than "Psychology as Philosophy," and "The Less-Than-Perfectly Strict Nomalism of the Mental" a less exciting thesis than "The Anomalism of the Mental."

9. The Arrhenius-Ostwald definition (according to which acids ionize in water to yield H+ ions) is the oldest, and too narrow for many uses. It agrees fairly closely with Bronsted-Lowry (according to which acids are proton donors) about acids, at least when behavior in water is concerned, but not so for bases. Bronsted-Lowry is broader and captures the "classic" acids, but is also somewhat "revisionist"—water is a Bronsted-Lowry acid, but is also the very paradigm of a neutral substance in the classical conception. Lewis (according to which acids are electron pair acceptors) agrees fairly closely with Bronsted-Lowry about bases, but is broader about acids. Usanovich (a complex matter having to do with forming salts with bases, giving up cations, adding to anions, and taking up free electrons) is the least widely used. Bronsted-Lowry is sometimes defended because of the interesting fact that strength of a Bronsted-Lowry acid is inversely related to strength of a Bronsted-Lowry conjugate base. (E.g., water is a weak Bronsted-Lowry acid, but OH− is a strong Bronsted-Lowry base.) Thus a *physical* definition of acid gets part of its rationale (as picking out an appropriate "natural kind") by revealing an additional *chemical* principle of similarity.

10. For this reason, this case is also of dubious intelligibility. Are we to imagine that we have discovered compelling reasons for us to think there are—at least, for us—no such things as reasons? See also Boghossian 1990.

11. It seems to me implausible that we are wrong about the existence of something playing the representing and action-guiding function of beliefs; and it seems to me quite plausible that anything playing these functions would ipso facto *be* belief. But I do not think the "implausible" can be coined into "conceptually impossible." There are, as noted earlier, paradoxes awaiting incautious formulations of the idea that there might not be beliefs. But precisely because commonsense psychology makes distinctive explanatory claims, it is possible that reality be otherwise. Even if we take much of the content of commonsense psychology to be analytically related to our mentalistic vocabulary, it is an empirical matter whether this analytic scheme is applicable to any given class of empirical objects, viz., Homo sapiens. I do not think, however, that we should underestimate the difficulties facing the supposition that the belief-desire theory picture is systematically in error. I am indebted here to conversations with Paul Boghossian, Frank Jackson, and Mark Johnston.

12. What does this loose-jointed discussion of possible revisionism with respect to our notion of rational agency presuppose about the semantics of our vocabulary of *belief, desire, rationality*, and their ilk? Certainly we would need some alternative to the view that our core folk psychology (if this could be identified), or orthodox rational decision theory, gives these terms strictly non-negotiable content. One alternative to such intransigent views is that these terms are holistically

tied to a large network of theory and application, concerned with explaining, evaluating, predicting, and guiding behavior. Their positions and roles in this web are revisable in response to multiple pressures, and so not knowable strictly a priori, even though some elements of position or role are more important than others. A second alternative is that these terms might be thought of possessing features akin to Kripke-Putnam natural kind terms — they are used to pick out phenomena with certain particular causal roles (e.g., in producing behavior) and with which we have associated certain important stereotypes, but it is left to psychological theory to discover their real nature. The tenability of either of these semantic alternatives — as well as the question of how much and in what ways they might differ from one another once fully spelled out — needs further scrutiny.

13. Interestingly for our purposes — because it is evidence of empirical tethering of the kind mentioned above — such work has produced not only the philosophical response of trying to defend the core theory by explaining away the apparent conflict with evidence, but also an effort by some philosophers and others to rethink the *normative* notions of what rationality and morality require to cohere better with this "revisionist" descriptive account.

14. Davidson again mentions the analogy with the transitivity of length: "If length is not transitive, what does it mean to measure a length at all?" (HEA, 273). If we ignore the epistemic flavor of this remark and treat it, as I believe Davidson intends it, as a remark about the constitutive role of transitivity in the concept of length, then it would be strained to ask whether the "empirical force" of an explanation involving length resides in individual attribution or general principle. The individual attribution would involve the general constitutive principle. Consider this explanation of why a new bolt will not fit flush in a threaded hole, even though the old one did: "It's half an inch longer than the old bolt". Should we say that explanation here resides in the attribution of relative length or in the intransitivity of length (if the new bolt is longer than the old one, and the old one was just as long as the hole, then the new bolt is longer than the hole)? How are we to imagine this to be a meaningful choice?

15. "Rational" in these sample "laws" is to be understood as Hempel does, not primarily as an evaluative term, but as standing for a disposition to act in characteristic ways — functionally akin to something like mathematical ability — which will manifest itself in certain circumstances if the agent is not upset, distracted, etc.

16. See, for example, some of the discussions of the belief-desire model as an empirical theory (and the recognized perils of interpretation) in Siegel, Siegel, and Andrews 1964; Fishbein and Ajzen 1975; and Eiser and van der Pligt 1988.

REFERENCES

Ajzen, Icek, and Martin Fishbein. 1980. *Understanding Attitudes and Predicting Social Behavior.* Englewood Cliffs, N.J.: Prentice-Hall.

Boghossian, Paul. 1990. "The Status of Content." *Philosophical Review* 99: 157–84.

570   **Peter Railton**

Considine, Douglas M., ed. 1976. *Van Nostrand Scientific Encyclopedia.* 5th ed. New York: Van Nostrand Reinhold.

Davidson, Donald. 1980. *Essays on Actions and Events.* Oxford: Clarendon.

Eiser, J. Richard, and J. van der Pligt. 1988. *Attitudes and Decisions.* London: Routledge.

Elster, Jon. 1986. "Introduction." In Jon Elster, ed., *Rational Choice.* Oxford: Basil Blackwell.

Fishbein, Martin, and Icek Ajzen. 1975. *Belief, Attitude, Intention, and Behavior: An Introduction to Theory and Research.* Reading, Mass.: Addison-Wesley.

Haugeland, John. 1982. "Weak Supervenience Revisited." *American Philosophical Quarterly* 19: 93–103.

Hempel, C. G. 1964. *Aspects of Scientific Explanation.* New York: Free Press.

Kuhn, Thomas S. 1970. *The Structure of Scientific Revolutions.* 2nd ed. Chicago: University of Chicago Press.

Lewis, David. 1983. *Philosophical Papers.* Vol. I. New York: Oxford University Press.

———. 1986. *Philosophical Papers.* Vol. II. New York: Oxford University Press.

Nisbett, Richard, and Lee Ross. 1980. *Human Inference: Strategies and Shortcomings of Social Judgment.* Englewood Cliffs, N.J.: Prentice-Hall.

Quine, W. V. 1960. *Word and Object.* Cambridge: MIT Press.

———. 1968. "Propositional Objects." In *Ontological Relativity and Other Essays.* New York: Columbia University Press.

Railton, Peter. 1981. "Explanation, Probability, and Information." *Synthese* 48: 233–56.

Ross, Lee, and Richard Nisbett. 1991. *The Person and the Situation: Perspectives of Social Psychology.* Philadelphia: Temple University Press.

Siegel, Sidney, Alberta Siegel, and Julia Andrews. 1964. *Choice, Strategy, and Utility.* New York: McGraw-Hill.

White, Morton. 1965. *Foundations of Historical Knowledge.* New York: Harper & Row.

Index